International Interdisciplinary Conference on Environmental Sciences and Sustainable Developments (IICESSD 2021)

IOP Conference Series: Earth and Environmental Science
Volume 1075

Palu, Indonesia
5-6 December 2021

Editors:

Tim Roberts
Muhammad Basir Cyio
Richard Gelderman
Miguel A. Altieri
Tomonori Kawakami
Bunyamin Maftuh
Suratman
F.X. Suryadi

ISBN: 978-1-7138-6431-8
ISSN: 1755-1307

Printed from e-media with permission by:

Curran Associates, Inc.
57 Morehouse Lane
Red Hook, NY 12571

Some format issues inherent in the e-media version may also appear in this print version.

Institute of Physics
Dirac House, Temple Back
Bristol BS1 6BE UK

Phone: 44 1 17 929 7481
Fax: 44 1 17 920 0979

techtracking@iop.org

Additional copies of this publication are available from:

Curran Associates, Inc.
57 Morehouse Lane
Red Hook, NY 12571 USA
Phone: 845-758-0400
Fax: 845-758-2633
Email: curran@proceedings.com
Web: www.proceedings.com

TABLE OF CONTENTS

AGRICULTURE FOR THE ENVIRONMENT AND SUSTAINABLE DEVELOPMENT

PUBLIC HEALTH FOR THE ENVIRONMENT AND SUSTAINABLE DEVELOPMENT

GREEN ENGINEERING FOR THE ENVIRONMENT AND SUSTAINABLE DEVELOPMENT

IOP Conf. Series: Earth and Environmental Science 1075 (2022) 011001 doi:10.1088/1755-1315/1075/1/011001

The Preface

It is indeed a great honor and a pleasure, on behalf of the organizing committee, to welcome you all to the First International Interdisciplinary Conference on Environment Sciences and Sustainability Development (IICSESSD) 2021, which was held on 5-6 December 2021 at Swiss-belhotel Convention Hall in Palu, the capital and a beautiful coastal city of Sulawesi Tengah Province, Indonesia.

As the conference is usually attended onsite by delegates, due to the growing concern regarding pandemic of covid-19, local goverment restrictions regarding mobilisation, and quarantine policy, the IICESSD 2021 was held virtually online via Zoom Webinar instead of postponed. The decision was a great that despite the approved budget and challenges the conference has to be cary on to contribute to the environmental science and sustainable development towards the Era of Societry 5.0 from new research ideas, methods and problem resolutions.

We have been lucky enough to have had distinguished speakers from a number of recognized worldwide universities:

Prof. Dr. Tim Roberts, Newcastle University, Australia
Prof. Dr. Muhammad Basir Cyio, Tadulako University, Indonesia
Prof. Richard Gelderman, Western Kentucky University, USA
Prof. Dr. Miguel A. Altieri, University of California, USA
Prof. Tomonori Kawakami, Toyama Prefectural University, Japan
Prof. Dr. Bunyamin Maftuh, University of Education Indonesia, Indonesia
Prof. Dr. Suratman, Gadjah Mada University, Indonesia
Dr. F. X. Suryadi, IHE Delft Institute, The Netherlands

Environmental science can be implicit by integrating them into a comprehensive study while at the same time, sustainable development refers to developing an economic system that uses natural resources in ways that do not deplete them for future generations – this is why environmental science to gether with sustainable development are referred inherently to as an interdisciplinary field because it concerns a complex include: green economics, energy consumption, waste production, threats to public health, practical education, poverty, social exclusion, management of natural resources, loss of biodiversity, and land use.

This conference will serve as an opportunity to disseminate or to share research results on environmental sciences and sustainable development that will give visibility to our studies and function, in turn, as means of transferring and communicating to others and to obtain various feedbacks through a combination of plenary, breakout or parallel sessions including (i) Agriculture for the environment and sustainable development, (ii) Public health for the environment and sustainable development, (iii) Green Engineering for the environment and sustainable development, (iv) Practical Education for the environment and sustainable development, and (v) Green economy, entrepreneurship and good governance for sustainable development.

IICESSD 2021 received 130 submissions, from various academic universities which were managed in EasyChair platform. Subsequently the editorial and review process with the assit of Grammarly and Turnitin Similarity Check, the conference committee accepted for presentation

112 papers which were presented into one of the named sessions for period of 15 minutes per paper, while the plenary speaker was 35 minutes including discussions and Q&A sessions.

Through this conference, we, the committee, hope to foster partnerships and build new synergies for a global environment and development.
I would like to express my appreciation to the Rector and Director of Postgraduate Tadulako University for tremendous financial support, to all the outstanding members of the organizing committee and reviewers who have worked very hard in the editorial process and constructing valuable recommendations for the authors to improve their work and to contribute their research results to the conference, and also special thanks go to the IOP Publishing EES that agreed with the works of this proceeding conference in an open-access volume of IOP Conference Series: Earth and Environmental Science.

I would like to thank you so much for your outstanding contribution and participation in the success of this event. We look forward to seeing all of you at the next IICESSD conference.

Professor Daud K. Walanda
IICESSD 2021 Chair

List of Editors
Professor Tim Roberts, The University of Newcastle
Professor F. X. Suryadi, IHE Delft Institute
Professor Tomonori Kawakami, Toyama Prefectural University
Professor Dr. Muhammad Basir Cyio, MS., ASEANEng., Tadulako University
Professor Mery Napitupulu, MSc., PhD., Tadulako University
Professor Dr. Alam Anshary, MS. ASEANEng., Tadulako University
Professor I Made Antara, MP., Tadulako University

ORGANIZING AND SCIENTIFIC COMMITTEE

Organizer
Postgraduate Universitas Tadulako

Conference and Program Chair
Professor Daud K. Walanda

Conference Vice-chair
Professor I Wayan Sutapa

Secretary
Dr. E. Fernilia Zahra M.Sc.

Treasurer
Dr. Yunidar Nur, M.Hum

Steering committee
1. Prof Dr. Shahabuddin, M.Si
2. Nur Edy, S.P., M.P., Ph.D
3. Dr. Misnah, S.Pd., M.Pd
4. Dr. Edhi Taqwa, S.E., M.Si
5. Dr. Dasa Ismaimuza M.Si
6. Dr. Muhammad Din, SE., M.Si.

Programme committee
1. Prof Dr. Ir. Effendy, M.Si
2. Prof Dr. Ir. Made Antara, M.P
3. Nur Edy, S.P., M.P., Ph.D
4. Prof Marsetyo, M.Agr.Sc., Ph.D.

Accomodation/venue committee
1. Dr. Siti Chacriah Ahsan, M.Si
2. Dr. Armin Muis, S.E., MP
3. Dr. H. Abu'Ijaiya SE., M.Si
4. Dr. Aminah, S.Pd., M.Pd., M.Ed

Technical committee
1. Dr. Ir. Anita Achmad, S.Kom, M.Cs
2. Dessy Santi, S.Kom. MT.
3. Ir. Syaifirl Handra, S.Kom., M.Kom
4. Syahrullah, S.Kom., M.Kom.

Peer Review Statement

All papers published in this volume have been reviewed through processes administered by the Editors. Reviews were conducted by expert referees to the professional and scientific standards expected of a proceedings journal published by IOP Publishing Publishing.

- **Type of peer review:** Single Anonymous
- **Conference submission management system:** Morressier
- **Number of submissions received:** 130
- **Number of submissions sent for review:** 112
- **Number of submissions accepted:** 94
- **Acceptance Rate (Submissions Accepted / Submissions Received × 100):** 72.3
- **Average number of reviews per paper:** 9
- **Total number of reviewers involved:** 16
- **Contact person for queries:**
  Name: Professor I Wayan Sutapa
  Email: wayan.sutapa@untad.ac.id
  Affiliation: Universitas Tadulako

IOP Conf. Series: Earth and Environmental Science 1075 (2022) 012001 doi:10.1088/1755-1315/1075/1/012001

The flowering of chrysanthemum (Chrysanthemum sp) growing under various concentrations of liquid organic fertilizer

R. Yusuf[1], A. Syakur[1], Y. Kalaba[2], R. Rostiati[1] and S' Ramadani[1]

[1]Laboratory of Horticulture Program Study Agroteknologi Faculty of Agriculture University of Tadulako, Palu
[2]Laboratory of Soil Science Program Study Agroteknologi Faculty of Agriculture University of Tadulako, Palu

ryusufus@yahoo.com

Abstract. Chrysanthemum is a mainstay commodity in the horticultural industry with bright market prospects. The chrysanthemum plant is one of the horticultural commodities in great demand because it has a variety of colors and shapes; besides that, ornamental plants help beautify the environment. The purpose of this study was to determine the effect of several concentrations of liquid organic fertilizer on the growth and flowering of chrysanthemum plants. This research was conducted at the Screen House of the Faculty of Agriculture, Tadulako University, Palu, Central Sulawesi. This study was arranged using a Randomized Block Design (RBD) consisting of 5 treatments, and each treatment was repeated three times so that there were 15 units of trials. The data obtained were analyzed using ANOVA analysis. The results showed that the liquid organic fertilizer (LOF) of 80 mL/1 liters of water significantly affected the vegetative phase, namely the number of leaves, plant height, and branches. The flowering has no significant effect on the number of flowers, flower diameter, and the time to flower blooming.

1. Introduction

Ornamental plants are horticultural commodities in high demand because they have various colors and shapes. Besides that, ornamental plants help decorate the environment. Ornamental plants are also helpful for satisfying spiritual needs and decorating the indoor [1,2].

Decorative plants include ornamental leaf plants and ornamental flower plants. One type of ornamental flower is a Crysanthemum sp [3]. The attractiveness of chrysanthemum plants includes various colors, types, and shapes. Chrysanthemum is also used in the health sector, consumed as herbal tea or medicinal tea. Although chrysanthemum is not native to Indonesia, it has been known for decades and is a mainstay commodity in the horticultural industry with bright market prospects. The various and beautiful shapes, types, and colors are increasingly making domestic and foreign demand increase from year to year [4]

Maintaining and increasing soil fertility by adding nutrients is necessary to maintain normal or increase crop production. Adding these nutrients allows achieving a balance between the nutrients that are lost and transported by the harvest, erosion, and another leaching. The act of returning/adding nutrients to soil plants is called fertilization. The type of fertilizer used must be as needed, thus requiring the correct method of diagnosis so that only those nutrients are added to plants and those that are lacking in the soil [5]

The use of organic fertilizers can be a solution in reducing the excessive use of inorganic fertilizers. Organic fertilizers can be in the form of solid or liquid. The advantage of liquid organic fertilizer is that the nutrients are available more quickly and easily absorbed by plant roots [6].

Liquid organic fertilizer also can improve the physical, chemical, and biological properties of the soil, help increase crop production, improve the quality of plant products, reduce the use of inorganic fertilizers, and as an alternative to manure [7].

Seaweed has long been used directly as a soil conditioner and fertilizer in various coastal areas of the world, and seaweed extract has also been widely marketed as an additive to plant fertilizers whose benefits and benefits have been widely reported. In addition, it contains many essential minerals from the sea needed by plants. Seaweed also has growth-promoting hormones that have been shown to increase plant growth and yields. Unlike chemical fertilizers, extracts made from seaweed are naturally biodegradable, non-toxic, non-contaminating, and safe for humans and animals [8].

The study aimed to determine the better concentration of LOF on the growth and flowering of chrysanthemum (Chrysanthemum sp.).

2. Material and methods

2.1. Study area

This research was conducted at the Screen House, Faculty of Agriculture, Tadulako University, Palu, Central Sulawesi. Research starts from January 16 to May 10, 2020.

2.2. Research tool and protocol

The tools used in the research are hoes, scissors, scales, cameras, glasses aqua, rope, bamboo, bucket with a lid on top, label, hand sprayer, ruler, and stationery. The materials used in the study were chrysanthemum plant seedling, granulated sugar, seaweed (Ulva sp), polybags, water, soil, and EM4. This study used a randomized block design (RBD) which consisted of 5 treatments, and each treatment was repeated three times so that there were 15 experimental units, namely S0 = No seaweed liquid organic fertilizer (control), S1 = 20mL LOF/1 L water, S2= 40 mL LOF/1 L water, S3= 60 mL LOF/1 L water, S4 = 80 mL LOF/1 L.

LOF preparation. The dry seaweed blender until smooth, then weigh 100 g. Then, prepare a 10 L bucket, put seaweed in it, add 3 L of water, four spoons of EM4, and four spoons of sugar, then stir and close tightly. Then open the bucket's lid every day and stir once a day so that it can be mixed evenly. After two weeks, the LOF is ready to be applied.

Preparation of growing media. The growing medium used in this study was soil. The soil used is cleaned first and then put into polybags of 15×30 cm. After that, the polybags that have been filled with planting media are arranged according to the research plan.

Seedling Preparation. Seedlings are placed in the middle of the polybag, which has been filled with growing media; a planting hole is made by digging the media base on the size of the root of a chrysanthemum plant. Then the seedling is planted in an upright position and then covered with soil. To maintain the humidity of the plants, container plastic is used to cover the plants. Seedlings are sown in the afternoon.

Care of plants. The care of chrysanthemums includes watering, weeding, fertilizing, controlling pests and diseases. Watering is carried out routinely every day, especially in the early growth phase. According to the weather condition, it is carried out twice a day, in the morning and evening. Weeding is done by pulling weeds in the planting area. The most important thing to note is when weeding so that the plant roots are not damaged or injured because it can facilitate disease attacks.

Liquid Organic Fertilizer Application. LOF application is carried out until the plant enters the generative period given at an interval of every seven days. The method of application is by giving seaweed LOF to each plant according to the dose of each plant according to the treatment by sprinkling them on the ground.

3. Results and discussion

Number of Leaves. The results of the 0.05 HSD test (Table 1) at the ages of 3, 4, 5, 6, 7, and 8 WAP (weak after planting) showed differences between treatments. At the age of 3, WAP with the symbol S3 gave the highest value on the number of leaves which was 9.33, significantly different from the S0 treatment. At the age of 4 WAP, the seaweed LOF treatment S4 gave the highest value, namely 18.33, which was significantly different from the treatment without seaweed LOF treatment, which was 16.00. At the age of 5 WAP, the treatment gave the highest value to the S4 symbol, which was 23.00, significantly different from the S0 treatment, 18.33. At the age of 6 WAP, the S4 treatment gave the highest value, which was 24.33, which was significantly different from the control treatment, which was 19.67. At the age of 7 WAP, the S4 treatment gave the highest value, which was 36.67, which was significantly different from the control treatment, which was 26.67. At 8 WAP, the S4 treatment gave the highest value, which was 38.33, which was significantly different from the control treatment, which was 32.67.

Table 1. The average number of leaves at the age of 3 WAP, 4 WAP, 5 WAP, 6 WAP, 7 WAP, and 8 WAP with the application of various concentrations of liquid organic fertilizer seaweed subsection. The paragraph text follows on from the subsection heading but should not be in italic.

Treatments	No. of leaves					
	3 WAP	4 WAP	5 WAP	6 WAP	7 WAP	8 WAP
S0	5,33[a]	16,00[a]	18,33[a]	19,67[a]	26,67[a]	32,67[a]
S1	7,00[a]	16,00[a]	19,33[a]	21,00[a]	26,33[a]	32,67[a]
S2	6,33[a]	16,33[a]	19,33[a]	20,67[a]	27,33[a]	34,00[a]
S3	9,33[b]	17,33[a]	20,33[a]	21,33[a]	28,67[a]	35,67[a]
S4	9,00[b]	18,33[b]	23,33[b]	24,33[b]	32,67[b]	38,33[b]
HSD 5%	2,69	1,72	2,82	2,59	2,18	3,65

Means followed by the same letter in the same column are not significantly different at the 0.05 HSD level.

Table 2. Average plant height at the age of 4 WAP, 5 WAP, 6 WAP, 7 WAP with the application of various concentrations of liquid organic fertilizer seaweed

Treatments	Plant high			
	4 WAP	5 WAP	6 WAP	7 WAP
S0	12,00 [a]	17,33 [a]	21,67 [a]	24,00 [a]
S1	11,33 [a]	17,33 [a]	22,00 [a]	25,00 [a]
S2	15,00 [a]	18,00 [a]	23,00 [a]	25,00 [a]
S3	16,00 b	19,00 [a]	23,00 [a]	24,33 [a]
S4	19,33 [b]	23,67 [b]	25,67 [b]	28,33 [b]
HSD 5%	3,72	2,30	2,77	2,46

References means followed by the same letter in the same column are not significantly different at the 0.05 HSD level.

Plant height. The results of the 0.05 HSD test (Table 2) at the age of 3, 4, 5, 6, 7, and 8 WAP showed differences between treatments. At 3 WAP and 8 WAP showed that the two treatments were not significantly different. At the age of 4 WAP, the S4 treatment gave the highest value for plant height, which was 19.33 cm, significantly different from the control treatment, which was 11.33 cm. At the age of 5 WAP, the S4 treatment gave the highest value for plant height, which was 23.67 cm, significantly different from the control treatment, which was 17.33 cm. At the age of 6 WAP, the S5 treatment gave the highest value, 25.67, significantly different from the control treatment, which was 21.67 cm. At the age of 7 WAP, the S4 treatment gave the highest value, which was 26.67 cm, which was significantly different from the control treatment, which was 24.00 cm.

Number of Branches. The results of the 0.05 HSD test (Table 3) at the age of 3, 4, 5, 6, 7, and 8 WAP showed differences between treatments. At the age of 3 WAP, the treatment with the symbol S4 gave the highest value on the number of branches, which was 4.33, significantly different from the control treatment without LOF, which was 2.67. At the age of 4 WAP, the S4 treatment gave the highest value in the number of branches is 5.00, significantly different from the control treatment, which is 3.67. At the age of 5 WAP, the S4 treatment gave the highest value on the number of branches, which was 8.67, significantly different from the S0 treatment, which was 6.00. At 6 WAP, the S4 treatment gave the highest value, namely 12.00, significantly different from the S0 treatment, which was 8.33. On age 7 WAP, the S4 treatment gave the highest score of 15.33, significantly different from the S1 treatment, 11.67. Moreover, at the age of observation 8 WAP treatment, S4 gave the highest value, which was 23.67, which was significantly different from the S1 treatment with a value of 20.33.

Table 3. Average number of branches at the age of 3 WAP, 4 WAP, 5 WAP, 6 WAP, 7 WAP, and 8 WAP with various doses of liquid organic fertilizer seaweed

Treatments	Number of Leaves					
	3 WAP	4 WAP	5 WAP	6 WAP	7 WAP	8 WAP
S0	2,67 a	3,67 a	6,00 a	8,33 a	11,67 a	21,67 a
S1	2,33 a	3,67 a	6,33 a	9,67 a	11,33 a	20,33 a
S2	3,00 a	3,00 a	6,33 a	9,00 a	12,33 a	22,00 a
S3	3,33 a	4,67 b	7,00 a	9,67 a	12,67 a	21,33 a
S4	4,33 b	5,00 c	8,67 b	12,00 b	15,33 b	23,67 b
HSD 5%	1,43	0,90	2,24	1,91	2,41	2,88

Means followed by the same letter in the same column are not significantly different at the 0.05 HSD level.

Flowering. The data from the observation of the number of flowers, flower diameter, and time to the flower booming showed that the treatment of seaweed liquid organic fertilizer had no significant effect on the number of flowers, diameter, and time to the flower booming (Figures, 1, 2, and 3).

Growth is a process in plant life. There will be a change in size; the plant will grow bigger. Genetic and other factors like the environment control the overall increase in size [9]. Based on the analysis of variance, the application of various concentrations of seaweed LOF had a significant effect on the growth components, namely plant height, number of leaves, and number of branches.

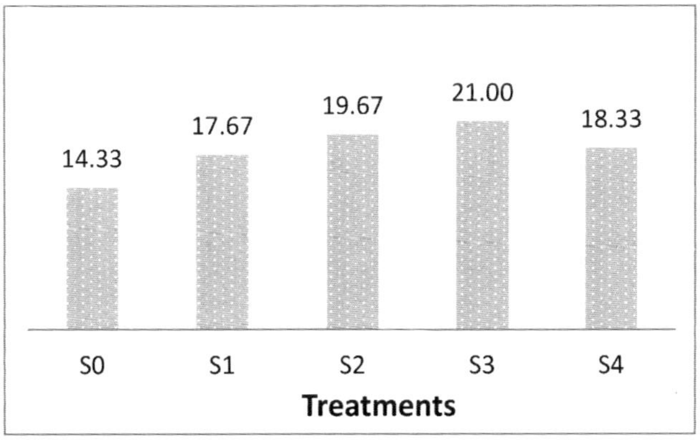

Figure 1. The number of chrysanthemum flowers on the application of seaweed LOF

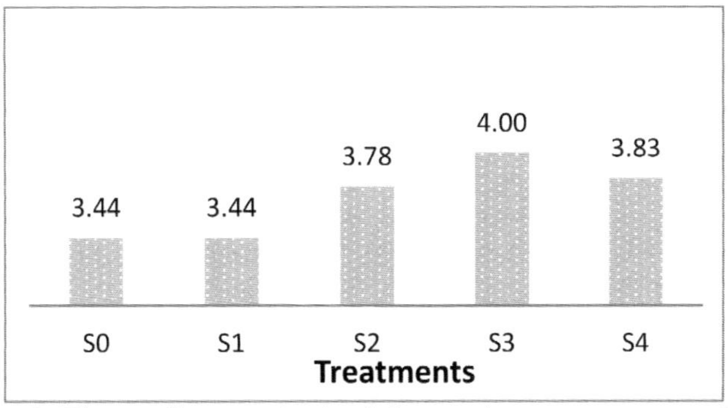

Figure 2. Flower diameter (cm) of chrysanthemum on seaweed LOF

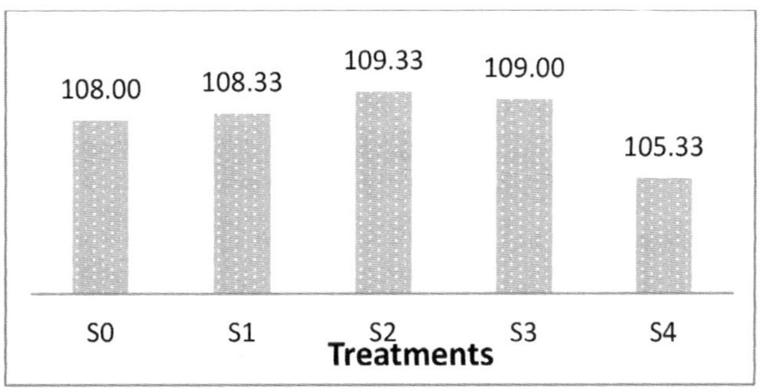

Figure 3. Blooming time (days) of chrysanthemum flowers on seaweed LOF administration

In the parameter of plant height, the best results on the growth of chrysanthemum plants, namely the S4 treatment, were significantly different from other treatments [10]. The application of liquid organic fertilizer can provide essential nutrients needed by plants for growth because seaweed LOF contains macro and micronutrients. The nutrients absorbed by plants will be utilized in metabolism to improve growth. Nitrogen nutrients are helpful for plant vegetative growth, namely the formation of new cells such as leaves, branches and replacing damaged cells [11].

Plant height growth can be supported by the availability of other macronutrients, such as carbon, oxygen, hydrogen, nitrogen, phosphorus, potassium, calcium, and sulfur, which are needed for the plant life cycle and produce other generations [12] [13]. The observations of the number of leaves of chrysanthemum plants showed that the S4 treatment had the highest value at each age of observation. This clearly shows that more leaves are formed when applying seaweed LOF with 80mL/L of water.

The higher the dose of LOF, the more nutrients are supplied for the growth of chrysanthemum plants. This is also suspected because the content of N, P, and K in the liquid organic fertilizer of seaweed has fulfilled the nutrient requirements needed by chrysanthemum plants [14][8]. LOF can provide added value to plants during plant growth and development; besides, this fertilizer is also helpful in improving soil and contains microorganisms that can reduce disease attacks on fertilized plants [14]. Plants can maximize the nutrients from fertilizers by minimizing leaching and evaporation. One of the efforts to avoid evaporation and leaching of fertilizers is repeated fertilization. Compounds contained in seaweed increase plant resistance to drought and insect attacks and improve soil structure. The function of N is to stimulate plant vegetative growth. Nutrient N is also helpful for forming chlorophyll in leaves which will later be helpful for the photosynthetic process [11]. The increased number of leaves was caused by higher nitrogen absorption. Plants use nitrogen compounds to form amino acids converted into proteins. Nitrogen is also needed to form essential chlorophyll, nucleic acids, and enzymes. Nitrogen is needed in relatively large amounts at every stage of plant growth, especially at the vegetative growth stage, such as shoot formation, stem, and leaf development [15].

The number of chrysanthemum branches showed a very significant effect; it was seen that the S4 treatment had the highest value in each age of observation. This clearly shows that the number of branches will be more formed when seaweed LOF is given with a concentration of 80mL/L of water. N is contained in protein and is helpful for the growth of leaf shoots; besides that, it can also fertilize parts of the leaf stem. Fertilizer containing N, P, K sufficient to meet the nutrient needs of plants is one of the essential factors needed by plants for growth [6] [11].

With the fulfillment of the nutrients needed by plants and supported by good environmental conditions, the results of photosynthesis will increase so that it can be used for the formation of flowers and fruit in the generative phase. Liquid organic fertilizer is a fertilizer that comes from nature [16]. It plays a role in improving the soil's physical, chemical, and biological properties because it contains nutrients that can increase plant growth. Organic fertilizers are fertilizers that increase the soil's biological, physical, and chemical activities so that the soil becomes fertile and suitable for plant growth [7].

Lack of nutrients in the soil can result in low crop production. If the nutrients in the soil are not available, then plant growth will be hampered, and production will decrease. Lack of one or more nutrients will result in improper plant growth, namely, there are abnormalities or deviations, and many plants die young [15].

4. Conclusions

Application of various concentrations of seaweed LOF significantly affected the growth components, namely plant height, number of leaves, and number of branches. The best treatment was giving a concentration of 80 mL LOF/1 liter of water. However, seaweed LOF showed no significant effect on the flowering parameter.

References

[1] P Abetz 1980 Seaweed extracts: have they a place in Australian agriculture or horticulture? *J Aust Inst Agric Sci* **46** 23–9

[2] Y P Situmeang I G M Arjana and K Siliani 2017 Use of Mulch and Organic Fertilizer on Chrysanthemum *Sustain Environ Agric Sci* **1** 39–45

[3] I G M Arjana Y P Situmeang and I Suaria 2015 Study of Development Potential Chrysanthemum in Buleleng Regency *Int J Adv Sci Eng Inf Technol* **5** 350-4

[4] D Bhattacharyya M Z Babgohari P Rathor and B Prithiviraj 2015 Seaweed extracts as biostimulants in horticulture *Scientia Horticulturae* **196** 39-48

[5] H Sokchea K Borin and T R Preston 2015 Carry-over effects of biochar on yield of Mustard Green vegetable (Brassica juncea) and on soil fertility *Livest Res Rural Dev* **27** 1-13

[6] J K Suge M E Omunyin and E N Omami 2011 Effect of organic and inorganic sources of

fertilizer on growth, yield and fruit quality of eggplant (Solanum Melongena L) *Arch Appl Sci Res* **3** 470–9

[7] E Nabti B Jha and A Hartmann 2017 Impact of seaweeds on agricultural crop production as biofertilizer *Int J Environ Sci Technol* **14** 1119–34

[8] N Vijayanand S S Ramy and S Rathinavel 2014 Potential of liquid extracts of Sargassum wightii on growth, biochemical and yield parameters of cluster bean plant *Asian Pacific J Reprod* **3** 150–5

[9] M D Maghfoer R Soelistyono and N Herlina 2014 Growth and yield of eggplant (Solanum melongena l) on various combinations of n-source and number of main branch *AGRIVITA* **36** 285–94

[10] P du Jardin 2015 Plant biostimulants: Definition, concept, main categories and regulation *Sci Hortic (Amsterdam)* **196** 3–14

[11] V Chouliaras M Tasioula C Chatzissavvidis I Therios and E Tsabolatidou 2009 The effects of a seaweed extract in addition to nitrogen and boron fertilization on productivity, fruit maturation, leaf nutritional status and oil quality of the olive (Olea europaea L) cultivar Koroneiki *J Sci Food Agric* **89** 984–8

[12] N Ozbay and A R Demirkiran 2019 Enhancement of Growth in Ornamental Pepper (Capsicum Annuum L) Plants with Application of a Commercial Seaweed Product, Stimplex® *Appl Ecol Environ Res* **17** 4361–75

[13] M Suman P D Sangma D R Meghawal and O P Sahu 2017 Effect of plant growth regulators on fruit crops *J Pharmacogn Phytochem* **331** 331–7

[14] R Yusuf P Kristiansen and N Warwick 2012 Potential effect of plant growth regulators in two seaweed products *Acta Hortic* **958** 133–8

[15] M Halpern A Bar-Tal M Ofek D Minz T Muller and U Yermiyahu 2015 The Use of Biostimulants for Enhancing Nutrient Uptake *Adv Agron* **130** 141-74

[16] R Yusuf A Syakur Y Kalaba and F Fatmawati 2020 Application of some types of local seaweed extract for the growth and yield of shallot (Allium wakegi) *AACL Bioflux* **13** 2203–10

Study of Phosphate Adsorption Using Ferrihydrite With Diffusive Gradient in Thin Films Method

Khairuddin[1*], Ruslan[1], Nindya Andarini[1], Widyastuti Oktriviani1, Meiva Toda'a[1], Van Hindersen Kaope[1]

[1]Department of Chemistry, Faculty of Mathematics and Natural Sciences, Tadulako University, Palu, Indonesia.

* *Coresponding author e-mail: kjunus24@gmail.com*

Abstract. The high input of phosphate into the aquatic system results in eutrophication which leads to algae blooming. Therefore, it is necessary to measure the concentrations of phosphate pollutants that occur in surface waters. Conventional analysis techniques so far only provide an overview of the pollution situation through instant sampling, so it may not represent the current water conditions. Pollutant concentrations fluctuate due to dynamic interactions of phosphate species in natural systems, hence species concentrations may change during sample storage, making accurate analysis difficult unless performed in situ. In this study, the Diffusive gradient in thin films (DGT) method was investigated using a ferrihydrite binding gel assembly on a polyacrylamide gel sheet. The ferrihydrite gel was eluted using 0.25 N H_2SO_4 to release the adsorbed phosphate. The performance test of the method on the absorption of phosphate by ferrihydrite was assessed by testing the effect of pH, contact time, and variation of initial concentration. Determination of the concentration of adsorbed phosphate using the phosphomolidenium blue method with a UV-Visible Spectrophotometer. Characterization of functional groups in polyacrylamide diffusing gel and ferrihydrite adsorbent gel using Fourier Transform Infra Red while the surface morphology of each gel was analyzed by Scanning Electron Microscope.

1. Introduction

Phosphorus (P) is an essential element for plants. Phosphorus is also a limiting nutrient on biological productivity in aquatic ecosystems. Phosphorus is present in waters in the form of orthophosphate species (H_2PO_4) which is considered the simplest form that can be absorbed biologically by aquatic biota [1]. In agricultural soil, phosphorus is a very important element because it plays a role in promoting root growth and development, and can also trigger flowering and fruit ripening in plants. However, if it is excessive, it can cause environmental pollution, therefore a method is needed that can measure phosphate levels in agricultural soil [2].
The most widely used phosphate measurement technique today is the adsorption technique because it is more effective, easy to prepare, and relatively inexpensive in cost compared to other methods. Phosphate adsorption is carried out with various inorganic and organic materials. Adsorbents with organic matter are very abundant in nature which come from various wastes that are not reused.
Various studies have been carried out to measure phosphate levels in aquatic samples. Methods that can be used to measure phosphate levels, one of which is the DGT (Diffusive Gradient in Thin Film) method. The DGT method is an excellent and easy in situ method for measuring unstable species in the aquatic environment. The basic principle of this method is the binding of analyte species to a binding gel or

adsorbent after diffusion through a hydrogel and then eluted with an acid to determine its concentration. In the DGT technique, the average solute concentration measured over a certain time can be known, sample contamination can be reduced, and it is possible to perform preconcentration without damaging the sample. The concentration of analyte species bound by binding gel in DGT can be considered the same as the concentration that may diffuse from water to aquatic biota so that this method can be used to predict the bioavailability of a labile metal, sulfate, and phosphate [3].

The method used in this study is the DGT (Diffusive Gradient in Thin Film) method. The DGT method itself is one method that can be used to measure phosphate levels. This DGT method is an excellent and easy in situ method for measuring unstable species in the aquatic environment. The basic principle of the DGT method is the binding of analyte species to an adsorbent after diffusion through a hydrogel and then eluted with an acid to determine its concentration. The advantages of this method include that the average solute concentration measured can be known within a certain time, preconcentration without damaging the sample and also reducing sample contamination. The concentration of analyte species bound by binding gel in DGT can be considered the same as the concentration that may diffuse from water to aquatic biota so that this method can be used to predict the bioavailability of a labile metal, sulfate, and phosphate [4].

In this study, ferrihydrite was used as a binding gel (adsorbent). Ferrihydrite is a metal oxide that is abundant in soil formed by protonation and release of Fe that comes out of primary or secondary minerals due to oxidation processes. The mechanism of absorption of anion (Phosphate) by iron oxide is through the replacement of one phosphate ion by two surface hydroxyl ions (or water molecules) of the iron oxide. Then the two oxygen atoms of the phosphate ion will coordinate with each of the Fe(3+) ions to produce complex bonds on the surface of the material. Ferrihydrite is found in almost all types of soil. In most soils, iron oxide is found in low concentrations because most of it is in the form of secondary minerals. The hydroxide form with a high specific surface area is an adsorbent complex that is highly reactive to various forms of agricultural chemicals and is a very important component in the soil environment. The ferrihydrite form has a high specific surface area, its presence as a soil component is a highly reactive adsorbent component to various ions in the soil [5].

2. Materials and Methods

2. 1 Materials

The materials used for the preparation of DGT gel were acrylamide (Sigma–Aldrich), ammonium persulfate (Sigma–Aldrich), tetramethylethylenediamine (Sigma–Aldrich), N,N'-methylenebisacrylamide (Sigma– Aldrich), membrane filter (Whatman). DGT probe (DGT Research, Lancaster University, UK), pH meter, Fourier Transform Infra Red Spectrophotometer (Prestige-21 Shimadzu), Scanning Electron Microscope. The other materials used were deionized water, sulfuric acid, ammoniummolybdate, potassium antimonyl tartrate, ascorbic acid. sodium nitrate. potassium dihydrogen phosphate, and iron nitrate.

2. 2 Methods

2.2.1 Preparation of N,N'-methylenebisacrylamide

A total of 2 grams of N,N'-methylenebisacrylamide was dissolved in methanol:water (1:9) in a 100 mL volumetric flask and then stir for ± 30 minutes.

2.2.2 Ferrihydrite Gel Preparation

Solution of 0.1 M $Fe(NO_3)_3.9H_20$ was added with 1 M NaOH while stirring and ensured that the pH was not more than 7. Water on top of the surface was removed using a pipette. The precipitate was washed two to three times and placed in the dark room at 4°C.

2.2.3 Binding Gel Preparation

For the manufacture of ferrihydrite gel, ferrihydrite was added acrylamide, aquademineral, 2% N,N' methylenebisacrylamide. Then added with Ammonium persulfate and TEMED in a container, and stirred until homogeneous (±15 – 20 seconds). Soon the solution was pipetted in a glass mold that had been cleaned with HNO_3. The gel solution that has been put in the glass mold is then heated in an oven at a temperature of 42-45 C for up to an hour until a gel is formed. The gel that has been formed is washed with aquademineralization and then soaked for 24 hours and replaced with water 3-4 times. The gel is then stored in aquademineral until it is used.

2.2.4 Diffusive Gel Preparation

The diffusive gel was made by adding acrylamide, aquademineral, 2% N,N' methylenebisacrylamide. Then added with ammonium persulfate and tetramethylethylenediamine (TEMED) in a container, and stirred until homogeneous (± 15 – 20 seconds). Soon the solution was pipetted to a glass mold that had been cleaned with HNO_3. The gel solution that has been put in the glass mold is then heated into an oven at a temperature of 42 - 45°C for up to one hour until the gel is formed (not liquid). Gels that has been formed is washed with aquademineral and then soaked for 24 hours and replaced with water 3-4 times. The gel was then stored in 0.01 M $NaNO_3$ until it was used.

2.2.5 Binding Gel Preparation

For the manufacture of ferrihydrite gel, ferrihydrite was added acrylamide, aquademineral, 2% N,N' methylenebisacrylamide. Then added with Ammonium persulfate and TEMED in a container, and stirred until homogeneous (±15 – 20 seconds). Soon the solution was pipetted in a glass mold that had been cleaned with HNO_3. The gel solution that has been put in the glass mold is then heated in an oven at a temperature of 42-45 C for up to an hour until a gel is formed (not liquid). The gel that has been formed is washed with aquademineralization and then soaked for 24 hours and replaced with water 3-4 times. The gel is then stored in aquademineral until it is used.

2.2.6 Assembly of DGT Components and Testing in Phosphate Solution.

For the use of the DGT device, a diameter of 25 mm of the diffusive and binding layer were cut using a DGT mold, and the Whatman filter membrane was immersed in aquademineralized. The DGT device was washed and rinsed with aquademineralized. The binding gel layer were placed first on the DGT probe with the ferrihydrite binding side facing up, then followed by placing the diffusive layer and filter membrane. DGT is closed properly.

2.2.7 Effect of pH on Phosphate Adsorption Experiments

Experiments of pH variation of phosphate solution was carried out by contacting the DGT device into a 3 mg/L phosphate solution with a pH variation of 3, 5, 7 and 9 for 4 hours, then eluted to determine the concentration.

2.2.8 Effect of Contact Time

The modified ferrihydrite and cellulose samples were used to adsorb 25 mL of 100 ppm phosphate solution at the optimum pH with various contact times of 15 minutes, 30 minutes, 45 minutes, 60 minutes, 120 minutes, 180 minutes and 240 minutes at a temperature of 25°C. The amount of phosphate ion that was not adsorbed was measured using a UV-Vis spectrophotometer at a wavelength of 828 nm.

2.2.9 Effect of Concentration

The modified ferrihydrite and cellulose samples were used to adsorb 25 mL of phosphate solution with various concentrations of 50 ppm, 75 ppm, 100 ppm, 125 ppm, 150 ppm, 175 ppm and 200 ppm. The process was carried out using a shaker at the optimum pH and time at a temperature of 25°C. The amount of phosphate ion that was not adsorbed was measured using a UV-Vis spectrophotometer at a wavelength of 828 nm.

2.2.10 Analysis of Phosphate

Measurement of the concentration of phosphate as P (P-PO_4) using a UV-Visible Spectrophotometer at a wavelength of 880 nm. The absorbance value was observed based on the formation of a phosphomolybdenum blue complex. Diffusive and binding layer were characterized using Shimadzu's Prestige-21 FTIR. The surface morphology of ferrihydrite gel was characterized by SEM.

3. Results and Discussion

3.1 Assembly of diffusive gel and binding gel

The formation of diffusive gels and binding gels is based on reactions of free radical polymerization. In this study, the diffusion gel used was an acrylamide polymer in the form of a hydrogel made of acrylamide, 0.5% N,N-methylenebisacrylamide crosslinker, ammonium persulfate as an initiator, and TEMED as a catalyst, while the binding gel was a diffusive gel with a ferrihydrite adsorbent added to bind the phosphate analyte. In the arrangement of the DGT device at the top, a cellulose nitrate filter membrane with a pore size of 0.45 μm is added which functions to filter the phosphate analyte that will enter the DGT device. Phosphate species that pass the filter are dissolved phosphate [3].
Ferrihydrite was synthesized by precipitation of $Fe(NO_3)_3.9H_20$ which was titrated with 1 M NaOH. Controlled precipitation during the preparation of ferrihydrite (FeOOH) from $Fe(NO_3)_3.9H_20$ and NaOH to avoid the formation of other iron oxides, such as geotite (α-FeOOH) and haematite (α-Fe_2O_3) [6]. The ferrihydrite binding gel was eluted using 0.25 M H_2SO_4 in order to release the phosphate anions bound to the ferrihydrite binding gel. The phosphate bonded to the iron atom will be replaced by SO_4^{-2}. The assembly of diffusive and binding gels must begin with the preparation of a gel solution first. To make a diffusive gel or binding gel mold, 5 mL of gel solution is needed. Each gel consists of a cross-linker, water, and acrylamide. After the gel solution is homogeneously mixed, the initiator is added. The initiator for this reaction is a 10% solution of ammonium persulfate. After the addition of the initiator, the catalyst was added. Stirring after adding TEMED is approximately 15-20 seconds. Stirring will cause collisions to accelerate polymerization. The diffusive gel solution that has been prepared is immediately inserted into the mold carefully until the mold is completely filled with gel, and no bubbles form. If bubbles form the pipetting is paused until the bubbles disappear so that all parts are equal. Transfer to the mold must be carried out immediately after stirring because polymerization proceeds very rapidly. If the stirring is too long and is not immediately inserted into the mold, the mixture will thicken so that it can no longer be pipette into the mold. If this has happened, then the mixture can no longer be used for the manufacture of diffusive gels.

3.2 Gel characterization with SEM

The characterization of ferrihydrite gel using a Scanning Electron Microscopy (SEM) tool aims to see the morphological appearance of the diffusive and binding gels. Binding surface of the ferrihydrite gel can be seen in figure 1.

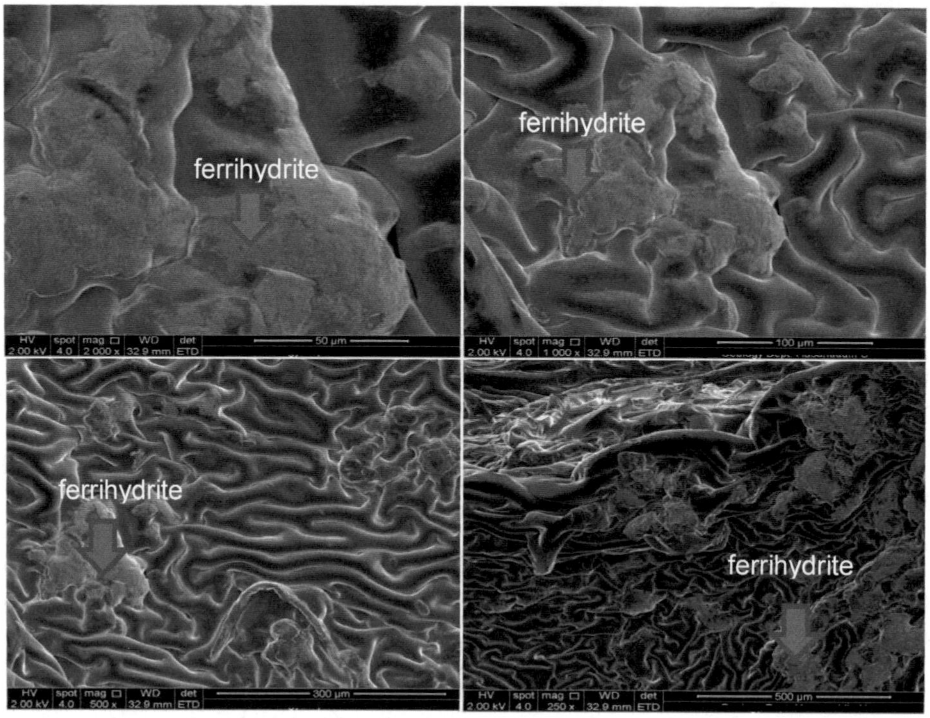

Figure 1. Surface morphology of ferrihydrite in binding gel

From the SEM image with magnification of 50 and 100 microns, it shows that the homogeneity of ferrihydrite binds well to the polyacrylamide matrix so that the concentration of each constituent compound in the manufacture of ferrihydrite gel by attaching it to the polyacrylamide gel is correct and the gel can be formed properly. The surface morphology image of the ferrihydrite gel with SEM showed that there was no chemical interaction between the ferrihydrite compound and the polyacrylamide compound and there was only physical mixing and there was no change in the structure of the acrylamide or ferrihydrite polymers [7].

3.3 Gel characterization by FTIR

To determine the functional groups contained in the diffusive and binding gels with FTIR characterization was carried out. FTIR spectrum of ferrihydrite and polyacrylamyde gels can be seen in figure 2.

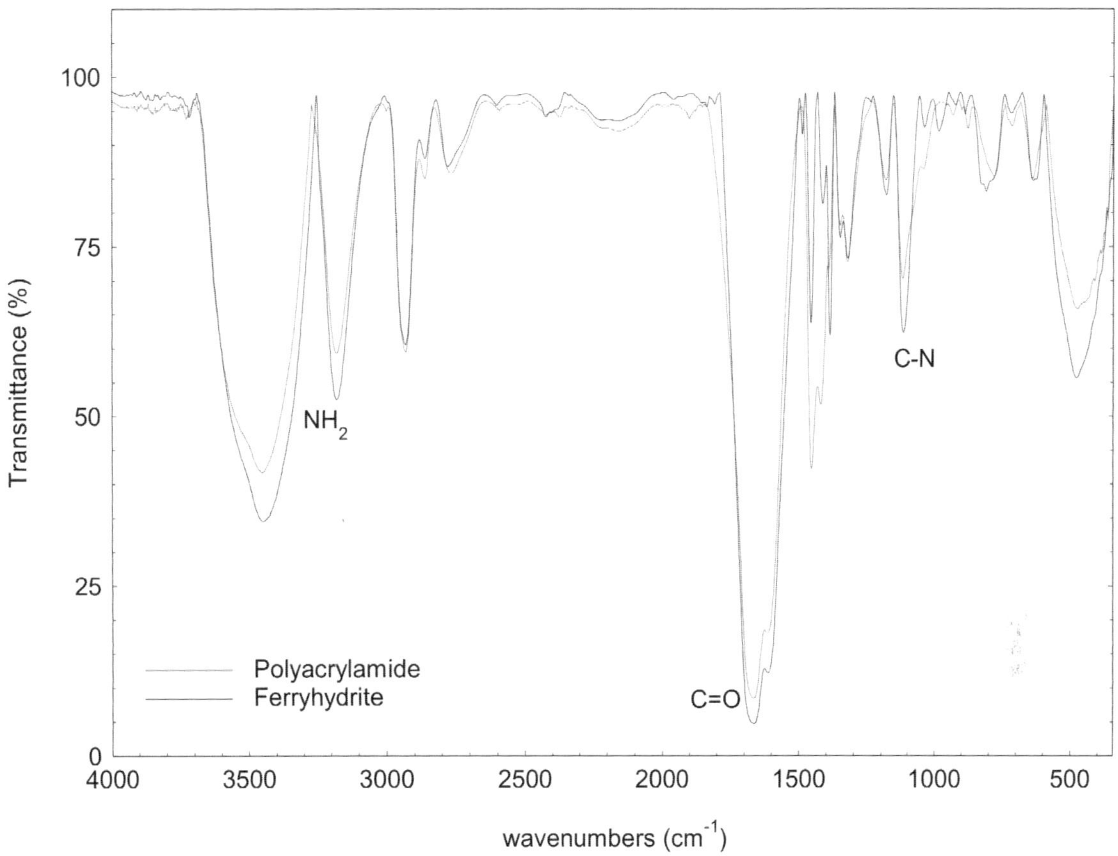

Figure 2. FTIR spectrum of polyacrylamide gel and ferrihydrite gel

Characterization with the FTIR instrument was used to ensure that polyacrylamide was successfully synthesized. There are two peaks at wave numbers 3450 cm^{-1} and 3180 cm^{-1} for the polyacrylamide gel sample, while also for the ferhydrite gel sample there are two peaks at wave numbers 3448 and 3178 cm^{-1}. This absorption wave number is typical for primary amines ($-NH_2$) in the wave number 3300–3600 cm^{-1} with a moderate intensity peaks form. Peaks at wave numbers of 1664 cm^{-1} for the polyacrylamide gel and 1662 cm^{-1} for the ferrihydrite revealed the absorption of carbonyl amide (-CONH-), usually found in the wave number 1690-1650 cm^{-1} with strong intensity forms. The peak at wave number 1182 cm^{-1} for the diffusion gel sample and 1180 cm^{-1} for the ferrihydrite sample indicates the absorption of aliphatic amines (C–N) is usually found at wave number 1250 cm^{-1} with moderate intensity peaks. The functional groups that appears is the polyacrylamide, so it can be reported that the polyacrylamide has been successfully synthesized. The functional groups that appear in the polyacrylamide and ferrihydrite are not too different, so there is no change in the polymer structure in the presence of ferrihydrite [8]. The SEM data also show that the interaction between ferrihydrite and polyacrylamide is a physical interaction.

3.4 Effect of pH variation on phosphate adsorption

This pH variation test aims to calculate the effect of pH on phosphate measurements. pH variations are carried out by adding HNO_3 to make an acidic pH or adding NaOH to make an alkaline pH. The phosphate ion concentration is determined after previously measuring the standard solution series so that a calibration curve is obtained as shown in the figure 3, and the concentration of phosphate absorbed by binding gel (C_{DGT}) is calculated using the following equation (1) bellow:

$$C_{DGT} = M\,\Delta g / (D.t.A)$$

Wherw C_{DGT} concentration of phosphate absorbed by binding gel, M mass of phosphate absorbed by binding gel, Δg the thickness of the diffusive gel plus the thickness of the filter membrane, D is diffusion coefficient of phosphate in the gel, t is deployment time, and A is exposure area.

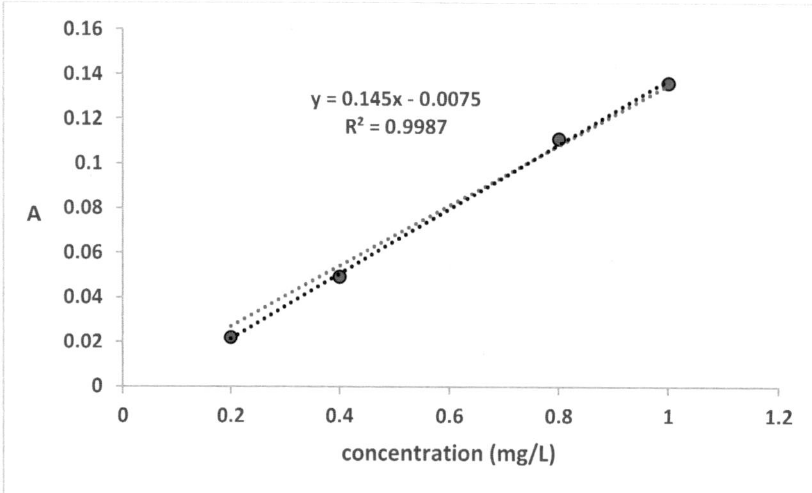

Figure 3. phosphate standard solution calibration curve

The results of pH effect variation on phosphate adsorption obtained data as shown in the figure 4.

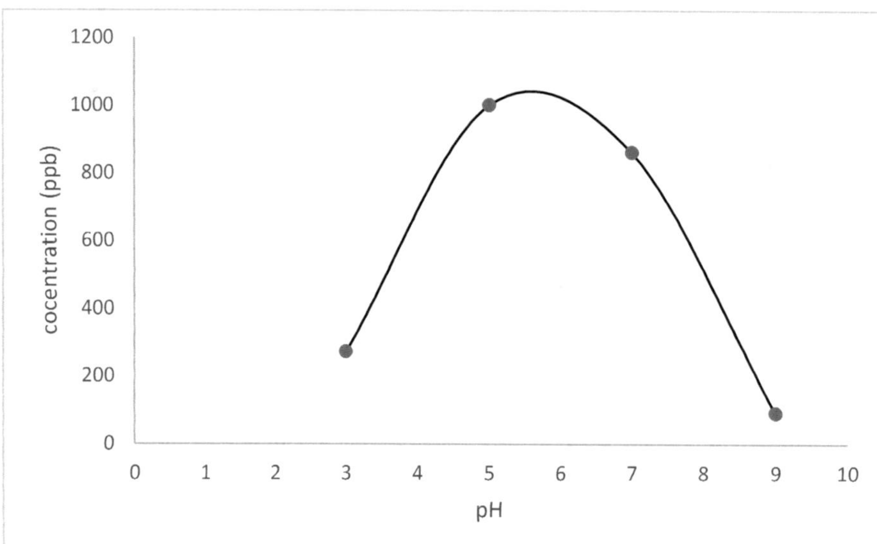

Figure 4. The effect of pH variations on the concentration of adsorbed phosphate.

Figure 4 shows that the concentration measured by DGT at pH 5 is 1.003 mg/L (1003 ppb), which is the optimum pH, while at pH 7 the phosphate concentration is 0.859 mg/L (859 ppb). At pH 3; 0.272mg/L (272 ppb) was obtained and the lowest was 0.093 mg/L (93 ppb) at pH 9. The low phosphate concentration at pH 9 was due to the presence of some Fe^{3+} ions forming $Fe(OH)_3$ precipitate. This phenomenon is caused by the presence of OH^- ions from the NaOH solution which is added when adjusting the pH of the solution to become more alkaline. It can be seen from the presence of a fine yellowish solid that is formed. This precipitation occurs because the K_{sp} value of $Fe(OH)_3$ is 6.3 x 10^{38}.

At pH 7-9, the solubility product value of the ions is greater than the K_{sp} value, so a precipitate of $Fe(OH)_3$ is formed. This causes the phosphate ion that can diffuse into the resin to decrease due to the larger size of $Fe(OH)_3$ compared to the size of the membrane pores, so that it cannot diffuse through the diffusive layer and into the gel layer binder. Phosphate adsorption on FeOOH is influenced by pH because of the contribution of H_2O and OH^- in the exchange of ligands. At high pH the ligand exchange is weak due to the presence of OH^-, which is known to be the strongest monovalent Lewis base. Therefore, phosphate is well absorbed by ferrihydrite at low pH. The ferrihydrite surface has the same amount of $FeOH^{2+}$ and FeO^- at high pH, while at low pH the ferrihydrite surface has a more positive charge and is easy to bind to phosphate [9].

3.5 Effect of Contact Time on phosphate ion adsorption

This test aims to determine the amount of phosphate that can be bound to ferrihydrite. In addition, it can also be determined the maximum ability of ferrihydrite in binding phosphate. The phosphate concentration tested was 0.5-5 ppm. The results obtained are as follows:

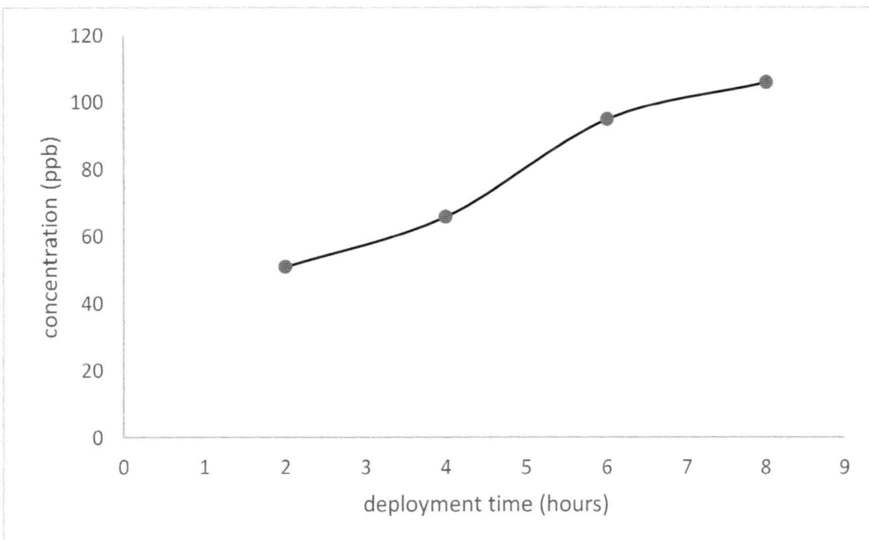

Figure 5. Effect of contact time variation on the concentration of adsorbed phosphate.

From Figure 5 above, it can be seen that the phosphate absorption increases with the increase in time used for contact immersion. The adsorption mechanism on ferrihydrite is generally dominated by ligand exchange in which two hydroxyl groups or water molecules are replaced by one phosphate anion to form a binuclear bidentate complex [10 - 11].

3.6 Effect of concentration variation on phosphate ion adsorption

This test aims to determine the amount of phosphate that can be bound to ferrihydrite. In addition, it can also be determined the maximum ability of ferrihydrite in binding phosphate. The phosphate concentration tested was 1-9 mg/L. The results obtained are as follows:

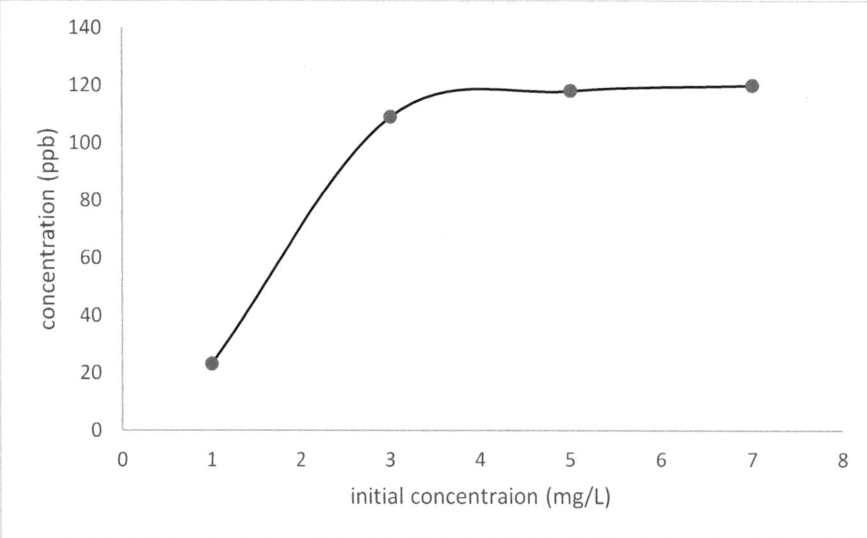

Figure 6. Effect of initial contact concentration on adsorbed phosphate concentration.

From Figure 6 regarding the test of the effect of concentration on the concentration of P above, it can be seen that with increasing phosphate concentration, the concentration of phosphate absorbed is higher. However, at a concentration of 3 – 7 mg/L, it has been sloping. This is probably caused by the binding capacity of ferrihydrite gel which has reached its maximum capacity so that its binding ability is no longer optimal. According to Khairuddin et al [12] that the greater the number of adsorbents used, the more adsorbate absorption, this happens because there are many empty spaces on the surface of the adsorbent and active groups that can bind ions. The results showed that the adsorbent dose given was not too large which could cause maximum absorption of the analyte that could be adsorbed. This is due to the large number of active groups capable of absorbing the adsorbate, especially in the treatment with a low adsorbent concentration while the higher number of active groups causes the adsorbents to not reach the equilibrium point.

4. Conclusion

Based on the experiments that have been carried out and the results of the analysis obtained, it is concluded that binding gel ferrihydrite can be used to adsorb phosphate ions. The ferrihydrite in the binding gel is uniformly dispersed. DGT using ferrihydrite binding gel was most effectively used in a solution with a pH of 5, deployment time of 8 hours, and initial concentration of 3 - 7 mg/L for phosphate measurement.

Acknowledgments

Thanks to the Indonesian Ministry of Education, Culture, Research and Technology for funding the PDUPT research grant, as well as thanks to the Chemistry Laboratory and Geology Laboratory of Hasanuddin University for FTIR and SEM analysis so that this research can be carried out smoothly.

References

[1] G. Elias, S. Díez, H. Zhang, and C. Fontàs, 2019 Chemosphere 245 125671
[2] R. Guibal, R. Buzier, S. Lissalde, and G. Guibaud, 2019 Science of The Total Environment 693 133537
[3] Zhang H, W Davison. 1995. Performance Characteristics of Diffusion Gradient in Thin Film for The in situ Measurement of Trace Metal in Aqueos Solution. Anal Chem. 67: 3391-3400
[4] M. Pesavento, G. Alberti, and R. Biesuz, 2009 Analytica Chimica Acta 631 129–141

[5] V. Smolíková, P. Pelcová, A. Ridošková, J. Hedbávný, and J. Grmela 2020 Analytica Chimica Acta 1102 36–45
[6] H. Du, N. Nie, W. Rao, L. Lu, M. Lei, and B. Tie, 2021 Environmental Pollution 290 118040
[7] M. Engel, J. S. Lezama Pacheco, V. Noël, K. Boye, and S. Fendorf 2020 Science of The Total Environment 750 141485
[8] H. Fu, H. He, R. Zhu, L. Ling, W. Zhang, and Q. Chen, 2020 Science of The Total Environment 764 142846
[9] C. Zhang et al., 2021 International Journal of Hydrogen Energy 46(34) 17720–17730
[10] Y. Wang, J. Wang, P. Li, H. Qin, J. Liang, and Q. Fan, 2021 Environmental Technology & Innovation 23 101615
[11] S. Lan et al., 2021 Science of The Total Environment 791 148225
[12] Khairuddin, Wahab A.W., Buchari, Raya I 2015 International Journal of Applied Chemistry 11(4) 567-577

Tensile Strength and Water Absorption of Oil Palm Mesocarp Fiber Reinforced Polyester Composites: Effect of Volume Fraction of Fiber

Bakri Bakri[1*], Naharuddin[1], Mustafa[1], Kristian Seleng[1], Sri Chandrabakty[1], Muhammad Iqbal[1], Anjar Asmara[1], Ardiansyah Hasan[1], Azizah Ayu Safitri [1]

[1]Mechanical Engineering Department, Tadulako University-Palu, Indonesia

*Corresponding author: bakri@untad.ac.id; bakriuntad@gmail.com

Abstract. Oil palm fiber has been developed for application of composite reinforcement. This fiber can be produced from the fruit of oil palm like mesocarp and empty fruit bunch fibers. In this paper, the oil palm mesocarp fiber used as reinforcement of composite was focused on the variation of volume fraction of fiber in the composite related to tensile strength and water absorption properties. The variation of volume fraction of fiber includes 10%, 20%, 30% and 40%. The unsaturated polyester was used as a matrix. Prior to the use of fiber, it was immersed in sodium bicarbonate ($NaHCO_3$) with 12wt % during 120 hours. Composite fabrication method is hand lay-up with compression molding. The composite samples were tested by the tensile and water absorption tests. Then, the fracture of tensile test was analyzed by scanning electronic microscopy (SEM). The result shows that the tensile strength of composite increases with increase in the volume fraction. Meanwhile, the percentage of water absorption decreases with rise in the volume fraction.

Keyword: Oil palm fruit fiber, polyester, tensile strength, water absorption, composite

1. Introduction

Natural fiber composites have been developed as an alternative to replace in synthetic fiber composite. They have advantages for application due to environmental friendly, no expensive, abundance and lightweight [1-3]. Some natural fiber have been used as reinforcement of composite such as ramie, coir, jute, flax, kenaf, oil palm fiber and sisal. Related to oil palm fiber, such natural fiber which has a potential for many applications because this fiber is abundant in the oil palm industry areas like on the Sulawesi island. For application in reinforcement composite, the content of fiber in composite is very important to investigation.

Poly(butylene succinate) biocomposites with oil palm mesocarp fiber reinforcement has been studied by Then et al. [4]. Their results show that the increase of the oil palm mesocarp fiber content in the composite induced the decrease of tensile strength of composite, but the tensile modulus of composite increases with increase the content of fiber in composite up to 50% fiber content, thereafter reduces with further enhance in fiber content. Lei et al. [5] investigated the various of ramie fiber volume fractions on the mechanical and physical properties of composite. The tensile strength, flexural, strength, flexural modulus and impact strength rise with increase fiber volume fraction (10%, 20% and 30%) in composite and the water absorption of composite rises with increase the volume fraction of fiber. Then, the effect of Alfa fiber volume fraction in polyester composite investigated Brahim and Cheikh [6] by that the tensile stress of composite increases with the increase of fiber content (12%-44%).

In this paper, the work focuses in the influence of oil palm mesocarp fiber volume fraction on tensile strength and water absorption of composite. The sodium bicarbonate solution is used for surface modification of fiber. The tensile fracture surface is also analyzed with scanning electronic microscopy.

2. Materials and Methods

2.1. Materials
The oil palm mesocarp fiber was obtained from fruit of oil palm which is the residual milling of palm fruit oil during production of crude palm oil. Sodium bicarbonate (NaHCO3) was used to treatment of fiber before composing with matrix. The matrix used is unsaturated polyester.

2.2. Methods

2.2.1. Surface Treatment of Fibers
The surface treatment of oil palm mesocarp fiber was conducted to improve the binding of fiber and matrix. 12 wt% NaHCO3 solution was used to immersion of fiber during 120 hours at room temperature [7, 8]. Fiber then was cleaned after immersion and dried during 24 hours in room temperature. After drying in room temperature, fiber is continued drying of fiber in the oven at 110ºC for 1 hour.

2.2.2. Fabrication of composites
After surface treatment of oil palm mesocarp fibers, composites were made with variations in fiber volume fraction, namely 10%, 20%, 30% and 40%. The unsaturated polyester is mixed with a catalyst, then poured into the mold along with the oil palm mesocarp fiber. The hand lay-up method was used followed by compression molding with a manual hydraulic press of up to 4 tons.

2.2.3. Tensile Testing
The tensile testing of composite is carried out using a Universal testing machine. ASTM 638-02 is the standard used in this test. Ten samples were tested for each different volume fraction to take the average value of tensile strength and tensile modulus

2.2.4. Scanning Electronic Microscopy
The tensile fracture surface of oil palm mesocarp fiber composite is captured by scanning electron microscopy (SEM). The SEM type for capturing image is SEM–JEOL JSM 6510 LA type.

2.2.5. Water Absorption Test
The water absorption test of composite was conducted based on the ASTM D 570-98 with the sample dimension is 76.2 long and 25.4 wide. The water absorption percentage of composite was calculated using Eq. (1).

$$\%w = \frac{W_t - W_o}{W_o} \tag{1}$$

where W_t is the wet weight sample after immersion, and W_o is the initial weight of sample.

3. Results and Discussion

3.1. Tensile strength and Tensile Modulus of Composites

The tensile strength of oil palm mesocarp fiber composite is pointed out in Figure 1. The tensile strength of composite increases with increasing the volume fraction of fiber. The highest tensile strength of composite with 40% volume fraction of fiber is 18.63 MPa followed by 30% volume fraction (17.19 MPa), 20% volume fraction (10.75 MPa) and 10% volume fraction (9.70 MPa).

The increase of the tensile strength with increase the volume fraction of fiber is due to the good interfacial strength between fiber and polyester matrix. The result of this study are in line with the study of Lei et al. [5]. The matrix may wrap and bind the fibers well so that they can maintain tensile loads.

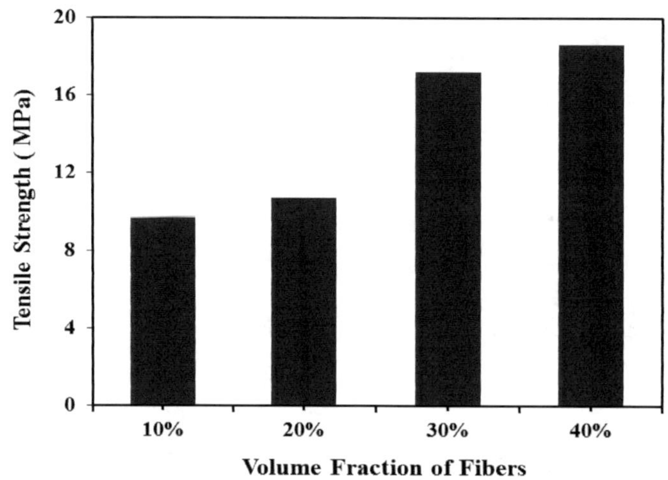

Figure 1. Effect of volume fraction of fiber on tensile strength of composite

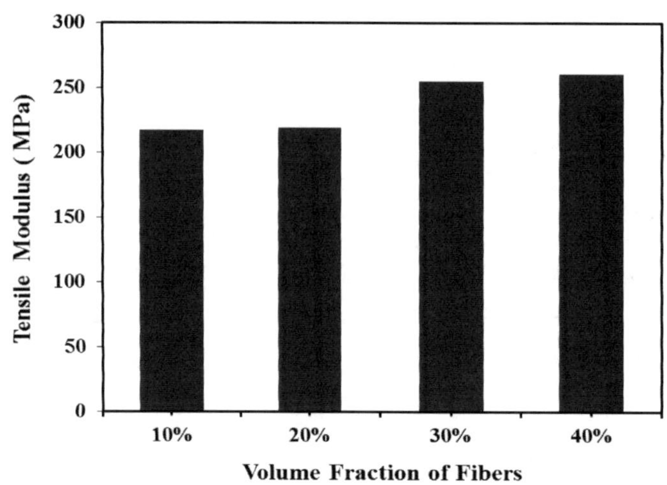

Figure 2. Effect of volume fraction of fiber on tensile modulus of composite

The effect of volume fraction of fiber on tensile modulus of oil palm mesocarp fiber composite were shown in Figure 2. It is displayed that the increase of volume fraction of fiber in composite increase the tensile modulus of composite. It is in line with the increase in the tensile strength of the composite. The tensile modulus of 40% volume fraction of fiber is 260.85 MPa followed by 30% volume fraction

(255.17 MPa), 20% volume fraction (219.52 MPa) and 10% volume fraction (217.52 MPa). The increasing of tensile modulus of composite with increase of volume fraction of fiber indicates that the stiffness of composite increase and also enhance the brittleness of composite [9].

3.2. Scanning Electronic Microscopy Analysis of Composite

The morphology of tensile fracture is analyzed by using Scanning Electron Microscopy as shown in Figure 3. It shows the breaking of fiber and matrix, debonding, void, fiber pull out and adhesion fiber-matrix in the composite. In Figure 3(a), (b), the breaking of matrix and debonding occur after tensile test. This indicates the poor adhesion between fiber and matrix. Fiber pull out from matrix also indicates the poor interfacial bonding of fiber-matrix as shown in Figure 3(c), (d). Good adhesion between fiber and matrix is shown in figure with fiber and matrix

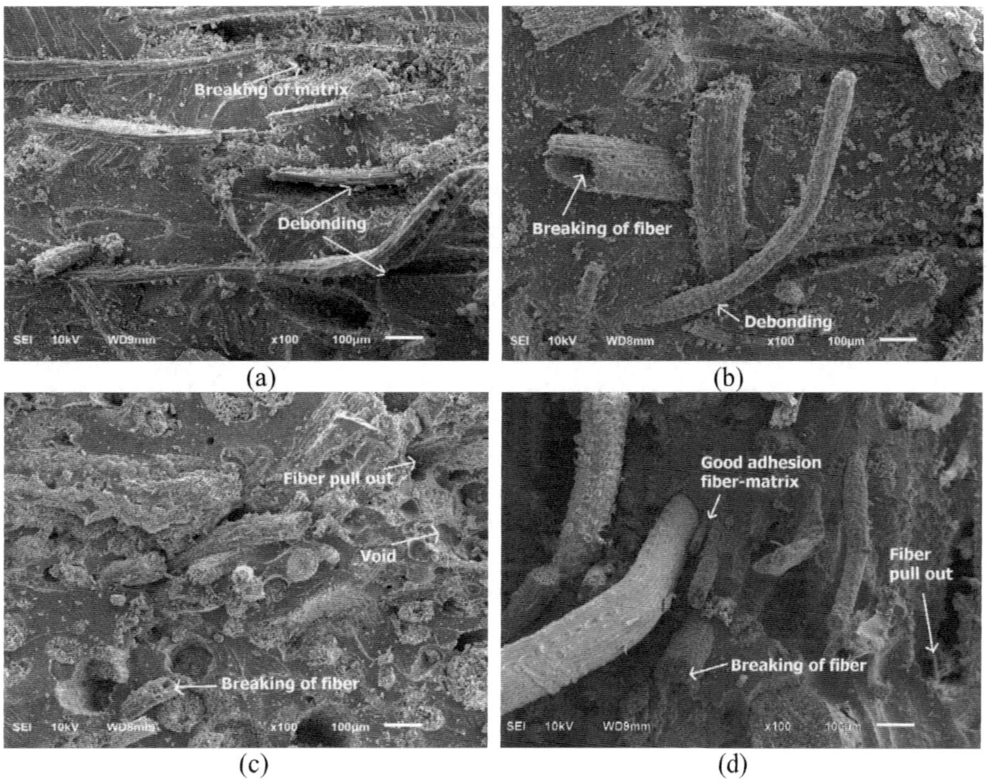

Figure 3. SEM images of tensile fracture of composite with volume fraction of fiber : (a) 10% , (b) 20% , (c) 30% and (d) 40%

3.3. Water Absorption of Composites

The immersion time effect of composite on water absorption is displayed in Figure 4. It can be seen that water absorption of composite enhances sharply until 48 hours after soaking for all volume fraction of fiber [10]. For 10% and 20% volume fraction of fiber, both volume fractions tended to be nearly constant after 48 hours of immersion. However, the 30% and 40% volume fractions tend to increase up to 360 hours of immersion. From Figure 4, it can be seen also that the increase in fiber content in the composite causes the increase in water absorption. This is due to fibers containing polar hydroxide groups which have the ability to absorb water [9,11].

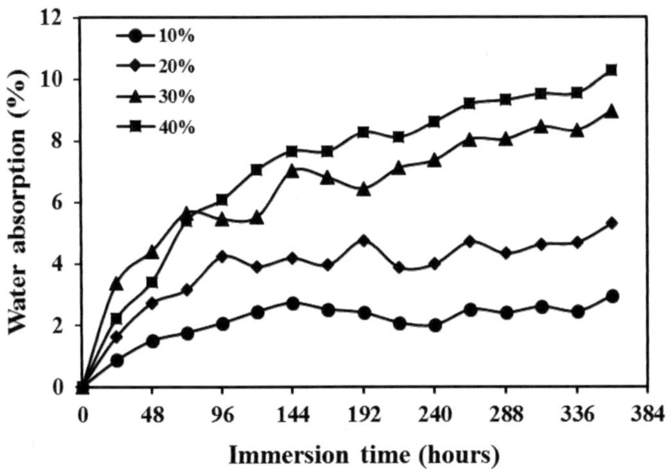

Figure 4. Effect of volume fraction of fiber on water absorption of composite

4. Conclusions

In this work, the tensile strength and water absorption of oil palm mesocarp fiber composite with different volume fraction of fiber were evaluated. The highest tensile strength of composite is 18.63 MPa at 40% volume fraction compared to 10%, 20%, and 30% volume fraction. Meanwhile, water absorption decreases with increasing volume fraction.

5. Acknowledgments

The authors are thankful to Engineering Faculty Tadulako University for funding support to oil palm fruit fiber composite research project. Also, the authors acknowledge the SEM Laboratory, Faculty of Mathematics and Sciences, The Bandung Institute of Technology, for SEM samples characterization

6. References

[1] Arif M F, Megat-Yusoff P S M, and Ahmad F 2009 *J. Reinf. Plast. Compos.* **29** 2105–18.
[2] Dhandapani S, Nayak S K, and Mohanty S 2015 *J. Elast. Plast.* **48** 456–479.
[3] Then Y Y, Ibrahim N A, Zainuddin N, Ariffin H, and Wan Yunus W M Z 2014 *Scie. Res. J.* **11** 1–12.
[4] Then Y Y, Ibrahim N A, Zainuddin N, Ariffin H, Chieng B W, and Wan Yunus W M Z 2015 *BioResources* **10** 2949–2968.
[5] Lei W, Lei W, and Ren C 2006 *Trans. Nonfer. Met. Soc. China* **16** s474–s477.
[6] Brahim S B and Cheikh R B 2007 *Compos. Scie. Tech.* **67** 140–147.
[7] Bakri B, Putra A E E, Mochtar A A, Renreng I, and Arsyad H 2018 *Inter. J. Auto. Mechan. Eng.* **15** 5562–72.
[8] Bakri B, Putra A E E, Renreng I, Arsyad H, and Mochtar A A 2019 *J. Phys. Conf. Ser.* **1242** 012056.
[9] Das G and Biswas S 2016 *IOP Conf. Ser. Mater. Scie. Engi.* **115** 012012.
[10] Renreng I, Bakri B, Chandrabakty S, and Naharuddin 2021 *J. Renew. Mater.* **9** 325-335.
[11] Raghu M J and Goud G 2020 *J. Appl. Agric. Scie. Tech.* **4**, 3–11.

Recovery and improvement of cocoa production with the use of nutrients from chitosan-tonnage blending ash

Suherman*, Sri Mulyani Sabang, Sitti Rahmawati, Supriadi

Departement of Chemistry, Faculty of Mathematic's and Natural Sciences, Tadulako University, Jl. Soekarno Hatta Km 9 Palu 94118, Central Sulawesi, Indonesia.

email: suherman@untad.ac.id

Abstract. Cocoa is one of the plantation crops that provides a high economic contribution to the Province of Central Sulawesi in particular and Indonesia in general. Cocoa is a non-oil and gas export commodity, therefore the government has determined cocoa as a strategic commodity that needs to be increased in productivity. The purpose of this study was to make a blending of chitosan-tonnage (chitosantonase) with a composition of 80%-90% and to characterize it and its optimal ratio to restore soil nutrients so that plants can be resistant to pests and vascular streak dieback (VSD). The cocoa plants used in this study came from Rantekala Village, SigiBiromaru Regency. The results showed that the diameter of cacao pods fed with blended chitosantonase ± 6 weeks was 23 cm, while only 14.5 cm were given ZA/NPK fertilizer. The length of the cocoa pods was 28 cm, while the cocoa pods fed with ZA fertilizer were only 21. The production of cocoa beans fed with blended chitosantonase nutrients increased by 60%. Mixing tonnage-chitosan can be used as nutrition from tonnage-chitosan mixing ash can increase the diameter of cocoa pods by about 37% and length by about 27%.

1. Introduction

Indonesia is the third-largest cocoa producer in the world after Ivory Coast and Ghana. Cocoa is one of the plantation crops that provide a high economic contribution to Central Sulawesi Province in particular and Indonesia in general and be the main income for local farmers. Cocoa is a non-oil and gas export commodity with high demand, therefore the government has determined cocoa as a strategic commodity that needs to be increased in productivity and developed as a source of income for Indonesia and the State of Sulawesi. The government continues to strive to increase crop production by establishing a revitalization of plantations, including an expansion of areas and crop rejuvenation or planting new cocoa seedlings.

The dilemma experienced by cocoa farmers was the level of cocoa crop productivity decreased after ± 10 years of age. The number and the size of the fruit begunbeganecrease, and also not resistant to pests, especially fruit borer (*Conopomorphacramerella*). Cocoa fruit production has decreased by 90%, so that in 2003 until 2008 cocoa fruit production only reached 625 kg/ha/year due to the attacks from these pests and *Vascular Streak Dieback* (VSD). Many efforts have been made by farmers and the government to overcome these pests and diseases, so that could increase cocoa productivity. The efforts that have been done have not been successful. Many cocoa plantations have been displaced, and also the farmers' income has gradually declined.

The research that needs to be followed up to get a solution to the problem of cocoa farmers is the improvement of cultivation techniques that cover various aspects including fertilization, pest control, pruning, and environmental sanitation in the cocoa plantation (Coffee and cocoa research center, 2003). It was identified in 2008, that the Indonesian cocoa farm consisted of 70,000 ha, with the condition of plants that are old, damaged, unproductive, and affected by pests and diseases, and also 145,000 ha of untreated cocoa plants so that those all need intensification [1]. Based on each plant's macro and micronutrient requirements, such as the nutrients absorbed by the roots, then transmitted to the leaves, flowers, and fruits. When the available nutrients are reduced, the fruit component of the plant did not get enough nutrients to maintain the fruit on the plant so it becomes vulnerable to pests, and the fruit becomes thin brown or black (tonnage).

The tonnage from cocoa pods contains minerals that can be converted into nutrients (although relatively low) that are needed by the cocoa plant itself but are inadequate. Therefore, the addition of other materials was needed to enrich the tonnage of cocoa fruits with a mineral content that can be converted into nutrients needed by cocoa plants to increase fruit production. Besides, it can also function against pests and vascular streak dieback (VSD). Chitosan is a new material containing hydroxyl groups (-OH) and amines (-NH). The group with minerals found in tonnage can form new minerals that function as nutrient elements needed by cocoa pods and can spread to cocoa pod scrapers and VSD disease. Based on this case, the problem of cocoa productivity can be overcome by providing nutrient elements from cocoa tonnage with chitosan. -NH which can form NH_4^+ or NO_3^-, these nutrient elements can stimulate photosynthesis, which can cause weak leaves, and control weeds [2].

Nutrients addition to cocoa plants are nutrients that function in plant physiology, both in the form of uncharged molecules, such as H_2O, O_2, and CO_2, and electrically charged ones like K+ and NO_3^-. These nutrients can be absorbed by plants in the form of gas or liquid. CO2, 02, and H2O through the leaves are absorbed in the form of gas, while in the form of liquid the process of absorption through roots starts from H_2O, O_2, and CO_2. Therefore there needs to be a synergy of absorption that occurs through the leaves and roots. Based on the environmental sanitation system of cocoa plantations, environmental cleanliness is needed both cleanliness near the roots/stems and cleanliness of the air around the plantation, the absorption process is as in Figure 1 and Figure 2.

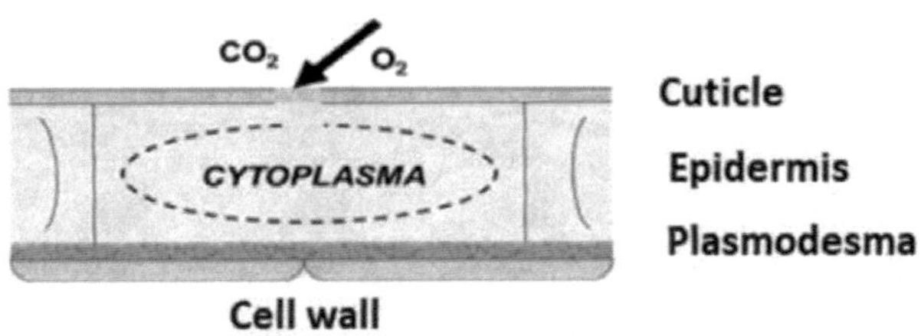

Figure 1. Absorption through leaf epidermal arrangement

The process of absorption of minerals/nutrients needed by plants through the roots are:

a. Active absorption, by the plasma of the outer cells and then passed from cell to cell through Plasmodesma to Endodermis and entering Xylem (Xymplast).
b. Passive absorption, where fluid moves within the cell wall and the space between cells until Endodermis, ions pass through the plasma to continue to the Xylem (Apoplast).

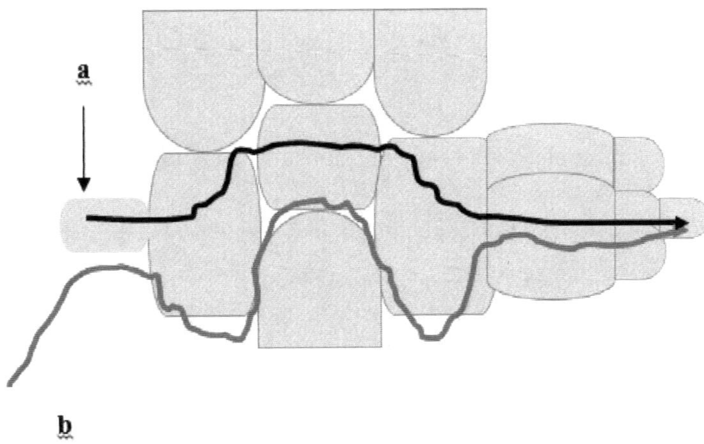

Figure 2. Mineral absorption through roots

Through the roots of nitric acid with amines first, form amino acids then proteins, and continue to form nucleates through DNA and RNA directly to the cell nucleus.

$NO_3^- + NH_4 \rightarrow NH_3 \rightarrow$ Protein $\rightarrow$ DNA & RNA (nucleic acids) to the cell nucleus. This shows vegetative growth. The ions that are formed to be absorbed by plants are subjected to soil pH. High pH will form OH^- so absorbed is NH_4^+, low pH will form H^+ so absorbed is NO_3^-. Similar ions will compete. Fertilizer N from Urea, namely CO $(NH_3)_2$ and ZA NH_4 $(SO_4)_2$, is absorbed by NH_4^+ by OH^- ions, which is good for soils with high pH. In contrast, the provision of nutrient elements from chichitosan–tonnage will compensate for the pH of the soil because there is a balance between the OH^- and NH^- groups owned by chitosan. Therefore, the nutrient elements of tonnage ash-chitosan could increase the absorption activity of roots to be transmitted to the stems, leaves, and fruit. Active absorbing roots will affect the formation of fruit on the stem.

The purpose of this research was to make the blending of tonnage ash, and cocoa pod ash with 80% - 90% chitosan and to characterize it as an optimal ratio to restore soil nutrients in plants so that plants become productive and resistant to pests and VSD. Furthermore, cocoa fruit is deposited as raw material for bispic products.

The study was urgent because the cocoa plant was an important commodity that contribute a lot to the economy of Central Sulawesi Province, and also Indonesia in general. Cocoa plantations are community plantations with a production of 0.87 tons/ha/ year, while the area of cocoa plantations is 162,444 ha. So, cocoa fruit products in this region per year were ± 147,574 tons [3]. The production was lower than the potential of cocoa products, which is 2 - 2.5 tons/ha/year [4,

5]. Based on this potential, cacao fruit products should be 324,888 - 406,110 tons/year. Thus, the potential for cocoa products (fruits) has not been reached above 50%/year. When assessed in terms of the economy, the economic contribution lost from the superior product of cocoa plants was very much for the Province of Central Sulawesi, even for Indonesia.

The main factor that caused the potential of cocoa products that have not been reached above 50% was the loss of soil fertility so that the nutrients needed for cocoa plants to produce fruit are very few. As a result, the cocoa plant was vulnerable to pests. Because of that, the cocoa fruit becomes stunted, and the seeds are hard and become black (tonnage). This result was not mean that nutrients run out/mineral content, but the nutrients they contain cannot compete to develop the fruit by challenging the pest attacks.

The tonnage of cocoa pods, skin, and seeds had a production value that still contains nutrients K, N, fiber, fat, and a number of organic acids. This content has the potential to be used as animal feed or raw material for compost / organic fertilizer to increase soil fertility [6]. Further explained that the cycle of formation of minerals in the soil is the availability/supply of nutrients or minerals in the soil for the purposes of growth and productivity of a plant. Nutrients absorbed by the roots are transported to the stem, then to the leaves and fruit. All parts are weathered in the soil by breaking down minerals and organic compounds. The results of the decomposition can function as a nutrient supply back to plants.

The nutrient supplements given to plants were not worse than the initial conditions. Therefore, the use of tonnage as a nutrient supplement in plants needs to be supplemented with chitosan, so that productivity increases, soil fertility is maintained, and plants and their products are resistant to pests.

Chitosan consists of organic compounds and their groups namely hydroxyl groups (-OH) and Amina groups (-NH). In the mineral N mineral polyblend from tonnage, the composition can be improved by the presence of bonds with –NH from chitosan, while the mineral K can form a base with a hydroxy group from chitosan. Likewise, the composition of organic fats and acids will be enriched with 2-amino-2-deoxy-D-glucose from chitosan. The increase in composition can double as a supply of minerals in plants (cocoa) and protection from fruit borer pests. In addition, chitosan polyblend with tonation ash does not damage soil structures, such as fertilizers and pesticides.

Based on this, chitosan-tonnage ash polyblend for cocoa plants is very suitable to be applied to supply the necessary nutrient elements for cocoa plants to increase their productivity. The supply of nutrient elements from these materials can increase the potential for cocoa fruit production so that the annual cocoa fruit production can reach 2 tons/ha. Achieving this has an effect on improving the economy of people in this region and the country.

Abundant cocoa products can be produced into cocoa flour. The flour can be used as raw material to produce bispik. Bispik was produced to fulfill the needs of community snacks that have nutritional value needed for the health of children and adults and even for parents. It also can increase the economic value of the Province of Central Sulawesi and also Indonesia, and become an Indonesian pride product. Because this product can be an export product.

2. Research methods

2.1. Source of research material

Chitosan from chitin with shrimp as raw material, tonnage was cocoa pods which have been infected with black fruit borers, 1 ha of cocoa plants from an unproductive farmer area, located in Rantekala Village, Gumbasa District, Sigi Biromaru Regency.

1. Chitosan used is derived from chitin which is deacetylation with 50% NaOII,
2. with a degree of deacetylation> 80%
3. Tonnage ash comes from black cacao tones that are grayed on the
4. temperature of 850 ˚C to 900 ˚C.
5. Chitosan polyblend with tonnage ash with a ratio (%) is 1: 3, 2: 2, and 3: 1, the polybags are capsules.
6. The productivity of cocoa plants, the treatment is that community cocoa farms that are less productive are supplied with polyblends from chitosan with tonnage ash, so that they become more productive.
7. The productivity of cocoa plants is compared with the previous results (previous production data were sourced from the data of the cocoa farmers' community) so that the most productive polybags are obtained.

2.2. Chitosan poliblend productivity test-tonnage ash

Third, the chitosan polyblock-tonation ash comparison was applied to less / unproductive cocoa plants. At first, there was a 1-hectare cocoa farm. Age and productivity considerations are the same and cocoa plants are evenly distributed. The area of the garden is divided by 4 ha each and given a code that is the KA, KB, KC, and KD. KA was treated with chitosan polyblend supplement: tonation ash with a ratio of % w/w was 1: 3, KB was given polyblend 2: 2, KC was given polyblend 3: 1, while KD was not treated with polyblend but was still given fruit fertilizer (control).

Furthermore, the development is observed every week, starting from the appearance of flowers, fruit, fruit development, and harvesting until the total yield is obtained (for 1 harvest). Each treatment was compared with the controls. Furthermore, the comparison of the most productive polybags or the greatest results obtained, the chocolate composition test of cocoa fruit (SNI) and determined as the recommended polybags to increase cocoa productivity and resistance to pests. Each application of polyblend added fruit fertilizer.

3. Results and discussion

The results of this study include;

1. The development of cacao fruit in the same two locations, namely the development of cacao fruit which is fertilized by providing additional fertilizer from blending chitosan-tonnage and cocoa plants that are only given fertilizer (ZA).
2. Weight (kg) of cocoa pods after being harvested and dried in both locations Information on these results is in Table 1.

Table 1. Results of observations of the development of cocoa fruit size

No	Time/week	Diameter (cm)		Length measure (cm)	
		Chitosan + tonnage	Fertilization	Chitosan + tonnage	Fertilization
1	1	7.5	6.2	8	7.1
2	2	9	6.7	15.5	9.4
3	3	16.5	9.5	18.0	13.5
4	4	21.5	12.5	22.5	16.5
5	5	22.5	13.7	27.5	19.0
6	6	23.0	14.2	28.0	20.5
7	7	23.0	14.5	28.1	21.0
8	8	23.0	14.5	28.1	21.0

Source: Measurement results from April to June 2019

Based on the results of measurements of the development of cocoa for 8 (eight) weeks or 2 months, information was obtained that cacao was fertilized and added with chitosan polyblend-ash tonnage fruit size is greater than that of cocoa plants which are only fertilized with N fertilizer namely urea or ZA fertilizer (Table 1). Based on the ability of N (Urae/ZA) fertilizer which is able to form NH_4^+ ions, which work actively in high soil pH conditions. But it is difficult to form ions at low pH, so cocoa plants are less productive. Unlike the nutrient elements owned by tonnage-ash ash, the NH group that is owned will be balanced by the hydroxy group (OH-), and vice versa. Because of that, the roots will still actively absorb nutrient elements that are available and needed by the cocoa plant itself. The activity of the roots absorbs available nutrient elements which are transmitted to the stems and fruit so that the plants become more productive and resistant to pest attacks.

But more important to note is the number or weight size of cocoa beans after being harvested dry. This is important because it provides a measure of economic value (selling value) that can increase cocoa farmers' incomes. This means that the welfare of farmers increases. This will have an impact on foreign exchange earnings. Production of dried cocoa beans per Ha. is a cocoa product that is given additional nutrient elements from the chitosan-gray tonnage ash one by one harvest is 160 kg. while the cacao fruit product which is only given fertilizer N (Urea/ZA) is 100 kg. This means the selection occurs by 60%. The thing that causes this is the addition of nutrient elements available in the soil needed by cocoa plants to supply the fruit, namely nutrient elements NH_4^+ and phosphorus (P) which effectively work at high pH. These nutrients are the constituents of ADP/ATP, in cocoa plants so that they can strengthen the stems and stimulate root development, and improve fruit quality. Thus the cocoa beans, the material content is denser/more (high concentration). While only fertilized with nutrient content from N (ZA/urea) fertilizer is only for growth, but not much supply to the fruit seeds. The reason is the supply of nutrient elements to the fruit will be disrupted in soil conditions whose pH is lower. The P nutrient is in the form of P_2O_5 molecule. Water solubility is 0.125 for Gafsa species and 99 is greater for TSP fertilizer types. his reaction in the soil for Gafsa species is neutral-base.

In addition to the phosphorus (p) nutrient elements, it also contains potassium (K) nutrient elements that function as stomatal activity, and enzymes, increasing disease resistance and weight size. As a result, the substance contained in the flesh of fruit seeds is lower, which causes a low weight size. Excess K nutrient elements can cause the oil content in the fruit to decrease so that the quality of the fruit increases. Nutrient K in the form of molecular K_2O is 49-53% for the type of ZK and as much as 52-60% for the MOP type. Both types are soluble in water and the reaction is neutral [7].

Cocoa plants need nutrients such as N, P, K, and Alkali metals, heavy locomotes in small amounts and limited, absorbed by the roots in the form of ions. Then it is continued throughout the plant. But the nutrients that have been absorbed by the roots for the whole network of cocoa plants are very lacking, so it needs to add nutrients to the cocoa plants. But keep in mind, that additional nutrients suitable for cocoa are for fruit growth and production. It is said that plant growth requires nutrient N, fruit production requires alkali metal nutrients, and for fruit quality organic nutrient elements are needed. Not only the growth and increase in production but the ability of cocoa plants to be able to withstand factors from within the soil and beyond. Because the maintenance of these plants requires a technique of accuracy and accuracy. This happens because the cacao plant is a productive plant, which is directly produced in the stem. This means that the absorption of the roots of the ions in the soil to be transported to the stem requires a large thrust. Therefore, continuous continuity is needed so that the fertility and productivity of cocoa remain normal.

Figure 3 development of cocoa fruit

However, what is happening (in Central Sulawesi Province) is that the condition of the cocoa plant has declined, namely the production of fruit has dropped to 50. Thus the yield of cocoa is only about 0.87 - 0.9 tons/ha/year [3]. Normally the yield of cocoa is 2.0 -2.5 tons/ha/year [4, 5]. The decline in cocoa production has an impact on the decline of the people's economy. Based on this hundreds of hectares. abandoned people's cocoa plants, the reason is the product produced is insufficient for maintenance costs. In fact, the production of dried cocoa beans kg/ha/year to 1.02 tons. means an increase of about 60%.

Figure 4. Normal and tonnage cocoa fruit conditions

According to Dirjenbun [8], The level of land suitability for cocoa plants was found to be of 4 types namely land suitable, quite appropriate, somewhat appropriate, and not appropriate. This type is based on agro-climate conditions such as altitude, soil fertility, and the availability of nutrient elements (in large amounts) needed by plants. Because it greatly affects the growth and productivity of cocoa for normal fruiting. It is important to open a cocoa plantation land first and foremost to be considered by the people who cultivate the land is the level of suitability and fertility/availability of nutrient elements in large numbers and can last long so that it requires physical tests and soil chemical tests, but this is very little who do. As a result of the age of the cacao plant more than 10 years, the yield of its products began to decline, thus its quality, the cacao plant, and its fruit range against pest attacks. The quality of cocoa beans can be seen in Table 2

The nutrition content test of the cocoa beans provides information that the cocoa beans are good for consumption to increase human body resistance, for physical growth, and for the development of thinking (for the brain). Therefore cocoa beans can add essential substances to the body under normal and healthy conditions. The quality of cocoa beans is classified into 2 namely; a quality I and quality 2. Physically the quality of cocoa fruit can be expressed by the number of seeds/100 gram samples of cocoa beans with a moisture content of 7%, and levels of foreign objects 0%.

Metal content such as K, Fe, Cu, and Zn are essential substances needed by the human body. potassium takes 3.3 grams, iron 3 mg, Cu 2 mg, and zinc 2.2 mg. This amount is needed for humans who want to diet. Based on these levels, the nutritional content of cocoa flour is good for consumption for dietary purposes [9]. Thus the levels of protein, fat, and carbohydrates are needed by the body in the appropriate amount. Information from table 2 about the levels of protein, fat, and carbohydrates shows the occurrence of balanced nutrition so that both are consumed and do not cause a negative impact on healthy growth and development. A balanced nutritional content of food is needed by humans.

Table 2. Nutrient content of cocoa beans

No	Parameters	Unit	Test Results

1	Potassium	mg/L	2.4
2	Iron	mg/L	0.8
3	Copper	mg/L	0.25
4	Zinc	mg/L	0.1
5	Protein	%	9.28
6	Fat	%	37.30
7	Carbohydrates	mg/100 gr	42.80

Source: Lab Analysis Results. Pend. Chemistry FKIP UNTAD, July 2019

The nutritional content of a food that needs to be considered is its calorific value related to the amount of energy (heat) required by 1 kg of water to raise the temperature of 1 °C (from 15 °C to 16 °C). The calorific value of a food item is related to the respiratory rate, which is a number that shows the ratio of the amount of CO_2 produced by an oxidation reaction to the volume of O_2 used [10]. The nutritional content of cocoa flour has the main components (as food) which are carbohydrates 42.89 mg/100 gr, 37.30% fat, and 9.28% protein. When compared with the main calorie values (carbohydrates, fats, and proteins) of food ingredients are respectively (calories/gram) 4.1; 9.4; and 4.1. The calorie value is safe for consumers.

The important thing to note is the suitability of the use of cocoa flour with human activities, namely light activities, moderate activities, and heavy activities. For lightworkers, the number of calories/day 2000-2400/day per person is needed, while for moderate workers the number of calories/day is 2700-3100, while heavy workers need calories/day: 3400 - 4800 [10]. The calorie value comes from the total food eaten.

4. Conclusion

The development of cacao fruit added by chitosan-tonation polybags is faster and greater than that which is only given fertilization. Production of dried cocoa beans added nutrient elements to fertilizer increased by about 60%. The nutritional content of dried cocoa beans as the main ingredient of the food is needed by consumers (humans), namely for endurance, growth, and body temperature regulators.

References

[1] Manggabarani A 2011 *Konsepsi Gerakan Peningkatan Produksi dan Mutu Kakao (Gernas kakao)* Seminar Evaluasi Pelaksanaan Gernas kakao BAPPENAS Jakarta
[2] Obidzinski K Andriani R Komarudin H & Andrianto A 2012 *Environmental and social impacts of oil palm plantations and their implications for biofuel production in Indonesia* Ecology and Society 17(1)
[3] BPS Sulawesi Tengah 2008 *Tanaman Kakao dan Produksinya* Palu
[4] Raharjo 1999 *Tanaman Kakao dan Potensinya* Sinar Mulya Surabaya
[5] Suhendi 2004 *Tanaman Kakao di Indonesia* Carnesius Jogyakarta
[6] Erlan 2010 *Komposisi Unsur Hara pada Daun dan Buah Kakao* Bionatura 8 (3)
[7] PPKS Medan 2010 *Peranan Unsur hara Pada Tanaman Kakao* Materi Pelatihan Jakarta

[8] Dirjenbun 2002 *Tanaman Kakao, kompetensi dan permasalahannya* Dinas Pertanian Tanaman Pangan Jakarta
[9] Darmono 1995 *Logam dalam sistem biologi mahluk hidup* UI – Press (UIP) Jakarta
[10] Wirahadikusumah M 1985 *Biokimia: Metabolisme energi, karbohidrat, dan lipid.* ITB Bandung

IOP Conf. Series: Earth and Environmental Science 1075 (2022) 012005 doi:10.1088/1755-1315/1075/1/012005

Arbuscular mycorrhizal fungi in cocoa plantation affected by different elevations and soil physical-chemical properties

N Edy[1,2*], E Zakaria[2], A Anshary[1,2], S Saleh[1,2*], I Lapanjang[1], HN Barus[1], I Lakani[1]

[1]Faculty of Agriculture Tadulako University, Palu Indonesia.
[2]Post Graduate Tadulako University, Palu Indonesia.

*Email: nuredy@untad.ac.id

Abstract. Arbuscular mycorrhiza fungi (AMF) play an essential role in various agroecosystems since they build symbiosis mutualism with plants. Understanding the differences in the distribution and diversity of AMF in cocoa plants with different elevations is important to explore the bio-ecological aspects of AMF. This study examined the differences in the number and types of AMF spores in cocoa plantations at three different elevations: <500, >500-1000, and >1000 masl. Environmental factors such as the physical and chemical properties of the soil were also investigated. This study found that the number of AMF spores was higher in the lowlands (<500 masl) than in the highlands. Rarefaction curves show that all soil samples taken are adequate for diversity analysis. Multivariate analysis with NMDS followed by the goodness of fit statistical test using the Bray Curtis method showed that a clear distribution pattern in the AMF community structure was significantly influenced by chemical and physical factors of the soil in the form of phosphorus, potassium, sodium, magnesium, total nitrogen, clay, and soil texture.

1. Introduction

Arbuscular mycorrhizae have considerable essential roles in soil, including helping organic decomposition, translocation of soil nutrients, increasing the ability of plant roots to absorb water, and maintaining the plant against plant pathogen and abiotic stresses [1–6]. Arbuscular mycorrhizae are obligate fungi, where their survival depends on their hosts [7–9]. Mycorrhizal infection begins from the germination of mycorrhizal spores in the soil. The spores, subsequently, form hyphae and appressorium. Then the extra-radical mycelia AMF connect the network between plant roots and the rhizosphere to absorb inorganic nutrients. Infected plant roots by AMF will develop arbuscules as a structure that functions to exchange nutrients between AMF and plants [10].

The role of mycorrhizae in different ecosystems has been widely studied [2]. However, information on the distribution and factors affecting the abundance and diversity of AMF in cocoa plantations at different elevations is still limited. This information is essential to explain the biology and ecology of AMF. Further, the AMF can be used to be developed as a plant symbiont agent that can help fertility and protect plants from pathogen disturbances and extreme environmental conditions, such as drought and salinity stresses.

The cocoa plantations are an agricultural ecosystems that are closely related to AMF [11–13]. Profitable yielding cocoa has been reported to be associated with AMF [14,15]. The elevation may affect different abundance and diversity of AMF [16–18].

Differences in the abundance and diversity of AMF can be influenced by environmental factors, including the number and type of plant vegetation, soil type, the proportion of soil physical structure content, and soil nutrients [19–21]. Previous studies found no significant difference in AMF at different altitudes based on plant diversity with a unimodal pattern [22]. However, few studies reported a different abundance of AMF with increasing or decreasing the elevations [16–18]. In the aspect of soil conditions, soil fertility directly related to the content of N and P is reported to have a significant effect on the presence of AMF [23–26].

This study aimed to examine the effect of the elevations of the cocoa plantation location, soil nutrient content, and soil physical structure on the abundance and diversity of AMF.

2. Methodology

2.1. Study sites and soil sampling

The study was conducted in Central Sulawesi, Indonesia, on three different elevations of cocoa plantations, located in Maku (0-500 m asl), Ue Rani (500-1,000 m asl), and Watumaeta (> 1,000 m asl). The soil cores (5 ′ 20 cm; diameter ′ deep). The soil was then transported to the plant pest and disease laboratory, Faculty of Agriculture Tadulako University.

2.2. AMF spore extraction

The AMF spore extraction was done by protocol pour-filtered followed by modification introduction by INVAM (https://invam.wvu.edu). Composite soil materials 200 g with 1 L water were passed on 200-, 125-, and 45-mm sieving, respectively. AMF determination was done on the genus level [11,27].

2.3. Soil physical and chemical analysis

Soil analysis include physical and chemical. Sand, mist, clay, and class texture were measured for soil physical analysis. Also, pH determined by electrometry method; P, K, Na, Ca, Mg, cation exchange capacity (CEC) were measured by using NH_4OAC 1N pH 7 protocol; total N was calculated by using Kjeldahl method, and C organic was measured by Walkey and Black protocol [28–31].

2.4. Data analysis

A rarefaction curve was applied to estimate the sufficiency of the soil sample used in this study for diversity analysis. Then, the Shannon diversity index was applied, and non-metric dimensional scaling (NMDS) was used to determine dissimilarities of AMF in different elevations of cocoa plantations with the Bray Curtis method. Those analyses were run on PAST version 4.08 [32]. Soil physical and chemical effect on AMF was used as environmental variables on NMDS then subjected to calculate the goodness of fit statistics (R^2) to determine the significant variable tested. The analysis was conducted in R (version R 4.11) and R Studio (version 1.4.1717) by using envit function Vegan package [33–35].

3. Results and discussion

3.1. Abundance and diversity of AMF in different elevations

A total of 2,266 spores were collected from the study sites. Table 1 shows that the higher the elevation, the lower the spores of AMF present. On the elevations of > 500 - 1,000 and > 1,000 m asl doesn't show big differences of AMF spores collected. Compared to the elevation of < 500 m asl, the spore increases twice.

The rarefaction curve shows that all samples collected from different elevations reached the plateau line at a chosen number of samples (Figure 1), which means that the samples are adequate for further diversity analysis. Similar to the abundance of AMF, the Shannon diversity index shows the same pattern. The lower elevation shows higher diversity than the lower elevation (Figure 2).

The hypothesis related to this study may cause by above-ground vegetation and soil climate condition [36–39]. Until now, information about AMF colonization's specifics in certain plants remains a question. A general point of view relates the between AMF and plants as different ecological strategies indicating functional groups form based on ecosystem conditions. The previous study has shown differences in the composition of the AMF community, for example, found in grasslands, vegetable crops, trees, forests or in certain plant species [36,40–43].

In addition, environmental climate conditions may also influence the abundance of AMF. Sosa-Hernández et al. [39], reported the abundant and unique community of AMF in subsoil adapted to the soil rhizosphere conditions.

Table 1. The abundance of AMF in different elevation of cocoa plantation

Elevation (m asl)	Number of AMF spore
< 500	1,175
> 500 - 1,000	496
> 1,000	595

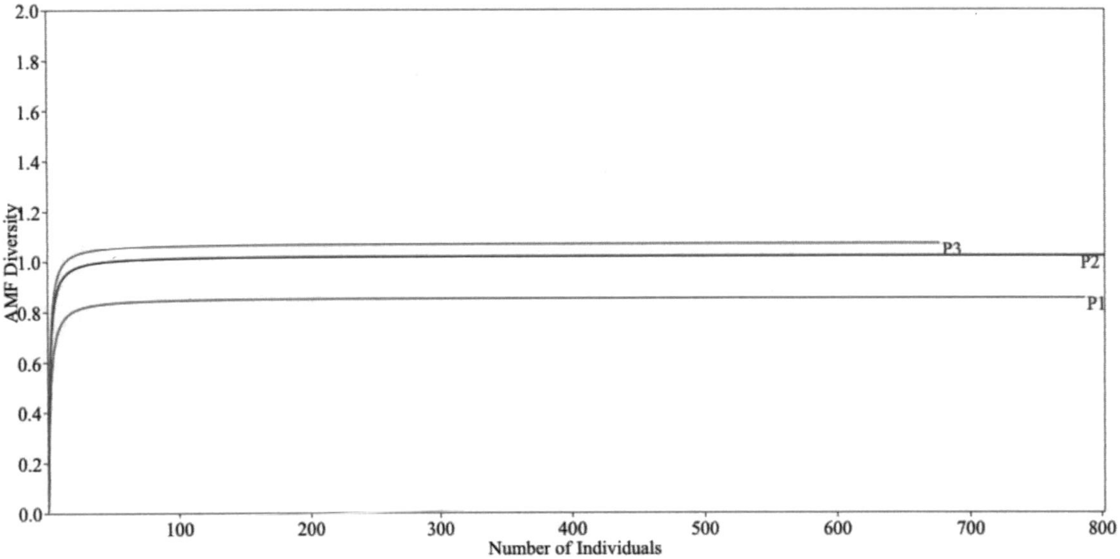

Figure 1. Rarefaction curve of sampling effort on AMF in different elevation of cocoa plantation. P1: elevation > 500 - 1,000 m asl; P2: elevation of > 1,000; P3 elevation of < 500 m asl.

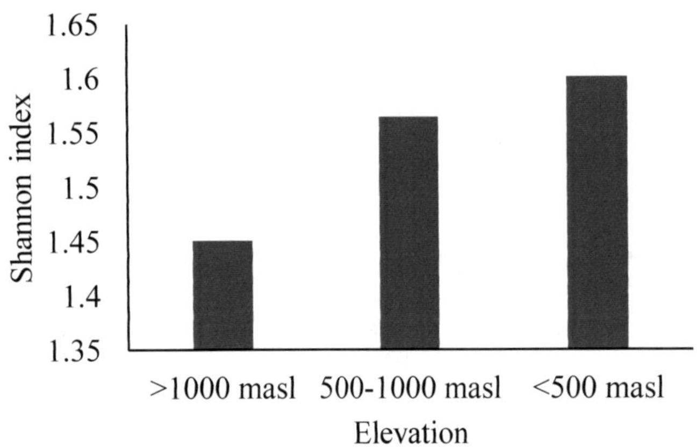

Figure 2. Shannon diversity index of AMF in different elevation of cocoa plantation

3.2. AMF communities affected by elevation and soil properties

The non-metric multidimensional scaling (NMDS) was applied to assess the community structure of AMF in different elevations. Environmental factors involved were pH, P, K, Na, Ca, Mg, CEC, N total, C organic, sand content, mist content, clay content, and soil texture (Figure 3). The NMDS stress was 0.07, an accepted value that explains low error distance between samples and low similarity in the group assessed. Additionally, the environmental factors determined showed that P, K, Na, Mg, N total, clay content, and soil texture were related to AMF community structures in different elevations (Table 2).

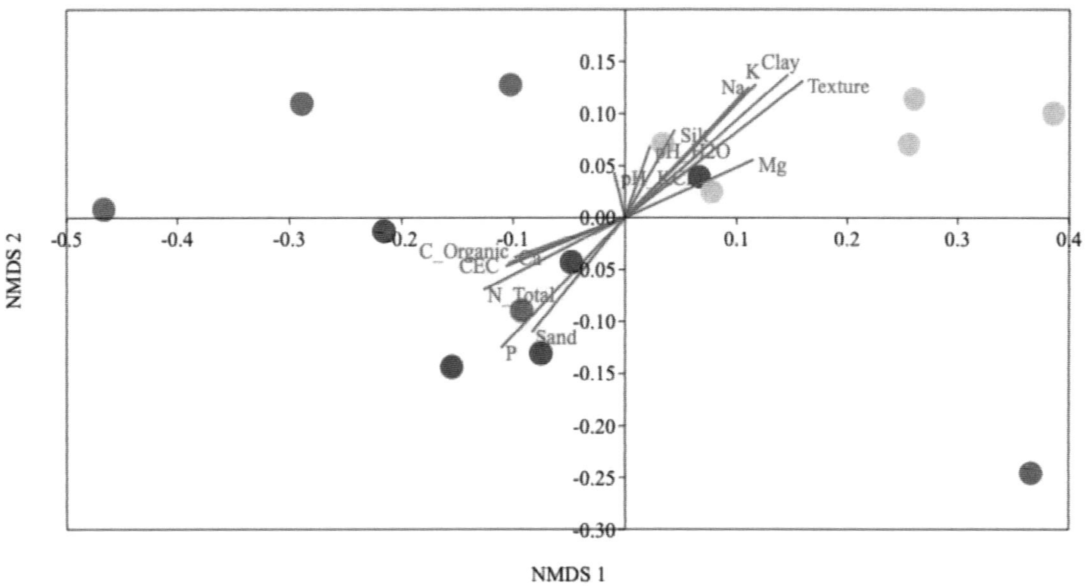

Figure 3. Ordination of NMDS of AMF communities using the Bray-Curtis dissimilarity to environmental factors that influence it. Red dots: elevation > 500 - 1,000 m asl; blue dots: elevation of > 1,000; green dots: elevation of < 500 m asl.

Tabel 2. Goodness of fit (R^2) statistics of AMF and environmental factors according to NMDS ordination. The significance is based on the 999 permutations.

Environmental Variable	R^2	*P*
pH H2O	0.0949	0.624
pH KCl	0.0629	0.789
P	0.4063	**0.021***
K	0.4397	**0.013** *
Na	0.4085	**0.017** *
Ca	0.2347	0.144
Mg	0.3124	**0.053***
CEC	0.2707	0.105
N Total	0.3778	**0.029** *
C Organic	0.2582	0.104
Sand	0.2795	0.100
Mist	0.1430	0.421
Clay	0.5979	**0.003** **
Soil texture	0.6581	**0.002** **

Mycorrhizal diversity in ecosystems is influenced by many factors, including biotic and abiotic factors. The role of AMF in phosphorous and nitrogen uptake has been studied broadly. However, limited studies have explicitly discussed the relation of AMF to K, Na, and CEC. A study related to available potassium concentration has been reported by Song et al.[44], which shows that available P and K are the most edaphic factors affecting the AMF communities in soil.

Related to sodium (Na) in soil, AMF has been reported to help the plant adapt to the saline condition [45–47]. A meta-analysis has proven the ability of AMF to absorb water and nutrients in soil conditions with saline conditions [48]. Thus, plants will be more efficient in carrying out photosynthesis and maintaining nutritional stability to form a resistance system to salinity stress. The ability of plants to resist salinity is influenced by the ability of AMF to produce antioxidant enzymes [49].

Soil CEC is one of the fertility indicators. Cation exchange capacity refers to clay content, type of clay, and organic compound content in the soil. CEC variation in different ecosystems is related to humus and clay proportions [50,51]. The higher CEC is related to soil fertility. Here in this study, CEC is one of the soil indicators significantly related to AMF community structure in different elevations of AMF. This study found that clay content and soil texture are significantly related to AMF community structure on soil physical properties. Soil texture, in this study sites, vary on sandy loam to clay sandy loam.

Conclusion

The AMF spore abundance and diversity was higher in the low elevation of cocoa plantations. Assessing the environmental factors related to the soil chemical and physical properties, P, K, Na, Mg, N total, clay content, and soil texture was significantly related to the community structures of AMF. Our results suggested that AMF triggered and contributed to the agricultural ecosystem of cocoa in different elevations. Thus, the AMF can be developed as potential agents to support the cocoa plantation ecosystem and production.

Acknowledgements

We thank the Ministry of Education, Culture, Research, and Technology Republic of Indonesia through Postgraduate Tadulako University research grants to support this study. Contract number: 453.ad/UN28.2/PL/2021.

Author Contribution

The study was developed by NE and IL. Fieldwork and AMF spore extraction by EZ. Soil analysis managed by NE. Data analysis by NE, AA, HNB, and IL. Draft preparation by NE, AA, and SS. Article final review by NE. Funding acquisition by NE and AA.

References

[1] Mackay J E, Cavagnaro T R, Müller Stöver D S, Macdonald L M, Grønlund M and Jakobsen I 2017 A key role for arbuscular mycorrhiza in plant acquisition of P from sewage sludge recycled to soil *Soil Biology and Biochemistry* **115** 11–20

[2] Bhale U N, Bansode S A and Singh S 2018 Multifactorial Role of Arbuscular Mycorrhizae in Agroecosystem *Fungi and their Role in Sustainable Development: Current Perspectives* ed P Gehlot and J Singh (Singapore: Springer) pp 205–20

[3] Borde M, Dudhane M and Kulkarni M 2017 Role of Arbuscular Mycorrhizal Fungi (AMF) in Salinity Tolerance and Growth Response in Plants Under Salt Stress Conditions *Mycorrhiza - Eco-Physiology, Secondary Metabolites, Nanomaterials* ed A Varma, R Prasad and N Tuteja (Cham: Springer International Publishing) pp 71–86

[4] Begum N, Qin C, Ahanger M A, Raza S, Khan M I, Ashraf M, Ahmed N and Zhang L 2019 Role of Arbuscular Mycorrhizal Fungi in Plant Growth Regulation: Implications in Abiotic Stress Tolerance *Frontiers in Plant Science* **10** 1068

[5] Debeljak M, van Elteren J T, Špruk A, Izmer A, Vanhaecke F and Vogel-Mikuš K 2018 The role of arbuscular mycorrhiza in mercury and mineral nutrient uptake in maize *Chemosphere* **212** 1076–84

[6] Zou Y-N, Wu Q-S and Kuča K 2021 Unravelling the role of arbuscular mycorrhizal fungi in mitigating the oxidative burst of plants under drought stress *Plant Biology* **23** 50–7

[7] Chandrasekaran M 2020 A Meta-Analytical Approach on Arbuscular Mycorrhizal Fungi Inoculation Efficiency on Plant Growth and Nutrient Uptake *Agriculture* **10** 370

[8] Wipf D, Krajinski F, van Tuinen D, Recorbet G and Courty P-E 2019 Trading on the arbuscular mycorrhiza market: from arbuscules to common mycorrhizal networks *New Phytologist* **223** 1127–42

[9] Keymer A, Pimprikar P, Wewer V, Huber C, Brands M, Bucerius S L, Delaux P-M, Klingl V, Röpenack-Lahaye E von, Wang T L, Eisenreich W, Dörmann P, Parniske M and Gutjahr C 2017 Lipid transfer from plants to arbuscular mycorrhiza fungi ed G Stacey *eLife* **6** e29107

[10] Smith S E and Read D J 2009 *Mycorrhizal symbiosis* (Amsterdam: Elsevier/Acad. Press)

[11] Edy N, Zakaria E K, Lakani I, and Hasriyanti 2019 Forest conversion into cacao agroforestry and cacao plantation change the diversity of arbuscular mycorrhizal fungi *IOP Conf. Ser.: Earth Environ. Sci.* **270** 012015

[12] Sitepu M E, Sari W P and Dwipa I 2021 Exploration and Identification of Arbuscular Mycorrhizal Fungi (AMF) in the Rhizosphere of Cocoa (Theobroma cacao. L) in West Sumatra *JERAMI Indonesian Journal of Crop Science* **4** 17–22

[13] Nurhalisyah, Pongpisutta R, Rattanakreetakul C and Kaewgrajang T 2563 Role of arbuscular mycorrhizal fungi (AMF) in cocoa (Theobroma cacao L.) seedlings growth *National Research Council of Thailand* **48** 923–32

[14] Nasaruddin, Syaiful S A, Farid B M, Ridwan I, Mantja K and Utami W 2020 Effectiveness of soil tillage and Arbuscular Mycorrhizal (AM) fungi inoculation on fruit development of the cocoa plant (Theobroma cacao L.) *IOP Conf. Ser.: Earth Environ. Sci.* **486** 012118

[15] Padjung R, Saad S H, Bahrun A H and Ridwan I 2019 Growth and development of Theobroma cacao seedlings as a response to different dosages of vermicompost and arbuscular mycorrhizal fungi *IOP Conf. Ser.: Earth Environ. Sci.* **343** 012017

[16] Zhao F, Feng X, Guo Y, Ren C, Wang J and Doughty R 2020 Elevation gradients affect the differences of arbuscular mycorrhizal fungi diversity between root and rhizosphere soil *Agricultural and Forest Meteorology* **284** 107894

[17] Liu M, Yue Y, Wang Z, Li L, Duan G, Bai S and Li T 2020 Composition of the arbuscular mycorrhizal fungal community and changes in diversity of the rhizosphere of Clematis fruticosa over three seasons across different elevations *European Journal of Soil Science* **71** 511–23

[18] Shi Z, Yin K, Wang F, Mickan B S, Wang X, Zhou W and Li Y 2019 Alterations of Arbuscular Mycorrhizal Fungal Diversity in Soil with Elevation in Tropical Forests of China *Diversity* **11** 181

[19] Frater P N, Borer E T, Fay P A, Jin V, Knaeble B, Seabloom E, Sullivan L, Wedin D A and Harpole W S 2018 Nutrients and environment influence arbuscular mycorrhizal colonization both independently and interactively in Schizachyrium scoparium *Plant Soil* **425** 493–506

[20] Xu X, Chen C, Zhang Z, Sun Z, Chen Y, Jiang J and Shen Z 2017 The influence of environmental factors on communities of arbuscular mycorrhizal fungi associated with Chenopodium ambrosioides revealed by MiSeq sequencing investigation *Sci Rep* **7** 45134
[21] Melo C D, Walker C, Krüger C, Borges P A V, Luna S, Mendonça D, Fonseca H M A C and Machado A C 2019 Environmental factors driving arbuscular mycorrhizal fungal communities associated with endemic woody plant Picconiaazorica on native forest of Azores *Ann Microbiol* **69** 1309–27
[22] Opik M, Vanatoa A, Vanatoa E, Moora M, Davison J, Kalwij J M, Reier U and Zobel M 2010 The online database MaarjAM reveals global and ecosystemic distribution patterns in arbuscular mycorrhizal fungi (Glomeromycota) *New Phytol.* **188** 223–41
[23] Mei L, Yang X, Zhang S, Zhang T and Guo J 2019 Arbuscular mycorrhizal fungi alleviate phosphorus limitation by reducing plant N:P ratios under warming and nitrogen addition in a temperate meadow ecosystem *Science of The Total Environment* **686** 1129–39
[24] Xiao D, Che R, Liu X, Tan Y, Yang R, Zhang W, He X, Xu Z and Wang K 2019 Arbuscular mycorrhizal fungi abundance was sensitive to nitrogen addition but diversity was sensitive to phosphorus addition in karst ecosystems *Biol Fertil Soils* **55** 457–69
[25] Han Y, Feng J, Han M and Zhu B 2020 Responses of arbuscular mycorrhizal fungi to nitrogen addition: A meta-analysis *Global Change Biology* **26** 7229–41
[26] Sheldrake M, Rosenstock N P, Mangan S, Revillini D, Sayer E J, Olsson P A, Verbruggen E, Tanner E V J, Turner B L and Wright S J 2018 Responses of arbuscular mycorrhizal fungi to long-term inorganic and organic nutrient addition in a lowland tropical forest *ISME J* **12** 2433–45
[27] Anon Methods | INVAM International Culture Collection of (Vesicular) Arbuscular Mycorrhizal Fungi | West Virginia University
[28] Gee G W and Bauder J W 1986 Particle-size Analysis *Methods of Soil Analysis* (John Wiley & Sons, Ltd) pp 383–411
[29] Tadesse T, Haque I and Aduayi E A 1991 *Soil, plant, water, fertilizer, animal manure and compost analysis manual* (International Livestock Centre for Africa)
[30] Walkley A and Black I A 1934 An examination of the degtjareff method for determining soil organic matter, and a proposed modification of the chromic acid titration method *Soil Science* **37** 29–38
[31] Kjeldahl J 1883 Neue Methode zur Bestimmung des Stickstoffs in organischen Körpern *Fresenius, Zeitschrift f. anal. Chemie* **22** 366–82
[32] Hammer Ø, Harper D A T and Ryan P D 2001 PAST-Palaeontological statistics *www. uv. es/~pardomv/pe/2001_1/past/pastprog/past. pdf, acessado em* **25** 2009
[33] Oksanen J, Blanchet F G, Kindt R, Legendre P, Minchin P R, O'Hara R B, Simpson G L, Solymos P, Stevens M H H and Wagner H 2013 Package 'vegan' *R Packag ver* **254** 20–8
[34] R Core Team 2020 R: A language and environment for statistical computing. R Foundation for Statistical Computing, Vienna, Austria. URL https://www.R-project.org/.
[35] RStudio Team 2019 RStudio: Integrated Development for R. RStudio, Inc., Boston, MA URL http://www.rstudio.com/.
[36] Davison J, García de León D, Zobel M, Moora M, Bueno C G, Barceló M, Gerz M, León D, Meng Y, Pillar V D, Sepp S-K, Soudzilovaskaia N A, Tedersoo L, Vaessen S, Vahter T, Winck B and Öpik M 2020 Plant functional groups associate with distinct arbuscular mycorrhizal fungal communities *New Phytologist* **226** 1117–28

[37] Oehl F, Laczko E, Oberholzer H-R, Jansa J and Egli S 2017 Diversity and biogeography of arbuscular mycorrhizal fungi in agricultural soils *Biol Fertil Soils* **53** 777–97
[38] López-García Á, Varela-Cervero S, Vasar M, Öpik M, Barea J M and Azcón-Aguilar C 2017 Plant traits determine the phylogenetic structure of arbuscular mycorrhizal fungal communities *Molecular Ecology* **26** 6948–59
[39] Sosa-Hernández M A, Leifheit E F, Ingraffia R and Rillig M C 2019 Subsoil Arbuscular Mycorrhizal Fungi for Sustainability and Climate-Smart Agriculture: A Solution Right Under Our Feet? *Frontiers in Microbiology* **10** 744
[40] Lukács A F and Kovács G M 2019 Effect of aboveground plant conditioner treatment on arbuscular mycorrhizal colonization of tomato and pepper *Horticultural Science* **46 (2019)** 208–14
[41] Trinchera A, Testani E, Roccuzzo G, Campanelli G and Ciaccia C 2021 Agroecological Service Crops Drive Plant Mycorrhization in Organic Horticultural Systems *Microorganisms* **9** 410
[42] Kokkoris V, Lekberg Y, Antunes P M, Fahey C, Fordyce J A, Kivlin S N and Hart M M 2020 Codependency between plant and arbuscular mycorrhizal fungal communities: what is the evidence? *New Phytologist* **228** 828–38
[43] Šmilauer P and Šmilauerová M 2000 Effect of AM symbiosis exclusion on grassland community composition *Folia Geobotanica* **35** 13–25
[44] Song J, Chen L, Chen F and Ye J 2019 Edaphic and host plant factors are linked to the composition of arbuscular mycorrhizal fungal communities in the root zone of endangered Ulmus chenmoui Cheng in China *Ecology and Evolution* **9** 8900–10
[45] Santander C, Ruiz A, García S, Aroca R, Cumming J and Cornejo P 2020 Efficiency of two arbuscular mycorrhizal fungal inocula to improve saline stress tolerance in lettuce plants by changes of antioxidant defense mechanisms *Journal of the Science of Food and Agriculture* **100** 1577–87
[46] Bencherif K, Dalpé Y and Lounès Hadj-Sahraoui A 2019 Arbuscular Mycorrhizal Fungi Alleviate Soil Salinity Stress in Arid and Semiarid Areas *Microorganisms in Saline Environments: Strategies and Functions* Soil Biology ed B Giri and A Varma (Cham: Springer International Publishing) pp 375–400
[47] Parvin S, Van Geel M, Yeasmin T, Lievens B and Honnay O 2019 Variation in arbuscular mycorrhizal fungal communities associated with lowland rice (Oryza sativa) along a gradient of soil salinity and arsenic contamination in Bangladesh *Science of The Total Environment* **686** 546–54
[48] Chandrasekaran M, Chanratana M, Kim K, Seshadri S and Sa T 2019 Impact of Arbuscular Mycorrhizal Fungi on Photosynthesis, Water Status, and Gas Exchange of Plants Under Salt Stress–A Meta-Analysis *Frontiers in Plant Science* **10** 457
[49] Chang W, Sui X, Fan X-X, Jia T-T and Song F-Q 2018 Arbuscular Mycorrhizal Symbiosis Modulates Antioxidant Response and Ion Distribution in Salt-Stressed Elaeagnus angustifolia Seedlings *Frontiers in Microbiology* **9** 652
[50] Oktaba L, Odrobińska D and Uzarowicz Ł 2018 The impact of different land uses in urban area on humus quality *J Soils Sediments* **18** 2823–32
[51] Yadav M and Kumar D 2021 Parameters of Soil Chemistry *Applied Soil Chemistry* (John Wiley & Sons, Ltd) pp 197–214

Liveweight gain, change in body dimension and condition score of Donggala bulls fed corn stover supplemented with different tree legume leaves

Marsetyo[1], Y Rusiyantono[1] and I W Sulendre[1]

[1]Department of Animal Science, Tadulako University, Palu Central Sulawesi, Indonesia

marsetyomarsetyo@yahoo.co.uk

Abstract. An experiment was done to examine the effect of the addition of different legume leaves on live weight gain, change in body dimension, and condition score of Donggala bulls fed corn stover (CS). Twenty-eight Donggala bulls (initial body weight (BW) 179.48±3.66 kg) were blocked by BW in a randomized complete block design with 4 treatments and 7 replicates. The treatments were CS ad libitum, CS plus Desmanthus virgatus (CSD), CS plus Gliricidia sepium (CSG), and CS plus Leucaena leucocephala (CSL). All supplements were offered at level 1% BW dry matter (DM)/day. The experiment was run for 16 weeks, with 2 and 14 weeks for the introduction and measurement period, respectively. The parameters observed include average daily live weight gain (ADG), wither height, heart girth, body length, and body condition score (BCS). It was observed that the addition of different tree legumes leaves lifted up significantly ($P<0.05$) ADG, change in heart girth, BCS but did not affect significantly ($P>0.05$) change of wither height and body length of Donggala bulls fed CS. The addition of Leucaena resulted in the highest increase ($P<0.05$) in ADG, change of heart girth, and BCS. It was concluded that supplementation of tree legume leaves increased the growth performance of Donggala bulls fed CS.

Keywords: Donggala bull, tree legume, growth, body dimension

1. Introduction

Donggala cattle (Bos indicus) have been recognized as an Indonesian native breed since 2014 by the Ministry of Agriculture of the Republic of Indonesia. These cattle are the predominant breed that is reared by smallholder farmers in Central Sulawesi province and surroundings, Indonesia. Physically Donggala cattle have many similarities with Ongole crossbreed cattle such as color, horn, hump, and wattle. This is because major genetic of Donggala cattle from Ongole cattle.

Nowadays many farmers in Central Sulawesi practice fattening Donggala bulls as part of their farming system. Farmers, in general, fed their bulls with agricultural byproducts such as corn stover, rice straw, or native grass as a single feed. With this practice, the average daily gain (ADG) of Donggala bulls is low. Marsetyo and Rusiyantono [1] reported that ADG of Donggala bulls fed CS was 0.34 kg/day. With this ADG performance, it takes a longer time (8-10 months) for Donggala bull to reach selling weight.

Supplementing cattle with concentrates that contained energy and a source of protein can increase ADG and reduce the time taken to attain market weight in finishing [2,3]. Marsetyo and Rusiyantono [1] reported that the addition of a mixture of rice bran and palm kernel meal (1:1) at level 1% live weight (W)/day of Donggala bulls fed corn stover increased ADG from 0.34 to 0.63 kg/day. Ba et al. [2] recorded that supplementation of Brahman crossbreed cattle with concentrate (mixture of cassava chips, rice bran, crushed rice grain, fishmeal, salt, and urea) at level 1.2%W/d, increased ADG from 0.09 kg/d (without supplement) to 0.79 kg/d. However, at the village level, concentrate ingredients are often not available for smallholder farmers. To some extent, the price of concentrate is not expensive, therefore farmers cannot afford this. The use of forage tree legumes and shrubs such as Desmanthus (Desmanthus virgatus) gliricidia (Gliricidia sepium) Leucaena (Leucaena leucocephala) leaves offer an alternative way of providing high-quality feed to fatten Donggala bulls in village systems, where these shrub and tree legumes are available. Previous studies [4,5] reported that supplementing Napier grass fodder with foliage from the tree legumes, Leucaena zeucocephala and Gliricidia sepium, improved the dry matter intake and live-weight gain of young cattle. The objective of this experiment was to test the influence of addition shrub and tree legume leaves on ADG, change in body dimension, and condition score of Donggala bulls fed corn stover.

2. Material and Methods

2.1. Location and Animals

An experiment was conducted at the experimental station located at Layana village, Mantikulore subdistrict, Sigi district, Central Sulawesi. Twenty-eight Donggala bulls, 1.5-2.5 years old and averaging 179.48±3.66 kg of live weight (LW) mean BCS 2.67 ±0.06 (scale 1-5) were used. The bulls were purchased from the local animal market around Sigi and Palu districts, Central Sulawesi province. Before the start of the experiment, the bulls were drenched with Ivomec (1 mL/50 kg live weight to control parasites.

2.2. Experimental design and diets

Twenty-eight Donggala bulls were divided into four groups based on unfasted weight and allocated to four dietary treatments in a completely randomized block design. The four experimental diets were as follows: corn stover (69 g CP, 658 g NDF, 449 g ADF, 16 g EE/kg DM) ad libitum (CS), CS plus 1% W DM/day of Desmanthus virgatus (18 g CP, 462 g NDF, 314 g NDF, 28 g EE/kg DM) (CSD), CS plus 1% W DM/day of Gliricidia sepium (21 g CP, 478 g NDF, 346 g ADF, 36 g EE/kg DM) (CSG), and CS plus 1% W DM/day Leuacena leucocephala (23 g CP, 397 g NDF, 244 g ADF, 42 g EE/kg DM)(CSL). Each dietary treatment was repeated 7 times (bulls). The whole CS was chopped into lengths 3-8 cm length. and offered ad libitum. CS and the legume forage were offered separately in two equal parts at 07:00 and 1300 h. The legume forage was harvested in the morning for the afternoon feed and the afternoon for the following day feed. The legume forage allocated each week was adjusted based on the latest bull live weight. The feeds for individual animals were weighed daily and adjusted to the actual DM content of the feeds. Refusals were removed and recorded before offering a new feed the next day. All bulls were confined in individual pens (1.6 x2m). The experiment consisted of a 2-week adaptation and an 8-week data collection period. Drinking water was on offer at all times.

2.2.1. Parameter measured.

Parameters measured include feed intake, live weight gain, heart girth, wither height, body length, and body condition score. The measurements of feed intake were done by reducing the amount of feed offered with the amount of feed refusal on the daily basis. Bulls were weighed every week. The ADG of each bull was determined by calculating the difference between the initial weight and the final weight of the bull divided by the number of days. Heart girth and body length were measured using a tape at week 1 and week 14. The measurement of wither height was done in weeks 1 and 14 using a measuring stick when a bull was standing as a method of [3]. The assessment of body condition score (BCS) of a bull was done in weeks 1 and 14 from behind, right, and left sides of the bull and followed with palpation of the tailbone and waist using hand. The value of BCS was determined using a single BCS unit 1-5 According to [6].

2.2.2. Feed and Statistical analysis.
DM of feeds was determined by drying the samples in an oven at 105°C for 24 h, ash by ashing at 550°C for 8 h, and crude protein (CP) by the official methods of the Association of Official Analytical Chemists [7]. Neutral detergent fiber (NDF) and acid detergent fiber (ADF) were determined by the method of [8].

Analysis of variance was carried out to test the effects of different legume leaves supplementation on feed intake, liveweight gain, change in wither height heart girth, body length and BCS of Donggala bull fed CS using Minitab 16 statistical software. Differences among treatment means were compared using Duncan's Multiple Range Test.

3. Results and Discussion
Data of feed DM intakes are presented in Table 2, none of the bulls consumed their total allocation of supplements. Bull received Desmanthus, Gliricidia and Leucaena consumed 94, 91, and 96% of the supplement allocation., respectively. The mean DM intake for seven unsupplemented bulls during the measurement period was 2.73 ± 0.08 %W/d. The daily basal DM intakes declined significantly ($P<0.05$) with the addition of legume leaves. Supplementation of low-quality diets have been reported to depress intake of the basal diet [9,10]. Supplementation with legume forage increased significantly ($P<0.05$) total DM intake. The highest total DM intake ($P<0.05$) was achieved by bull received CSD and CSL and no differences between both treatments.

Table 1. Effect of Desmanthus, Gliricidia, and Leucaena leaves supplementation on average daily live weight gain, change of wither height, heart girth, body length, and body condition score of Donggala bulls given fed corn stover as a basal diet

Parameters	Dietary treatment			
	CS	CSD	CSG	CSL
Supplement DM intake (%W/d)	0.00 ± 0.00	0.94 ± 0.01	0.91 ± 0.01	0.96 ± 0.01
Basal diet DM intake (%W/d)	2.73 ± 0.08^a	2.37 ± 0.06^b	2.34 ± 0.06^c	2.41± 0.04^b
Total DM intake (%W/d)	2.73 ± 0.08^a	3.32± 0.06^c	3.25 ± 0.07^b	3.36 ± 0.04^c
Daily liveweight gain (kg/d)	0.32 ± 0.01^a	0.45 ± 0,02^b	0.54 ±0.03^c	0.62 ± 0.04^d
Wither height (cm. week 1)	109.5±1.21	113.4±2.21	116.3±1.63	117.6±1.85
Wither height (cm. week 14)	111.2±1.14	115.2±1.57	118.0±1.71	119.3±1.94
Change in wither height (cm)	**1.70±0.16**	**1.80±0.15**	**1.70±0.17**	**1.70±0.18**
Heart girth (cm. week 1)	136.2±0.75	137.6±0.71	134.7±0.84	134.6±0.77
Heart girth (cm. week 14)	138.1±1.53	141.1±2.21	139.3±2.54	139.7±2.28
Change in heart girth (cm)	**1.90±0.16** ^{a}	**3.50±0.21**^{b}	**4.60±0.24**^{b}	**5.10±0.26**^{c}
Body length (cm. week 1)	116.7±2.56	117.6±2.51	119.5±2.37	118.4±2.32
Body length (cm. week 14)	117.1±2.32	118.0±2.24	119.9±2.26	118.9±2.46
Change in body length (cm)	**0.40±0.06**	**0.40±0.09**	**0.50±0.13**	**0.50±0.15**
Body condition score (week 1)	2.58±0.17	2.47±0.19	2.58±0.26	2.63±0.29
Body condition score (week 14)	3.11±0.16	3.32±0.27	3.48±0.22	3.75±0.32
Change in body condition score	**0.53±0.08**^{a}	**0.85±0.09**^{b}	**0.90±0.09**^{b}	**1.12±0.08**^{c}

CS = corn stover ad libitum, CSD = CS + Desmanthus (1% W DM/d), CSG = CS+Gliricidia (1% W DM/d, CSL = CS + Leucaena (1% W DM/d)

Means with different superscripts in the same row are significantly different (P<0.05).

The ADG, body dimension, and BCS of Donggala bulls supplemented with different legume leaves are presented in Table 2. Supplementation with different legume leaves substantially increased (P<0.05) ADG, change of heart girth, and BCS of Donggala bulls. However, the greatest ADG, changes of heart girth, and BCS of Donggala bulls were found with Leucaena supplementation. These differences in ADG, changes in heart girth, and BCS were associated with CP intake. Leucaena contained a higher CP content than Desmanthus and Gliricidia. This finding is in agreement with the previous finding of [4,11] who reported a higher ADG of Leucaena than Gliricidia supplemented cattle. From this study, it was suggested that there is a strong relationship between change in heart girth and ADG of the Donggala bull. The previous study [12] documented that the best parameter to predict the live weight of Brown Swiss and crossbred cattle is heart girth. The change in heart girth could be a result of the growth of chest bone and muscle around the chest.

No significant effect (P>0.05) was detected of the effect of different legume leaves supplementation on wither height and body length of Donggala bulls fed CS. The bull is already mature therefore no development of the bone to increase the wither height and body length.

4. Conclusion
Growth performance, heart girth, and body condition score of Donggala bull fed low-quality forage can be increased by the inclusion of legume leave such as Desmanthus, Gliricidia, and Leucaena. The greatest change was associated with Leucaena supplementation.

Acknowledgment
Thanks are due to the Ministry of Education, Culture, Research, and Technology of the Republic of Indonesia through the Applied Research scheme, for financial supports of this experiment. The authors also directed technicians and students of the Department of Animal Science, Tadulako University for assistance throughout the work.

References
[1] Marsetyo and Y Rusiyantono 2019 IOP Conference Series: Earth Environ Sci 387 012005
[2] Ba NX, Van NH, Ngoan LD, Leddin CM, Doyle PT 2008 AJAS 21:1736–44
[3] Quang, DV, Ba, NX, Doyle PT, Hai, DV, Lane, PA, Malau-aduli, A EO, Van, NH and Parsons, D, 2015 J Anim Sci Tech 57:1-8
[4] Abdulrazak, SA, Muinga, RW, Thorpe, W and Orskov ER, 1997 Livest Prod Sci 49:53-62
[5] Muinga, RW, Topps, JH, Rooke, JA, Thorpe, W, 1995 Anim Sci 60: 13-23
[6] Affandhy L, Yusran M A, dan Winugroho M 2001 Proc National Seminar Anim Husb Vet Puslibangnak, Departemen Pertanian 17-18 September 2001 Bogor, Indonesia, pp147-154
[7] AOAC 1990 Official Methods of Analysis15th Edition Association of Official Analytical Chemistry, Arlington, VA
[8] Van Soest P J, Robertson J B, and Lewis B A 1991 J Dairy Sci 74:3583-97
[9] Mosi, AK, and Butterworth, MH, 1985 Anim Feed Sci Tech 12: 241-251
[10] Getachew, G, Said, AN, Sundstol, F, 1994 Anim Feed Sci Tech 46: 97-108
[11] Simons, AJ, and Stewart, JL, 1994 Gliricidia sepium - a multipurpose forage tree legume In eds RC Gutteridge and HM Shelton, Forage Tree Legumes in Tropical Agriculture CAB International, Wallingford, pp 30-48
[12] Oskaya S and Boskurt Y 2009 The accuracy of prediction of body weight from body measurements in beef cattle Arch Tierz 4: 371-77

Effects of formaldehyde treated-soybeal meal on sheep performance

M S Arifuddin[1], A Yakupitiyage[2], S Tumwasorn[2], G K Hansen[2], P Parkpian[2]

[1]Department of Animal Husbandry, Faculty ofAnimal Science and Fisheries, University of Tadulako
[2]School of Environment, Resources and Development, Asian Institute of Technology (AIT) Bangkok, Thailand

sadik_arifuddin@yahoo.com

Abstract. The study aims at investigating the effects of protecting soybean meal proteinby using formaldehyde on live weight gain, dry matter consumption and feedefficiency of local rams. The research was conducted during 10 weeks period, using a completely randomized design (CRD) with 4 treatments within 4 replicates. Sixteen fat tail sheep, approximately 10 – 14 months of age with body weights ranging from 9 – 12 kg were used in this experiment. Soybean meal was treated with various concentration of formaldehyde, i.e 0 g, 5 g, 10 g and 15 g formaldehyde/ kg soybean meal. Untreated soybean meal was used as the treatment control. Maize forage (zea mays) was given ad-libitum as basal diet. Treatment with highest formaldehyde levels enhanced live weight gain (19.27 kg), dry matter consumption (412.80 g/d) and feed efficiency (35.52 %). The result indicated that, formaldehyde treatment of soybean meal protected dietary protein from rumen micro-organisms. Further investigation the optimum level of formaldehyde is warranted.

Keywords: Formaldehyde, soybean meal, maize forage.

1. Introduction

Small ruminant production throughout Indonesia is primarily limited by the quantity and quality of feed available throughout the year. Regardless the region, sufficient and quality of feed for desired reproduction and growth is lacking. Several attempts have been made to develop a feeding strategy for ruminants in the tropics such as protein supplement, including feeding grain or molasses.

The quality and quantity of feed as well as the feeding management are undoubtedly important factors affecting production of small ruminant. Sheep like other ruminants consume forage as natural feed. However, forage in the tropic generally is low, with low crude protein and high the crude fibre. Therefore, ruminants in the tropic like sheep mostly lack protein, not energy, because the requirement of energy for maintenance and growth is normally met with forage [1].

Dietary protein at ruminants function as by pass and nitrogen source that used by microbe proliferation with deaminase and transanimasi process. Availability of protein in the ruminant like sheep to fulfil their production, growth, and respiration process, is effected by the availability protein in the quality and quantities with activities of microbial rumen. Frequently, the high quality protein is consume by the rumen microbial protein and is not avoided by degradation. Therefore, to protect or minimize protein the high quality from degradation microbial rumen and directly to abomasum (by pass protein) can be used some substance in the diet small ruminant [2,3].

Soybean meal as a source ruminants protein with crude protein content 44%, can be used as protein concentrate supplement for sheep. The use of protein concentrate supplement is not efficient without the protected degradation rumen of microbial in the rumen ruminants. To be able to increase the proportion of dietary protein reaching the intestines (by-pass protein), without any reduction in the time spent in the rumen and thus exposure to microbial attack, many protecting techniques have been developed, encapsulation of amino acid [4] and formaldehyde or other aldehyde [5,3].

2. Materials and method

2.1. Site of the study

The study will be conducted in the District Donggala of Palu in the province of Central Sulawesi Indonesia, between March and May 1998. Central Sulawesi is one of the area in Indonesia which has the high potential and more suitable for sheep production. The population of sheep in this area in 1995 was 20,430 [6]. With a prodeminant breed of the fat tail sheep. Like other sheep, Donggala fat-tail sheep can contribute to the meat requirement of Indonesia which in term can be a good income supplemented to farmers.

2.2. Materials

2.2.1. Animals and experimental design. Sixteen fat-tail sheep, approximately 10 to 14-months-age with body weight ranged from 9 to 12 kg were used in this experiment. A 10-weeks experimental period was adopted. For the first 14 d, dietary adaptation was carried out. Adaptation was performed either on protein concentrate and maize forage. Twelve sheeps were given formaldehyde-treated soybean meal and other sheep, were given soybean meal as control. When the sheep finish consuming soybean meal, maize forage as the basal feedstuff was given ad-libitum. Completely Randomised Design (CRD) was used during the experiment period, with 4 treatments and 4 replications. Sheeps were housed individually in platform floor (100 x 150 cm) in a room at an ambient temperature between 30 – 32°C. The treatments were as follows :

R0 = 0 g formaldehyde / kg soybean meal + maize forage (zea mays) ad-libitum.
R1 = 5 g formaldehyde / kg soybean meal + maize forage (zea mays) ad-libitum.
R2 = 10 g formaldehyde / kg soybean meal + maize forage (zea mays) ad-libitum.
R3 = 15 g formaldehyde / kg soybean meal + maize forage (zea mays) ad-libitum.

2.2.2. Diets. Formaldehyde with different concentration (as treatments) were mixed into soybean meal at rate as previously described. Liquid formaldehyde was taken as required and sprayed onto soybean meal and then mixed homogeneously. The homogeneously mixture was placed in the plastic bag and stored for 1 week prior to feeding the sheep. The sheep were fed at rate of 1% of body weight (BW) per day at 08.00 am. In case of the feed did not consumed totally until 12.00 am, the uneaten feed were the given directly to the mouth individually by hand. Fresh maize forage (zea mays) were given ad-libitum after 12.00 am. These maize forage (zea mays) was previously chopped into 1 – 2 cm in length. Fresh drinking water was freely available.

2.3. Data collection

2.3.1. Live weight gain (LWG). After acclimation period, the experimental animals were weighed for initial weight and 1 week there after. The following formula was given for live weight gain:

$$W = W\_t - W\_0$$

Where: *W* = *live weight gain (kg)*
W_t = *Final weight (kg)*
W_0 = *Initial weight (kg)*

2.3.2. Dry matter consumption (DMC). Dry matter consumption that was generated from soybean meal and forage intake was measured daily by recording the amount of offered feed and refusal feed. Feeds were sampled by taking 5 % of feed twice a day (at 08.00 and 12.00 am) for later analysis.

Determination of dry matter consumption (g/d) for both soybean meal and maize forage (Zea mays), was calculated by formula:

Dry Matter Consumption = (A x %DM) + (B x %DM)
Where A = Amount of soybean meal consumed (g/d)
B = Amount of maize forage consumed (g/d)
DM = Dry matter

2.3.3. Feed effiency. Feed efficiency was calculated at the end of experiment throught the following formula:

Feed effiency (%) = weight gain / Dry matter consumption x 100 %

2.4. Data analysis
The data were analysed by using analysis of covariance (Steel and Torrie, 1991), based on the model as follow:

$$Y\_ij = \mu + T\_i + B\_j + b\_xy\,(x - x) + E\_ij$$

Where Y_ij = observated yield for treatment i in block j
μ = mean yield
A_i = treatment effect on yield for treatment i
B_j = block effect on yield for block j
b_xy (x – x) = regression of final weight (y) on initial weight (x)
E_ij = error effect

The least aquare mean (LSM) test was used for further statistical analysis. The data of feed efficiency were transformated to arcsin before analysed.

3. Results and discussion

3.1. Result

3.1.1. Live weight gained (LWG). The least square mean (LSM) of various concentration formaldehyde-treated soybean meal on live weight gain (LWG) were presented in Table 6

Table 1. Least square mean and standard error of live weight gain (LWG) at various concentration formaldehyde-treated soybean meal.

Treatment	Total weight	Live weight gaint
	Initial weight	Final weight
R012.93	18.20	16.88±0.49^{a}
R110.40	15.88	16.81±0.46^{a}
R210.40	16.35	17.29±0.46^{a}
R312.08	19.83	19.27±0.44^{b}

R0 = 0.0 g formaldehyde/kg soybean meal; R1 = 5 g formaldehyde/kg soybean meal -; R2 = 10 g /kg soybean meal and R3 = 15g /kg soybean meal. LWG = live weight gain [a,b] Means in the same row with different superscripts differ (P<0.05)

Analysis of covariance indicated that formaldehyde-treated soybean meal significantly (P<0.05) influenced live weight gain. Least significant different (LSD) test revealed that the live weight gain of treatment R3 (19.27) was significant different (P<0.05) higher than treatment R0, R1 and R2. However, there was no significantly different (P>0.05) within R0, R1 and R3 (Table 1).

3.1.2. Dry matter consumption (DMC), and Feed Efficiency (FE). The effects of formaldehyde-treated soybean meal protein at various concentration on dry matter consumption and feed efficiency are presented in Table 2.

Table 2. Adjusted mean dry matter consumption (DMC), and feed efficiency (FE) of sheep whit formaldehyde-treated soybean meal protein

Parameters	Treatment			
	R0	**R1**	**R2**	**R3**
DMC (g/d)	413.74±28.21^{a}	350.74±13.47^{b}	359.42±16^{bc}	412.80±57.99^{a}
FE (%)	22.84±2.20^{a}	27.95±3.28^{b}	29.63±2.14^{bc}	33.52±4.67^{c}

R0 = 0.0 g formaldehyde/kg soybean meal; R1 = 5 g formaldehyde/kg soybean meal -; R2 = 10 g /kg soybean meal and R3 = 15g /kg soybean meal. LWG = live weight gain [a,b,c] Means in the same row with different superscripts differ (P<0.05).

Based on the covariances analysis indicated the formaldehyde-treated soybean meal significantly influenced (P<0.05) the dry matter consumption. The highest dry matter consumption was achieved by sheep on R3 group. There was no significant effect of DMC of sheep on R1 compared to R3 group (P<0.05). Similar trend was found between R2 and R3 (P<0.05).

The mean of value of feed efficiency is presented in table 2. Analysis of covariances indicated that formaldehyde influenced (P<0.05) the feed efficiency. The highest value of feed efficiency was found on R3 group (P<0.05). meanwhile, no significant (P<0.05) different was found in other comparison of treatments.

3.2. Discussion

Protein is a main nutrient required for normal growth of animal. [7] stated that there are two sources of protein (amino acids) to the host, i.e., microbial protein (ruminal microbe synthesis) and by-pass protein (undegraded protein). However, during the catabolism of this nutrient, there is a minimum amount of protein absorbed in the rumen has to be met by the host. In ruminants, feed consumption is known to be controlled to a large extent by the amount of digesta that can be accommodated in the rumen and the rate of clearance of indigestible material from the rumen.

The significant effect of formaldehyde inclusion on LWG could be explained as follow; firstly, ingested protein might have passed through to the intestine without protein degradation in the rumen, or the formaldehyde effectively protected the protein from microbial degradation in the rumen. Secondly formaldehyde might have acted as a disinfectant allowing protein being coagulated [8] and lowering the surface area of protein due to fibre-protein forming [4] which cause limited change of microbial to reach these protein. Therefore, there is a high amount of protein is available at post-rumen.

The efficacy of formaldehyde to increase the body weight of animals were reported by number of authors. [9] found that formaldehyde treatment on a groundnut meal diet with 12% CP gave 9% increase in weight gains of calves. Similar responses were reported by Broderick (1994) that increasing the level of formaldehyde-treated casein gave a linear drop in both the rates of rat growth and release of amino acids during in vitro pepsin pancreatic digestion. Faichney (1970) found that live weight gain of lambs fed formaldehyde treated casein increased from 154.0 to 165.5 g/d.

Generally, the increasing of body weight is a result of the increase feed consumption. However, the current result showed that the higher feed consumption (treatment R0, control) did not enhance live weight gain. This indicated that formaldehyde act as a protecting agent so that there was higher amount of protein by-pass to the intestine, rendering in the higher weight gain of sheep feed formaldehyde treated soybean meal.

Analysis of covariance indicated that formaldehyde-treated soybean meal significantly ($P<0.05$) influenced the dry matter consumption (DMC). Control treatment (R0) had a higher DMC compared to other treatments. This is possibly due to higher feed degradation in the rumen. Perdock and Leng, (1990) reported that dry matter consumption was directly affected by degradation process the rumen indicated by the highest ammonia concentration due to increasing microbial activity. The other reason is that small particle size which allow to increase the surface area of the soybean meal, and ultimately has an effect on the DMC. Increasing DMC as a result of grinding (reduction of particle size) have been reported by Weston, (1996). And found that a reduction of particle size of the diets reduced the mean retention time in tractusdigestivus, allowing more feed to be ingested.

There was a decrease DMC of sheeps fed formaldehyde treated soybean meal (table 2). Leng and Nolan (1982) found that steer fed with wheat treated formaldehyde had decrease feed intake. Church, (1988) noted that feed consumption was effected by several factors such as feed palatability, temperature, disease, humidity and body weight.

It was observed that the feed efficiency was enhanced by increasing the dose rate of formaldehyde (table 2). FE was higher in high dose rate of formaldehyde than the lower doses. The positive impact of formaldehyde utilization related to feed efficiency were reported for lambs, steers and dairy cows [10].

4. Conclusion

It can be concluded that formaldehyde significantly influenced the live weight gain and feed efficiency and the highes dose rate of 15 g formaldehyde per kg soybean meal produced the best performance of sheep.

References

[1] Leng, R A (1993) Quantitative ruminant nutrition – A green science Australian Journal of Agricultural Research,44 :363 – 80
[2] Arora, SP (1989) PencernaanMikrobapadaRuminansia GadjahMada University Press, Yogyakarta, Indonesia
[3] Siregar, S B (1994) Feeding Ruminant PenebarSwadaya, Bogor Indonesia
[4] Singh, VB and Mehra, R (1987) Protected proteins and protected amino acids in ruminant nutrition, In :Advanced Animal Nutrition for Developing Countries, pp : 205 – 18, Editor : Sing, UB, Indo Vision Private limited, India
[5] Broderick, GA (1978) In vitro procedures for estimating rates of ruminal protein degradation and proportions of protein escaping the rumen undergraded Department Of Animal Science Journal Nutrition,108 : 181 – 90
[6] Statistic Central Sulawesi, (1995) Sulawesi Tengah DalamAngka Kantor BPS statistic Propinsi Sulawesi Tengah, Indonesia
[7] Orskov, ER (1992) Protein nutrition in ruminants Second edition Academic press, London
[8] Martindale, (1979) The Extra Pharmacepoea 27 th Ed Pharmaceutial press, London
[9] Faichney, CJ Tagari, H Tleki E and Boston, RC (1994) Nitrogen Transaction in the Rumen of Sheep given a Barley Straw diet Supplemented with Untreated of formaldehyde-treated Sunflower Seed Meal Australian Journal of Agricultural Research, 45 : 1203 -1214
[10] Broderick, G A (1994) Quantifying forgae protein quality In : GC Fahey (ed) Forage quality, evaluation and utilization American Society of agronomy, Wisconsin, USA Pp 200-28

Multiple Biota Cultivation (Gracilaria sp and Chanos-chanos) Development Model as a Pillar of Milk Fish-Agar Agroindustry and Its Applications

Zakirah Raihani Ya'la[1], Dwi Sulistiawati[2], Nasmia[1], Desiana Trisnawati Tobigo[1]

[1] Aquaculture, Animal Husbandry, and Fisheries faculty Tadulako University, Palu Soekarno-Hatta Street Indonesia

[2] Animal Husbandry and Fisheries Faculty Tadulako University, Palu Soekarno-Hatta Street Indonesia

zakirahraihaniyala@gmail.com (the corresponding author's e-mail)

Abstract. The general objective of the research is to formulate a development model for the cultivation of multiple biota (Gracilaria sp and Chanos chanos) as a pillar as the main support for the implementation of the development of the milkfish- agar agroindustry in the province of Central Sulawesi. The research was conducted in 3 districts producing seaweed and milkfish, namely, Morowali Regency, Parimo Regency and Donggala Regency. Each district is represented by 2 sub-districts which are the largest producers of seaweed so that there are 6 sub-districts. This research using a survey method by taking samples directly in the field by simple random sampling and purposive sampling, GIS methods for land suitability and SWOT analysis. Seaweed processing technology and marketing analysis supported SWOT analysis as requirements for development model of dual biota cultivation. The results showed that The suitability of the land is mostly suitable for the cultivation of milkfish and seaweed in Donggala Regency (716.09 Ha), Parimo Regency (674.91 Ha) and Morowali Regency (118.33ha). supported by water quality condition in 3 Regency which is quite optimal for the land suitability for cultivation of seaweed and milkfish, as well as factors that support the development model of milkfish-agar agroindustry.

1. Introduction

The income gap in Central Sulawesi province has not been fully evenly distributed. As a result, it also affects the economic and social capabilities of the community. The number of poor people in Central Sulawesi Province in 2010 amounted to 474,990 people (18.07%) decreased to 423,630 people (17.89%) in 2011 [1]. Poverty rate in Central Sulawesi during the period 2016-2020 has a tendency decreased even though in 2017 there is an increase. In terms of number, population the poor in Central Sulawesi for the period of September 2020 as many as 403.74 thousand people [2]. Therefore, the government cooperates with various elements of the education world to improve the human development index (HDI) as a benchmark for quality of life in Indonesia [3]. Seaweed is one of the fishery commodities that can be used to encourage people's economic growth. Seaweed cultivation does not require high technology, investment tends to be low, absorbs considerable labor and generates relatively large profits.

The development of the business is expected to reduce the unemployment rate (pro job), increase people's income (pro growth) and in turn can reduce poverty (pro poor) [4]. Seaweed is one of the leading commodities that has a prospective market. The world's demand is high enough to cause production from nature is insufficient, so cultivation efforts must be made [5]. The target is carried out by developing 15,000 hectares of cultivation land until 2009 with a production target of 0.4 million tons for Gracilaria sp and 1.5 million wet tons for Eucheuma sp. Total expected production of 1.9 million tons or equivalent to 186,332 dry tons, while the expected export amounted to $ 111,501,000. The reality in the field shows that seaweed cultivation activities have absorbed a lot of labor, helped poverty alleviation efforts and in turn opened up opportunities for economic growth of the community. At least, the margin of eucheuma sp seaweed cultivation business ranges from Rp 20,500,000/ KK / 0.5 Ha / Year while Gracilaria sp ranges from Rp 24,700,000 / KK / Ha / Year. The margin can increase if coupled with increased production per unit area, expansion of area, improvement in quality and potential for increasing world demand.

Central Sulawesi Province with a coastal length of around 4,013 km covering Tomini Bay, Tolo Bay and Makassar Strait has a potential for seaweed commodity development of about 65,426 ha [6]. The dominant species cultivated is Eucheuma cottonii producing karaginophytes, while agarophyte producers such as Gracilaria verrucosa cultivated in ponds are being developed [7]. In 2009 the development of seaweed cultivation business in Central Sulawesi is predicted to reach an area of 7,851.12 Ha, in 2010 it can reach an area of 11,776.68 Ha, and in 2011 it can reach an area of 15,702.24 Ha.

One of the other fishery commodities that have high economic value in addition to shrimp to be cultivated in ponds is Milkfish (Chanos chanos Forsk) Milkfish is a fishery commodity that has a fairly good and savory taste so it is very popular with the public. In addition, the price is also affordable by all levels of society. Milkfish is classified as high protein and low fat fish [8]. Milkfish farming activities in ponds have been developed for quite a long time, this is supported by the potential of excellent natural resources, especially the availability of Milkfish seeds (Nener) both naturally and from hatchery, but production and productivity are relatively low [9]. This low production and productivity is partly due to the limited knowledge and skills of farm farmers about Milkfish farming techniques [10].

Milkfish and seaweed biologically have properties that can synergize so that this kind of double biota cultivation can be developed because it is one form of polyculture cultivation that is environmentally friendly [11]. Seaweed is a supplier of oxygen through photosynthesis during the day and has the ability to absorb excess toxic nutrients and pollutants in the waters [12]. While the fish as plankton eaters are controlling against the excess plankton in the waters [13]. Shrimp droppings, fish and other organic materials are a source of nutrients that can be utilized by seaweed and phytoplankton for growth [14]. Such a relationship can balance aquatic ecosystems, so it needs to be researched about the model of developing double biota biota fish and seaweed.
Central Sulawesi does have a fairly long coastline, which is more than 4,000 km. The length of this coastline has the potential for seaweed development. Seaweed is one of the flagship programs in Central Sulawesi. The Ministry of Marine Affairs and Fisheries even made Central Sulawesi as one of the areas that will be used as a seaweed center in Indonesia.

Cultivation of Gracillaria verrucosa, and other types of seaweed, can be done monoculturally and polyculture with fish (finfish) moored [15]. By using polyculture cultivation systems can increase the efficiency of farm land use and farmers' income on an ongoing basis. This cultivation is based on the principle of natural balance. Seaweed serves as an oxygen producer and shelter for fish and shrimp from predators and as a biological filter. Fish and shrimp remove dirt that can be used as nutrients by seaweed. Seaweed absorbs dissolved CO_2 from the breathing of fish and shrimp. In general, the presence of seaweed in shrimp ponds / Milkfish has a positive impact [16].

Seaweed is one of the fishery commodities that also acts as a biofilter, because in its growth seaweed absorbs nutrients (ammonia, nitrate, and nitrite) from aquatic media by diffusion through its thallus wall. The function of biofilters in pond areas is needed, especially those whose water sources are very murky due to mud or other particles. Cultivation activities that use biofiltration, the content of organic matter and ammonia in the maintenance plot are relatively lower compared to maintenance ponds that do not use biofiltration systems. This is because seaweed is able to absorb ammonia, nitrate and phospat ions. In addition, seaweed also has the ability to absorb elements or other compounds such as heavy metals. It was also explained by [17] that seaweed as a aquatic plant can absorb the degradation of water organic matter that will be needed for growth, thereby reducing the risk of increasing water organic matter that will be used to maintain Windu Shrimp. In line with the government program, the

planning of the development of seaweed processing industry can be used as a driving force for the regional economy by preparing investment opportunity information, especially in seaweed and Milkfish producing areas such as Bangkep, Donggala, Morowali, and Parimo.
The amount of land potential and production of Milkfish in Central Sulawesi until it has not been qualified, this is due to the lack of attention from both the government and private sector about this commodity. The production of Milkfish has not been well exposed so until now the latest amount of Milkfish production. Some of the largest Milkfish producing districts in Central Sulawesi include Donggala, Parimo and Morowali regencies.

The general purpose of research is to formulate a model of development of dual biota cultivation (Gracilaria sp and Chanos chanos) as a pillar as the main support for the implementation of milkfish-agar agroindustry development in Central Sulawesi Province. The specific objectives include: estimating the land potential and production of seaweed and the latest Milkfish in Morowali, Parimo, and Donggala regencies Analyzing the management of the correct management of seaweed and Milkfish cultivation as the main supporters of the development of milkfish-agar agroindustry, isolating and identifying bacteria in Gracilaria seaweed affected by ice, analyze the economic feasibility of dual biota cultivation of seaweed and Milkfish, as well as the economic viability of milkfish-agar agroindustrials, analyze the marketing of seaweed and presto processed production, formulate a model of development of double biota cultivation of seaweed and Milkfish as pillars of the formation of milkfish-agar agroindustry business, assist seaweed processing training into agar (sheets, stems and flour) and training in making presto Milkfish in the 3 districts above.

2. Study Site and Methods

The research was conducted in 3 districts producing seaweed and milkfish, namely, Morowali Regency, Parimo Regency and Donggala Regency. Each district is represented by 2 sub-districts which are the largest producers of seaweed so that there are 6 sub-districts. Achieving the target of this research using a survey method by taking samples directly in the field by simple random sampling and purposive sampling, GIS methods for land suitability and environmental carrying capacity analysis, SWOT analysis and direct interviews with the help of structured questionnaires. The expected outputs include land suitability analysis, land carrying capacity analysis, analysis of seaweed processing technology, seaweed marketing analysis as requirements for development model of dual biota cultivation (Gracilaria sp and Chanos-chanos) as pillars of milkfish-agar agroindustry and their applications in Central Sulawesi Province.

3. Results and Discussion

3.1. Land Suitability for Donggala Regency

The physical characteristics of the waters play an important role in determining the suitability of the area for seaweed cultivation and are interrelated, where this research was carried out through an ecological approach to carry out seaweed and milkfish cultivation. Marine organisms have environmental requirements in order to live and grow properly. The more suitable the conditions of the aquatic environment, the better the growth of an organism. Seaweed is one of the marine organisms that requires an environmental habitat to grow and reproduce. Seaweed growth is highly dependent on oceanographic factors such as physical, chemical and biological parameters.

At the study location, namely Lalombi Village, the suitable cultivation area was 518.29 Ha, and there was no unsuitable and unsuitable land, in Surmana Village it showed a suitability level of 197.80 Ha (suitable) and 11.45 Ha. Not suitable. The value of this suitability level indicates that the management of the cultivation of multiple biota (Gracilaria sp and Chanoschanos) is adequate, especially water quality monitoring. (**Figure 1.**)

IOP Conf. Series: Earth and Environmental Science 1075 (2022) 012008 doi:10.1088/1755-1315/1075/1/012008

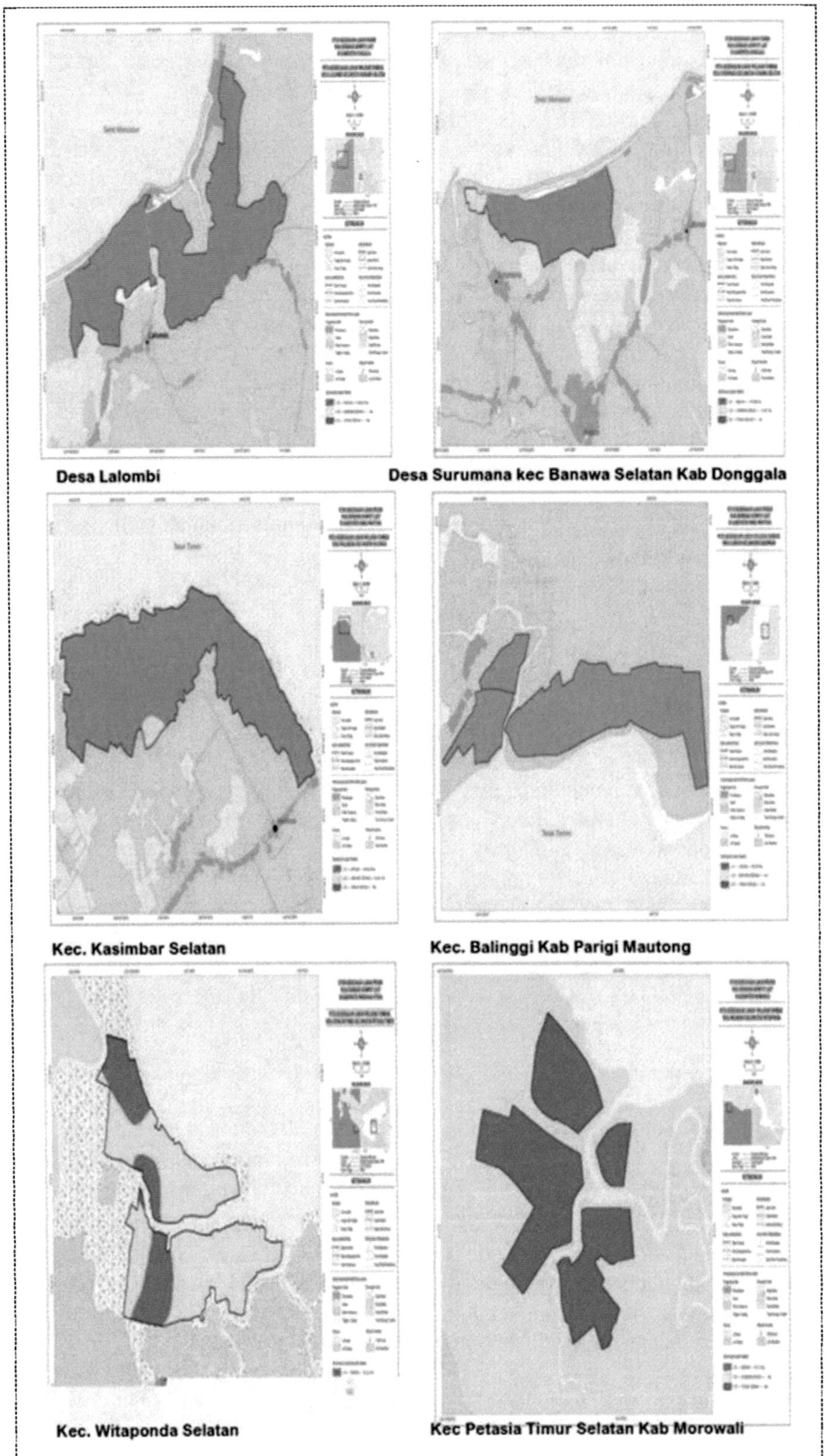

Figure 1. Land Suitability for seaweed and milkfish.

Pests and diseases of this dual biota cultivation are more in Donggala Regency than Parimo Regency. Seaweed pests commonly encountered are mullet and sea cucumber larvae (Holothuria sp.). Other pests include porcupine fish (Siganus sp). While the diseases that attack seaweed are bacterial, fungal and ice-ice and mildew. Ice-ice disease (some people call it white spot) is a major obstacle to the cultivation of Gracilaria verrucosa seaweed

Symptoms shown in seaweeds affected by this disease include: slow growth, changes in the color of the thallus to pale or not bright, and some or all of the thallus on some branches turns white and rots. The disease is mainly caused by environmental changes such as current, temperature and brightness. Very high water clarity and low solubility of nitrate nutrients in water are also the causes of the emergence of this disease.

3.2. Land Suitability for Parigi Mautong Regency

Marine waters in Kasimbar and Balinggi sub-districts have potential natural resources for the development of marine aquaculture, one of which is the cultivation of seaweed and milkfish. To support these activities, it is necessary to carry out a land/location suitability analysis where in determining the location of seaweed cultivation it is necessary to pay attention to several requirements as indicators that support these activities.

The suitability analysis carried out is based on the limiting factors/parameters according to their utilization in terms of the ecological aspect. The initial criteria that are generally compiled are from ecological prerequisites, then the results of the GIS analysis are in the form of location and area in accordance with the required criteria which ultimately determine the carrying capacity of the land. This analysis is intended to assess the feasibility or suitability of the water areas of Kasimbar and Balinggi sub-districts. The results of the analysis are grouped into three categories, namely not suitable (Sl), conditionally appropriate (S2) and very suitable (S3). Based on the results of the land suitability analysis, the suitable cultivation area in Kasimbar District is around 644.56 Ha and the unsuitable area is 26.18 Ha. In Balinggi sub-district the suitable land area is 30.35 ha and there is no unsuitable land and unsuitable land. Gracilaria verucosa seaweed cultivation is quite easy, inexpensive and relatively short. In addition, the cultivation of this type of seaweed can also help increase the value of pond productivity because it can be done in polyculture with milkfish and shrimp. Ecologically, this seaweed helps maintain the quality of pond water as well as a place for the growth and development of fish and milkfish.

3.3. Land Suitability for Morowali Regency

Pond waters in Witaponda and Petasia sub-districts have natural resource potential, especially for the development of seaweed and milkfish aquaculture, although the overall level of physical and chemical suitability of waters in Gracilaria sp cultivation shows that the class is not suitable at 231.12 Ha, the appropriate class is 82, 22 Ha (District of Petasia). This shows that pond management is still carried out in a traditional manner and there is a lack of monitoring of water quality both at the time of planting seaweed and during its maintenance (Figure 1).

Based on the results of surveys and interviews, almost all seaweed aquaculture ponds are not controlled for water quality. At the opening of the pond, most of the cultivators did not do liming and fertilizing. Gracilaria sp requires nutrients for its growth such as nitrogen, phosphate, magnesium and potassium. The nutritional quality of pond water affects the growth and development of cultivated Gracilaria sp in Witaponda District, the appropriate class is 36.11 ha and the non-conforming class is not suitable and does not exist.

Problems that arise in polyculture cultivation in aquaculture areas are usually caused by internal factors in the pond itself, namely the environmental conditions of the pond which have decreased its carrying capacity as a result of pond management that does not consider the carrying capacity of the existing environment. Based on the problems that arise, it is suspected that the dominant factor causing the failure of polyculture cultivation, seaweed, milkfish and shrimp is the environmental factor of aquaculture, especially living media water that is not suitable for cultivation in ponds, because water sources as a living medium for shrimp with low quality are widely used for cultivation. cultivation.

The mindset of farmers who have to change the water in their plots is between 10 – 30% every day without considering the quality of the water added. Farmers entering the water is actually waste from the surrounding ponds, because the disposal/replacement of water is not the same, so that the effort to replace water does not improve the quality of water in the pond. Water with a high level of turbidity is often forced to irrigate ponds, because there is no settling basin. Particles carried by this turbid water settle in ponds, without realizing it causes pollution, siltation, interferes with shrimp respiration and limits primary productivity.

The abandonment of traditional ponds and intensive ponds that are not operated is a strong indication that there has been a decline in the quality of the aquaculture environment which is characterized by crop failure. One of the causes of the decline in pond water quality during operations is the high concentration of organic waste and nutrients as a consequence of aqua-inputs in shrimp farming which results in residual feed and faeces dissolved into the water [18] [19] [20] [21].

3.4. Agribusiness management in the coastal area of Central Sulawesi Province

Analysis of the strategy for developing agribusiness management in the coastal area of Central Sulawesi Province was carried out using a SWOT analysis (strengths, weaknesses, opportunities, and threats). This analysis was carried out by applying the criteria for conformity with quantitative data and a description of the situation (internal and external factors) obtained with the FGD technique. The weighting and scoring in the SWOT analysis was carried out based on the results of interviews with the RRA technique, which was then justified by researchers in the form of weights and scores [22].

3.5. SWOT Strategic Elements

1) Strength:
   - S1: Land Availability Is Still Wide
   - S2: There is Investment from the Community
   - S3: Seaweed Production Potential
   - S4: Availability of Local Labor related to the poor
   - S5: Cultivation Work Ethic
   - S6: Availability of Cultivation Institutional Facilities
2) Weaknesses:
   - W1: Limited Production Facilities/Supporting Infrastructure
   - W2: Lack of Market Information Facilities
   - W3: Lack of Knowledge of Post-Harvest Processing Technology
   - W4: Low HR Quality
3) Opportunity:
   - O1: High Market Demand
   - O2: Capital Support from Local Government and Partner Companies
   - O3: There are Educational Institutions that Support Aquaculture
4) Threats:
   - Q1: The Middlemen Who Dominate the Market
   - Q2: Competition with Products from Outside the Region
   - Q3: Land Use Conflict
   - Q4: Extreme Oceanographic Conditions in Certain Seasons

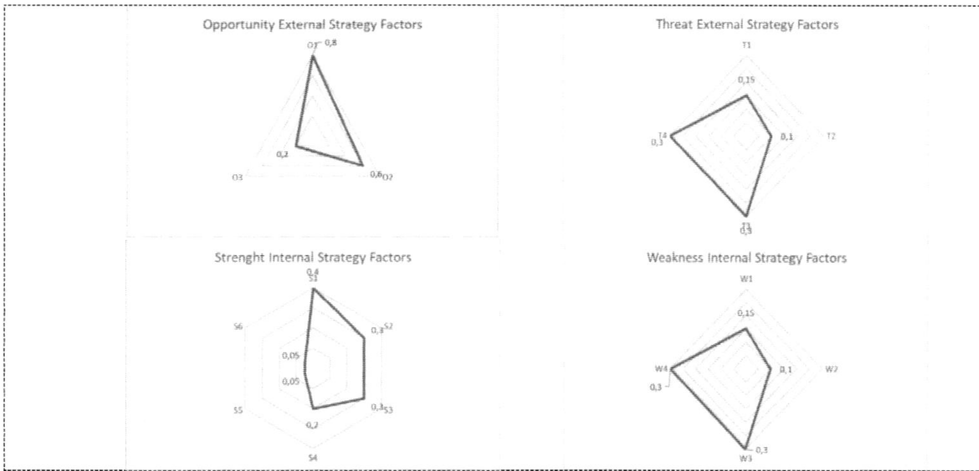

Figure 2. The strategy for developing Milkfish-Agar Agroindustry in Central Sulawesi

The results SWOT analysis from the weighting of the influencing factors, the results show that external factors (opportunities and threats) have a greater influence than internal factors (strengths and weaknesses), on the development of the milkfish-agar agroindustry, with a ratio of 2.75: 2.15. Based on the EFAS and IFAS matrices mentioned above, the SWOT matrix model obtained strategies that were grouped into 4 categories, namely:

i) SO strategy, namely the use of elements of coastal area strength to take advantage of existing opportunities;
ii) WO strategy, which is to improve existing weaknesses in coastal areas by taking advantage of available opportunities,
iii) ST strategy, namely the use of existing strengths to avoid or minimize the impact of external threats;
iv) WT strategy, namely defense tactics directed at reducing internal weaknesses to face external threats

Table 1. SWOT Matrix of Aquaculture Area Development Strategy

MATRIKS SWOT	STRENGTH (S) S1: land availability is still wide S2: there is investment from the community S3: feasibility of seaweed cultivation S4: local labor available S5: cultivation work ethic S6: available means of cultivation institutions
OPPORTUNITIES (O) O1: high market demand O2: capital support from local government and partners O3: there are educational institutions that support seaweed cultivation	**SO STRATEGY** 1) increasing the scale of aquaculture business by utilizing investments from partners or local governments; 2) empowerment of local workers as a side or main job in aquaculture;
THREATH (T) T1: middlemen who dominate the market T2: competition with products from outside the region T3: land use conflicts T4: extreme oceanographic conditions	**ST STRATEGY** 1) development of a marketing system that can drive the local economy;

MATRIKS SWOT	**STRENGTH (S)** S1: land availability is still wide S2: there is investment from the community S3: feasibility of seaweed cultivation S4: local labor available S5: cultivation work ethic S6: available means of cultivation institutions
OPPORTUNITIES (O) O1: high market demand O2: capital support from local government and partners O3: there are educational institutions that support seaweed cultivation	**SO STRATEGY** 1) increasing the scale of aquaculture business by utilizing investments from partners or local governments; 2) empowerment of local workers as a side or main job in aquaculture;

The strategies above are then sorted according to ranking based on the total scores of the constituent elements, as presented in the following table.

Table 2. Determination of Priority Strategies for Development of milkfish-agar Agroindustry

SWOT ELEMENTS		RELATED	SCORE	RANK
Strategy 1	1) Development of industrial cluster pattern that utilizes investment from banks and financial institutions	S1,S2,S3,O1,O2	2,40	1
Strategy 2	2) development of cultivation techniques and post-harvest;	W3,W4,O1,O3	1,60	2
Strategy 3	3) increasing the capacity of coastal human resources;	W3,W4,O2, O3	1,40	3
Strategy 4	4) empowerment of local workers as a side job or main	S4,S5,O1	1,05	4
Strategy 5	5) development of facilities and infrastructure for cultivation and agroindustry	W1,O1	0,95	6
Strategy 6	6) development of a marketing system that can drive the local economy;	S2,S6,T1,T2	0,60	7
Strategy 7	7) development of access to information	W1,W2,T1,T2	0,50	8

After paying attention to all the potential resources and activities of coastal seaweed cultivation in supporting the milkfish-agar agroindustry in Central Sulawesi Province and combined with the factors from the SWOT analysis, a work program plan and strategic plan in the development of the agroindustry are prepared.

3.6. Community Empowerment

The implementation of training assistance in the manufacture of agar sheets, agar flour as well as the manufacture of market-worthy presto milkfish to the community has been carried out in Donggala Regency located in 2 villages (Lalombi Village and Surumana Village), and Parimo Regency in Kasimbar District, Balinggi District Morowali Regency with 30 participants each any training assistance. It was carried out in August-September 2021 according to an agreement with the aquaculture community.

4. CONCLUSION

Based on the results of the analysis and discussion, it can be concluded that the potential of seaweed and milkfish is quite large in 3 regencies, namely Donggala Regency, Parimo Regency and Morowali Regency. Seaweed production has decreased, especially in Donggala Regency there is no cultivation activity because in recent years there has been a decline in selling prices. Generally, the condition of water quality parameters in 3 regencies is quite optimal for the cultivation of seaweed and milkfish. The suitability of the land is mostly suitable for the cultivation of milkfish and seaweed in Dongaala Regency and Parimo Regency which are 716.09 Ha and 674.91 Ha respectively and 11.45 Ha and 36.11 Ha are not suitable. In Morowali Regency, Petasia Subdistrict, the less suitable land is 231.12 Ha, the appropriate class is 82.22 Ha, the Witaponda Subdistrict is 36.11 Ha. Donggala Regency has an effective area of 436.524 Ha, of which 291.06 Ha is used for other purposes. Parimo Regency has an effective area of 420,654 Ha, of which 280,436 Ha is used for other purposes. Morowali Regency has an effective area of 209.67 Ha, which for other uses is 139.78 Ha. Milkfish-agar agroindustry development will require government policies to deal with price management selling seaweed to increase public interest in cultivating seaweed and need collaboration between the government, the private sector and the community in particular cultivators in order to build milkfish-agar agroindustry in Central Sulawesi Province.

5. Acknowledgments
We are grateful for the support and grants from the Ministry of Research and Technology the Republic of Indonesia (Stranas 2014, MP3i 2015, Insinas 2016), The Coordinating Ministry for Human Development and Culture the Republic of Indonesia (Mental Revolution Movement Program 2021)

6. References

[1] Sari N R and Pujiyono A 2013 Analysis of economic growth and income inequality between provinces in Indonesia in 2004-2010 Faculty of Economics and Business
[2] BPS Provinsi Sulawesi Tengah 2021 Sulawesi Tengah dalam Angka 2020 Palu
[3] Septiarini M M 2016 Achievement of Human Development Based on Islamic-Human Development Index (Maqashid Syariah) Regency/City in East Java Period 2010-2014 Airlangga University
[4] Aziz A 2011 Analysis of the Urgency of Indonesia's Foreign Loans in the Context of Financing the State Budget Deficit. *Economics and Finance Studies* 15(1) 47
[5] Anggita L L 2020 Analysis of Competitiveness and Factors Affecting Exports of Indonesian Footwear Products to the United States Islamic Economic Perspective UIN Raden Intan Lampung
[6] Gazali M 2016 Consumer Preferences for Seafood Products (Case Study of Seafood Restaurant Mjm Palu City Central Sulawesi *Journal of Tropical Fisheries* 3(1)
[7] Iskandar K Hendri M Agussalim A 2019 Seaweed Growth Rate Turbinaria Sp. Using the Long Line Method in Lampung Bay Waters Sriwijaya University
[8] Susanto E 2017 Pengolahan Bandeng (Chanos-chanos Forks) Duri Lunak t: https://www.researchgate.net/publication/278000982
[9] Fauzi 2020 Feasibility analysis of milkfish (Chanos chanos Forskall) nursery in Gandapura District, Bireuen Regency. Arwana: Scientific Journal of Aquatic Studies Program 2(2) 112–117
[10]Husain N Rustam Rauf A 2020 Strategy for the Development of Sustainable Pond Cultivation Businesses in Lawallu Village, Barru Regency. Journal Of Indonesian Tropical Fisheries (JOINT-FISH): Journal of Aquaculture, Capture Fisheries Technology and Management, Marine Science, 3(2), 138–150
[11] Fidyansari D Rafli W 2015 Comparative Analysis of Seaweed Farming between Milkfish and No Milkfish Integration Systems in Balandai Village Bara District Palopo City Perbal: *Journal of Sustainable Agriculture* 4(1)
[12] Mulyani S Indrawati E 2021 Seaweed Cultivation Potential Waters in Sinjai Regency South Sulawesi. Almaida's heirloom

[13] Mira F M Reswati E (2014 Application of Blue Economy Principles in Coastal Communities in Brebes Regency, Central Java. *Marina Socio-Economic Marine and Fisheries Scientific Bulletin* 9(1) 17–23
[14] Wandira A W Sunaryo Sedjati S 2018 Seaweed Gracilaria sp. as Bioremedian In Polyculture Cultivation System With Mangrove Crab (Scylla serrata) *Journal of Marine Research* 7(2) 113–124
[15] Reksono B Hamdani H 2012 Effect of stocking density of Gracilaria sp. on the growth and survival of milkfish (Chanos chanos) in polyculture system cultivation *Journal of Marine Fisheries* 3(3)
[16] Ichdayati L I Hartoyo S Syaukat Y Kuntjoro S U 2013 The effect of pond pollutants on the technical efficiency of milkfish production in Karawang Regency. Journal of Indonesian Agribusiness (Journal of Indonesian Agribusiness) 1(2) 107–124
[17] Supito D A Maskar J Damang S 2008 Teknik Budidaya Udang Windu Intensif Dengan Green Water System Melalui Penggunaan Pupuk Nitrat Dan Penambahan Sumber Karbon. *Media Budidaya Air Payau* 7 38-53
[18]Boyd C E Massaut L and Weddig L J 1998 Towards reducing environmental impacts of pond aquaculture *INFOFISII Internasional* 2 /98 p: 27 – 33
[19]Boyd C E 1999 Management of shrimp pond to reduce the eutrophication potential of effluents. *The Advocate* December 1999:12-14
[20]Horowitz A and S Horowitz 2000 Microorganism and feed management in aquaculture. *Global Aquaculture Alliance The Advocate* Vol 3 Issue 2 April 2000 p; 33 – 34
[21] Montoya R and M Velasco 2000 the Role of bacteria on nutritional and management strategic in aquaqlture systems *Global Aquaculture Alliance The Advocate* Vol 3 Issue 2 April 2000 p 35–38
[22] Rangkuti F 2006 Analisis Swot Teknik Membedah Kasus Bisnis. Gramedia Pustaka Utama. Jakarta

Growth and survival of Banggai cardinalfish (*Pterapogon kauderni*, Koumans 1933) reared with artificial microhabitat

I K S Artayasa, M S Manabanti, Karimullah, D E Sari, S Ndobe

Aquaculture Study Program, Department of Fisheries and Marine Science, Faculty of Animal Husbandry and Fisheries, Universitas Tadulako, Palu 94118, Indonesia

Corresponding author: samndobe@yahoo.com

Abstract. The Banggai cardinalfish (*Pterapogon kauderni*), endemic to the Banggai Archipelago (Central Sulawesi), has been declared the Indonesian National Marine Ornamental Fish Mascot. Sea urchins (*Diadema* sp.) are a key *P. kauderni* microhabitat. The decline of urchin populations due to harvesting for human consumption has adversely affected *P. kauderni* populations. The purpose of the study was to develop artificial microhabitat from environmentally friendly materials (coir and cement) and to determine the effectiveness of this artificial microhabitat in terms of fish growth and survival. The research was conducted in July 2021 at the Water Quality and Aquatic Biota Laboratory, Tadulako University. Two treatments were used (*P. kauderni* with live urchin microhabitat = BBA; *P. kauderni* with artificial microhabitat = BBB) with 8 replicates. T-test analysis showed net growth in weight differed significantly ($p<0.05$) between treatments (higher in BBA), as did survival rate (higher in BBB), while net growth in length did not differ significantly. Based on the results, it was concluded that the sea urchin-like artificial microhabitat developed could be effective as a surrogate *P. kauderni* microhabitat, at least in a controlled environment, and shows promise for supporting the conservation of *P. kauderni*. The next stage will be trials in the natural environment.

1. Introduction

The Banggai cardinalfish (*Pterapogon kauderni*) is a marine fish endemic to the Banggai Islands, Central Sulawesi Province and a few nearby islands, traded internationally as an ornamental fish [1]. Due to its unique colour, life history and attractive body shape, the Banggai cardinalfish is favoured by overseas ornamental fish enthusiasts and can sell for as much as 21-25 USD [2,3].

Under Decree of the Minister of Maritime Affairs and Fisheries/KEPMEN KP Number 2 of 2021 [4], *P. kauderni* has been declared the National Mascot for marine ornamental fish. However, in 2000 *P. kauderni* was already considered endangered [5] and an assessment under the International Union for Conservation of Nature (IUCN) Red List criteria assigned *P. kauderni* the Endangered (EN) status [6]. Factors causing the decline in *P. kauderni* populations include overexploitation for the ornamental fish trade as well as habitat degradation due to human activities and climate change [6–10].

In the wild, *P. kauderni* makes use of benthic organisms as microhabitat, especially sea urchins (predominantly *Diadema* sp.) and sea anemones [8,11,12]. One important role of these microhabitats is that they serve as a refuge from predation, including cannibalism of newly recruited juvenile *P. kauderni* (called recruits) by adult fish [12]. The Government of Indonesia has recognized the importance of these microhabitats for *P. kauderni* conservation [3]. Sea anemones are mainly used as microhabitat by juvenile *P. kauderni*, while *P. kauderni* of all sizes use *Diadema* sp. sea urchin microhabitat as shelter [8,12–14]. According to Adel et al. (2016), *Diadema* sea urchins are not only the main microhabitat of *P. kauderni* but are also exploited by coastal communities around the Banggai Islands [15]. There are strong indications that loss or reduction of microhabitat due to over-exploitation of *Diadema* sp. is a major driver of declines in the abundance of *P. kauderni* populations [7,16,17].

In view of this situation, innovative solutions are needed to avoid or reverse declines in the abundance of wild *P. kauderni* populations. Research on the use of different *P. kauderni* microhabitats found that growth and survival were highest when *Diadema* sp. was used as a microhabitat [18]. This prompted the idea that one possible solution would be to create an environmentally friendly artificial microhabitat from natural materials which would resemble *Diadema* sp. sea urchins. Artificial urchin-like microhabitat could be made from palm fibres (coir) to form the spines with cast cement to form the body. However, this concept needed to be tested in terms of effectiveness as a microhabitat for *P. kauderni*, especially juvenile fish (recruits) that are particularly vulnerable to predation.

The purpose of the study was to examine and analyse the effectiveness of artificial microhabitat made from palm fibres (coir) and cast cement in supporting the growth and survival of *P. kauderni* maintained in controlled conditions (*ex-situ*). The benefit and potential usefulness of the research are that it is expected to support the conservation of the endangered Banggai cardinalfish *P. kauderni*, as the national ornamental fish mascot by providing artificial microhabitat from environmentally friendly materials which could serve as an alternative to the natural microhabitat *Diadema* sp.

2. Material and methods

2.1. Time and place
This research under the Student Creativity Research (PKM) programme was carried out from June to September 2021. The initial phase was the development of the artificial microhabitat units. The experimental component was carried out over 30 days in July 2021 at the Water Quality and Aquatic Biota Laboratory, Department of Fisheries and Marine Affairs, Faculty of Animal Husbandry and Fisheries, Tadulako University, Palu City, Central Sulawesi Province, Indonesia.

2.2. Equipment and materials
Equipment used in the study included 16 aquaria (50×25×40.5 cm) with aeration equipment (aerator, aeration hoses and stones) and filtration equipment (cloth filtration bags, aquarium filter sets), water quality measurement equipment (pH meter, thermometer, DO meter, refractometer), and a digital camera. Equipment for measuring the experimental fish comprised digital scales (precision 0.001 g) and a fish ruler (precision 0.5 mm). Miscellaneous equipment included buckets and styrofoam containers, scoop nets, scissors, electrical leads, builder's trowels, masking tape and stationery. Materials used during the study included experimental animals (i.e. *P. kauderni* and *Diadema* sea urchins), sea water and fresh water, filter foam, palm fibre (coir) and cement as well as feed for the experimental animals: *Artemia* sp. (live feed for *P. kauderni*) and seagrass (feed for *Diadema* sp.).

2.3. Study design and layout
The study used an experimental method with a completely randomized design consisting of 2 treatments with 8 replicates so that there were 16 experimental units. The treatment replicate placements were randomized using the randomisation function in Microsoft Excel 2010 (Figure 1). The two treatments were: BBA = live sea urchin microhabitat (*Diadema setosum*); and BBB = artificial sea urchin-like microhabitat (from coir and cement). Each experimental unit (aquarium) was filled with seawater and stocked with 1 microhabitat (*D. setosum* or artificial microhabitat) and 3 *P. kauderni*.

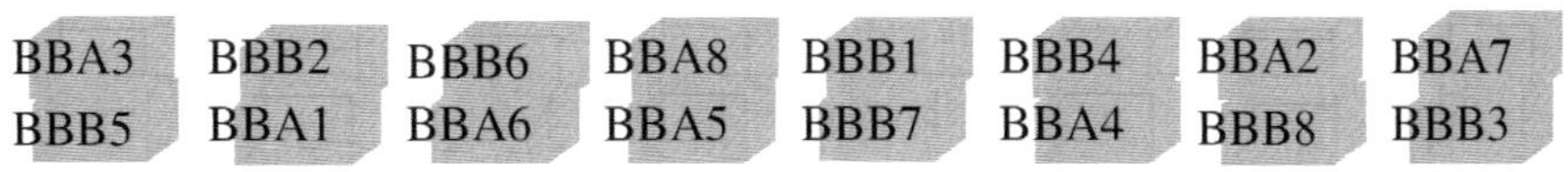

Figure 1. Experimental layout: BBA = microhabitat *Diadema setosum*; BBB = artificial microhabitat

2.4. Source and care of experimental animals

The experimental fish were 48 individuals captured from the introduced *P. kauderni* population [19] in Palu Bay, Palu City, Central Sulawesi Province, Indonesia with a standard length (SL) of 20-33 mm and a weight (W) of 0.5-1.5 g. Based on the classification in Ndobe et al. (2020) [17], *P. kauderni* in the SL range 18-35 mm are considered juveniles, with smaller individuals (SL < 18 mm) classified as recruits and larger individuals (SL > 35 mm) classified as adult. Before stocking, the experimental fish were acclimatized for 24 hours then released gently into the experimental units (aquaria). Live food (*Artemia* sp.) was cultivated and provided 3 times a day to satiation. The sea urchins (*D. setosum*) in treatment BBA were also collected from Palu Bay and fed daily with seagrass (*Thalassia hemprichii* and *Enhalus acoroides*). Each aquarium was siphoned once a week to remove faeces and other debris.

2.5. Manufacturing the artificial microhabitat

The urchin-like artificial microhabitat units were manufactured from coir fibres and cement (Figure 2). The coir palm fibres were cut into lengths similar to *Diadema* sp. spines with pointed ends. Cement and sand were mixed in a ratio of 1:1 to which water was added until the texture was suitable for moulding. The mixture was shaped into hemispheres and the coir fibres pushed into the cement while it was still moist. Each unit had 40 long coir fibres (similar to primary spines) measuring 26 cm in length and 80 short fibres (similar to secondary spines) measuring 5 cm in length. The artificial microhabitat units were then dried for 48 hours and soaked before use to remove the odour of the cement.

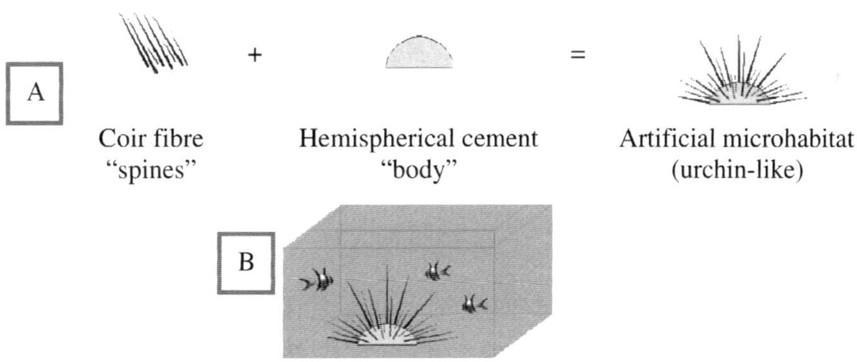

Figure2. Manufacture and use of artificial microhabitat units: (A) Artificial microhabitat construction; (B) Artificial microhabitat and *P. kauderni* in an aquarium (experimental unit)

2.6. Data collection

Data were collected on *P. kauderni* growth and survival. The length (to the nearest 0.5. mm) and weight (to the nearest 0.001 g) of the experimental fish were measured at the beginning and end of the study. Survival was calculated by comparing the number of fish stocked at the beginning and end of the study.

For each experimental unit, net weight gain (in g) was calculated as:

$$W = \overline{Wt} - \overline{W0}$$

where:

W = net weight gain (g)
$\overline{Wt}$ = mean final weight of the experimental fish (g)
$\overline{W0}$ = mean initial weight of the experimental fish (g)

Net growth in length was calculated as:

$$L = \overline{Lt} - \overline{L0}$$

where:
L = net growth in length (g)
$\overline{Lt}$ = mean final length of the experimental fish (g)
$\overline{L0}$ = mean initial length of the experimental fish (g)

Survival rate was calculated as:

$$SR\ (\%) = \frac{Nt}{N0} \times 100$$

where:
SR = survival rate (%)
Nt = number of fish placed in the unit at the start of the study
N0 = number of live fish in the unit at the end of the study

Water quality is a factor that can affect the survival of *P. kauderni*. Water quality data collected comprised: temperature (thermometer), pH (pH meter), salinity (refractometer) and dissolved oxygen (DO meter). Water quality parameters temperature, pH and salinity were measured twice daily, in the morning and evening. Dissolved oxygen was measured once a week in the morning and evening.

2.7. Data analysis

Data were tabulated and analysed in Microsoft Excel 2010. T-test analysis was applied at the 95% confidence level ($\alpha = 0.05$) to determine whether *P. kauderni* growth and survival were significantly different between the two treatments. The length-weigh relationship ($W = aL^b$) for the final weight was calculated to obtain the growth pattern and enable the calculation of the relative condition factor K. For isometric growth, $K = W/aL^3$ otherwise $K = W/aL^b$ [20]. For each data set, the F-test for equal variance was applied to each data set to determine whether t-test assuming equal or unequal variance was applied. Water quality data were analysed descriptively.

3. Results and discussion

3.1. Net weight gain

The mean net weight gain of the surviving *P. kauderni* varied considerably between replicates of each treatment (Figure 3A) as did the total net weight (biomass) gain per experimental unit (Figure 4B).

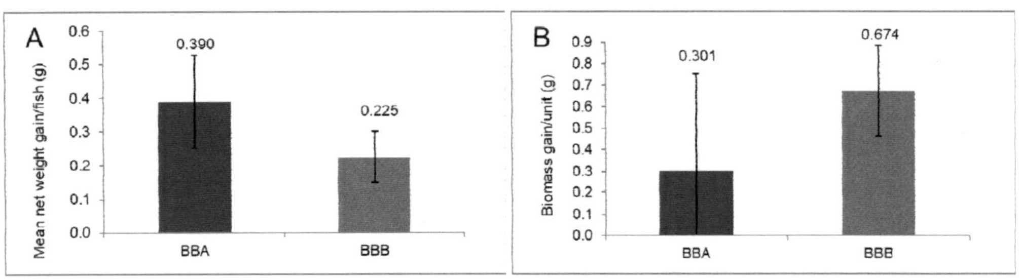

Figure 3. Mean net weight gain of surviving fish (A) and biomass gain per unit (B) of *P. kauderni* reared with natural Diadema *setosum* microhabitat (BBA) and artificial microhabitat (BBB)

The F-test did not indicate unequal variance in mean individual weight gain ($p = 0.067$) between treatments. The t test assuming equal variance showed that the mean individual net weight gain was significantly different ($p < 0.05$) between treatments with higher weight gain under the BBA treatment (natural *D. setosum* microhabitat) than the BBB treatment (artificial urchin-like microhabitat). On average, individual weight gain was 42% lower in the BBB treatment units than the BBA treatment units. With respect to biomass, the F-test showed unequal variance in total net weight (biomass) gain per experimental unit ($p = 0.04$). The t-test assuming unequal variance showed a significant difference in biomass gain between the two treatments, with biomass gain considerably higher for the BBB treatment than the BBA treatment.

The relatively slow growth in the BBB treatment could be related to disease. Some *P. kauderni* in this treatment group showed symptoms of disease including damaged or ragged fins and/or white spots. Adriany and Koesharyani (2017) found the bacteria *Alcaligenes faecalis*, *Micrococcus luteus*, *Acinetobacter* spp., *Pleisiomonas shigelloides*, *Yersinia enterocolitica*, and *Vibrio alginolyticus* in *P. kauderni* specimens sampled from the catch of ornamental fishermen [21]. Meanwhile the same study found *Vibrio alginolyticus* and *V. shigelloides* infecting *P. kauderni* samples collected from collectors (middlemen/intermediary traders) in Luwuk City, Banggai District. Factors that can cause fish reared or held in a controlled environment (e.g. aquaria) to be attacked by disease include a lack of care during capture and subsequent handling; fish that are injured will be more susceptible to infection with pathogens and are more likely to suffer with disease [21,22].

It is also possible that in the BBA treatment there was a beneficial effect from the symbiosis between *P. kauderni* and the sea urchins which may have promoted faster growth. For example, the fish may have obtained additional nutrition from the sea urchin faeces if these became suspended. Another consideration is that stocking density can be negatively correlated with fish growth and mean individual weight gain, even though biomass increase can be higher at higher densities [23]. Within the BBA replicates, all units with mortality (and hence a lower stocking density for much of the study period) had higher mean individual net weight gain (mean 0.456 g) than those where all three fish survived (mean 0.280 g), while the latter were not significantly different (t-test, $p = 0.33$) from the BBB treatment units (mean 0.225 g). However, all these potential explanations require further research.

3.2. Net increase in length, length-weight relation and relative condition factor

The mean net increase in length of *P. kauderni* was on average 0.04 mm (13%) less under the BBB treatment than the BBA treatment (Figure 4A). The F-test did not indicate unequal variance, and the t-test (assuming equal variances) showed the between-treatment difference was not significant ($p = 0.77$). The combined length-weigh relationship (Figure 4B) indicates an isometric growth pattern (b not significantly different from 3).

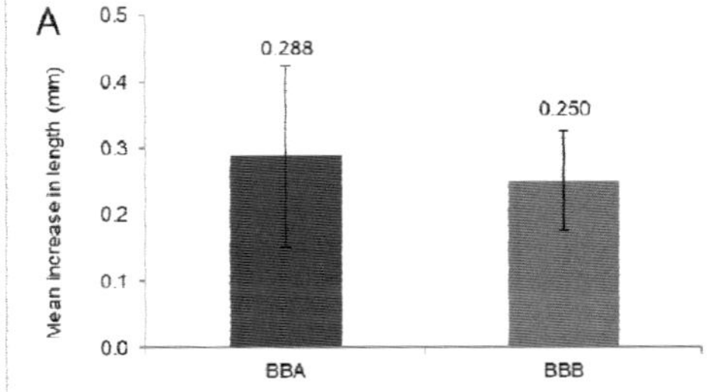

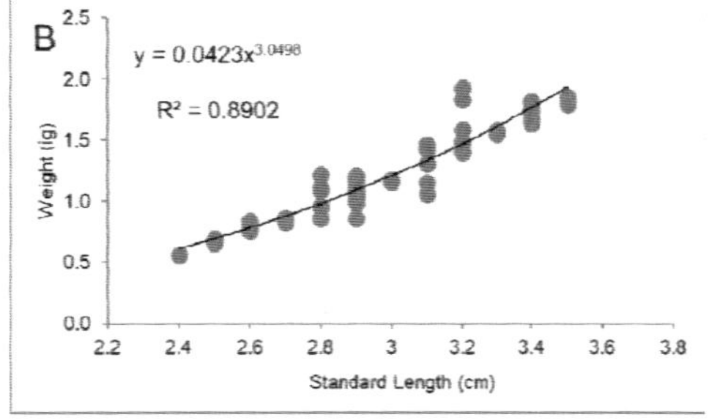

Figure 4. Net increase in length of *P. kauderni* reared with natural *Diadema setosum* microhabitat (BBA) and artificial microhabitat (BBB) (A); length-weight relationship for both treatments combined at the end of the experimental period (B)

The relative condition factor K based on isometric growth was not significantly different between treatments (p = 0.22) with a mean close to 1 for both treatments. Based on these analyses, even though the BBA treatment using *Diadema setosum* resulted in a slightly higher net increase in length growth, both treatments resulted in similar growth rates and growth patterns. *Diadema* sp. is a natural microhabitat of wild *P. kauderni* and it could therefore be expected that the experimental fish could readily adapt to this microhabitat. Meanwhile, apart from the similar shape of the artificial habitat, colour of the coir fibres was similar to that of the black spines of *Diadema* sp. sea urchins, which may have aided adaptation. Rahman and Safir (2018) found that *Diadema* sp. microhabitat gave a better growth performance in captive *P. kauderni* compared to the other two main microhabitats observed in the wild (sea anemones and hard corals), while growth in all three microhabitats was significantly higher than in the control (no microhabitat) [18]. Despite the lack of statistical significance, the slightly lower growth in treatment BBB indicates that the adaptation of *P. kauderni* to the artificial microhabitat may be less than perfect, and could have resulted in somewhat slower growth.

3.3. Survival rate

The survival rate of *P. kauderni* was not normally distributed, with only two values (step function): 100% and 66.7% (2 out of 3 fish surviving). All mortality occurred in the BBA treatment (5 out of 8 replicates) with 100% survival in the BBB treatment and a mean of 83.33% in the BBA treatment. The t-test indicated significant between treatment difference (p = 0.033).

Despite indications of disease in some individuals, the artificial urchin-like microhabitat out-performed the natural sea urchin microhabitat in terms of juvenile *P. kauderni* survival, with no mortality in the BBB treatment. This indicates that artificial sea urchins could be considered an effective or feasible replacement for natural sea urchins as microhabitat for the captive husbandry and grow-out of *P. kauderni*. One reason for the higher survival rate could be water quality, with the water remaining clearer and presumably more stable, as it was not affected by organic inputs related to the sea urchins and their feed. In addition, the artificial microhabitat was similar in shape and colour to the natural *Diadema* sp. microhabitat, which is likely why the *P. kauderni* in the BBB treatment seemed attracted to their artificial urchins and used them as a refuge.

The lower survival rate in the BBA treatment was suspected to be due to the *Diadema setosum* sea urchins eating the juvenile *P. kauderni* in the experimental units, because the bodies of the fish were not found. During an experiment on the thermal tolerance of *P. kauderni* and *Diadema* sp. in the context of climate change, *Diadema setosum* were observed piercing *P. kauderni* with their spines and consuming the injured or dead fish [24]. In that case, all fish predated were already showing signs of stress or disease (e.g. emaciation and/or unusual behaviour and colouration) and predation appeared more frequent at higher water temperatures. In this study, the water temperature did not exceed the historical temperature range reported for the natural habitat of *P. kauderni* [8]. However, the experimental fish could have been stressed due to other factors and/or similar disease symptoms to those seen in some surviving fish. This study reinforces the need for further study of the nature of the symbiotic relationship between *P. kauderni* and *Diadema* sp.

Water quality is a factor that may have affected the *P. kauderni* in the BBA treatment, specifically the effects of the release of large volumes of faeces by the *Diadema* urchins. These faeces caused cloudiness or turbidity, which may have affected the vision of the fish. While ammonia was not measured in this study, faeces and feed residues are also sources of ammonia; excessive ammonia concentrations can cause fish disease, including nervous system and behavioural disorders, and even death [25]. Furthermore, *P. kauderni* tends to be sedentary in nature [26], and observations during a sea urchin species preference trial indicate that individuals tend to remain closely associated with their sea urchin microhabitat [27]. It is possible that the movements of the *Diadema* sea urchin within the BBA treatment units, especially when they climbed the walls of the aquaria or were manipulating and feeding on the seagrass, may have resulted in a mismatch between the behaviour of *P. kauderni* and their host sea urchin. This could result in intentional or unintentional injury and subsequent predation of the fish by the urchin.

3.4. Water quality

The water quality data for the *P. kauderni* experimental rearing units during the study are shown in Table 1. The water quality parameters temperature, salinity, and DO remained similar between the two treatments throughout the study, while pH was lower (7.7) in the BBA treatment with *Diadema setosum* microhabitat than in the BBB treatment with artificial microhabitat (8.0).

Table 1. Mean values of water quality parameters during the study on *P. kauderni* reared with natural sea urchin microhabitat (BBA) and artificial microhabitat (BBB)

Treatment	Temperature (°C)		Salinity (ppt)		DO (ppm)		pH	
	a.m.	p.m.	a.m.	p.m.	a.m.	p.m.	a.m.	p.m.
BBA								
mean	28.4	28.5	31.1	30.9	6.3	6.4	7.7	7.7
SD	0.58	0.50	0.64	0.58	0.56	0.52	0.11	0.11
BBB								
mean	28.4	28.5	31.1	31.0	6.6	6.6	8.0	8.0
SD	0.59	0.50	0.64	0.65	0.61	0.64	0.11	0.10

The water quality parameter data in Table 1 show that the ranges recorded in the mornings and afternoons remained within reasonable limits for supporting the growth and survival of *P. kauderni*. The temperature recorded remained within the temperature range of 26-30°C which is within the range reported from *P. kauderni* habitat in Palu Bay and in the endemic distribution of *P. kauderni* [8] and below the temperature at which adverse effects on *P. kauderni* have been reported [24]. Salinity range was 29-33 ppt, with mean values slightly above the optimum 27-30 ppt recommended by Madinawati et al. (2009) for rearing juvenile *P. kauderni* [28]. However, salinity remained well within the range reported from *P. kauderni* endemic and introduced habitat [8]. DO varied between 4.6 and 7.7 ppm, within the levels recommended for captive fish welfare [25] and similar to or higher than levels at which *P. kauderni* has been successfully bred in captivity [29].

The pH in all experimental units remained within the range 7.3-8.1, which is within the range recommended for ornamental fish in general [25], the range reported for *P. kauderni* habitat [14,30], and for successful husbandry of captive *P. kauderni* [28]. However, the pH was consistently lower in the BBA treatment than the BBB treatment units. As noted previously, the BBA treatment water was also generally more cloudy or turbid than in the BBB treatment units, even though all the aquaria in this study were equipped with filters and siphoning of faeces and other debris was carried out more frequently for the natural sea urchin treatment (BBA) than the artificial urchin-like treatment (BBB). Both pH and turbidity can be affected by the faeces excreted by the sea urchins serving as microhabitat and residue from their feed; this contamination could account for the difference in pH as well as potentially contributing to the mortality in the BBB treatment.

4. Conclusion

Based on the results of the study, it can be concluded that the artificial sea urchins developed can be used effectively to replace natural sea urchins as the microhabitat of *P. kauderni* in a controlled environment (*ex-situ*). Thus, it is hoped that this research can support the conservation of *P. kauderni* as the national ornamental fish mascot by providing artificial microhabitat made from environmentally friendly materials. The next stage will be to test the feasibility and effectiveness of using the artificial microhabitat *in situ*, to assist the recovery of a depleted *P. kauderni* population at a site where *Diadema* urchin populations have been severely depleted or extirpated.

Acknowledgments

The authors are grateful to the Directorate General of Higher Education, Ministry of Research and Technology of the Republic of Indonesia for funding this research under the Student Creativity

Programme and to the Aquaculture Study Program at Tadulako University for the use of research facilities at the Water Quality and Aquatic Biota Laboratory.

References

[1] Ndobe S, Yasir I, Moore A M, Biondo M V and Foster S J 2018 *A study to assess the impact of international trade on the conservation status of Pterapogon kauderni (Banggai cardinalfish)* (Gland: International Union for Conservation of Nature)

[2] Hopkins S, Ako H and Tamaru C S 2005 *Manual for the Production of the Banggai Cardinalfish, Pterapogon kauderni, in Hawai'i* (Hawai'i: Rain Garden Ornamentals)

[3] Suwardi., Arbi U Y, Wiadnyana N N, Ndobe S, Reksodihardjo G, Lestari Y T, Julianto E, Nelly E and Suharti S R 2019 *Pedoman Monitoring Banggai Cardinalfish (Pterapogon kauderni)* (Jakarta: Ministry of Marine Affairs and Fisheries)

[4] MMAF 2021 Keputusan Menteri Kelautan dan Perikanan Republik Indonesia Nomor 2 Tahun 2021 tentang Maskot Ikan Hias Nasional

[5] Allen G R 2000 Threatened fishes of the world: *Pterapogon kauderni* Koumans, 1933 (Apogonidae) *Environ. Biol. Fishes* **57** 142

[6] Allen G R and Donaldson T J 2007 *Pterapogon kauderni IUCN Red List Threat. Species 2007* e.T63572A12692964

[7] Ndobe S, Moore A, Salanggon A I M, Muslihudin ., Setyohadi D, Herawati E Y and Soemarno . 2016 Pengelolaan Banggai cardinalfish (*Pterapogon kauderni*) melalui Konsep Ecosystem-Based Approach [Banggai cardinalfish (*Pterapogon kauderni*) Management - an Ecosystem-Based Approach] *Mar. Fish. J. Mar. Fish. Technol. Manag.* **4** 115

[8] Ndobe S, Soemarno, Herawati E Y, Setyohadi D, Moore A, Palomares M L D and Pauly D 2013 Life History of Banggai Cardinalfish, *Pterapogon kauderni* (Actinopterygii: Perciformes: Apogonidae), from Banggai Islands and Palu Bay, Sulawesi, Indonesia *Acta Ichthyol. Piscat.* **43** 237–50

[9] Kasim K, Hartati S T and Thordarson G 2014 Impact of fishing and habitat degradation on the density of Banggai cardinal fish (*Pterapogon kauderni*, Koumans 1933) in Banggai Archipelago, Indonesia *Indones. Fish. Res. J.* **1** 29–36

[10] Moore A M, Ndobe S, Yasir I, Ambo-Rappe R and Jompa J 2019 Banggai cardinalfish and its microhabitats in a warming world: A preliminary study *IOP Conf. Ser. Earth Environ. Sci.* **253** 012021

[11] Vagelli A A 2011 *The Banggai Cardinalfish: Natural History, Conservation, and Culture of Pterapogon kauderni* (Chichester, UK: John Wiley & Sons, Ltd.)

[12] Ndobe S, Widiastuti I and Moore A 2013 Sex Ratio dan pemangsaan terhadap rekrut pada ikan hias Banggai cardinalfish *(Pterapogon kauderni) Prosiding Konferensi Akuakultur Indonesia 2013 9-20*

[13] Vagelli A A 2004 Ontogenetic shift in habitat preference by *Pterapogon kauderni*, a shallow water coral reef apogonid, with direct development *Copeia* **2004** 364–9

[14] Ndobe S, Madinawati and Moore A 2008 Pengkajian *ontogenetic shift* pada ikan endemik *Pterapogon kauderni J. Mitra Bahari* **2** 32–55

[15] Adel Y, Yonvitner and Rahardjo M F 2016 Management of the Banggai cardinalfish (*Pterapogon kauderni*, Koumans 1933) using an ecosystem approach (Banggai Island case study, Banggai Laut District) *J. Ilmu Pertan. Indones.* **21** 186–94

[16] Ndobe S, Moore A M and Jompa J 2017 Status of and threats to microhabitats of the endangered endemic Banggai Cardinalfish (*Pterapogon kauderni*) *Coast. Ocean J.* **1** 73–82

[17] Ndobe S, Handoko K, Wahyudi D, Yasir M, Irawati Y, Tanod W A and Moore A M 2020 Monitoring the endemic ornamental fish *Pterapogon kauderni* in Bokan Kepulauan, Banggai marine protected area, Indonesia *Depik J. Ilmu-Ilmu Perairan, Pesisir dan Perikan.* **9** 18–31

[18] Rahman S A and Safir M 2018 Performa Pertumbuhan dan Kelangsungan Hidup Ikan Capungan Banggai (Pterapogon kauderni) pada Mikrohabitat yang Berbeda *Octopus* **7** 1–6

[19] Moore A and Ndobe S 2007 Discovery of an introduced Banggai Cardinalfish population in Palu Bay, Central Sulawesi, Indonesia *Coral Reefs* **26** 569–569
[20] Froese R 2006 Cube law, condition factor and weight-length relationships: History, meta-analysis and recommendations *J. Appl. Ichthyol.* **22** 241–53
[21] Adriany D T and Koesharyani I 2017 Infeksi penyakit ikan Banggai cardinal (*Pterapogon kauderni*) dalam rantai perdagangan. *J. Ris. Akuakultur* **12** 283–94
[22] Smith K F, Schmidt V, Rosen G E and Amaral-Zettler L 2012 Microbial diversity and potential pathogens in ornamental fish aquarium water *PLoS One* **7** e39971
[23] Khatune-Jannat M, Rahman M M, Bashar A, Hasan N, Ahamed F and Hossain Y 2012 Effects of stocking density on survival, growth and production of Thai climbing perch (*Anabas testudineus*) under fed ponds *Sains Malaysiana* **41** 1205–10
[24] Moore A M, Ndobe S, Yasir I and Jompa J 2019 Disasters and biodiversity: case study on the endangered endemic marine ornamental Banggai cardinalfish *IOP Conf. Ser. Earth Environ. Sci.* **253** 012036
[25] Yavuzcan Yildiz H, Robaina L, Pirhonen J, Mente E, Domínguez D and Parisi G 2017 Fish welfare in aquaponic systems: its relation to water quality with an emphasis on feed and faeces—A Review *Water* **9** 13
[26] Kolm N, Hoffman E A, Olsson J, Berglund A and Jones A G 2005 Group stability and homing behavior but no kin group structures in a coral reef fish *Behav. Ecol.* **16** 521–7
[27] Ndobe S, Moore A and Jompa J 2018 A Tale of Two Urchins - Implications for *In-Situ* Breeding of the Endangered Banggai Cardinalfish (*Pterapogon kauderni*) *Aquac. Indones.* **19** 65–75
[28] Madinawati, Ndobe S and Gamgulu A 2009 Pertumbuhan ikan kardinal Banggai (*Pterapogon kauderni*) yang dipelihara pada salinitas yang berbeda dalam wadah terkontrol [Growth of Banggai cardinalfish *Pterapogon kauderni* reared at different salinity in a controlled system] *J. Akuakultur Indones.* **8** 193–8
[29] Pietoyo A, Andayani S and Soeprijanto A 2016 Latent and eggs production of Banggai cardinal (*Pterapogon kauderni*, Koumans 1933) on various salinity levels: conservation efforts *J. Exp. Life Sci.* **6** 16–8
[30] Muslihuddin, Salanggon A-I, Ndobe S, Moore A and Ederyan 2012 Monitoring fish home restoration modules in Central Sulawesi *J. Mitra Bahari* **6** 36–52

Assessment of forest road network alternatives in forest harvesting activity

A Malik[1*], N A Santi[2]

[1]Faculty of Forestry, Tadulako University, Palu, Indonesia 94118
[2]Department of Forest Management, Faculty of Forestry and Environment, IPB University, Bogor, Indonesia 16680

*Email: amalik@untad.ac.id

Abstract. One of timber harvesting purposes is moving wood from the forest to sawmills. Road network planning requires consideration for alternatives. This paper purposed to determine decision-making in the selection of alternative road networks with utility analysis by considering economic, ecological, and social criteria. The indicator used were the length of the road, open area, cost of road construction, felling, skidding, and empowerment of forest communities. The analysis showed that the first alternative was the best forest road network of the three alternatives designed. The results are used in the decision-making plan for implementation in the next rotation.

1. Introduction

FAO [1] explains the definition of forest as an area with an area of at least 0.5 hectares containing tree stands with a minimum height of 5 meters and having a canopy cover percentage of >10%, or a collection of trees that can reach this threshold. The existence of the forest must be maintained and managed with sustainable forest management. The existence of the forest can also be known continuously if monitoring and regulation of the existence of the forest are carried out.

Forest harvesting activities can be defined as a series of forestry activities aimed at converting trees and biomass into forms that can be transferred to other locations so that they are beneficial to the life and economy of the community [2]. This is necessary because forests have an important role in human life. In addition to providing wood, forests, especially tropical natural forests, have biodiversity that is able to provide solutions to current problems in the fields of food, health, and energy [3]. Timber forest products and non-timber forest products (NTFPs) will be more useful if they can be removed from the forest properly and correctly. In order for sustainable forest management and maximum utilization of forest products to be achieved, the infrastructure for access to and from the forest in the form of main roads and branch roads must be able to facilitate well, so that forest management activities which include harvesting of forest products, evaluation, monitoring, and supervision can be carried out properly.

Elias [2] argued that forest area clearing (PWH) is a forestry activity carried out to facilitate forest management activities by facilitating facilities and infrastructure. The infrastructure provided is in the form of a road network, log pond, log yard, main basecamp, branch basecamp, forest development basecamp, wood stockpile (TPK), wood collection point (TPN), bridges and culverts, control towers, and others. The infrastructure built must be permanent to lead to sustainable forest management.

PWHs that fall into the good category are PWHs that can serve all parts of the forest equally in the long term. In addition, they also have low forest openness. Planning activities, forest management, PWH, and an integrated timber harvesting system will be able to achieve efficient forest management and minimize environmental damage. PWH planning is an activity that must be able to design, design, and plan PWH well, starting from determining a good area to be used in the construction of PWH facilities and infrastructure, making road alignments to selecting alternative road plans.

Road network planning is carried out to facilitate forest product harvesting activities. A new road network planning approach has been intensively developed for timber harvesting activities. A road network with good quality will reduce production costs so that transportation can be carried out more smoothly and continuously, as well as increase the efficiency of forest product output, and minimize forest damage [4]. This is necessary because forests have an important role in human life. In addition to providing wood, forests have biodiversity that is able to provide solutions to problems in the fields of food, health, and energy [3]. Forests are also able to reduce erosion [5-7] and are able to absorb carbon dioxide in the atmosphere [8]. in road network planning.

Utility analysis is a decision-making method to determine the optimum road network for harvesting forest products. The utility analysis determines the optimum road network using economic, social, and ecological criteria. This paper aims to determine alternative decision-making in the selection of road networks based on utility analysis.

2. Materials and Methods

2.1 Study site

Assessment of alternative roads was carried out on tree distribution maps and plot topography 36 RKT 2002 RKL IV Administrator PT. Inhutani I Labanan, East Kalimantan. The total area of the cutting plots is 85.1 Ha. In addition, there are protected plants. The purpose of managing the forest area is as a production forest.

2.2 Materials and tools

The materials used are cutting plots 36 RKT 2002 RKL IV Administrator PT. Inhutani I Labanan, East Kalimantan. The scale used on the map is 1: 2000. The tool used is a curvimeter to measure the length of the road (main roads, branches, branches, and skids), and an excel program for making decision-making models.

2.3 Data collection

An assessment of the best alternative road network as part of a forest harvesting plan was made using three assessment criteria, namely economic, ecological, and technical aspects. The purpose of the economic aspect is to maximize the profits obtained by the company with the minimum possible costs. The costs incurred are the activities of clearing forest areas in the construction of a road network in terms of construction and operations. Indicators used in the assessment of alternative road networks on the economic aspect include road construction costs, skidding costs, and income from felled trees. Assessment of the ecological aspect, namely the minimal impact of damage to forest ecosystems caused by PWH activities. Assessment indicators include road length, open area due to road construction, and stand damage due to road construction. Assessment of technical aspects is important to consider in sustainable forest management because forest sustainability will be influenced by its technical value.

Road types are divided into corridor roads, main roads, branch roads, branch roads, and skid trails. The type of road will determine the costs incurred in making the road [2]. In practice, making forest roads requires tools, human resources, and no small amount of money. So that careful planning is needed so that it can reduce road construction costs. The road that has been made on the map with a scale of 1:2000 is measured using a curvimeter. The length of this road is also to determine the amount of costs that must be incurred by the company to make the road. The length of the road desired by the management is a short road but can reach the entire area to be harvested so as to facilitate the process of moving wood out of the forest.

The utility analyst method is used by selecting the forest road network with. using a scale of 1–5. The total utility value for each alternative road network is obtained by adding the utility value of the overall indicator which has been changed from a nominal scale to an interval scale, the scale value used is 1-5 [9]. Forest road network planning must be made to minimize construction and maintenance costs, while the road specification design must be in accordance with environmental requirements. Ecological, economic, and social aspects are taken into consideration in the criteria and indicators for the assessment of the PWH plan for the development of harvesting facilities and infrastructure using utility analysis (Table 1).

Table 1 Preference criteria and indicators used in the selection of alternative road networks

Criteria	Indicators	Unit	Preference
Technic	Main road length	km	Minimum
	Branch road length	km	Minimum
	Road density	M/ha	Minimum
	Open area by main road	Ha	Minimum
	Open area by branch road length	Ha	Minimum
	Vcorr	-	Minimum
	E (percent of PWH)	%	Maximum
Ecology	Number of protected trees damaged	Tree	Minimum
	Number of core trees damaged	Tree	Minimum
Economy	Main road construction costs	Rp	Minimum
	Branch road construction costs	Rp	Minimum

3. Results and Discussion

3.1 Road Quality

Forest road network planning aims to design an alternative road by considering the low total costs of construction, maintenance, and transportation costs, but the alternatives made must comply with specifications, environmental requirements, and operator safety [10]. The main factor that is considered in planning the road network is the length of the forest road (Figure 1).

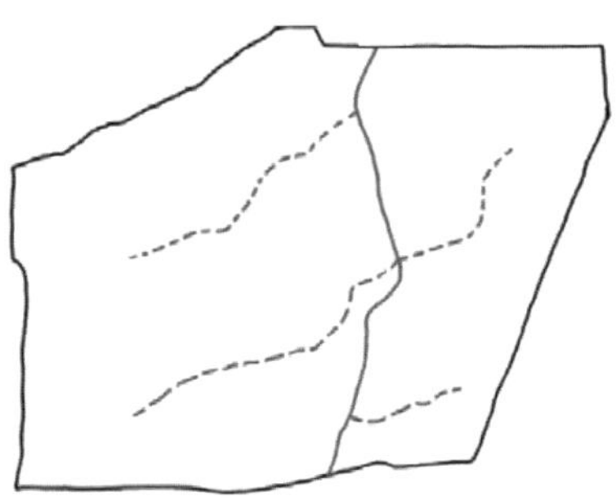

Status Quo (SQ)

Alternative 2

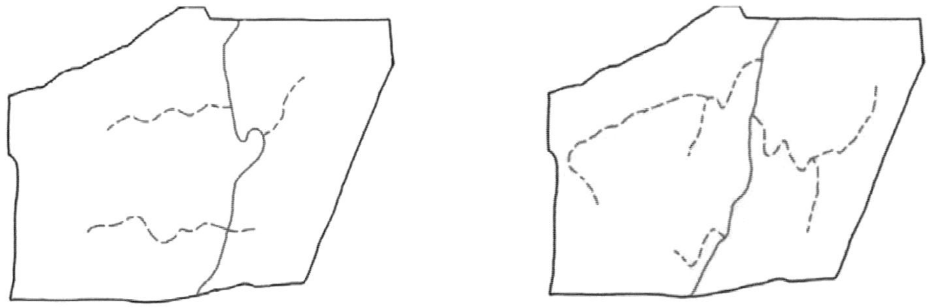

Alternative 1 Alternative 3

note – : Boundaries –: Main road - -: Branch road

Figure 1. Alternative of Forest Road

Table 2 Indicator quantification of three alternative forest road networks in PWH activities

Criteria	Indicators	Unit	Alternative			
			SQ	A1	A2	A3
Technic	Main road length	Meter	910	1 040	980	980
	Branch road length	Meter	1 740	1 220	1 642	2420
	Road density	M/ha	31.14	26.56	30.81	39.95
	Open area by main road	M^2	9 100	10 400	9 800	9800
	Open area by branch road length	M^2	13 920	9 760	13 136	19 360
	Vcorr		1.22	1.30	1.37	1.27
	E (percent of PWH)	%	81.96	76.92	72.99	78.51
Ecology	Number of protected trees damaged	Tree	12	29	27	5
	Number of core trees damaged	Tree	9	20	19	7
Economy	Main road construction costs	million	1 001	1 144	1 078	1 078
	Branch road construction costs	million	1 740	1 220	1 642	2 420

3.2 Best Road Network

Sustainable management of forest resources can be achieved through a well-planned road network. In order to minimize the environmental impact of forest roads, forest area managers must design a road network that can reach the entire area but is also environmentally friendly [12-14]. The planning of a multifunctional forest road network is one of the important things to meet the objectives of the sustainable forest concept. Damage to forest road construction and nature protection must be considered in designing forest road networks. Various criteria from economic, ecological, social, and institutional aspects are taken into consideration in making decisions on the selection of alternative road networks.

Decision-making is needed in choosing several alternatives so as to provide a decision on the optimum benefit. Alternative forest road network planning is designed to make assessments and sensitivity analyses in the decision-making process in good forest clearing activities [15]. The alternative used in determining the best PWH in this paper is SQ. A1, A2, and A3 using utility analysis on PWH activities (Table 3).

Table 3 Utility analysis and scoring of three alternative forest road networks in PWH activities

Criteria	Indicator	Alternative			
		SQ	A1	A2	A3
Technic	Main road length	5	1	3	3
	Branch road length	3	5	4	1
	Road density	4	5	4	1
	Open area by main road	5	1	3	3
	Open area by branch road length	3	5	4	1
	Vcorr	5	3	1	4
	E (percent of PWH)	5	3	1	4
Ecology	Number of protected trees damaged	4	1	1	5
	Number of core trees damaged	4	1	1	5
Economy	Main road construction costs	5	1	3	3
	Branch road construction costs	3	5	4	1
	Total Score	46	31	29	31

Note: 1(Bad), 2 (Quite bad), 3(Quite good), 4 (Good), 5 (Excellent)

The highest score of the four alternatives designed is SQ, while the alternative that has the lowest score is A2. The first alternative SQ is the best alternative which is selected according to established economic, ecological, and social criteria (Table 3). Alternative SQ is considered good in the construction of PWH facilities as seen from the minimal costs incurred and the impact of environmental damage caused. The best decision as a goal achievement is PWH activities that are able to provide maximum benefits, provide benefits to the community from a social perspective and have a minimal impact on environmental and ecosystem damage.

4. Conclusion

Alternatives SQ, A1, A2, and A3 were assessed using utility analysis based on ecological, economic, social, and institutional criteria. The best alternative chosen in decision-making based on its utility value is SQ in the first rank, A1, A3 in the second rank, and A2 in the third rank. The best forest road network is a road network that has short forest roads, minimal area openness due to road construction, minimal road construction costs, the optimal number of trees harvested, and maximum community empowerment around the forest.

ORCID iDs

A Malik https://orcid.org/0000-0002-0184-0058

References

[1]. Food and Agriculture Organization 2010 Global Forest Resources Assessment 2010 Main Report FAO Forestry Paper 163 Rome FAO.
[2]. Elias 2012 Pembukaan Wilayah Hutan Edisi II Bogor (ID) IPB Press
[3]. Sukara E 2014 Tropical forest biodiversity to provide food, health and energy solution of the rapid growth of modern society *Procedia Environmental Sciences* **20** 803-8
[4]. Ismail M H, Jusoff K, and Azizon M. 2008 Forest road assessment in Ulu Muda *Modern Applied Science* **2** 100–8
[5]. Zimmermann A, Francke T, and Elsenbeer H 2012 Forests and erosion: Insights from a study of suspended-sediment dynamics in an overland flow-prone rainforest catchment *J Hydrology* **428–429** 170-1

[6]. Poulsen J R, Clark C J, and Palmer T M 2013 Ecological erosion of an Afrotropical forest and potential consequences for tree recruitment and forest biomass *Biological Conservation* **163** 122-30 http://dx.doi.org/1-.1016/j.ecoleng.2008.07.006
[7]. Marden M, Herzig A, and Basher L 2014 Erosion process contribution to sediment yield before and after the establishment of exotic forest: Waipaoa catchment, New Zealand *Geomorphology* **226** 162-74
[8]. Couture S, and Reynaud A 2011 Forest management under fire risk when forest carbon sequestration has value *Ecological Economy* **70** 2002-11
[9]. Budiaman A, and Heryana 2013 Assessment of forest road network alternatives for pine resin extraction *Jurnal Manajement Hutan Tropica* **19** 23-30 doi:10.7226/jtfm.19.1.23
[10]. Akay A E 2006 Minimizing total costs of forest roads with computer-aided design model *Sadhana* **31** 621–33
[11]. Abdi E, Majnounian B, Darvishsefat A, Mashayekhi Z, and Sessions J 2009 A GIS-MCE based model for forest road planning *J Forest Science* **55** 171-6
[12]. Hayati E, Majnounian B, and Abdi E 2013 An expert-based approach to forest road network planning by combining delphi and spatial multi-criteria evaluation *Environment Monitoring and Assessment* **185** 1767-76 doi:10.1007/s10661-012-2666-1
[13]. Caliskan E 2013 Planning of forest road network and analysis in mountainous area *Life Science J* **10** 2456-2565
[14]. Gumus S, Acar H H, and Toksoy D 2008 Functional forest road network planning by consideration of environmental impact assessment for wood harvesting *Environmental Monitoring and Assessment* **142** 109-116 doi:10.1007/s10661-007- 9912-y
[15]. Enache A, Stampfer K, Ciobanu V, Branzea O, and Duta C 2011 Forest road network planning with state of the art tools in a private forest district from lower Austria *Agricultural Food Engineering* **4** 33-40

IOP Conf. Series: Earth and Environmental Science 1075 (2022) 012011 doi:10.1088/1755-1315/1075/1/012011

Performance analysis of breadth-first search and depth-first search on MANET for health monitoring system

Alamsyah[1], A Amir[1], M Subito[1], R Fauzi[1], Amirullah[2]

[1]Department of Electrical Engineering, Faculty of Engineering, Tadulako University, Palu, 94119, Indonesia
[2]Electrical Engineering Study Program, Faculty of Engineering, Universitas Bhayangkara Surabaya, Surabaya, 60231, Indonesia.

alamsyah.zakaria74@gmail.com

Abstract. The development of mobile ad-hoc network (MANET) technology is increasing and is directly proportional to the need. MANET is a wireless network consisting of several nodes moving freely and without using the infrastructure. MANET can be applied in various fields such as the military, natural disasters, and health services. The health sector, especially the health monitoring system, gets more attention because it involves maintaining patient safety. However, wireless devices installed in health services have not been able to reach patients. This condition causes the required time and energy consumption to increase. Therefore, the breadth-first search (BFS) and depth-first search (DFS) methods are indispensable in determining the quality of service (QoS) in MANET. QoS parameters analyzed include packet delivery ratio (PDR), throughput, packet loss, delay, and energy consumption. This study aims to choose a health data traffic route that can improve QoS when transmitting data. The simulation results using the 2018 version of MATLAB show that BFS has a better performance than DFS on delay and energy consumption parameters. Meanwhile, DFS has better performance than BFS on PDR, throughput, and packet loss parameters.

1. Introduction

The development of technology is currently increasing and is directly proportional to the need for facilities and infrastructure. The availability of facilities ranging from fiber-optic networks to 5G-based mobile phone networks requires infrastructure to support information delivery. Conditions like this certainly require network technology that can send information to its destination without being connected to existing infrastructure such as access points or routers. One of the network technologies that can use to overcome infrastructure limitations is the use of MANET technology. The MANET concept [1, 2] is a wireless network consisting of several nodes that can move freely. MANET is widely used in various applications because it is fast, able to manage topological changes independently [3], without using fixed infrastructure [4-6], and can be applied to emergency conditions such as natural disasters, agriculture, and health.

The health sector, especially the vital sign monitoring system [7, 8], needs more attention because it successfully distributes information and public services to maintain patient safety. Based on the regulation of the Minister of Health of the Republic of Indonesia number 4 of 2018, which states that the provision of health monitoring facilities is essential to improve patient health services. However, the existing health monitoring system in hospitals or health centers still has weaknesses. One of the problems is that the installed wireless device has not been able to reach the patient. This condition causes data transmission to be not optimal because it affects long delays and increases energy consumption. Therefore, route selection [9] is considered to improve optimal network quality and reduce energy consumption [10].

This study aims to choose a health data traffic route that can improve QoS when transmitting data. Meanwhile, the urgency of this research is to obtain the selection of health data traffic routes that can apply to hospitals, health centers, and other health services. Choice of the proposed health data traffic route using the BFS and DFS methods

2. Review of Literature

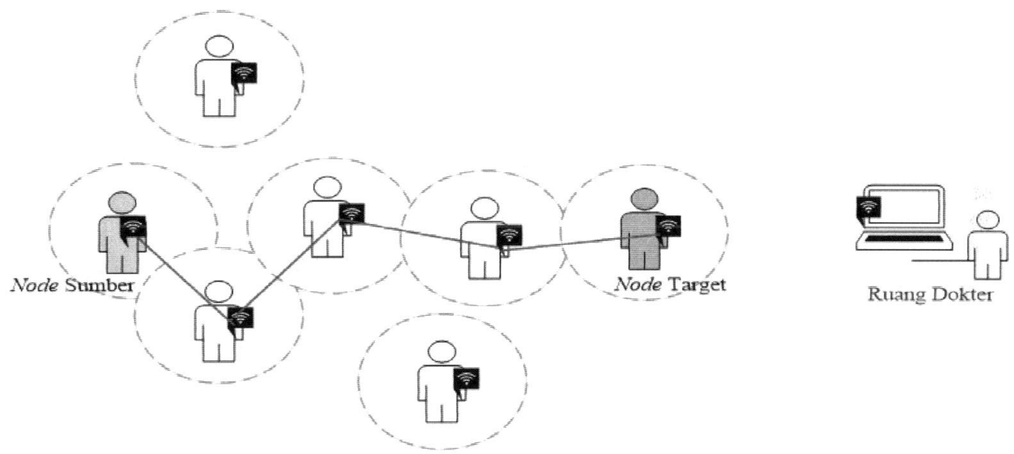

Figure 1. Route selection from node source to server

Figure 1 shows that the nodes are selected based on the distance between the nodes with each other. The source node searches for the destination node through alternative routes in a network. Furthermore, the target node that has been found will respond to the destination node based on the path that has been selected. Nodes send data to target nodes to communicate with each other without using an infrastructure called MANET. This concept collects several ad hoc devices that phrase in a wireless network to form a dynamic network, stand-alone, and do not depend on existing infrastructure

2.1. Route Selection of Data Traffic

Route selection is a mechanism for sending data packets that are transmitted from one network to another. Route selection [11] aims to determine the number of movements originating from the source node to the destination node using several available route options. Factors that influence route selection include travel time, distance, and cost. Route selection can be said to be ideal if it accommodates all the factors that influence its use. However, factors that are often considered in choosing a route are travel time and distance.

The data traffic route is sending data sent from the source node to the destination node via wired or wireless communication media. The selection of data traffic routes related to the exchange of data is called routing. The routing determines the appropriate course, gathering information, repairing broken ways, and even identifying and preventing attacks. Routing consists of the process of selecting and improving the path from the source node to the destination. To carry out the process of determining the optimal route, need route discovery and route selection processes are carried out. Route discovery aims to find alternative ways to send information from the source node to the destination and select the route as the data delivery route.

2.2. *Health Monitoring*

Health monitoring is the monitoring of a person to identify changes in health status. Health monitoring includes a person's medical history or lifestyles such as smoking, alcoholic beverages, and related health factors. Health monitoring is essential and mandatory for every human being, especially for the elderly. The element of old age causes a decrease in the physical condition and is susceptible to disease.

2.3. *Route Selection of Data Traffic*

Vital signs are various physiological statistical measures that are used to help determine a patient's health status. The examination is more directed at patients who are medically unstable or have a dangerous body condition. Vital sign monitoring is an examination related to vital signs to assess the patient's physical health. It is essential to monitor vital signs to monitor the condition of patients under treatment [12]. Eight vital signs can monitor patient health, body temperature, heart rate, blood pressure, and breath.

2.4. *Topology MANET*

Every network that is built always uses a specific network topology. Topology is essential so that can know the communication routes between nodes. MANET is a wireless network with a dynamic topology. MANET applies point-to-point or ad-hoc network topology classification in building data communication networks. Modeling with a point-to-point concept causes MANET to be classified as a mesh-network topology. The role of MANET is to provide temporary telecommunications network infrastructure for emergency conditions.

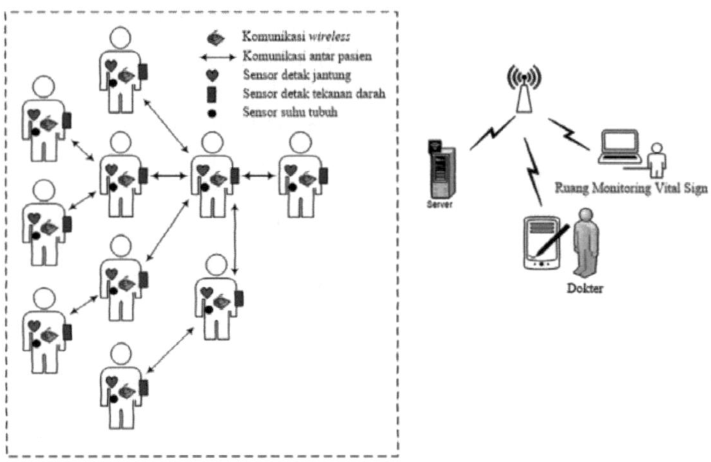

Figure 2. Dynamic topology model for health monitoring

2.5. *Breadth-First Search (BFS)*

BFS is an algorithm that performs a broad search [13] by visiting nodes in preorder. This process continues until all paths are seen so that we can obtain a solution. This procedure is guaranteed not only to find a solution. However, found a solution with the shortest path from the goal. The advantage of the BFS algorithm is that it can find a solution as the shortest path and find the minimum solution if there is more than one solution. BFS will not be deadlocked. The disadvantage of the BFS algorithm is that it requires a lot of memory to store all nodes in one tree.

The working principle of BFS is to enter the end node into a queue and then take the node from the beginning of the line. Check if the starting node is a solution. If the node is a solution, the search is complete, and results are returned. If the node is not a solution, enter all nodes that are neighbors to that node. If all nodes have been checked and the queue is empty, the search is completed by returning the solution results.

2.6. Depth First Search (DFS)

DFS is a deep search algorithm that starts from the initial node and continues by visiting only the leftmost child nodes at the next level. The way the DFS algorithm works is to insert the root node into a stack [14]. Then take the first node at the topmost level. If the node is a solution, the search is complete, and results are returned. If the vertex is not a solution, add all the vertices neighboring that vertex to the stack. When all nodes have been checked, and the queue is empty. The search is completed by returning the result of a solution not found. The search is repeated from the starting node of the line.

The advantage of the DFS algorithm is that it uses less memory to store nodes on the active path and find solutions without having to test more in the state space. The DFS algorithm can find a solution quickly if the answer is at the deepest and leftmost levels. The disadvantage of the DFS algorithm is that DFS may not see the expected resolution and only find one solution in each search.

2.7. Quality of Service (QoS)

QoS [15] refers to the ability of network technology to measure network performance on available services. The primary purpose of QoS is to know network services, capabilities, so that information exchange is achieved. QoS is designed to help end-users be more productive in getting a reliable performance. The QoS parameters used in this study include packet delivery ratio (PDR), packet loss, throughput, delay, topology control (TC), and energy consumption.

3. Method

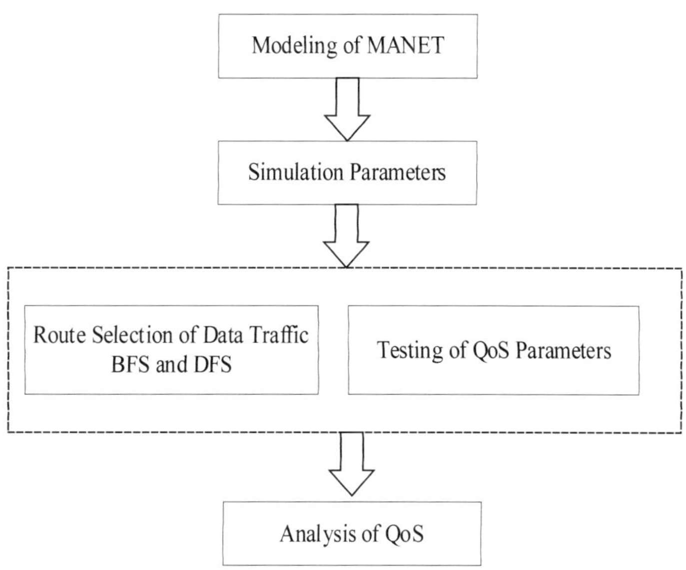

Figure 3. Stages of research

The methodology used is simulation-based research using the 2018 version of MATLAB. Figure 3 shows the stages of the study, starting from MANET modeling, simulation parameters, selecting data traffic routes using BFS and DFS, testing, and analyzing.

Communication for sending data packets uses the concept of user datagram protocol (UDP). UDP selection is based on considering that the data transmission process occurs without going through a complex series of operations. The simulated data traffic route searches are BFS and DFS. The scenario of giving the number of nodes varies from 25 to 100 nodes. The purpose of providing the same number of nodes is to ensure the accuracy of the resulting simulation. The simulation speed is 20 m/s with a duration of 200 seconds, and the simulation area is 800 x 800 m2. After the simulation parameters are entered, it is continued to test the QoS parameters. The calculated QoS values include PDR, throughput, packet loss, delay, and energy consumption.

4. Results and Discussion

4.1. Connectivity and Distance Matrix

Table 1. Connectivity matrix between nodes

	N_A	N_B	N_C	N_D	N_E	N_F	N_G	N_H	N_I	N_J
N_A	1	1	0	0	0	1	1	0	0	0
N_B	1	0	1	0	0	1	1	1	0	0
N_C	0	1	0	1	0	0	1	1	1	0
N_D	0	0	1	0	1	0	0	1	1	1
N_E	0	0	0	1	0	0	0	0	1	1
N_F	1	1	0	0	0	0	1	0	0	0
N_G	1	1	1	0	0	1	0	1	0	0
N_H	0	0	1	1	0	0	1	0	1	0
N_I	0	0	1	1	1	0	0	1	0	1
N_J	0	0	0	1	1	0	0	0	1	0

Table 2. Distance matrix between nodes

	N_A	N_B	N_C	N_D	N_E	N_F	N_G	N_H	N_I	N_J
N_A	0	60	0	0	0	64.12	82.07	0	0	0
N_B	60	0	56.14	0	0	90.51	60.13	82.07	0	0
N_C	0	56.14	0	64.12	0	0	82.07	56	84.85	0
N_D	0	0	64.12	0	60	0	0	87.73	64.12	84.89
N_E	0	0	0	60	0	0	0	0	90.51	62.03
N_F	64.12	90.51	0	0	0	0	60.13	0	0	0
N_G	82.07	60.13	82.07	0	0	60.13	0	60	0	0
N_H	0	0	56	87.73	0	0	60	0	60.13	0
N_I	0	0	84.85	64.12	90.51	0	0	60.13	0	62.03
N_J	0	0	0	84.89	62.03	0	0	0	62.03	0

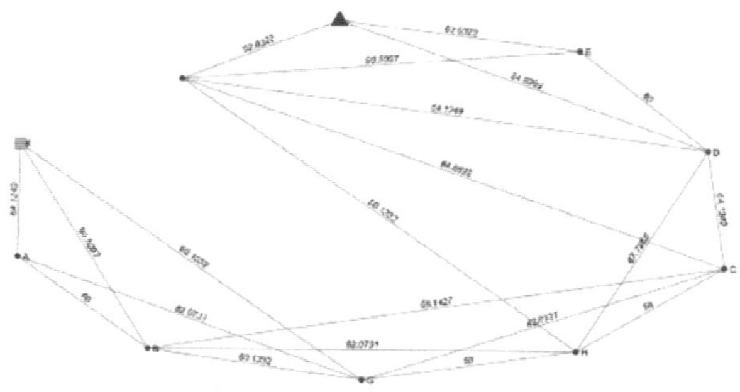

Figure 4. Display node connectivity on BFS

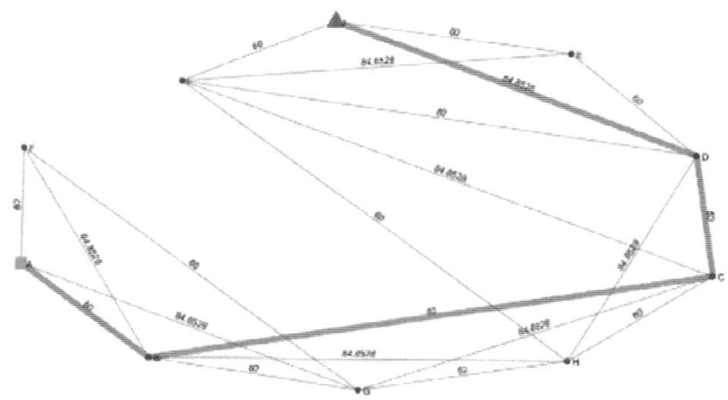

Figure 5. Display node connectivity on DFS

Tables 1 and 2 show the connectivity matrix with a value of 1 or 0. If it is worth 1, it indicates that there is a connected route. While a value of 0 means that there are no connected routes. Furthermore, the connectivity of nodes on BFS and DFS can be seen in Figures 4 and 5. Node n_A as a source node will send information to node n_J as a target. In the simulated case, all nodes will send to the target node in turn. Before sending the data, there is a choice of routes available to the target node.

4.2. Node Marking and Determination of Destination Node

After building the connectivity and distance matrix, mark each node with a level number starting from the source node (N1). The marking process applies BFS and DFS search algorithms from source to destination nodes. The process is carried out as shown in Figures 6 and 7.

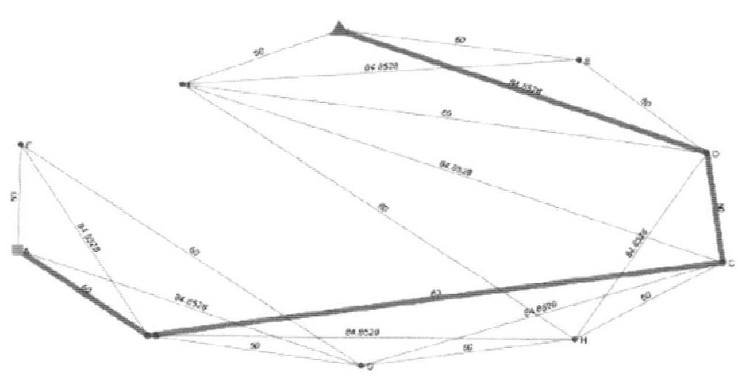

Figure 6. Marking of BFS node

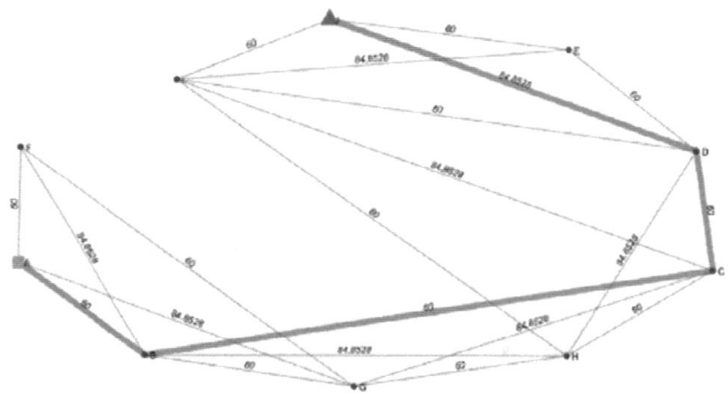

Figure 7. Marking of DFS node

4.3. Calculation of BFS and DFS Data Traffic Routes Based on QoS

The BFS and DFS algorithms can use determine the data traffic route to reach the destination node. The selection of BFS and DFS algorithms applied to the MANET concept is based on two variables. The two variables are the distance of the relay to the destination node and the number of connectivity of the initial node to the next node. The shorter the length generated from the destination node, the better the quality of the information signal that will receive will be. The number of connectivity between nodes with the next level above allows connectivity to reach the target node. Furthermore, the results of the calculation of BFS and DFS data traffic routes based on QoS parameters such as packet delivery ratio (PDR), throughput, packet loss, and delay can be seen in Table 3.

Table 3. Results of the calculation of BFS and DFS data traffic routes based on QoS

Method	Node	Quality of Service				
		PDR (%)	Throughput (Kbps)	Packet Loss (%)	Delay (ms)	Energy Consumption (Joule)
BFS	10	79,20	66,04	20,79	104990	0,30
	20	80,69	66,20	19,31	112161	0,44
	40	73,41	60,38	26,59	110625	0,43
	60	88,31	75,07	11,69	89112	0,63
	80	86,09	67,95	13,91	135979	0,39
	100	91,38	73,18	8,62	126503	0,54
DFS	10	81,86	68,26	18,14	104990	3,16
	20	81,94	67,23	18,06	112161	7,76
	40	83,35	68,17	16,65	114210	8,92
	60	90,25	76,72	9,75	89112	13,54
	80	88,38	69,70	11,62	136491	21,16
	100	92,49	73,92	7,51	127783	14,27

4.3.1. Packet Delivery Ratio (PDR)

PDR is a QoS parameter that indicates the success rate of data traffic routes. Table 3 shows that PDR performance on DFS is better than BFS on all nodes. The increase in the PDR value is caused by the ability of DFS to optimize the node selection process. The average PDR for DFS is 86.38%, and BFS is 79.86%.

4.3.2. Throughput

Throughput is the total number of data packets that are successfully received in a unit of time and describes the data speed conditions. Table 3 shows that the throughput value of DFS is better than BFS for all numbers of nodes. The increase in DFS throughput performance is due to the maximum amount of data sent from the destination node being successfully received at the destination node. The average throughput performance on DFS is 70.67 Kbps, and BFS is 68.14 Kbps.

4.3.3. Packet Loss

Packet loss is measured as the percentage of packets lost concerning packets sent between source nodes to destination nodes. Table 3 shows that DFS has better packet loss performance than BFS for all number of nodes. DFS packet loss performance is better because the amount of data lost on the target node is less. The average version of packet loss on DFS is 16.82%, and BFS is 13.62%.

4.3.4. Delay

Delay is the delay time for all packets successfully sent from the source node to the destination node. Table 3 shows that DFS has the same delay performance as DFS at 10 to 60 nodes. However, at 80 to 100 nodes, the delay performance on BFS is better than DFS. The average delay performance on BFS is 113228 milliseconds, and DFS is 114125 milliseconds.

4.3.5. Energy consumption

Energy consumption is the amount of energy required by a node to send and receive packets. Table 3 shows that energy consumption performance on BFS is better than DFS for all nodes. The decrease in consumption generated by BFS is due to the ability to route data traffic which takes less time than DFS. The average energy consumption performance on BFS is 0.46 joules, and DFS is 11.47 joules.

5. Conclusion

Each data traffic ute selection is simulated using the same parameters to ensure the accuracy of the resulting simulation. This study aims to analyze the performance of data traffic route selection using BFS and DFS based on QoS. The analyzed QoS parameters consist of PDR, throughput, packet loss, delay, and energy consumption. The simulation results show that DFS has better performance than BFS for PDR, throughput, and packet loss parameters. The increase in PDR value, throughput, and packet loss is caused by the ability of DFS to choose a node that is more optimally accepted at the destination node. Meanwhile, BFS has better performance than DFS for delay parameters and energy consumption. The decrease in delay and energy consumption in BFS is caused by the time required at the destination node being further minor than DFS.

References

[1] Sanaa J H Al-Shakarchi and Raaid Alubady, 2021 A Survey of selfish nodes detection in manet: solutions and opportunities of research IEEE Babylon International Conference on Information Technology and Science (BICITS), Babil, Iraq

[2] Alamsyah, I Ketut E Purnama, Eko Setijadi, Mauridhi H Purnomo, 2018 Minmax algorithm for MPR selection in improving the performance of OLSR protocol on manet IEEE International Conference on Intelligent Autonomous Systems (ICoIAS), Singapore

[3] Ibrahim Alameri, Jitka Komarkova, Mustafa K Ramadhan, 2021 Conceptual analysis of single and multiple path routing in MANET network IEEE International Conference on Information and Digital Technologies (IDT), Zilina, Slovakia

[4] Ankita A Mahamune and M M Chandane, 2021 Evaluating Routing Protocols for Mobile Ad Hoc Networks Under Varying Network Scenarios IEEE International Conference on Intelligent Communication Technologies and Virtual Mobile Networks (ICICV), Tirunelveli, India
[5] L Maheshavel and C Senthilkumar, 2021 Mitigation of black hole attack in MANETs: A sequence
no-based approach Journal of Physics: Conference Series, pp 1-8
[6] Umang R, Balachandra P, B B Sundaram, Manish K Mishra, Dhanabal T, 2021 Investigation of DSR protocol performance in Wireless MANET using Correlation Method Journal of Physics: Conference Series, pp 1-11
[7] Alamsyah, M Subito, A Amir, 2021 Design system body temperature and blood pressure monitoring based on internet of things International Conference on Information and Communications Technology (ICOIACT), Yogyakarta, Indonesia
[8] Alamsyah, M Subito, M Ikhlayel, E Setijadi, 2020 Internet of things-based vital sign monitoring system International Journal of Electrical and Computer Engineering (IJECE), vol 10, no 6, pp 5891–5898
[9] K Gunasekaran , G Suresh, V Kavitha, S Lekashri, 2020 Resourceful and reliable routing selection scheme in MANET IEEE International Conference on Smart Structures and Systems (ICSSS), Chennai, India
[10] S Pariselvam, T Ramya, T Rajasri, C Aisha, 2021 Survey on energy-efficient routing protocols for MANET IEEE International Conference on System, Computation, Automation and Networking (ICSCAN), Puducherry, India
[11] Prabhdeep S and Heena D, 2015 Node mobility based route selection in AODV for use in MANETs IEEE International Conference on Computing Communication Control and Automation, Pune, India
[12] Pradeep K, Mohammed O A, Prabaakaran K, Aparna B, Princy A, Raksha R, 2021 Coma patient's health monitoring and observatory system using internet of things Journal of Physics: Conference Series, pp 1-6
[13] Rashmi D, Somesh K Dewangan, 2014 An enhanced light-weight proactive source routing protocol using BFS and DFS for MANET International Journal of Engineering Trends and Technology, vol 15, no 5, pp 261-267
[14] Balram S and Ravindar S, 2016 Comparative analysis of scalability and energy efficiency of ordered walk learning routing protocol IEEE Proceedings of the International Congress on Information and Communication Technology, Advances in Intelligent System and Computing, pp 265-272
[15] Kashif Nisar, Ibrahim A Lawal, Usman I A Malik, Abubakar A Mu'azu, BS Chowdhry, Sohrab K, Shuaib M, 2020 QoS analysis of the MANET routing protocols concerning delay, throughput, & network load: challenges and open issues IEEE International Conference on Application of Information and Communication Technologies (AICT), Tashkent, Uzbekistan.

The Addition of Seaweed Extract (*Sargassum* sp) as Fertilizer on Culture Media on The Total Density of *Skeletonema costatum* Cells

Nasmia[1], Kusmadi L. Binangkari[1], Rusaini[1], Eka Rosyida[1], Muh. Saleh Nurdin[1], Zakirah Raihani Yala[1]

[1]Aquaculture Study Program, Department of Fisheries and Marine Science, Faculty of Animal Husbandry and Fishery, Tadulako University, Palu 94118, Indonesia

nasmia68@gmail.com

Abstract. Microalgae *Skeletonema costatum* is a natural food that contains nutrients needed for aquaculture. Seaweed contains several nutrients and has body regulatory substances, namely auxin, gibberellin and cytokinin which are useful for increasing the growth of microalgae cells, including *S. costatum*. The study aims to examine the addition of *Sargassum sp.* on culture media on cell density of *S. costatum*. The benefit of this study is to obtain scientific information regarding the addition of *Sargassum sp.*, thus, it can be a natural fertilizer that can be used in the economical cultivation of *S. costatum*. This study consisted of 5 treatments, namely the concentration of *Sargassum sp.* extract in media (control) 0 mg/500 mL, 25 mg/500 mL, 50 mg/500 mL, 75 mg/500 mL and a concentration of 100 mg/500 mL, these treatments were repeated three times. *S. costatum* cell density was checked every 6 hours. The results showed that the use of extracts of *Sargassum sp.* 100 mg/500 mL had the highest cell density of *S. costatum* ($2.08x10^7$ cells/mL), followed by 75 mg/500 mL ($2.00x10^7$ cells/mL), 50 mg/500 mL ($1.99x10^7$ cells /mL, 25 mg/500 mL ($1.97x10^7$ cells/mL), and 0 mg/500 mL ($1.05x10^7$ cells/mL).

Keywords: concentration, cultivation, microalgae, nutrient

1. INTRODUCTION

Skeletonema costatum is a type of phytoplankton used in the maintenance of fish, shrimp, and crab larvae [(1)], and for the cultivation of artemia [(2,3)]. *S. costatum* has the primacy that it has its own autolysis enzyme, thus, it is easily digested by larvae and does not contaminate the culture media [(4)] and contains quite complete nutrients [(1)]. *S. costatum* contains 31% protein, 1.3% fat and 21.5% carbohydrates, fatty acids, free amino acids, ß-1, glucans and polysaccharides [(5,6,7)]. Various kinds of inorganic compounds as macro nutrients (N, P, K, S, Na, Si, and Ca) and micro nutrients (Mg, S, Fe, Mn, Zn, Co, Mo and Cu) are needed in synthetic media for microalgae growth [(8)], including *S. costatum*.

One of the natural resources in the waters that contain macromineral and micromineral elements is seaweed. According to [(9)] stated that *Sargassum sp.* contains K, Ca, P, Na, Fe, and iodine as well as vitamins A, B1, B2, B6, B12, and vitmain C [(10)], reported that the nutrient in the extract of *Sargassum sp.* consisted of N 35.59 mg/L, P 3.81 mg/L, K 3.22 mg/L, Ca 2.02 mg/L, Mg 13.31 mg/L, Fe 2.17 mg/L, S 6.67 mg/l, Cu 0.19 mg/L and Zn 19.62 mg/L. Further, [(11)] stated that *Sargassum sp.* contained growth hormones including auxin, gibberellins and cytokinins [(12)] which became body regulators to stimulate plant growth that functions as fertilizer. Because of its growth-stimulating activity, algae (seaweed) are used as a biostimulant in plant production [(13,14,15)].

Increasing the number of phytolankton cell density can be enriched with nutrient content through fertilizer application. The use of fertilizers in *S. costatum* culture media is very important to increase its productivity and good biomass growth. Cultivating *S. costatum* growth is strongly influenced by the nutrients in the environment where it lives, therefore fertilizer is needed in the culture media to support the availability of both macro and micro nutrients. Based on the above, this study was conducted on the use of *Sargassum sp.* extracts as a natural fertilizer in culture media and to determine the best concentration for the total density of *S. costatum*. The results of this study are expected to be used as an environmentally friendly organic fertilizer and to get the best dose for the growth of *S. costatum* cell density.

2. MATERIAL AND METHOD

2.1 Material

The study took place in Aquaculture Laboratory, Faculty of Animal Husbandry and Fisheries, Tadulako University. The test organism used was the microalgae *S. costatum* from the Windu Shrimp Cultivation Installation, Barru Regency, South Sulawesi.

2.2 Research Procedures

2.2.1 Sterilization of tools and materials

Sterilization of equipment such as scale pipettes, jars and dropper is carried out by dry sterilization. The sterilized tools are wrapped in paper and then placed in the oven at 180°C for ±1 hour. Materials used such as sea water are sterilized using an autoclave. Before being autoclaved, the seawater was diluted to 25 ppt. Furthermore, the sea water is inserted into the Erlemeyer flask and then covered by aluminum foil. Seawater was autoclaved at 121°C for 15 minutes.

2.2.2 *Sargassum* sp. extract making

Sargassum sp. were taken from Kolami village, District Walea Islands, Tojo Una-Una Regency, Central Sulawesi Province. The samples of *Sargassum sp.* were washed and dried at room temperature. Making *Sargassum sp.* extract were conducted at the Nutrition Laboratory of the Faculty of Animal Husbandry and Fisheries. *Sargassum sp.* extract were dissolved using distilled water, and homogenized using a vortex [(10)].

2.2.3 Spreading *Skeletonema Costatum* Seed

The maintenance container was placed on a culture rack that has been equipped with aeration and 600 lux of lighting from fluorescent lamps. Pure cultures of *Skeletonema costatum* were stocked in 34.48 mL containers with a density of 1.45×10^6. The number of cultured *S. costatum* cells could be determined by the following equation:

$$V1 \times V2 = N1 \times N2$$

Notes:
V1 = *S. costatum* volume in the initial spreading (mL)
V2 = Medium water volume (mL)
N1 = The total amount of *S. costatum*
N2 = The desired amount of *S. costatum* (Sel/mL)

2.3 Research Design

This study was designed with 5 treatments and 3 replications, so the number of experimental units was 15 units. The treatment used was the administration of *Sarggasum sp.* with different concentrations, namely: A. without extract *Sargassum sp.* (control), B. *Sargassum sp.* extract. (25 µg/ 500 mL), C. *Sargassum sp.* Extract (50 µg/ 500 m)L, D. *Sargassum sp.* extract (75 µg/ 500 mL) and E. extract of *Sargassum sp.* (100 µg/ 500 mL).

2.4 Observed Variables

2.4.1 *Skeletonema costatum* density

S. costatum culture was carried out until it reached the death phase and the density was observed every 3 hours. Calculation of the population density of *S. costatum*, using the big block calculation formula [(16)].

$$JS = N \times 10^4 \text{ sel/mL}$$

Notes:
JS = *S. costatum* density (sel/ mL)
N = The number of phytoplankton cells in the big box
10^4 = Hemocytometer big box volume

2.4.2 Water quality

The water quality variables observed during the study are listed in Table 1.

Tabel 1. Water quality variables observed during the study

Water Quality	Tools	Observation Time
Temperature	Thermometer	Every day
Salinity	Refractometer	Every day
pH	Litmus paper	Every day
Dissolved Oxygen	DO meter	Beginning and end
Nitrogen	Spectrophotometer	Beginning and end
Phosphate	Spectrophotometer	Beginning and end

2.5 Data analysis

Cell density of *S. costatum* and water quality were analyzed descriptively using graphs and tables.

3. RESULT AND DISCUSSION

3.1 *Skeletonema Costatum* density

The results showed that the addition of *Sargassum sp.* on culture media can increase the density of *S. costatum*, where the highest treatment was given the concentration of *Sargassum sp.* 100 µg/500 mL (E) of 2.08 $x10^7$ cells/mL, followed by treatment with an extract concentration of 75 µg/500 mL (D) is 2.00 x 10^7 cells/mL, extract concentration 50 µg/500 mL (C) of $1.99x10^7$ cells/mL, extract concentration of 25 µg/500 mL (B) was $1.97x10^7$ cells/mL, and the lowest was in treatment A without *Sargassum sp.* (control) is $1.05x10^7$ cells/mL (Figure 1).

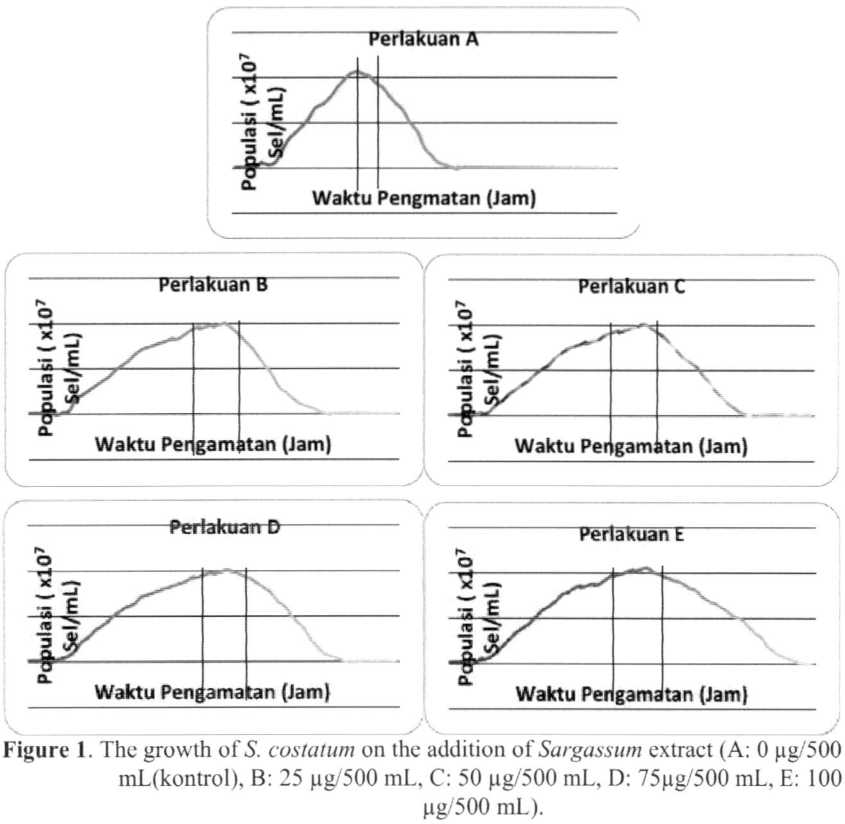

Figure 1. The growth of *S. costatum* on the addition of *Sargassum* extract (A: 0 µg/500 mL(kontrol), B: 25 µg/500 mL, C: 50 µg/500 mL, D: 75µg/500 mL, E: 100 µg/500 mL).

The results of the study (Figure 1) show that the addition of *Sarggasum sp.* on culture media of *S. costatum* gave a better cell number density than that without *Sargassum sp.* (control) because the extract of *Sargassum sp.* added nutrients to the culture medium of *S. costatum* to perform cell division. [17] stated that the culture of *S. costatum* requires nutrients to support its growth. The nutrients needed consist of macro and micro elements, namely macro elements such as N, P, K, S, Na, Si and Ca, while the micro elements are Fe, Zn, Mn, Cu, Mg, Mo, Co, B [18]. Liquid extract of *Sargasum sp.* contains micro and macro nutrients as well as growth-promoting hormones [19,20], dan [21] reported that the liquid extract of seaweed contained growth-promoting substances auxin at 127.48 ppm, gibberellins 131.11 ppm, cytokinins-kinetin 68.77 ppm, and cytokinins zeatin 82.41 ppm, as well as macronutrients potassium (K) of 345, 29 mg/100 g, nitrogen (N) 0.78%, phosphorus (P) 55.39 mg/100 ml.

The comparison of the time achieved from each growth phase of *S. costatum* in each treatment are shown in Table 2.

Table 2. The comparison of the time achieved from each growth phase

Treatment (µg/500 mL)	Growth Phase (Time)				
	Adaptation	Exponential	The Deflation in Growth Rate	Stationary	Death
A (0)	0-21	21-45	45-48	48-60	60-99
B (25)	0-18	18-57	57-69	69-93	93-138
C (50)	0-18	18-57	57-69	69-93	93-141
D (75)	0-18	18-57	57-75	75-96	96-147
E (100)	0-18	18-51	51-69	69-93	93-168

The results of Table 2 show that the adaptation phase in treatment A occurred at 0-21 hours of observation, treatments B, C, D and E saw the adaptation phase at 0-18 hours of observation. Treatment A (control) showed that *S. costatum* needed a relatively longer adaptation time, compared to other treatments because the nutrient concentration in this treatment was lower than other treatments with the addition of extracts of *Sargassum sp.* The adaptation phase of treatments B, C, D and E lasted for 18 hours faster than treatment A because there were sufficient nutrients to stimulate *S. costatum* to do cell division. Acording to [22] that, in the adaptation phase, microalgae are still in the process of adjusting to their environment. [23] stated that the adaptation phase may require a shorter or longer time, depending on the microalgae before the inoculum and the nutrient content of the culture medium. Microalgae need nutrients for the growth process. [24] stated that mineral nutrients are needed by plants and microalgae for their growth process. According to [17] growth is a process of changes that occur in organisms, whether it is increasing in length or increasing in weight or increasing the number of *S. costatum* cells.

The exponential phase in treatment A occurred at 21-45 hours of observation, and treatments B, C, D at 18-57 hours of observation, and treatment E (100 µg/500 mL) showed an exponential phase at 18-51 hours of observation. In the exponential phase, the number of cells increases rapidly. The peak of population increase occurred in that phase and also this phase shows that the utilization of nutrients has been carried out optimally by *S. costatum.* The population growth curve shows that treatment E in this phase experienced a very high increase in cell density (2,08 x10^7 sel/mL), and the time that occurred in this phase was longer than the control treatment (0 µg/500 mL). It is influenced by the nutrients in *Sargassum sp.* includes macro and micro elements that are able to meet the nutritional needs of *S. costatum* cells. Population growth was also showed in treatments B, C, and D, the exponential phase was faster and lasted longer, namely 18-57 jsi because the nutrients were met for *S. costatum* to perform cell division. In treatment A (control), to reach the exponential phase, *S. costatum* took longer and this phase lasted 21-45 faster, due to the limited nutrients in the media. Acording to [22], the exponential phase occurs after the cell adjusts to the environment and begins to develop to enter the exponential phase. According to [23] the exponential growth rate of microalgae varies which is influenced by environmental conditions and the composition of the culture medium as well as genetics. *Sargassum sp*. contains various ingredients needed for plant growth, namely the presence of hormones such as gibberellins, cytokinins and auxin, and contains nutrients (N, P, K, Mg and Ca).

The growth deflation phase in treatment A started at 45-48 hours, treatments B and C at 57-69 hours of observation, D treatment at 57-75 hours and treatment E at 51-69 hours of observation. The growth of *S. costatum* cells deflated because the nutrients contained in the maintenance medium were decreasing. According to (25) the deflation in growth rate was caused by decreased nutrients in the culture media and also the limited incoming light caused by the density of *S. costatum*. While the stationary phase in treatment A (control) occurred at 48-60 hours of observation, treatments B, C and E at 69-93 hours of observation and treatment D at 75-96 hours of observation.

The death phase of treatment A (control) started at 60-99 hours of observation, treatment B at 93-138 hours of observation, treatment C at 93-141 hours of observation, treatment D at 96-147 hours of observation and treatment E at 93-168 hours of observation. The death phase occurs after the growth deflation phase, where the number of cells decrease due to cell death. This is because the nutrients available in the culture media are decreasing, so they cannot meet the needs of *S. costatum* cells and the competition between individual cell took place in utilizing nutrients, space, light, and other supporting factors. The results showed (Table 2) that the higher the concentration of *Sargassum sp.* used the longer the time needed to reach the death phase compared to those without *Sargassum sp.* (26) stated that the death phase in microalgae is influenced by nutrient depletion.

3.2 Water Quality

The growth of *S. costatum* is not only influenced by the availability of nutrients but also by environmental factors in the growth media. The results of the water quality variable at the time of *S. costatum* culture showed a reasonable range for *S. costatum* culture as shown in Table 3.

Table 3. Water Quality Parameters

Parameter	Range
pH	6-8
Temperature	29°C - 30°C
Salinity	30-32 ppt
Dissolved Oxygen	6-7.9 mg/L

The degree of acidity (pH) during the study ranged from 6-8. This pH value is still within the normal range for *S. costatum* culture. According to (26), the optimal pH for microalgae culture ranges from 7-8. The same thing was stated by (4) that acidic waters with a pH of less than 6.0 cause microalgae not to live properly and when the pH is less than 4.0 it can cause the death of microalgae. Continued (27) that the degree of acidity (pH) affects the density of *S. costatum*, because enzymes in cells are very sensitive to changes in pH so that it affects cell activity.

Temperature directly affects the efficiency of photosynthesis and is a very influential factor in plant growth. The temperature range obtained during the research ranged from 29-30°C. These results indicate that the temperature used is still in a suitable condition for *S. costatum* culture. According to (16), *S. costatum* can grow in a temperature range of 15°-34°C, but for optimal growth requires a temperature range of 25°-27°C.

The salinity obtained at the time of the study ranged from 30-32 ppt, and the results showed that the salinity used during the study was in fairly good condition. According to (28), salinity affects the growth of *S. costatum* and the optimum salinity is 32 ppt. Media salinity is related to the ability of microalgae to maintain osmotic pressure between the protoplasm and its environment (4). Too high or too low salinity rage can lead to the impairment of cell activity.

Dissolved oxygen is needed by *S. costatum* for respiration. Dissolved oxygen during the study was in a suitable condition (6-7.9 mg/L) for the growth of *S. costatum.* [16] revealed that *S. costatum* can grow well in the dissolved oxygen range of 6-7 mg/L. Dissolved oxygen can be increased by the process of photosynthesis in the waters carried out by microalgae and decreases when the microalgae stop carrying out the photosynthesis process.

The results of the analysis of nitrogen and phosphate content in each treatment with the addition of *Sargassum sp.* can be seen in Table 4.

Table 4. The results of the analysis of nitrogen and phosphate

Treatment	Nitrogen (mg/L)		Phosphate (mg/L)	
	Beginning	Ending	Beginning	Ending
A (control)	1,039	1,090	0,185	0,097
B	1,059	1,111	0,198	0,128
C	1,068	1,148	0,210	0,161
D	1,090	1,184	0,221	0,193
E	1,111	1,228	0,232	0,226

Based on the results of the study, the nitrogen concentration increased at the end of the study, it was suspected that the nitrogen content increased because *S. costatum* had completely died so that the nitrogen concentration increased. Thus, in the culture media, the nitrification process occurs. [29] stated that the amount of sediment in the waters can increase the total nitrogen.

The results showed that the phosphate content decreased during the study. This occurred due to the phosphate content in the culture media had been well utilized by *S. costatum* when it performs cell division.

4. CONCLUSION

The addition of *Sargassum sp.* extracts in *S. costatum* culture media can be used as a safe and environmentally friendly fertilizer and a concentration of 100 µg /500 mL gives the highest density level of 2.08 x10^7 cells/mL. Water quality during the study showed a reasonable range for the cultivation of *S. costatum*.

ACKNOWLEDGEMENTS

The authors would like to thank the Directorate General of Higher Education, Ministry of Education and Culture of the Republic of Indonesia for funding assistance through the National Priority Research Masterplan for the Acceleration and Expansion of Economic Development. Thank you to the Chancellor of Tadulako University, the Dean of Fapetkan, and all teaching staff of the Aquaculture study program, as well as all staff of the Fisheries and Biotechnology Laboratory of Tadulako University, Faculty of Animal Husbandry and Fisheries, Tadulako University Palu.

REFERENCES

[1] Azmi, K.A., Arsad, S., and Sari, L.A. 2020. The effect of commercial nutrients to increase the population of *Skeletonema costatum* on laboratory and mass scales. 2nd International Conference on Fisheries and Marine Science. IOP Conf. Series: Earth and Environmental Science, 441.

[2] Lestari, D.P., Arning, W., Ekawati, Maftuch. 2014. Dried *Skeletonema costatum* in Feed Formulation for the Growth of Vaname Shrimp (*Litopenaeus vannamei*). J.Exp. Life Sci. Vol. 4 (2).

[3] Herawati V.E., Johannes H, Budi P. 2012. The effect of essential amino acid profile, fatty acid profile and to growth of *Skeletonema costatum* using technical media culture guillard and double walne. J Coast Dev. 10:48-54.

[4] Supriyantini, E., 2013. Effect of salinity on nutritional content of *Skeletonema costatum*. Buletin Oseanografi Marina. Vol. 2:51-57.

[5] Widiastuti, R., Hutabarat, J., Herawati. V.E. 2016. Effect of feeding different natural (*Skeletonema costatum* and *Chaetoceros gracilis*) to absolute biomass growth and nutrient content of local *Artemia* sp. J. Aquacult. Manage. Technol. 1(1). 236-248.
[6] Monkonsit, S., Powtongsook, S., Pavasant, P. 2011. Comparison between airlift photobioreactor and bubble column for *Skeletonema costatum* cultivation. Eng. J. 15(4). 53-64.
[7] Granum, E., S., Kirkvold, S. M. Myklestad. 2002. Cellular and extracellular production of carbohydrate and amino acid by the marine diatom *Skeletonema costatum*: diel variations and effects of N depletion. Mar. Ecol. Prog. Ser. 242. 83-94.
[8] Amit, Dahiya, D., Ghosh, U.K., Nigam, P.S., and Jaiswal, A.K. 2021. Food Industries Wastewater Recycling for Biodiesel Production through Microalgal Remediation. Sustainability. 13, 8267.
[9] Handayani, T., Sutarno, Setyawan, D.A. 2004. Nutritional composition analysis of seaweed *Sargassum crassifolium* J. Agardh. Biofarmasi. Vol. 2(2):45-52.
[10] Nasmia, Natsir, S., Rosyida, E., Rusaini, dan Ya'la, R.Z. 2017. Toxicity of liquid extract of seaweed *Sargassum* sp. on the growth of microalgae *Skeletonema costatum*." AACL Bioflux 10(2): 247-253.
[11] Montano N.E., and Topas L.M., 1997. Plant growth hormonal activities of aqueous extracts from Phillipnes seaweeds. SICEN Leaflet 2, Marine Science Institure, University of Phllpines.
[12] Chojnacka, K., Saeid, A., Witkowska, Z., and Tuhy, L. 2012. Biologically active compounds in seaweed extracts the prospects for the application, The Open Conference Proceedings Journal. Vol.3(1), pp. 20–28.
[13] Khan, W., Rayirath, U.P., Subramanian, S. 2009. Seaweedextracts as biostimulants of plant growth and development. Journal of Plant Growth Regulation. Vo. 28(4), pp. 386–399.
[14] Rathore, S.S., Chaudhary, D.R., Boricha, G.N. 2009. Effect of seaweed extract on the growth, yield and nutrient uptake of soybean (Glycinemax) under rainfed conditions. South African Journal of Botany. Vol. 75(2); pp. 351–355.
[15] Sivasamkari, S., Venkatesalu, V., Anantharaj, M., and Chandrasekaran, M. 2009. Effect of seaweed extracts on the growth and biochemical constituents of *Vigna sinensis*. Bioresource Technology. Vol. 97(14): pp. 1745–1751.
[16] Isnansetyo, A. dan Kurniastuty. 1995. Phytoplankton and Zooplankton Culture Techniques: Natural Feed for Hatchery of Marine Organisms. Kanisius Publishers, Yogyakarta. 116 p.
[17] Fauziah dan Hatta, M. 2015. The effect of vermicompost with different doses in cultured *Skeletonema costatum*. Acta Aquatica. Vol. 2(1): 11-17
[18] Cahyaningsih, S., Achmad, N., Sugeng, J.P. 2005. Kultur Murni Phythoplankton. Departemen Kelautan dan Perikanan Budidaya Balai Budidaya Air Payau Situbondo.
[19] Widanarto. A. 2013. Pemanfaatan silase limbah kepala ikan kuniran (*Uphuneus sulphureus*) dan limbah alginate dari rumput laut coklat (*Sargassum* sp.) sebagai pupuk organik. Sekolah Tinggi Perikanan Jakarta. p.90.
[20] Winda, H., Lubis, A., dan Supriadi. 2013. Utilization of *Sargassum polycystum* waste from the pharmaceutical industry as liquid fertilizer and its effect on the chemical properties of ultisol soil and the growth of mustard plants. Journal Agroekoteknologi. Vol.1(3): 820-832.
[21] Basmal, J., Kusumawati, R., and Utomo, B.S.B. 2015. Quality of *Sargassum* Sap Liquid Extracted using Potassium Hydroxide as Material for Fertilizer. JPB marine and fisheries. Vol. 10(2): 143–153
[22] Lee, S.H., Ahn, C.Y., Jo, B.H., Lee, S.A., Park, J.Y., An, K.G., and Oh, H.M. 2013. Increased microalgae growth and nutrient removal using balanced N:P ratio in wastewater. Journal of microbiology and biotechnology. 23(1); 92-98.
[23] Madigan, T.D., Martinko, J.M., and Parker, J. 2013. Biology of Microorganism 13th Edition. San Francisco: Pearson/Benjamin Cummings.

[24] Grobbelaar, J.U. 2013. Chapter Inorganic Algal Nutrition. In Book: Handbook of Microalgal Culture. pp.123-133.

[25] Utami, Yuniarti, Kiki. 2012. Pertumbuhan *Chlorella* sp. yang dikultur pada perioditas cahaya yang berbeda. Fakultas perikanan dan ilmu kelautan, Universitas Padjajaran, Bandung.

[26] Armanda, T.D. 2013. The growt of diatom *Skeletonema costatum* (Greville) cleve of Jepara's isolates cultures in f/2 and conway culture. Journal Bioma. Vol. 2(1).

[27] Nurjamila, Tenriware, Arbit, N.I.S. 2021. *Skeletonema costatum* density on different pH level. SIGANUS. Journal of Fisheries and Marine Science. Vol. 2.(2).

[28] Rudiyanti, S. 2011. Growth of *Skeletonema costatum* at various levels of media salinity. Jurnal Saintek Perikanan. Vol. 6(2): 69 -76.

[29] Fatih, R.A. 2008. Kajian kandungan nitrogen pada kolom perairan dan sedimen akibat aktivitas kerambajaring apung di waduk Cirata, Jawa Barat. Program Studi Manajemen Sumberdaya Perairan Fakultas Perikanan dan Ilmu kelautan Institut Pertanian Bogor

The use of MSG (*monosodium glutamat*), aloe vera, and papaya leaf (*carica papaya L*) as an alternative of corrosion inhibitor

Sitti Rahmawati[1]*, Qonita Kurnia Anjani[2], Magfirah[1], Selmiati[1]

[1]Departement of Mathematics and Natural Sciences Education, Faculty of Teacher Training and Education, Tadulako University, Jl. Soekarno Hatta Km 9 Palu 94118, Indonesia
[2]Department of Pharmaceutics, Faculty of Pharmacy, Hasanuddin University, Jl. Perintis Kemerdekaan Km. 10, Makassar, Indonesia 90245

*E-mail: sittirahmawati.q3a@gmail.com

Abstract. Corrosion is a complicated problem in everyday life. Corrosion causes many losses because it reduces the life span of metal-based goods or buildings. One solution is to use organic corrosion inhibitors. Organic corrosion inhibitors are an alternative choice because they are more environmentally friendly. Heteroatomic organic compounds containing nitrogen, oxygen, or sulfur in molecules such as those found in MSG, papaya leaves and aloe vera can be used as organic corrosion inhibitors, because they can be well adsorbed on metal surfaces. Amino acids such as glutamic acid which have two carboxylic groups and one amine group are expected to be able to inhibit corrosion by adsorbing metal (iron) surfaces. This study aims to determine the use of MSG, aloe vera and papaya leaves as alternative materials for organic corrosion inhibitors. The results showed that the rust level using MSG solution was 8.33%, aloe vera extract was 9.09%, papaya leaf extract was 8%, and the container with a mixture of aloe vera extract, papaya leaf extract and MSG was added 4.35 %. This shows that aloe vera extract, papaya leaf extract and MSG can be used as an alternative to organic corrosion inhibitors.

1. INTRODUCTION

Indonesia that is under the expanse of the equator, causes a very significant temperature fluctuation during the day or night. Coupled with the high quantity of rainfall causes the effect of corrosion in metallic object is higher. Corrosion of metals is a common thing to happen. But there are many negative impact caused by the corrosion process. Natural process in the metal due to the contact with moist air (O_2 and water) can not be eliminated but can be slowed.

Nail is one example of iron metal that is found often in daily life. One disadvantage of nail (iron metal) is easy to experience corrosion and easily observed the corrosion process. Corrosion causes many disadvantages because it reduces the age range of goods and buildings that use iron or steel. Therefore it is needed a way to prevent such corrosion. Corrosion prevention process can be done, such as by coating on a metal surface, cathodic protection, the addition of corrosion inhibitors and others [1]. The use of inhibitors is one of the most effective way to prevent corrosion, because of its relatively low cost and simple process.

Corrosion inhibitor is defined as a substance which when added in small amounts into the environment will decrease the attack of environment corrosion against the metal. Corrosion inhibitors are generally derived from organic compounds and inorganic that contains groups which have free electron pair, such as nitrite, chromate, phosphate, urea, phenylalanine, imidazolin, and amine compounds. One alternative is the extract of natural substance, especially compound which contains atom N, O, P, S, and atoms that have free electron pair. Elements that contain this free electron pair can later serve as ligand that will form complex compound with metal. One alternative material that contains a pair of free electron that are easily found and accessible to the society is MSG, aloe vera and papaya leaf.

MSG (Monosodium Glutamate) contains amino acids such as glutamic acid that has two carboxylic groups and one amine group which is expected to do inhibition in corrosion by adsorbed the surface of iron metal or carbon steel. In the presence of these groups, it can be assumed that amino acid compound and its derivative are potential candidate as a corrosion inhibitor. Some research has involved several amino acids as inhibitor [2 3], cysteine [4, 5], and tryptophan [6]. Other amino acids and their derivatives that have been studied to give the efficiency of inhibition (EI) is good but economically is still not profitable [7, 8]. Glutamic acid is an intermediate compound in the manufacture of mono sodium glutamate (MSG) which is relatively cheap in the market, so that when glutamic acid is potential proven as a corrosion inhibitor, automatically mono sodium glutamate becomes a candidate of corrosion inhibitor candidates which are economically relatively cheap and improve the economic value in MSG production generally.

In addition to MSG, natural ingredients that can be easily obtained by the society and economically relatively cheap are aloe vera and papaya. Aloe vera contains aloin, aloenin, aloesin and amino acids. Papaya leaf contains N-acetyl-glukosaminida, benzyl isothiocyanate, amino acids [9]. This research aims to "*Harnessing the MSG, Aloe Vera and Leaf Papaya (Carica papaya L) as an Alternative Corrosion Inhibitors*".

2. RESEARCH METHODS

2.1 Materials
Basic materials used in the research are MSG, Aloe Vera, papaya leaf, nail, gelatin and aquades. While the tool used are petri dish, spatula, digital balance, Bunsen burners, beakers, mortar / pestle, measuring cups, pipette drops, funnel and filter paper.

2.2 Procedures
Procedures in this research are as: Weigh the petri dish and nail, and then record the results. Smoothes the papaya leaf with mortar. Weigh the papaya leaf with balance, and wrote the result. Add 40 ml of aquades in papaya leaf that has been mashed, stir, and then filter to separate the pulp of papaya leaf. Repeat the step of (b) to (d) on aloe vera. Take the aloe vera and papaya leaf that have been filtered in 30 ml into each beaker that already contains with the 30 ml of gelatin solution. Heat the aloe vera and papaya leaf that have been mixed with the gelatin solution until boiling. Repeat the step of (c) on MSG. Dissolve the MSG with 30 ml of water. Take all MSG solution into a beaker that contains 30 ml of gelatin solution. Repeat the step of (g) on MSG. Make the variable of control. Take the papaya leaf solution that has been heated for 40 ml into the petri dish. Repeat the step of (m) on aloe vera and MSG. Mix the 12 ml of papaya leaf solution, 12 ml of aloe vera solution, and 16 ml of MSG solution into one petri dish. Weigh the whole petri dish that already consist of solution and nail. Take the whole petri dish into refrigerator for ±30 minutes until the solution of each variable becomes chewy. Put the whole petri dish in a place with an open air conditions. Observe the physical changes of the nails in each petri dish every day for 6 days. Weigh the whole petri dish and calculate the mass difference on day-6 [10].

2.3 Data Analysis
Rust levels are analyzed using the gravimetric methods. Gravimetric is one of the main branch of analytical chemistry. The step of gravimetric measurement is weighing. Analyte is physically separated from all other components of the sample and of the solvent. Its products are usually very slightly soluble substances, which can be weighed [11]. In this research, the corrosion on a nail is based on the reaction :

$$Fe(OH)_2 \xrightarrow{O_2\ ,\ H_2O} Fe(OH)_3 \rightarrow \underset{rust}{Fe_2O_3\ x\ H_2O_{(s)}}$$

The steps :
The container is weighed (container mass) : m_w gr

Iron (nail) that has not rusted yet ($Fe_{(s)}$) + container : m_{wp0} gr
Nail (iron) mass : ($m_w - m_{wp0}$) gr
After treatment in each time interval :
Iron (nail) that has rusted + container : m_{wpt} gr
Iron rust mass ($Fe_2O_3.\, xH_2O_{(s)}$) : $m_{wpt} - (m_w - m_{wp0})$ gr

According to the procedure in the gravimetric method, a sediment is weighed and from this value the weight of the analyte in the sample can be calculated, in this case iron rust. So the percentage of iron rust in this research can be calculated by gravimetric analysis:

$$\%\, \boldsymbol{iron\, rust}\, (\boldsymbol{Fe_2O_3\, x\, H_2O}) = \frac{\boldsymbol{m_{wpt}} - (\boldsymbol{m_w} - \boldsymbol{m_{wpo}})}{(\boldsymbol{m_w} - \boldsymbol{m_{wpo}})}\, \boldsymbol{x}\, 100\, \%$$

3. RESULTS AND DISCUSSION

In the event of corrosion, the metal is oxidized while the oxygen (air) is reduced. Generally, metal corrosion is form of the oxide or carbonate in the form of solids. From the research result for 5 days can be seen the significant changes in nail of each cup. These changes can be seen from Table 1.

Table 1 Research Result Data

No	Variable Type	Observation of day- 1	2	3	4	5
1	Control Variable I	Half of the nail surface has rusted	Whole nail surface has been covered by rust	Rust increases	Rust increases	Constant
2	Control Variable II	Half of the nail surface has rusted	Whole nail surface has been covered by rust	Rust increases	Rust increases	Constant
3	MSG solution	Gelatin color is red around the nail	Gelatin color is dark red around the nail	Gelatin color is a little brown around the nail and there is a part of nail that has rusted	Rust increases and the nail color changes become rather darker	Rust increases and a half of nail surface has rusted
4	Aloe vera Solution	-	Gelatin color is red around the nail	Half of nail surface has rusted	Rust increases	Rust increases and the whole of nail surface has rusted
5	Papaya Leaf Solution	-	-	-	Gelatin color is dark around the nail	There is a part of nail that has rusted
6	The Mixed Solution of MSG, Aloe vera, and Papaya Leaf	-	-	-		

The nail changes can be observed qualitatively as can be seen in Table 1. On the first day, the nail on the control variables has been partially covered with rust. In the cup of aloe vera, papaya leaf, and a mixture of nail conditions have not changed yet. On day 4, the cup of papaya leaf solution and the cup of the three mixed solution has the color changing around the nail become rather dark. This color changing is not rust, but the peeling of nail layer that have been oxidized by the compound in papaya leaf. On day 5, all the nails on the cup have rusted at different levels. The mass data of the nail and its container that is measured by the researcher at the beginning and the end of the research can be seen in Table 2.

Table 2 The Mass Data of the Nail and Its Container at the Beginning and the End of the Research

No.	Variable Type	m_w(gr)	m_{wp0}(gr)	m_p(gr)	m_{wpl0}(gr)	m_{wpt}(gr)	m_k(gr)
1.	Control Variable I	47,1	49,4	2,3	82,1	82,5	0,4
2.	Control Variable II	47,4	49,6	2,2	81,6	82,0	0,4
3.	MSG Solution	48,0	50,4	2,4	86,0	86,2	0,2
4.	Aloe vera Solution	48,0	50,2	2,2	87,3	87,5	0,2
5.	Papaya Leaf Solution	46,2	48,7	2,5	86,1	86,3	0,2
6.	The Mixed Solution of MSG, Aloe vera, and Papaya Leaf	81,5	83,8	2,3	122,7	122,8	0,1

Description:
m_w = container mass
m_{wp0} = the mass of nail that has not rusted yet + container
m_p = nail mass
m_{wpl0} = the mass of nail that has not rusted yet + container + solution
m_{wpt} = the mass of nail that has rusted + container + solution
m_k = rust mass

Based on data in the table above can be determined percentage of the iron rust in each container by using the gravimetric method, that is the data in the table above is put into the following equation:

$$\%\ iron\ rust(Fe_2O_3\ x\ H_2O) = \frac{m_{wpt} - (m_w - m_{wpo})}{(m_w - m_{wpo})} x100\ \%$$

The example of container 1:

$$= \frac{0,4}{2,3} x100\ \%$$

$$=17,39\%$$

The results of calculations using these equations can be seen in Table 3.

Table 3 Level of Nail Rust from Each Container or Petri Dish

No.	Variable Type	% rust
1.	Control variable I	17.39%
2.	Control variable II	18.18%
3.	MSG solution	8.33%
4.	Aloe vera solution	9.09%
5.	Papaya leaf solution	8.00%
6.	The mixed solution of MSG, aloe vera, and papaya leaf	4.35%

Calculation result data of rust level in Table 3 shows that there is a significantly different level of rust, rust level of the control variable, 17-18%, mixed containers of 9.09% aloe vera, 8% papaya leaf, and 8.33% for MSG, and the container that contains the mixture of MSG, aloe vera, papaya leaves at 4.35%. This is because the presence of amine groups and carboxylic groups on the MSG can inhibit corrosion by being absorbed on the nail surface, so the level of corrosion is much smaller than the control variable.

Similarly to the addition of papaya leaf extract, the corrosion level is also much smaller than the control variable. This is because the papaya leaf contains N-acetyl-glukosaminida, benzyl isothiocyanate, and amino acid, the presence of nitrogen atom groups will give its free electron pair to donate an electron to the metals of Fe^{2+} so formed complex compounds that can be an inhibitor in corrosion.

The addition of aloe vera extract also causes the corrosion level to be much smaller than the control variable because aloe vera contains aloin, aloenin, aloesin, and amino acid, which contains atom N, O, P, S, and atoms that have free electron pair [12]. Elements that contain this free electron pair can later serve as ligands that will form a complex compound with metal. Sudrajat and Ilim [13] suggest that the extract of tobacco leaf, aloe vera, papaya leaf, tea leaf, and coffee can effectively reduce the corrosion rate of *mild steel*.

A mixture of MSG solution, papaya leaf solution, and aloe vera is the best corrosion inhibitor from the whole cup that was observed with a rust level of 4% is much smaller than the rust level of 17-18% on the control variable. This can be explained that MSG with the $-NH_2$ groups, and carboxylate groups, papaya leaf that also contains amino acid and has a group of atoms that can give its free electron pair to donate an electron to the metals of Fe^{2+}, and aloe vera contains aloin, aloenin, aloesin, and amino acid. These natural materials contain atoms N, O, P, S, and atoms that have free electron pairs that give their free electron pair to donate electrons to the metals of Fe^{2+}.

The mixture of three materials will increase the number of atoms that has a group of the atom that can give its free electron pair to donate an electron to the metal of Fe^{2+} plus the presence of amino acids contained in MSG, papaya leaf, and aloe vera. Amine groups and carboxylic groups that are contained in the amino acid can do inhibition in corrosion by adsorbing the nail surface.

4. CONCLUSION

Based on the research result of MSG, papaya leaf, and aloe vera as the corrosion inhibitor can be summed up some of the following:

1. MSG, papaya leaf, and aloe vera can be used as corrosion inhibitor alternative
2. From the research results, the rust level that uses the MSG solution is 8.33%, for aloe vera extract is 9.09%, for papaya leaf extract is 8%, and for the container that is added the mixture of aloe vera extract, papaya leaf extract, and MSG is 4.35%.

REFERENCES

[1] Achmad H 1992 Penuntun belajar kimia dasar, elektrokimia dan kinetika dasar Bandung: PT. Citra Aditya Bakti 59 105

[2] Wahyuningrum D A Shobarudin R Brahma S Achmad B Ariwahjoedi Ciptati 2006 The synthesis of some acetyl substituted histidine compounds containing imidazole ring framework and the investigation of their inhibitor activity towards carbon steel in 1% NaCl solution Proceeding of Indonesian Symposium on Science and Technology on Chemistry (ISSToC) Serpong Indonesia ISSN: 1907-1485

[3] Mateo E Marti Ch Methivier P. Dubot and C M Pradier 2003 Adsorption of (S)-Histidine on Cu(110) and Oxygen-Covered Cu(110), a Combined Fourier Transform Reflection Absorption Infrared Spectroscopy and Force Field Calculation Study J Phys Chem B 107 10785-10792

[4] Brolo AG P Germain and G Hager 2002 Investigation of the adsorption of l-cysteine on a polycrystalline silver electrode by surface-enhanced raman scattering (SERS) and Surface Enhanced second harmonic generation (SESHG) J Phys Chem B 106 5982–87

IOP Conf. Series: Earth and Environmental Science 1075 (2022) 012013 doi:10.1088/1755-1315/1075/1/012013

[5] Matos J B L P Pereira S M L Agostinho O E Barcia G G O Cordeiro and E D'Elia 2004 Effect of cysteine on the anodic dissolution of copper in sulfuric acid medium J Electroanal Chem 570 91–94
[6] El-Shafei A A S A Abd El-Maksoud A S Fouda 2004 The role of indole and its derivatives in the pitting corrosion of al in neutral chloride solution Corros. Sci 46, 579–590
[7] Rajendran S B V Apparao N Palaniswamy V Periasamy and G Karthikeyan 2001 Corrosion Inhibition by Strainless Complexes, Corros. Sci 43 1345-1354
[8] Stupnisek-Lisac E A Gazivoda and M Madzarac 2002 Evaluation of Non-Toxic Corrosion Inhibitors for Copper in Sulphuric Acid.Electrochim Acta 47 4189-4194
[9] Andrade Silva M 1943. Histamine and proteolytic enzymes, Liberation of histamine by papain. Journal of Biological Chemistry 149: 9-7
[10] Ketis Ni Ketut Deana Wahyuningrum Sadijah Achmad and Bunbun Bundjali 2010 Efektivitas asam glutamat sebagai inhibitor korosi pada baja karbon dalam larutan NaCl 1% Jurnal Matematika dan Sains 15 No 1 5
[11] Petrucci and Harwood 2007 General Chemistry Macmillan Publishing Company
[12] Wahyuningrum D Achmad S Nugraha A dan Damanik R 2004 Sintesis senyawa ester histidin dan aplikasinya sebagai inhibitor korosi Proceeding Seminar MIPA IV, ISSN 979-368-8-02-5
[13] Sudrajat and Ilim 2006 Studi Penggunaan Inhibitor Organik yang Mengandung Nitrogen dari Ekstrak Bahan Aam terhadap Laju Korosi Baja Lunak dengan Metode Gravimetri. Universitas Lampung. Bandar Lampung

Etnochemistry Study of Medicine Plants for Liver Disease in The Community of Balane Village

Purnama Ningsih*, Ijirana Ijirana, Sri Mulyani and Frischilia Febby Patanda

Chemistry Education, Department of Mathematics and Natural Science Education, Faculty of Teacher Training and Education, Tadulako University

*purnamaningsih@yahoo.com

Abstract. Based on their ethnochemical expertise, the study intends to find the types, sections of plants, processing methods, and ways of use used to treat the liver ailment in the community of Balane village, Kinovaro District, Sigi Regency, Central Sulawesi. A qualitative research method was used, which included interview techniques. To treat liver disorders, the Balane community uses six distinct types of herbs. Cakar ayam (*Selaginella doederleinii Hieron*), daun sendok (*Plantago major L.),* jarong (*Stachytarpheta indica (L.) Vahl*), pegagan (*Centella asiatica (L.) Urban.*), sambiloto (*Andrographis paniculate (Burm. F) Wallich*), temulawak (*Curcuma zanthorrhiza Roxb.*). Plant parts that are used as medicine include herbs, leaves, rhizomes, and grains. Boiling is used to prepare the medicinal herbs, and the treatment is used for 1-3 months or until the patient notices a difference or recovers.

1. Introduction

Indonesia is an archipelagic country with the world's second most excellent tropical forest and biodiversity, and it is considered one of the world's second "mega biodiversity" countries after Brazil [1]. According to estimates, Indonesia's forests contain 30,000 species of medical plants, 940 of which have been designated as medicinal, and 78% of them are still obtained via direct extraction from the forest.

Sulawesi is one of Indonesia's largest and most important islands. It is part of the Wallacea biogeographic region, which is home to a diverse range of endemic flora and animals. On the other hand, the Sulawesi region is home to several ethnic or indigenous groups comprised of a diverse range of ecologists. Similarly, health awareness is a defining characteristic of each ethnic group in Indonesia.

Several villages in Central Sulawesi, including distant indigenous groups, were identified using data from the Central Statistics Agency [2]. This comprises the Balane people, as well as the tolare/da'a, rai, and ledo tribes. Small indigenous groups as a communal group follow a pattern of conduct that is closely tied to the customs passed down from generation to generation by their forefathers [3]. Several examples have chemical substances, despite the fact that the adopted lifestyle is still pure and regarded as 'outdated.' One example of their way of life is that they use the plants that grow around them as medicine to treat fevers, stomach-aches, and coughs.

Plants are still used as medicine by the communities of Balane Village. It stands apart from other civilizations because of its diverse user base. The people of Balane Village have remained steadfast in their management of plants as a medicine until now. Liver illness is one of the diseases that the residents of Balane Village still believe must be treated traditionally. In addition to the information disclosed above, it is necessary to reveal the pattern of life behavior associated with ethnochemistry in the Balane community in terms of treatment. Thus, this study, will investigate which medicinal plants are used by the people of Balane Village to treat liver disease, which parts of the plant are used as medicine, and how medicinal plants are processed and used by the people of Balane Village to treat liver disease.

The liver is a vital organ for humans, according to [4]. This organ is located just below the diaphragm in the right abdominal cavity. The liver performs various functions, including acting as an antidote to and neutralizer of toxins, regulating hormone circulation, and controlling the composition of blood, which contains fat, sugar, protein, and other substances. Additionally, the liver produces bile, a substance that aids in fat digestion. Liver disease is a malfunction of any of the liver's functions. The liver is responsible for critical roles in the body, and their loss can result in significant damage. Although the liver is the only organ in the body capable of quickly replacing damaged cells, if these cells are lost, the liver may not meet the body's requirements. Because the liver disease may be asymptomatic, it is frequently referred to as a silent killer. According to [5], liver disease is a leading cause of death worldwide, increasing the mortality rate from time to time. Globally, estimates place the number of patients with chronic liver disease (not all types of liver disease) at 884 million, with an average annual death rate of 2 million.

2. Methods

This research uses ethnographic methodologies to conduct a descriptive qualitative investigation. The goal of descriptive qualitative research is to describe or gather knowledge from a large, broad, and in-depth set of study data of traditional medicine for liver disease in the people of Balane Village in the Sigi area of Central Sulawesi. Interview guidelines are developed to confirm the findings acquired both in theory and in the field. Documentation procedures are employed to record the plants used as medicine, and interviews with informants are among the research instruments used.

3. Result and Discussion

Patients with liver problems in the area provided data for the study, as did sandro (shaman). In Balane village (Figure 1), there are six different types of herbs that are utilized to treat liver illness. Table 1 shows the sorts of plants utilized by the people in Balane village, Kinovaro sub-district, Sigi district, Central Sulawesi, to treat liver illness.

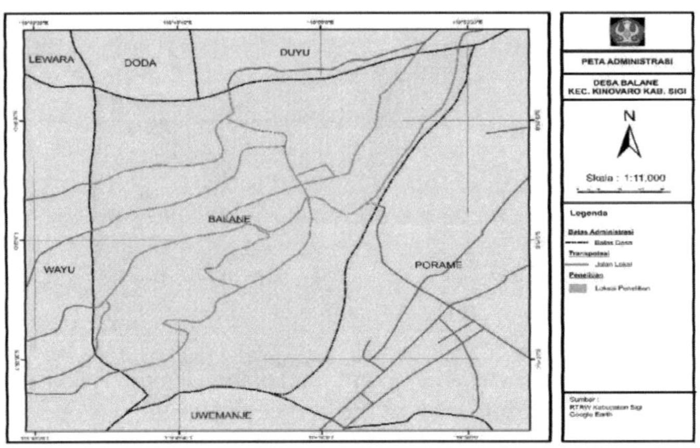

Figure 1. Balane Village Map (Geospatial, 2019)

Table 1. Medicinal plants for liver disease

No	Name of Plants	Family	Species
1	Cakar Ayam	Selaginellaceae	*Selaginella doederleinii Hieron.*
2	Daun Sendok	Plantaginaceae	*Plantago major L.*
3	Jarong	Verbenaceae	*Stachytarpheta indica (L.) Vahl.*
4	Pegagan	Mackinlayaceae	*Centella asiatica (L.) Urban.*
5	Sambiloto	Acanthaceae	*Andrographis paniculate (Burm. F) Wallich*
6	Temulawak	Zingiberaceae	*Curcuma zanthorrhiza Roxb.*

The grains, leaves, stems, roots, and rhizomes are the plant parts used as therapeutic substances. With a ratio of 42.9 percent, the leaves are the most commonly utilized as a treatment to treat liver illness. This is due to the fact that the leaf is an important part of the plant, and most plants have several leaves. The leaves are also quite easy to find and always available, and retrieving them is less difficult than retrieving other plant parts. The leaves are usually thin, broad, and high in chlorophyll, a green chemical [6].

Boiling is the method employed in the Balane village community to manage plants used as therapeutic components. By drinking the boiled water from the stew, internal diseases such as liver disease may usually be treated [7].

3.1. Cakar Ayam (Selaginella doederleinii Hieron)

Selaginella doederleinii Hieron (Figure 2) is a fern that grows in the wild along river banks and on moist rocks. The components in *Selaginella doederleinii*, especially biflavanone, tetrahydro robustaflavone, and trimethyl ether, are cytotoxic elements that can attack cancer cells in humans, according to [8]. Alkaloids, saponins, phytosterols, and flavonoid glycosides are also found in the *Selaginella doederleinii Hieron* plant. Alkaloids and flavonoid glycosides, according to [9], act as hepatoprotectors [10] One of the active compounds in the *Selaginella doederleinii Hieron* plant is flavonoid, which acts as an antioxidant in the lipid peroxidation process. Saponins have a direct influence on mending impaired liver function in their work. Similar to alkaloids, if the concentration is high and the administration duration is extended, fat buildup will promote cell expansion.

Figure 2. *Selaginella doederleinii Hieron*

3.2. Daun Sendok (Plantago major L)

Plantago major L (Figure 3) leaves contain a variety of active chemicals that are known to have hepatoprotective properties. The leaves have anti-inflammatory, antiproliferative, anti-apoptotic, and antioxidant properties in addition to hepatoprotective properties [11-12]. Ursolic acid, apigenin, lutheolin, baicalein, aucubin, and scutellarin are some of the active chemicals found in *Plantago major L* leaves [11]. Compounds in the leaves with antioxidant properties will protect liver cells from ciprofibrate-induced oxidative stress reactions and enhance the number of antioxidants inhibited by omeprazole, allowing the liver to repair damaged tissues. The anti-inflammatory impact of the ethanol extract of *Plantago major L* leaves can limit the inflammatory reaction created by omeprazole, while the antiproliferative effect inhibits hepatocyte cell hyperproliferation caused by ciprofibrate induction, preventing liver mass and volume gain. Similarly, ethanol extract of *Plantago major L* leaves can repair damage to hepatocyte cells, enhancing the histological image of the liver. Compound's hispidulin, baicalein, oleanic acid, and lutheolin have a major role in the antioxidant effect of ethanol extract of spoon leaves, which is useful in preventing liver cell damage caused by oxidative stress reactions. Lutheolin aids in the production of antioxidants such as vitamin A, vitamin C, and beta-carotene [13]. Baicalein and hispidulin have the effect of reducing glutathione depletion [14] and inhibiting iNOS gene expression and inducing metalloproteins that act like glutathione in protecting the body from toxic substances [15].

Figure 3. *Plantago major L*

3.3. Jarong (S. indica (L.) Vahl.)

Chemical substances including as terpenoids, flavonoids, glycosides, and tannins are found in *S. indica (L.) Vahl* (Figure 4). Almost all plants contain flavonoid chemicals, including the *S. indica (L.) Vahl* [16]. This plant can be found in abundance in Indonesia. [17] discovered that employing the leaf soxhletation method, 95 % ethanol extract had a hepatoprotective effect.

Figure 4. *Stachytarpheta indica (L.) Vahl*

3.4. *Pegagan (Centella asiatica (L.) Urban.)*

Several bioactive components in *Centella asiatica (L.) Urban* (Figure 5) plant include asiaticoside, tankunicide, isotankunicide, madecasoside, brahmoside, brahminoside, brahmic acid, madasiatic acid, meso-inositol, senteloside, carotenoids, hydrocotyledon, vellarin, tannins and mineral salts such as potassium, sodium, magnesium, calcium, and iron [18]. *Centella asiatica (L.) Urban* plant, which has asiaticoside in the form of glycosides and is frequently employed in traditional medicinal herbs or herbal medicine, also contains asiaticoside in the form of glycosides. The triterpenoid group includes asiaticoside, asiatic acid, madekasida, and madecasoside, whereas the steroid group includes sitosterol and stigmasterol, and the saponin group includes vallerin brahmoside.

Centella asiatica (L.) Urban. has a variety of active compounds, including 1) triterpenoid saponins, 2) triterpenoid genin, 3) essential oils, 4) flavonoids, 5) phytosterols, and other active components, according to [19]. Triterpenoids and saponins, which comprise 1) asiaticoside, 2) senteloside, 3) madecoside, and 4) asiatic acid, as well as volatile oils, flavonoids, tannins, phytosterols, amino acids, and carbohydrates, are the most important active constituents. The entire bioactive content of the gotu kola plant is an antioxidant that helps the human body's immune system function better.

Many factors determine the active ingredient in *Centella asiatica (L.) Urban.* The content of chemicals generated in plants is influenced by the type of soil or growing environment, according to [20]. The most important chemicals in the *Centella asiatica (L.) Urban* plant are triterpenoids. Triterpenoids have a soothing impact and help to improve brain function. This molecule can also help to rejuvenate blood arteries, allowing for better blood flow to the brain. Asiaticoside is a triterpenoid that strengthens and increases the regeneration of skin cells, stimulates blood cells and the immune system, and serves as a natural antibacterial. Brahmoside is a molecule that helps blood flow through the brain and is a vital protein for brain cells.

Figure 5. *Centella asiatica (L.)*

3.5. *Sambiloto (Andrographis paniculate (Burm. F) Wallich)*

Deoxyandrographolide, andrographolide, neondrographolide, dehydroandrographolide, and homoandrographolide are some of the active lactone compounds found in *Andrographis paniculate (Burm. F) Wallich* [21]. Flavonoids, alkanes, ketones, aldehydes, minerals, and resins are among the chemical substances found in it. Andrographolide compounds act as hepatoprotectors or liver protectors, with the mechanism of action of these compounds being able to reduce lipids, boost glutathione levels, and increase antioxidant enzyme activity [22].

Figure 6. *Andrographis paniculate (Burm. F) Wallich*

3.6. Temulawak (Curcuma zanthorrhiza Roxb.)

Curcuma zanthorrhiza Roxb (Figure 7) is a kind of curcuma. Curcumin, demethoxycurcumin, and bisdemethoxycurcumin are all found in Roxb. Curcumin has a component in the antioxidant process that has activity as an antioxidant that counteracts free radicals. Curcumin has a phenolic hydroxy group and a diketone group in its structure. In the first phase of the antioxidant mechanism, the phenolic hydroxy group acts as a free radical scavenger. The phenolic hydroxy group acts as an oxidant, and by a hydrogen transfer mechanism, the radical is transformed into a stable state. The reactive group on DPPH (1,1-diphenyl-picrylhydrazyl) is a nitrogen group that will pair with a hydrogen atom on the antioxidant to generate a stable DPPH radical (1,1-diphenyl-2picrylhydrazine), which will inhibit the increase in peroxidation by preventing free radicals from forming. the liver's lipids [23].

Figure 7. *Curcuma zanthorrhiza Roxb*

Alkaloids, flavonoids, glycosides, saponins, hispidulin, baicalein, oleanic acid, lutheolin, triterpenoids, asiaticoside, brahmoside, andrographolide, and curcumin are some of the chemical components used in traditional medicine to treat liver illness. Alkaloids and flavonoid glycosides that function as hepatoprotectors [9] are the most important chemical compounds in the healing process of liver disease, hepatoprotectors are compounds or substances that are effective in protecting cells while repairing liver tissue damaged by toxic substances. Saponins also have a direct effect on repairing damaged liver function [24] with the mechanism of action as antioxidants to protect the liver by inhibiting the activity of cytochrome P450 inhibitors in the liver and inhibiting the increase in AST (Aspartate Transaminase) and ALT (Aspartate Transaminase/Aspartate Transaminase/Aspartate Transaminase (Alanine Transaminase) [25].

4. Conclusion

1. People in Balane Village, Kinovaro Subdistrict, Sigi Regency, Central Sulawesi, use six different plants that can be used as medications to treat liver illness, including cakar ayam (*Selaginella doederleinii Hieron*), daun sendok (*Plantago major L.*), jarong (*Stachytarpheta indica (L.) Vahl*), pegagan (*Centella asiatica (L.) Urban*.), sambiloto (*Andrographis paniculate (Burm. F) Wallich*), and temulawak (*Curcuma zanthorrhiza Roxb.).*
2. The people of Balane Village, Kinovaro Subdistrict, Sigi Regency, Central Sulawesi, use the rhizome, grain, herbs, and leaves of therapeutic plants. The leaves are the most commonly employed plant portion for the treatment of liver illness, accounting for 42.9% of all use.
3. People in Balane Village, Kinovaro Subdistrict, Sigi Regency, Central Sulawesi, grow medicinal herbs for liver sickness, which they process directly by boiling and then consume.

Acknowledgments
The authors acknowledge Balane village community who were involved in this study.

References

[1] Ersam, T. 2004. Keunggulan Biodiversitas Hutan Tropika Indonesia Kimia VI. http://www.its.ac.id/personal/files/pub/764-beckers-chem-Kimia%20ITS%20TE%2004.pdf.[09 Sep 2019].
[2] Statistik B P 2020 Badan Pusat Statistik Diambil Dari https://www.bps.go.id
[3] Permana S 2016 Antropologi Perdesaan Dan Pembangunan Berkelanjutan Deepublish
[4] Pusporoni. E. Dkk (2019). Klasifikasi Pasien Penderita Penyakit Liver dengan Pendekatan Machine Learning. INFERENSI. Institut Teknologi Sepuluh September. Surabaya. Hlm. 25-32. Vol. 2(1).
[5] Widodo, Harto, Abdul Rohman, and Sismindari Sismindari. "Pemanfaatan Tumbuhan Famili Fabaceae Untuk Pengobatan Penyakit Liver Oleh Pengobat Tradisional Berbagai Etnis Di Indonesia." Media Penelitian Dan Pengembangan Kesehatan 29, no. 1 (2019): 65-88. doi:10.22435/mpk.v29i1.538.
[6] Tjitrosoepomo, G. 2001. Morfologi Tumbuhan, Cetakan 13, Gadjah Mada University Press, Yogyakarta
[7] Amrul, I. (2015). Pemanfaatan Tumbuhan Obat Pada Masyarakat Ternate, Maluku Utara, Fakultas Kehutanan, Institut Pertanian Bogor, Bogor.
[8] Lee. N.Y. 2008. Identification of a new cytotoxic biflavanone from Selaginella doederleinii. Chem Pharm Bull: 56(9): 1360-61.
[9] Prastiwi. R. 2011. Efek Hepatoprotektor Brotowali (Tinospora Cordifolia Miers) Terhadap Virus Hepatitis B [Skripsi]. UGM: Yogyakarta
[10] Achmadi S.S., Iskandriati D., Amzu, A.A., Batubara, I. (2003). Kajian Aktivitas Hepatoprotektor Saponin Dan Segi Budaya Akar Kuning (Arcengelisia Floava (L) Merr) [Skripsi]. UGM: Yogyakarta.

[11] Ozbek, H., Erten, R., Oner, A., Cengiz, N., Yilmaz, O., & Turel, I. (2009). Hepatoprotective and anti-inflammatory activities of Plantago major L. Indian Journal of Pharmacology, 41(3), 120. doi:10.4103/0253-7613.55211.

[12] Mao-ye, W., An, Li-guo. 2011. Effects of Plantago Major L. Sedds Extract On Endurance Exercise Capacity In Mice. J Medical Plants Res., 5(9):1659-63.

[13] Seelinger, G., Merfort, I., & Schempp, C. (2008). Anti-Oxidant, Anti-Inflammatory and Anti-Allergic Activities of Luteolin. Planta Medica, 74(14), 1667-1677. doi:10.1055/s-0028-1088314

[14] Lu, J., Wu, D., Zheng, Y., Hu, B., Zhang, Z., Ye, Q.,. Wang, Y. (2010). Ursolic Acid Attenuates D-Galactose-Induced Inflammatory Response in Mouse Prefrontal Cortex through Inhibiting AGEs/RAGE/NF- B Pathway Activation. Cerebral Cortex, 20(11), 2540-2548. doi:10.1093/cercor/bhq002

[15] Chang, W., Chen, C., & Lu, F. (2002). Different Effects of Baicalein, Baicalin and Wogonin on Mitochondrial Function, Glutathione Content and Cell Cycle Progression in Human Hepatoma Cell Lines. Planta Medica, 68(2), 128-132. doi:10.1055/s-2002-20246.

[16] Chowdhury, R., Rashid, M., Khan, O., & Hasan, C. (2003). Ipolamiide and α-spinasterol from Stachytarpheta urticaefolia. Biochemical Systematics and Ecology, 31(10), 1209-1211. doi:10.1016/s0305-1978(03)00073-5

[17] Gayatri., Remesh., Sumalatha., Venkates, W., Vidyadhara. (2011). Hepatoprotective Activity Of Ethanolic Extract Of Stachytarpheta Indica On Wistar Rats. Pharmacy Globale.(2), 26.

[18] Dalimartha, S., 2006. Atlas Tumbuhan Obat Indonesia. Trubus Agriwidya, Jakarta.

[19] Winarto, W.R. dan M. Surbakti. 2003. Khasiat dan Manfaat Pegagan. Agromedia Pustaka, Jakarta.

[20] Bermawie, N., S. Purwiyanti, dan Mardiana. (2008). *Keragaan sifat morfologi, hasil, dan mutu plasma nutfah pegagan (Centella asiatica (L.) Urban)*. Bul. Penel. Tan. Rempah dan Obat XIX (1): 118.

[21] Prapanja EP, Adi ML. (2003). Khasiat Dan Manfaat Sambiloto Raja Pahit Penakluk Aneka Penyakit, Agromedia, Jakarta.

[22] Sagita, A. (2006). Pengaruh Ekstrak Andrographis Paniculate (sambiloto) Terhadap Kadar Serum Oksaloasetat Transaminase Pada Tikus Wistar Yang Diberi Paracetamol. Universitas Diponegoro, Semarang.

[23] Rosiyani, L. (2010). Evaluasi Perubahan Metabolit Pada Temulawak Dengan Waktu Tanam Berbeda.. Institut Pertanian Bogor. Bogor.

[24] Achmadi S.S., Iskandriati D., Amzu, A.A., Batubara, I. (2003). Kajian Aktivitas Hepatoprotektor Saponin Dan Segi Budaya Akar Kuning (Arcengelisia Floava (L) Merr) [Skripsi]. UGM: Yogyakarta.

[25] Suminar, B., Iramanida, L., Yulin. 2006. Promoting Selected Non-Timber Forest Products Based On Community Participation Approach To Support Sustainable Forest Management In East Kalimantan, Biopharmaca Research Center, Institut Pertanian Bogor. Bogor

IOP Conf. Series: Earth and Environmental Science 1075 (2022) 012015 doi:10.1088/1755-1315/1075/1/012015

The Relationship Between Pesticide Exposure and Hypertension Incidence on Paddy Farmers in Dolago Padang Village, Central Sulawesi

NA Salikunna[1], Kurniawan A[1], Fitriana Y[1], Ramadhan MZ[1]

[1]Faculty of Medicine, Tadulako University, Palu, Central Sulawesi, Indonesia

nurasmarsalikunna82@gmail.com

Abstract Hypertension is one of the leading causes of death in the world, including Indonesia. Pesticide exposure is one of the risk factors for hypertension. This study aims to determine the relationship between the length of work of farmers, the amount types of pesticides used, the time of spraying pesticides and the use of PPE on the incidence of hypertension. The research subjects were 80 paddy farmers who participated in this study, voluntarily. Blood pressure have been measured by sphygmomanometer. The other data has been collective by interview. The statistical analysis test used the chi square test. The results showed that there was a relationship between the length of work of farmers, the amount types of pesticides used, the time of spraying pesticides with the incidence of hypertension (p value < 0.05). There is no relationship between the use of PPE with the incidence of hypertension (p value > 0.05). The conclusion of this study is that there is a relationship between pesticide exposure and the incidence of hypertension in farmers in terms of length of work, the amount types of pesticides used and the time of spraying pesticides.

1. Introduction

Hypertension or commonly referred to as blood pressure is a condition in which a person's blood pressure rises above normal. Hypertension occurs due to increased blood pressure pumped by the heart. A person is said to be hypertension if his blood pressure is more than 140/90 mmHg (1). Hypertension can increase the risk of heart, kidney, brain and other diseases. This can make hypertension the main cause of premature death in the world. Hypertension is a burden for low- and middle-income countries, where risk factors increase in these conditions [2]. Several risk factors for hypertension are genetics, age, physical activity, obesity, alcohol consumption, high salt intake, unhealthy diet and exposure to environmental pollution [3,4]. One of the environmental pollutions is pesticides exposure[5,6]. Indonesia is one of the countries with a high prevalence of hypertension. The prevalence of hypertension in Indonesia has increased from 25.8% in 2013 to 34.1% in 2018. The prevalence of hypertension in Central Sulawesi in the age group 18 and years in 2018 was 29.75% and 25.10% in Parigi Moutong [7].

Indonesia is an agricultural country where most of the population lives from farming or farming so that the agricultural sector is a very important sector in supporting the national economy, where the agricultural sector is one of the main jobs for some Indonesian people to be able to live and fulfill their lives for daily needs [8]. Likewise with Central Sulawesi, the majority of the population's occupations are farmers[9]. In carrying out their work, farmers usually use pesticides to eradicate pests.

Pesticides are combinations of various materials that function to eradicate, or control every type of pest or insect on plants that have the potential to damage the plant. Pesticides can be chemicals, biological agonists (eg viruses and bacteria), antibacterial agents, fungicides or other substances. Pests can be in the form of insects that become vectors of disease transmission or insects that can damage plants, causing disturbances [10].In addition to eradicating pests in agriculture, pesticides can have an impact on human health, either directly or indirectly. Farmers are a group that is at high risk related to pesticide exposure [10]. Pesticide exposure is associated with several diseases such as chronic kidney disease, insulin resistance, neurological disease and dysfunction, metabolic disorders, idiopathic preterm birth, gestational hypertension and hypertension[5,11–16]. Therefore, this study aims to determine the relationship between the length of work of farmers, the amount of pesticides used and the time of spraying pesticides, the use of PPE on the incidence of hypertension in paddy farmers.

2. Methods

Eighty paddy farmers participated voluntarily in the study after receiving explanation from the researcher. Ethical clearance was published by the Faculty of Medicine of Tadulako University for this study with the following registration number: 2317/UN.28.1.30/KL/2021. All the paddy farmers involved in this study were in good health and had no history of cardiovascular diseases, metabolic diseases and chronic kidney disease. Blood pressure checks using a sphygmomanometer and a stethoscope. The other data is done by interview. Statistical analysis using chi square test.

3. Results

Table 1. Characteristics of Subjects Based on this Study

Characteristics	n (%)
Age group	
50-60 year	30 (37.5)
40-49 year	26 (32.5)
30-39 year	18 (22.5)
<30 year	6 (7.5)
Gender	
Male	65 (81.2)
Female	15 (18.8)
Blood Pressure status	
Normal (blood pressure < 140/90 mmHg)	10 (12.5)
Hypertension (blood pressure ≥140/90 mmHG)	70(87.2)

Table 2. Relationship Between Variables with Hypertension Incidence of study

Variable	Hypertension		Odd Ratio	p-value
	No	Yes		
Length of work				
a. ≤ 5 years	5	0	2.275	0.001
b. >5 years	5	70		
Number of types of pesticides				
a. ≤ 3 types	1	0	4.389	0.008
b. > 3 types	9	70		
The time of spraying pesticides				
a. Morning	8	70	4.588	0.001
b. Afternoon	2	0		
Personal protective equipment				
a. Good	0	6	0.655	
b. Poor	10	64		0.336

4. Discussions

The results showed that there was a relationship between the incidence of hypertension and the length of work (p value = 0.001), the number of types of pesticides used (p value = 0.008), and the time of pesticide spraying (p value = 0.001). however, it was not related to the completeness of the use of personal protective equipment (p value = 0.336). There were 75 farmers with a length of work > 5 years and 70 of them suffer from hypertension. Farmers who had length of work (> 5 years) had a 2,275 times risk of experiencing hypertension compared to farmers who had < 5 years of length work . There were 79 farmers who used pesticides > 3 types of pesticides and 70 of them suffer from hypertension. Farmers who used pesticides > 3 types had a 4,389 times risk of developing hypertension compared to farmers who used pesticides < 3 types. There were 78 farmers who spray in the morning (high risk) and 70 of them suffer from hypertension. Farmers who did the spraying time in the morning had 4,588 times the risk of experiencing hypertension compared to farmers who do the spraying time in the afternoon. Farmers who used incomplete PPE are 74 people and 64 of them suffer from hypertension. Farmers who used personal protective equipment (PPE) spraying <5 types had a 0.622 times risk of experiencing hypertension compared to farmers who used personal protective equipment (PPE) sprayers> 5 types.

Farmers who work for long periods of time will cause pesticides to be exposed for longer in the body, which will accumulate the amount of toxins that enter the body and increase the risk of negative effects from pesticides. Exposure to pesticides can enter the body through the skin surface, inhalation or contamination of food. The toxicity of a compound is influenced by the dose and frequency of exposure. The longer we are exposed to pesticides, the greater the risk for our bodies to experience pesticide poisoning. There are two types of toxicity, some are acute and some are chronic. Acute toxicity is the ability of a substance to cause harmful effects that develop very rapidly after exposure, i.e. hours or days. while chronic toxicity is the ability of a substance to cause adverse health effects for the body, caused by prolonged exposure[10]

One of the most common types of chronic toxicity is cholinesterase inhibition. Cholinesterase is an enzyme produced in the body that regulates the transmission of nerve impulses. If cholinesterase is not present in the body, then the impulses from one nerve to the next will continue to flow. Many pesticides work against or suppress cholinesterase pests. However, these chemicals can also be toxic to humans in some situations. Long-term exposure to cholinesterase inhibitory chemicals can be harmful to the body, for example types of pesticides such as organophosphates and carbamates. Because the longer a person works as a farmer, the longer he or she has been exposed to pesticides. Increased cardiac output and peripheral resistance can occur, one of them as a result of sympathetic nerve activity. Preganglion of this sympathetic nerve, has a neurotransmitter in the form of acetylcholine (ACh). Such stimulation can cause disruption of the function of the sympathetic system in regulating blood pressure [16,17].

Excessive pesticides will increase the concentration of the mixture and produce vapors that can enter through the respiratory system, thereby increasing the risk of exposure to pesticides. The best time to spray pesticides is in the afternoon, from 16.00 - 17.00 PM when the air temperature is <30° and humidity is 50-80%[18]. If spraying is carried out in the morning or at noon, the possibility of farmers being exposed to pesticides is higher because when the environmental temperature increases it can cause farmers to sweat more and the skin pores open, so that the absorption of pesticides into the skin will be easier [19].

5. Conclussions

The incidence of hypertension in paddy farmers is related to the length of work, the number of types of pesticides used, the time of spraying pesticides.

References

[1] Armstrong C, Senior Associate Editor A. *JNC 8 Guidelines for the Management of Hypertension in Adults.* 2014.
[2] WHO. *Hypertension.* 2021.
[3] Whelton PK, Einhorn PT, Muntner P, Appel LJ, Cushman WC, Roux AVD, et al. Research needs to improve hypertension treatment and control in African Americans. *Hypertension.* 2016;68:1066–72. https://doi.org/10.1161/HYPERTENSIONAHA.116.07905.
[4] Mills KT, Stefanescu A, He J. The global epidemiology of hypertension. Nature Reviews Nephrology 2020;16:223–37. https://doi.org/10.1038/s41581-019-0244-2.
[5] Ledda C, Fiore M, Santarelli L, Bracci M, Mascali G, D'agati MG, et al. Gestational Hypertension and Organophosphorus Pesticide Exposure: A Cross-Sectional Study. *BioMed Research International.* 2015;2015. https://doi.org/10.1155/2015/280891.
[6] Goncharov A, Pavuk M, Foushee HR, Carpenter DO. Blood pressure in relation to concentrations of PCB congeners and Chlorinated pesticides. *Environmental Health Perspective.* 2011;119:319–25. https://doi.org/10.1289/ehp.1002830.
[7] Tim Riskesdas. *Laporan Riskesdas 2018 Provinsi Sulawesi Tengah.* Lembaga Penerbit Badan Penelitian dan Pengembangan Kesehatan : Jakarta.2018
[8] Ministry of agriculture of the Reppublic of Indonesia. *History of the ministry of agriculture.* Jakarta.2021
[9] Pusdatin (Pusat data dan informasi) Kementrian Pertanian. *Statisitik Ketenagakerjaan Sektor Pertanian.*Jakarta.2020.
[10] Jabbar A. PESTICIDE POISONING IN HUMANS. n.d.1992
[11] Seesen M, Lucchini RG, Siriruttanapruk S, Sapbamrer R, Hongsibsong S, Woskie S, et al. Association between organophosphate pesticide exposure and insulin resistance in pesticide sprayers and nonfarmworkers. *International Journal of Environmental Research and Public Health.* 2020;17:1–13. https://doi.org/10.3390/ijerph17218140.
[12] Tyagi V, Mustafa MD, Sharma T, Banerjee BD, Ahmed RS, Tripathi AK, et al. Association of organochlorine pesticides with the mRNA expression of tumour necrosis factor-alpha (TNF-α) & cyclooxygenase-2 (COX-2) genes in idiopathic preterm birth. *Indian Journal of Medical Research.*2016;143:731–8. https://doi.org/10.4103/0971-5916.191986.
[13] Javeres MNL, Habib R, Laure NJ, Shah STA, Valis M, Kuca K, et al. Chronic exposure to organophosphates pesticides and risk of metabolic disorder in cohort from pakistan and

cameroon. *International Journal of Environmental Research and Public Health*.2021;18:1–13. https://doi.org/10.3390/ijerph18052310.

[14] Suarez-Lopez JR, Amchich F, Murillo J, Denenberg J. Blood pressure after a heightened pesticide spray period among children living in agricultural communities in Ecuador. *Environmental Research*. 2019;175:335–42. https://doi.org/10.1016/j.envres.2019.05.030.

[15] Valcke M, Levasseur ME, Soares Da Silva A, Wesseling C. Pesticide exposures and chronic kidney disease of unknown etiology: an epidemiologic review. Environmental Health: A Global Access Science Source 2017;16. https://doi.org/10.1186/s12940-017-0254-0.

[16] Kamel F, Hoppin JA. Association of pesticide exposure with neurologic dysfunction and disease. *Environmental Health Perspectives*.2004;112:950–8. https://doi.org/10.1289/ehp.7135.

[17] Crane AL, Abdel Rasoul G, Ismail AA, Hendy O, Bonner MR, Lasarev MR, et al. Longitudinal assessment of chlorpyrifos exposure and effect biomarkers in adolescent Egyptian agricultural workers. *Journal of Exposure Science and Environmental Epidemiology*. 2013;23:356–62. https://doi.org/10.1038/jes.2012.113.

[18] Litbang Pertanian. *Teknik Penyemprotan Pestisida*. Badan Penelitian dan Pengembangan Pertanian Kementerian Pertanian : Jakarta. 2016

[19] Sungkawa, H. B. *Hubungan Riwayat Paparan Pestisida dengan Kejadian Goiter Pada Petani Hortikultura di Kecamatan Ngablak Kabupaten Magelang*. Thesis. Pascasarjana Universitas Diponegoro : Semarang. 2008

Effect of adding squid meal as an attractant to freshwater eel (*Anguilla marmorata*) feed on elver growth, feed palatability, efficiency and conversion

Samliok Ndobe*, Muamar, Eka Rosyida, Irawati Mei Widiastuti, Kasim Mansyur, Fadly Y. Tantu

Aquaculture Study Program, Department of Fisheries and Marine Science, Faculty of Animal Husbandry and Fisheries, Universitas Tadulako, Palu 94118, Indonesia

*Corresponding author's e-mail address: samndobe@yahoo.com

Abstract. Squid meal can serve as an attractant to increase feed palatability and fish growth. The purpose of this study was to determine the optimum dose of squid meal as an attractant in freshwater eel (*Anguilla marmorata*) elver feed based on feed palatability and elver growth. This research was conducted from December 2020 to February 2021 at the Water Quality and Aquatic Biology Laboratory, Tadulako University, Palu, Indonesia. The completely randomized design (CRD) comprised four treatments with five replicates. Commercial feed was used with four squid meal attractant doses: 0% (A); 1% (B); 3% (C); 5% (D). Variables were palatability (speed of finding and consuming the feed, morning and afternoon), growth (net weight gain; net increase in length), feed efficiency, feed conversion ratio (FCR), and survival rate (SR). FCR did not vary significantly between treatments. The best results were obtained under treatment C with a uniformly high palatability (feed finding speeds of 0.12 s/cm in the morning and 0.06 s/cm in the afternoon; 100% feeding response); net weight gain 3.50 g; net growth in length 4 cm; feed efficiency 89.9%; SR 85.56%. It was concluded that 3% is the optimum dose for enriching elver feed with squid meal as an attractant.

1. Introduction

The freshwater eels of the genus *Anguilla* are in fact catadromous fish that spend some part of their life cycle in marine, brackish and freshwater environments [1]. Anguillid eels spawn in the deep sea; planktonic eggs and leptocephalus larvae are borne by currents to coastal areas where the larvae metamorphose into glass eels which migrate to brackish and freshwater bodies, where they become benthic elvers then yellow eels and finally silver eels with maturing gonads which migrate back to their ancestral spawning grounds [2–4]. The largest of the anguillid eels, the giant mottled eel *Anguilla marmorata* (author) has a wide distribution in the tropical Indo-Pacific, spreading north-south from Japan to northern Australia and west-east from eastern Africa and Madagascar to Tahiti [1]. Although this species likely has more than one spawning ground, there is strong evidence of an *A. marmorata* spawning ground in the North Equatorial Current region of the western North Pacific Ocean from which larvae disperse to areas including Japan, parts of Indonesia including Sulawesi, and Brunei [1,2,5,6].

Anguillid eels are high value fish popular for their tasty meat, high levels of essential fatty acids and health benefits [3,4,7–9]. Over the past two decades, global recorded freshwater eel production has ranged from around 220,000 to 280,000 tons/year [10,11]. Overexploitation and habitat loss or degradation have brought the European eel *Anguilla anguilla* to the brink of extinction, as reflected in the IUCN Red List category of Critically Endangered (CR) and CITES Appendix II listing [8,12,13]. The Japanese eel *A. japonica* and American eel *A. rostrata* are also considered overexploited and listed as Endangered (EN) in the IUCN Red List [4,8,14,15]. This situation has increased the interest in and market demand for tropical anguillid eels, in particular *A. marmorata* [7,8,16–19]. In recent decades, *A. marmorata* has become a valuable fisheries commodity traded internationally, with Japan as one of the main markets [17,18], and growing demand from farms in China as well as Taiwan and other Asian countries [11,19]. In 2018 *A. marmorata* was considered the most commonly exported anguillid eel species from Indonesia, especially Sulawesi [19].

Despite advances in eel reproductive biology, including successful captive gonad maturation and spawning, full-cycle aquaculture is still impractical, and all commercial eel aquaculture relies on wild-caught seed, mostly glass eels but also elvers [4,8,16,20,21]. While eel aquaculture originally focused on temperate species, mainly *A. anguilla* and *A japonica*, several tropical anguillid eel species, including *A. marmorata*, are also increasingly sought after both at consumer-ready sizes and as glass eels or elvers for grow-out aquaculture [4,8,16–19,22]. Glass eels caught in Indonesia have been (and still are) exported, legally or illegally, mostly destined for industrial scale eel farms in other Asian countries [18,22,23]. However, Indonesian eel aquaculture (grow-out) has been increasing, with drivers including regulations intended to discourage export, growing awareness of the market opportunities, research and outreach activities, government support and (primarily overseas) investment [7,23–25].

Indonesian freshwater eel production capacity has been estimated at around 120,000 tons, but in 2020 the recorded production was just 19,351 tons, around 16% of the estimated potential volume [26]. Despite doubts regarding the accuracy of freshwater eel capture and aquaculture statistics [25,27], this represents a substantial increase from 4,407 tons reported in 2017 [23]. Despite the high value of eel products, there are still constraints on eel culture in Indonesia. These include the slow growth of wild- caught eel seeds [28], and hence the long time for glass eels to reach marketable size [29]. For example, it can take around 8-9 months of maintenance for eels to reach a weight of 120 g, a size considered fit for consumption [29]. The most difficult stages in eel aquaculture are the glass eel to elver stage and the elver to fingerling stage [23], one reason being that at these stages the young eels are very sensitive to water quality and can be vulnerable to disease [20,28,30].

One reason for the slow growth is thought to be due to the adaptation from natural feed to artificial feed, especially as digestibility and/or palatability of artificial feed is often low [30,31]. In order to support the growth and success of anguillid eel farming, balanced feed formulations are needed which fulfil the needs of the eels at each life stage from capture to marketable size, including high protein content [10,20,23,28,32]. From the elver stage onwards, artificial eel feed is generally provided in the form of a moist paste or dough which can be shaped into pellets and can be readily eaten by the eels [20,29–32]. However, this feed has the disadvantage that left-overs often settle to the bottom of the pond where they are not eaten and affect water quality. Leftover or uneaten feed can be due to several factors including a lack of stimulation to feed so that the young eels are slow to respond to the feed or do not have an appetite for the feed [29].

One way to overcome the challenges with artificial feed and accelerate the adaptation of the young eels to artificial feed is to stimulate feeding by increasing the attractiveness and palatability of the feed [20]. Ingredients which can provide a stimulus for the fish to feed are called attractants [33]. The attractant gives off an appropriate signal to enable the fish to recognize the artificial feed pellets as a food source [20,33]. Attractants are generally produced from ingredients rich in free amino acids (alanine, glycine, histidine and proline), which also function as dietary components to stimulate growth and act as energy sources [29,34]. However, dosage is important; the effects can be negligible if the dose is too low [34], while excessive doses or inappropriate attractants can be counter-productive [29,31]. Squid meal is one potential feed ingredient that can be used as an attractant as it has a relatively high free amino acid content with around 619-928 mg of glycine and betaine in 100 g of squid meat [35].

Research on elver feed attractants in 2013 [35] tested the effect of several ingredients when added as attractants at a dosage of 2%; these were squid meal (genus *Loligo*), shrimp meal (genus *Acetes*) and a mixture of squid and shrimp meal. The squid meal attractant gave the best results in terms of feed consumption, daily growth rate and feed efficiency. Based on the situation description above, the purpose of this study was to determine the optimal squid meal attractant dose to increase feed palatability and promote growth of freshwater eel (*Anguilla marmorata*) elvers.

2. Materials and Methods

2.1. Location, time, materials and equipment

The experimental research took place from 31 December 2020 to 24 February 2021 (60 days) at the Water Quality and Aquatic Biology Laboratory, Aquaculture Study Program, Universitas Tadulako in Palu, Central Sulawesi, Indonesia. The experimental fish were freshwater eel (*Anguilla marmorata*) elvers from the Palu River with weight 0.50-2 g and length 6-7.7 cm. The main consumable materials were fresh squid (*Loligo* sp), standard high protein formulated feed in moist (paste/dough form), and water.

The main items of equipment used were aquaria (experimental units) equipped with a recirculation and filtering system to maintain water quality. The filtration tank contained layers of filter foam, zeolite and activated charcoal. Other equipment used included digital scales (precision 0.01g), callipers (precision 0.01 mm), a digital camera with video capacity, a stop-watch and a hand counter, a digital pH meter and thermometer, a DO meter, titration equipment (ammonia), and general maintenance equipment such as feeding trays and scoop nets.

Each unit contained 15 L of clean water, aerated for 24 hours before the elvers were stocked at a density of 18 elvers/container. Prior to stocking, the elvers were acclimatized for 20-30 minutes to adjust temperature in the transport medium and experimental unit. Twice-daily water changes (10% by volume) were carried out in the morning and evening. The light was dimmed in the laboratory in order to stimulate the primarily nocturnal eels to feed.

2.2. Squid meal preparation and feeding pattern

Fresh squid (*Loligo* sp.), still complete with ink sacks, were cleaned, de-boned and steamed for 20 minutes. The flesh was then finely chopped, drained and placed on aluminium foil in an oven at 60-70°C for 4 hours or until dry. The dried squid were reduced to fine powder (meal) in a blender, sieved to remove any coarse fragments, and stored in an airtight container until used.

Squid meal was added to the feed at the appropriate dose for each experimental treatment as needed. Feed was provided at a daily rate of 5% of the total body weight of the elvers [29]. The feed was divided between two daily feeding times with 40% provided in the morning at 07.00 and 60% in the afternoon at 16.00 local time (GMT + 8). After each weekly elver measurement session, an adjustment was made to the amount of feed provided. A proximate analysis of each treatment was carried out to measure the effect of squid meal attractant enrichment on the protein content of the feed.

2.3. Experimental treatments, design and parameters

A completely randomized design was used with 4 treatments and 5 replicates per treatment. The treatments were: A = control (no squid meal); B = 1% squid meal; C = 3% squid meal; and D = 5% squid meal. Water quality parameters pH and temperature were measured daily in the morning, afternoon and evening. Dissolved oxygen and ammonia concentrations were measured once a week.

Feed palatability was evaluated based on the response rate (FP, expressed in s/cm) at which the elvers found and began to consume the feed (morning and afternoon). The feeding process was recording on video using a digital camera for ± 15 minutes after feed was placed on the feeding tray in each aquarium. Based on previous studies [36,37], observations of elver response to the modified test feed comprised: (1) the time (T_f) taken for the elvers to find and start consuming the test feed, measured in seconds; (2) the distance (D_f) from which elvers approached the feed, measured in cm. The mean values of T_t and D_t were calculated for each unit and treatment, giving the response rate FP as T_t divided by D_t. Lower values of FP indicate greater attraction and therefore higher palatability. The proportion of elvers in each unit responding to the feed, responding to and eating the feed, and not responding to the feed was recorded and expressed as a percentage (%).

The weight and length of the elvers were measured every 7 days and mortality was noted. Elver growth was evaluated based on mean net gain in weight (g) and length (cm) over the 60 day experimental period. The survival rate (SR) was also calculated.

Feed efficiency (FE) and feed conversion ratio (FCR) were calculated for each experimental unit based on [37] using the following equations:

$$FE = ((Wt+D)-W0) \cdot Wf^{-1} \cdot 100\%$$

$$FCR = Wf \cdot ((Wt+D)-W0)^{-1}$$

where

Wt = final weight of all live elvers (g)
D = weight of all elvers which died during the study period (g)
Wo = initial weight of all elvers (g)
Wf = weight of all feed consumed (g)

2.4. *Data analysis*

Data were tabulated and analysed in Microsoft Excel 2010. The null hypothesis (H_0) was that enriching feed with squid meal as an attractant at different dosages has no effect on feed palatability and the growth of freshwater eel (Anguilla marmorata) elvers. The alternative hypothesis (H_1) was that enriching feed with squid meal as an attractant at different dosages affects feed palatability and has a significant effect on the growth of freshwater eel (*Anguilla marmorata*) elvers. To test this hypothesis, between treatment data on elver growth and survival rate were subjected to analysis of variance (ANOVA) followed by post-hoc pairwise Tukey honestly significant difference (HSD) tests, both at the 95% confidence level ($\alpha = 0.05$). Furthermore, it was hypothesized that there is an optimal dose which maximises feed palatability and growth of freshwater eel (*Anguilla marmorata*) elvers.

3. Results and Discussion

3.1. *Water quality*

The water quality parameter values remained within suitable limits for tropical elver husbandry (Table 1). This demonstrates that the installation and protocols used were able to maintain water quality under all treatments. The filter foam played a role in filtering suspended solids from water, such as uneaten feed residue and faeces, while the zeolite and active carbon helped to absorb toxic substances such as ammonia and nitrite. This is similar to the results from rearing the Indonesian shortfin eel *A, bicolor bicolor* in a recirculation system with physical, chemical and biological filtration systems [38].

Table 1. Mean values of water quality parameters in the experimental units

Parameter	Treatment (% squid meal)				Optimum Range [reference]
	A (0)	B (1%)	C (3%)	D (5%)	
Temperature (°C)	27.6	27.7	27.72	27.56	21-31 [39]
pH	7.7	7.8	7.62	7.58	6.0-8.0 [30,40]
DO (mg/L)	7.72	7.7	7.6	7.7	>3.0 [41]
Ammonia (mg/L)	0.15	0.15	0.15	0.15	>0.1 [42]

3.2. *Feed palatability and protein content*

The proximate analysis showed that feed enrichment with squid meal had an effect on the nutritional composition of the feed, with the protein content increasing in line with the amount of squid meal added. In terms of feeding behaviour, the elvers responded differently to each of the treatments. Feed palatability parameters and protein content for the four experimental treatments are shown in Table 2.

Table 2. Mean values of *A. marmorata* elver feed palatability parameters and nutritional content for *squid* meal attractant treatments

Ppalatability/nutritional content parameter (unit)	A (0% squid meal)		B (1 % squid meal)		C (3 % squid meal)		D (5 % squid meal)	
	a.m.	p.m.	a.m.	p.m.	a.m.	p.m.	a.m.	p.m.
FP (s/cm)	0.24	0.14	0.17	0.10	0.12	0.06	0.24	0.18
No response (%)	0		0		0		0	
Approach only (%)	23.19		1.81		0		8.69	
Approach and feed (%)	77.39		98.88		100		92.19	
Protein (%)	34.32		35.81		38.11		41.05	

The sense of smell in fish, including eels, has several functions including migration, avoiding predators, reproduction, and seeking and locating food [3]. Fish appetite is controlled by the hypothalamus which is stimulated by the level of gut fullness as well as various metabolic factors and feeding behaviour is initiated by the presence of appetite, and the level of response to stimuli to eat depends on the level of satiety as well as the attractiveness of the food [43–45]. Responses to feeding-associated stimuli can also be caused by habit, or actions that are carried out repeatedly, while the fish's response to eating stimuli is divided into five stages: the fish are aware of and detect the presence of food or stimuli; they determine the direction of movement; move from the initial position to the desired destination; begin to recognize and make decisions whether to accept or reject stimuli; and finally eat or do not eat the food. [37].

The data in Table 2 show that under all treatments, the elvers showed awareness of and approached the feed faster in the afternoon than in the morning. Treatment C (3% squid meal) had the highest response in terms of feeding behaviour (100%) as well as the fastest rate of response and approach (0.06 s/cm, in the afternoon), while the control (treatment A) had the lowest response rate. The elvers in treatment D (5% squid meal) took the longest time to find and consume the test feed, indicating that this level of enrichment may be excessive and could have had a negative impact on the smell and taste of the feed. Other studies have found that, although an appropriate dosage of attractant can increase the speed of the response and feed consumption [35,37], an excess of attractant can make the aroma of the feed less pleasant and reduce the attraction of the fish [29,31].

Substances providing the aroma in fish feed attractants include proteins such as glycine and alanine which can stimulate the sense of smell of fish and shrimp [3,36]. In particular, glycine in the attractant will be captured by the eel's olfactory and gustatory receptors and the signals sent interpreted as food aroma. These amino acids are present in squid meal, which has been found to contain 2.54% glycine and 2.66%, making it a good attractant [35]. The fast response of *A. marmorata* elvers to treatment C indicates that 3% attractant gave the best aroma and taste or in other words the optimal palatability for this species.

3.3. *Weight gain and increase in length*

The net weight gain (Figure 1A) and increase in length (Figure 1B) of *A. marmorata* elvers under the four experimental treatments shows a significant between treatment difference in growth (ANOVA, $P < 0.05$). Treatments A and D did not differ significantly from each other (Tukey HSD test, $P > 0.05$) while treatments B and C each differed significantly from all other treatments (Tukey HSD test, $P < 0.05$). Growth in weight and length was highest in treatment C followed by treatment B and lowest in treatments A and D.

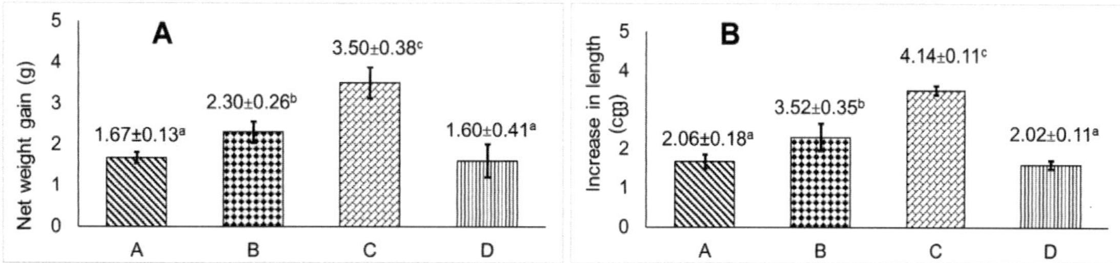

Figure 1. Net growth of *A. marmorata* elvers: a. weight gain and b. increase in length. Squid meal attractant treatments: A = 0%; B = 1 %l; C = 3 %; D = 5%.

The higher net weight gain in treatment C (3% squid meal attractant) was likely due to a high level of feed consumption and effective utilisation of the relatively high protein content (38.11%). Higher increases in weight related high levels of feed consumption and feed protein content utilization have been reported in *A. marmorata* under various culture conditions [46]. Conversely, the lower weight gain in other treatments were likely due to lower levels of consumption, most likely caused by the difference in the aroma of the feed which stimulates the fish to eat, and may also have been related to utilization of protein in feed. Growth in fish is caused by an excess of energy sourced from the feed consumed over the energy needed for other processes [29,45]. Adding the optimal dose of attractant to feed can increase the intake as well as the speed of nutrient absorption, providing additional nutrients for protein and energy metabolism, while faster and more voracious feeding responses can reduce the amount of waste as well as the time the feed is in contact with the water, thereby reducing the amount of nutrients becoming dissolved or suspended in the water.

Despite the highest protein content (41.09%), treatment D (5% attractant) did not significantly increase growth relative to the control, with lower indicators of palatability than treatments B and C. This could be related to a possibly lower intake as well as to energy used for excretion. Research on the shortfin eel *A. bicolor* found that excess protein content in the feed can be utilized for processes related to amino acid transformation and the excretion of nitrogenous compounds, as the processes of excretion and catabolism require a lot of energy [47]. These processes are influenced by various factors including the protein content, balance of amino acids, and the energy ratio of the feed, and can result in a decrease in growth with increasing protein content above a certain threshold. In addition to the palatability factor, this phenomenon could help to explain why growth performance of the elvers increased up to 3% squid meal (treatment C) and then declined at 5% attractant (treatment D).

3.4. Survival rate

In anguillid eel culture, survival rates of 70-90% are considered high [3]. The mortality data recorded during this experiment show that the survival rates of *A. marmorata* elvers were quite high (over 85%) under all treatments. The between treatment difference were not significant (ANOVA, $P > 0.05$). This indicates that the survival rate was not directly affected by feed, and that the culture conditions provided were adequate.

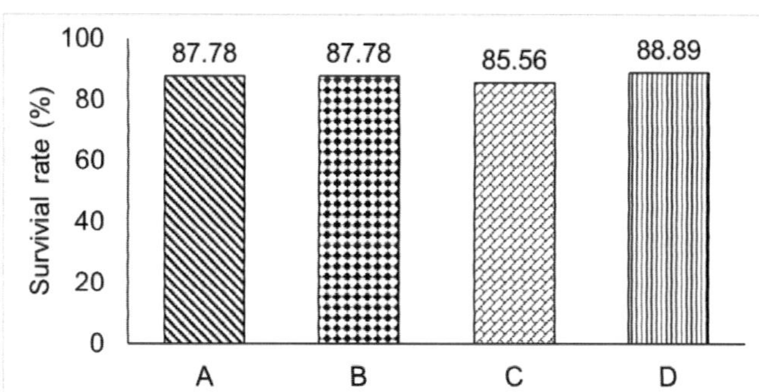

Figure 2. Survival rate of *A. marmorata* elvers.
Squid meal attractant treatments:
A = 0%; B = 1%;
C = 3%; D = 5%.

3.5. *Feed efficiency (FE) and feed conversion ratio (FCR)*

The feed conversion value (FCR) is an illustration of the level of feed utilization used by fish; in general, the higher the FCR value, the lower the quality of the feed and vice versa. FCR tends to have a negative correlation with growth rate, although this is not always the case, and the correlation can be positive or insignificant [48]. Meanwhile, feed efficiency (FE) is widely viewed as a major key to reducing production costs and to achieving sustainability in aquaculture [49]. Feed efficiency (FE) differed significantly (ANOVA, $P < 0.05$) between all treatments (Figure 3a); conversely, there was no significant difference (ANOVA, $P > 0.05$) in feed conversion ratio (FCR) between any of the treatments (Figure 3b).

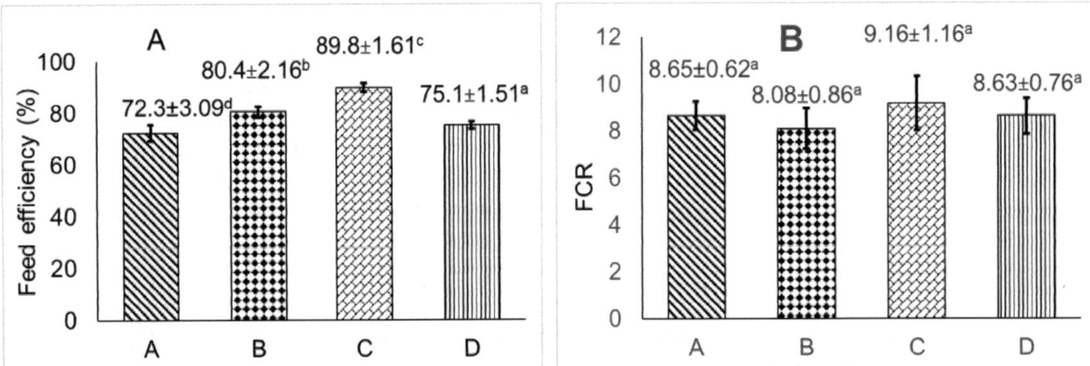

Figure 3. Feed efficiency (a) and feed conversion ratio (FCR) (b) of *A. marmorata* elvers. Squid meal attractant treatments: A = 0%; B = 1 %l; C = 3 %; D = 5%.

Feed efficiency was highest under treatment C (addition of 3% squid flour attractant) followed by treatments B (addition of 1% squid flour attractant), D (addition of 5% squid flour attractant) and finally A (control with the lowest feed efficiency). These data confirm the net growth results and further reinforce the conclusion that treatment C provided the best performance. The low FE value in treatment A (without attractant) may indicate a lack of nutrients in terms of energy and/or protein for growth, with energy possibly largely used for respiration and protein breakdown. The poor performance of treatment D (addition of 5% squid flour attractant) may be related to an increase in fat content, which can reduce the fish's body's ability to break down protein into energy.

4. Conclusion

Enrichment of freshwater eel (*Anguilla marmorata*) elver feed with squid (*Loligo* sp.) meal at doses of 1%, 3% and 5% did affect the feed palatability. The attractant had a significant effect on elver growth but not on survival rate, and on feed efficiency but not on feed conversion ratio. The 3% squid meal attractant gave the highest palatability, growth rates (net weight and length) and feed efficiency. The addition of 5% squid meal gave no improvement over the control and may have a negative effect on palatability. Survival rate was high under all treatments and, together with water quality data, indicate the experimental conditions were adequate. Based on the results of this study, the optimal dosage of squid meal as an attractant in *A. marmorata* elver feed is 3%.

Acknowledgments

The authors thank all who assisted in any way with the research and preparation of this paper. In particular, the authors are grateful to Abigail Mary Moore for assistance with English language proof reading and editing.

References

[1] Aoyama J 2009 Life History and Evolution of Migration in Catadromous Eels (Genus *Anguilla) Aqua-BioScience Monogr.* **2** 1–42

[2] Arai T 2020 Ecology and evolution of migration in the freshwater eels of the genus *Anguilla* Schrank, 1798 *Heliyon* **6** e05176
[3] Tesch E W 2003 *The Eel* ed J E Thorpe (Oxford: Blackwell Science Ltd)
[4] Arai T 2014 How have spawning ground investigations of the Japanese eel *Anguilla japonica* contributed to the stock enhancement? *Rev. Fish Biol. Fish.* **24** 75–88
[5] Tsukamoto K, Chow S, Otake T, Kurogi H, Mochioka N, Miller M J, Aoyama J, Kimura S, Watanabe S, Yoshinaga T, Shinoda A, Kuroki M, Oya M, Watanabe T, Hata K, Ijiri S, Kazeto Y, Nomura K and Tanaka H 2011 Oceanic spawning ecology of freshwater eels in the western North Pacific *Nat. Commun.* **2** 179
[6] Zan N D, Sarbini A, Taha H, Tan I V, Azri A, Kahar R, Metali F, Ahmad N and Arai T 2020 Occurrence and ecological implication of a tropical anguillid eel, *Anguilla marmorata*, in Brunei Darussalam, Borneo Island *Zoologia* **37** 1–7
[7] Muthmainnah D, Honda S, Suryati N K and Prisantoso B I 2016 Understanding the Current Status of Anguillid Eel Fisheries in Southeast Asia *Fish People* **14** 19–25
[8] Arai T 2015 Save Our Eels: Protection or Extinction? *Fish People* **13** 12–8
[9] Nafsiyah I, Nurilmala M and Abdullah A 2018 Nutrient Composition of Eel *Anguilla bicolor bicolor* and *Anguilla marmorata J. Pengolah. Has. Perikan. Indones.* **21** 504–12
[10] Monticini P 2014 *Eel (Anguilla spp.): Production and Trade according to Washington Convention Legislation (GLOBEFISH Research Programme 114)* vol 114 (Rome: Food and Agriculture Organization of the United Nations)
[11] Shiraishi H and Crook V 2015 *Eel market dynamics: An analysis of Anguilla production, trade and consumption in East Asia* (Tokyo: TRAFFIC Japan Office)
[12] Pike C, Crook V and Gollock M 2020 European eel *Anguilla anguilla IUCN Red List Threat. Species 2020* e.T60344A152845178
[13] Silfvergrip A M C 2009 *CITES Identification Guide to the Freshwater eels (Anguillidae) with Focus on the European eel Anguilla anguilla* (Bromma: Swedish Environmental Protection Agency)
[14] Pike C, Kaifu K, Crook V, Jacoby D and Gollock M 2020 Japanese Eel *Anguilla japonica* (amended version of 2020 assessment) *IUCN Red List Threat. Species 2020* e.T166184A176493270
[15] Jacoby D, Casselman J, DeLucia M and Gollock M 2017 American eel *Anguilla rostrata* (amended version of 2014 assessment) *IUCN Red List Threat. Species 2017* e.T191108A121739077
[16] Arai T 2016 Current status and concerns on tropical anguillid eel stocks *Sci. Bruneiana* **15**
[17] Crook V and Nakamura M 2013 Assessing supply chain and market impacts of a CITES listing on *Anguilla* species *TRAFFIC Bull.* **25** 24–30
[18] Nijman V 2015 CITES-listings, EU eel trade bans and the increase of export of tropical eels out of Indonesia *Mar. Policy* **58** 36–41
[19] Gollock M, Shiraishi H, Carrizo S, Crook V and Levy E 2018 Status of non-CITES listed anguillid eels. CITES AC30 Doc. 18.1 Annex 2 176 pp
[20] Cuvin-Aralar M L, Aya F, Romana-Eguia M R and Logronio D J 2019 *Nursery Culture of Tropical Anguillid Eels in the Philippines: Aquaculture Extension Manual No. 65* (Tigbauan: Southeast Asian Fisheries Development Center)
[21] Righton D, Piper A, Aarestrup K, Amilhat E, Belpaire C, Casselman J, Castonguay M, Díaz E, Dörner H, Faliex E, Feunteun E, Fukuda N, Hanel R, Hanzen C, Jellyman D, Kaifu K, McCarthy K, Miller M J, Pratt T, Sasal P, Schabetsberger R, Shiraishi H, Simon G, Sjöberg N, Steele K, Tsukamoto K, Walker A, Westerberg H, Yokouchi K and Gollock M 2021 Important questions to progress science and sustainable management of anguillid eels *Fish Fish.* **22** 762–88
[22] Richards J L, Sheng V, Yi C W, Ying C L, Ting N S, Sadovy Y and Baker D 2020 Prevalence of critically endangered European eel (*Anguilla anguilla*) in Hong Kong supermarkets *Sci.*

Adv. **6** 2–7
[23] Muthmainnah D, Suryati N K, Pamungkas Y P and Mulyani Y S 2020 Western Region of Indonesia: the nucleus of anguillid eel fisheries and trade *Fish People* **18** 37–42
[24] Serdiati N, Ndobe S, Moore A and Wahyudi D 2013 Species composition of glass eels (*Anguilla* spp.) recruiting to the Palu River, Central Sulawesi *J. Iktiologi Indones.* **13** 109–23
[25] Honda S, Muthmainnah D, Suryati N K, Oktaviani D, Siriraksophon S, Amornpiyakrit T and Prisantoso B I 2016 Current status and problems of the catch statistics on anguillid eel fishery in Indonesia *Mar. Res. Indones.* **41** 1–14
[26] Ahmad A and Budi W 2021 *5 Rahasia Jitu Budidaya Ikan Sidat* (Bogor: Sidat Labas)
[27] Haryono, Krismono, Fahmi M R, Riyanto M, Lubis S B, Agung M F, Darmasyah S and Setiono 2020 *Panduan Pendataan dan Monitoring Populasi Sidat* (Jakarta: Ministry of Marine Affairs and Fisheries R I)
[28] Affandi R, Budiardi T, Wahju R I and Taurusman A A 2013 Pemeliharaan Ikan Sidat Dengan Sistem Air Bersirkulasi [Eel Rearing in Water Recirculation System] *J. Ilmu Pertan. Indones.* **18** 55–60
[29] Yudiarto S, Arief M and Agustono 2019 Pengaruh Penambahan Atraktan Yang Berbeda Dalam Pakan Pasta Terhadap Retensi Protein, Lemak Dan Energi Benih Ikan Sidat (*Anguilla bicolor*) Stadia Elver [Effects of addition different attractants in pasta feed against retention of protein, fat and energy eel fish seed (*Anguilla bicolor*) stadia elver *J. Ilm. Perikan. dan Kelaut.* **4** 135
[30] Soeprijanto A, Guntur G and Fakhri M 2018 Application of Probiotic and Fermented Feed in the Nursery of *Anguilla bicolor J. Perikan. Univ. Gadjah Mada* **20** 19
[31] Perdana A A, Suminto and Chilmawati D 2016 Performa efisiensi pakan pertumbuhan dan kualitas nutrisi elver sidat (*Anguilla bicolor*) melalui pengkayaan pakan buatan dengan minyak ikan - Feed efficiency performance, growth and quality nutrition of elver eels (*Anguilla bicolor*) enrichment artificial feed with fish oil *J. Aquac. Manag. Technol.* **5** 26–34
[32] Djajasewaka H and Tahapari E 1999 Kebutuhan protein optimal pada pakan benih ikan sidat (*Anguilla bicolor*) *J. Penelit. Perikan. Indones.* **5** 64–8
[33] Khasani I 2013 Atraktan pada pakan ikan: jenis, fungsi, dan respons ikan *Media Akuakultur* **8** 127
[34] Pratama R H, Tarsim T and Yudha I G 2019 Efektifitas penambahan asam amino pada pakan untuk pertumbuhan ikan sidat, *Anguilla bicolor* (McCelland, 1844) *e-Jurnal Rekayasa dan Teknol. Budid. Perair.* **7** 835
[35] Kurniawan A 2013 *Palatability and Growth of eel Anguilla marmorata with Attractant Squid Meal and Shrimp Meal - Palatabilitas dan Pertumbuhan Sidat Anguilla marmorata dengan Pemberian Atraktan Tepung Cumi dan Tepung Udang Rebon* (Institut Pertanian Bogor)
[36] Nunes A J P, Sá M V C, Andriola-Neto F F and Lemos D 2006 Behavioral response to selected feed attractants and stimulants in Pacific white shrimp, *Litopenaeus vannamei Aquaculture* **260** 244–54
[37] Arditya B P, Subandiyono S and Samidjan I 2019 The effects of various sources of dietary attractant on the feed response, total feed consumption, and growth of snakehead fingerlings (*Channa striata*) *Sains Akuakultur Trop.* **3** 70–81
[38] Scabra A R, Budiardi T and Djokosetiyanto D 2016 Production performance of *Anguilla bicolor bicolor* with the addition of calcium carbonate (CaCO3) into culture media *J. Akuakultur Indones.* **15** 1–7
[39] Usui A 1974 *Eel culture.* (London, UK: Fishing News Books)
[40] Ritonga T P T 2014 *Respons benih ikan sidat (Anguilla bicolor bicolor) terhadap derajat keasaman (pH)* (Institut Pertanian Bogor)
[41] Herianti I 2005 Rekayasa lingkungan untuk memacu perkembangan ovarium ikan sidat (*Anguilla bicolor*) *Oseanologi dan Limnol.* **37** 25–41
[42] MMAF 2011 *Panduan budidaya ikan sidat Anguilla spp.* (Jakarta: Ministry of Marine Affairs

and Fisheries R I)

[43] Volkoff H, Canosa L F, Unniappan S, Cerdá-Reverter J M, Bernier N J, Kelly S P and Peter R E 2005 Neuropeptides and the control of food intake in fish *Gen. Comp. Endocrinol.* **142** 3–19

[44] Soengas J L, Cerdá-Reverter J M and Delgado M J 2018 Central regulation of food intake in fish: an evolutionary perspective *J. Mol. Endocrinol.* **60** R171–99

[45] Canosa L F and Bertucci J I 2020 Nutrient regulation of somatic growth in teleost fish. The interaction between somatic growth, feeding and metabolism *Mol. Cell. Endocrinol.* **518** 111029

[46] Harianto E, Supriyono E, Budiardi T, Affandi R and Hadiroseyani Y 2020 Production performance and physiology response of *Anguilla bicolor bicolor* rearing with a wet, damp and dry system *J. Iktiologi Indones.* **20** 117

[47] Nawir F, Utomo N B P and Budiardi T 2015 The growth of eel fed with different protein level and protein-energy ratio *J. Akuakultur Indones.* **14** 128

[48] Besson M, Komen H, Rose G and Vandeputte M 2020 The genetic correlation between feed conversion ratio and growth rate affects the design of a breeding program for more sustainable fish production *Genet. Sel. Evol.* **52** 5

[49] Verdal H, Komen H, Quillet E, Chatain B, Allal F, Benzie J A H and Vandeputte M 2018 Improving feed efficiency in fish using selective breeding: a review *Rev. Aquac.* **10** 833–51

Characterization and evaluation of sweet basil (Ocimum basilicum L.) oil as larvicidal and repellent against the yellow fever mosquito

R P Ludia[1], S Shahabuddin[2,3] dan S H Gatot[2,3]

[1] Post Graduate Student, Tadulako University, Palu, 94118, Indonesia
[2] Agrotecnology Study Program, Agriculture Faculty, Tadulako University, Palu, 94118 Indonesia
[3] Post Graduate Program, Tadulako University, Palu, 94118, Indonesia

shahabsaleh@gmail.com

Abstract. Dengue Hemorrhagic Fever (DHF) is one of the leading health problems in Indonesia and other countries. However, a study on sweet basil oil and its potency to control Aedes aegypti, or Yellow fever mosquito (YFM), as the vector of DHF, is rarely investigated. This research aims to study the chemical characteristics of sweet basil oil and evaluate its potency as insecticides against Aedes aegypti. A spectrophotometer gas chromatography to identify its compounds followed by the standardized toxicity and repellency test methods were conducted. The results showed that the main compounds of sweet basil oil are linalool (31,78 %), thymol (11,60%), and cineol (0,99%). The sweet basil oil showed high toxicity for the larvae of Aedes aegypti with LC50 and LC90 were 72,08 ppm and 121,47 ppm. The protection effect of sweet basil oil increases following the oil concentration. However, at 30 % concentration, this essential oil could protect from Aedes aegypti up to six hours with repellency levels of 81,3 %. Therefore, this study supports the effective and environmentally friendly method of managing the YFM using sweet basil oil.

Keywords: Sweet basil oil, botanical insecticides, Aedes aegypti

1. Introduction

Dengue Haemorrhagic Fever (DHF) is a disease caused by the dengue virus and transmitted by the Aedes aegypti and A. albopictus mosquitoes as the primary vectors. Globally, this disease infects an average of 390 million people each year. Meanwhile, Indonesia is the 2nd country with the most prominent DHF cases among 30 countries in endemic areas [1].

One of the efforts to reduce the transmission of the dengue hemorrhagic fever virus is to control the Aedes mosquito as a vector of the disease. However, the control techniques need to consider the effectiveness and environmental safety. The study shows that chemical insecticides hurt beneficial insects, especially bees and aquatic organisms [2]. Its efficacy has also decreased due to the increased resistance of A. aegypti to several insecticides such as DDT, permethrin, deltamethrin, and temephos groups [3,4]. Therefore, more environmentally friendly control techniques are needed, such as using natural ingredients as botanical insecticides.

Indonesia has many plants species that potential to be developed as botanical insecticides. At least 25 species have been identified that can be used as larvicides to control A. aegypti [5]. However, many plants might be used to control Aedes aegypti. One of them is the sweet basil (Ocimum basilicum L.), which has insecticidal properties against several types of insects [6.7.8]. Sweet basil is an aromatic herb used extensively to add a fragrant aroma and flavor to food. The leaves might be used fresh or dried for use as a spice. It is essential oils of new leaves and flowers that can be used as aroma additives in cosmetics, pharmaceuticals, and food [9].

The sweet basil grows wild on the roadside and community yards in Central Sulawesi. It has not been studied for its chemical characteristics and potential as a botanical insecticide to control A. aegypti. Thus, this study will characterize and examine the possibility of sweet basil oil as larvicidal and repellent against A. aegypti.

2. Materials and Methods

2.1. Plant collection and extraction

This research was conducted from February to June 2020. The sweet basil samples were collected from Labuan Panimba Village, Donggala regency, Central Sulawesi, Indonesia. O. basilicum grows wild on roadsides and yards in several sites at the Palu valley. This plant is an erect herb, very fragrant, with a height of 0.6-1.6 m, rectangular stems, and brown-green (Figure 1). Plant collection is carried out on days when the plant conditions are still fresh. The essential oil was extracted from sweet basil leaves using steam distillation. This method extracts oil by converting volatile liquids and condensing vapors into liquids [10].

Figure 1. Sweet basil (*O.basilicum*) that grows at several sites in Palu Valley

2.2. Analysis of the chemical compound

According to the Laboratory of the Research Institute for Spices and Medicinal Plants (Balittro), West Java, Indonesia, analysis of the chemical compounds of extracted sweet basil oil was carried out using a gas chromatography spectrophotometer the procedure of Sastrohamidjojo [11].

2.3. Larvacidal and repellency test

A. aegypti larvae were obtained from the Institute for Health Research and Development (BP4B2) Donggala, Central Sulawesi. The larvacidal activity of sweet basil oil against third instar larvae was tested at four different extract concentrations, namely: 180 ppm, 150 ppm, 20 ppm, 80 ppm, and 60 ppm. The negative and positive control groups used the aqua dest (200 ml) and temephos (1 ppm). Each treatment has three replicates. A total of 25 larvae were tested at each concentration by putting the larvae into a 200 ml of extract solution. Observation of mosquito larvae mortality was carried out at 24 hours. The lethal concentrations (LC50 and LC90) were calculated by probit analysis [12].

For the repellency test, five concentrations of sweet basil oil dissolved in virgin coconut oil (VCO) were tested: 10%, 15%, 20%, and 25%. The repellency test followed the WHO procedure [13]. The test was carried out by applying sweet basil oil on the volunteer's arm. According to the treatment concentration, the right forearm was smeared with VCO as a control, and the left forearm was smeared with essential oil. Both arms were exposed to a mosquito cage (93 cm x 48cm x 48 cm) containing 25 female mosquitoes (3-5 day-old). The number of mosquitoes perched and bites on both arms were counted for 5 minutes every hour until 6 hours after application. Then calculate the repellent activity by using this formula: Percentage of protection (100%) = [(Control - Treated) / Control] ×100 [14]. The repellency test was repeated three times for each sweet basil oil concentration to the same respondents.

3. Results and Discussion

3.1. Chemical composition

The results of gas chromatography (GC) analysis showed that the main volatile compounds of sweet basil oil were; linalool (31.78%), thymol (11.60%), and cineole (0.99%). This is slightly different from the research results by Rodríguez-González et al. [8]. They reported that the main volatile compounds in sweet basil oil were Citral (0.70), Citronellol (11.50), Citronellal (34.00), Geraniol (22.00), Limonene (3.50 %). Meanwhile, Silva [15] recorded 19 volatile compounds of sweet basil oil dominated by linalool (79.34 %), 1,8-cineole (6.02%), and geraniol (9.86%). Linalool is the primary compound of the sweet basil. This secondary metabolite is a terpenoid that exhibits toxicity to various pests by interrupting the nerve system of insects [15].

3.2. Larvicidal activity and toxicity

The 24 h larvicidal effect of O. basilicum oil against A. aegypti is presented in Table 1. The essential oil of this plant exhibited significant larvicidal activity. The mortality of *A. aegypti* larvae increased following the increase of the crucial concentration. The highest mortality (100%) was obtained from a concentration of 180 ppm and temephos. However, larval mortality at concentrations of 120 and 150 ppm of sweet basil oil was not significantly different from both treatments. This indicates the high potential of this essential oil as a natural larvacide against *Aedes aegypti.*

Table 1. Mortality of *A. aegypti* larvae 24 hours after exposure to sweet basil oil at different concentration

Sweet basil oil concentration	Larva mortality (%)
180 ppm	100 ± 0.0a
150 ppm	98.6 ± 2.3a
120 ppm	97.3 ± 2.3a
80 ppm	56.0 ± 16.0b
60 ppm	32 ± 6.9c
Aquadest (200 ml)	0 ± 0d
Temephos (1 ppm)	100 ± 0a

The high toxicity of sweet basil oil against A. aegypti is also shown by LC50 and LC90 values of 72.07 ppm and 121.47 ppm, respectively (Table 2). This toxicity value is higher than that reported by Govindarajan et al. [6], which were 11.97 ppm and 21.17 ppm for A. albopictus, respectively. Besides exhibiting high toxicity to A. aegypti, sweet basil oil also had a significant toxic effect against late third-stage larvae of Culex tritaeniorhynchus and Anopheles subpictus [6].

Table 2. Toxicity of sweet basil oil on *A.aegypti*

Toxicity level	Toxicity value	LCL	UCL
LC_{50} (ppm)	72,07	63,41	79,75
LC_{90} (ppm)	121,47	105,41	157,66

LCL = Lower confidence level, UCL = Upper Confidence level

3.3. Repellent activity

The results showed that the protective effect of the sweet basil oils in the first hour for all treatments was very high (98-100%). This protective power decreased with the time of testing. However, until the 6th hour of testing, the repellency ability of sweet basil oil was still relatively high, namely 67.6% and 81.3% at concentrations of 20% and 25% (Figure 2).

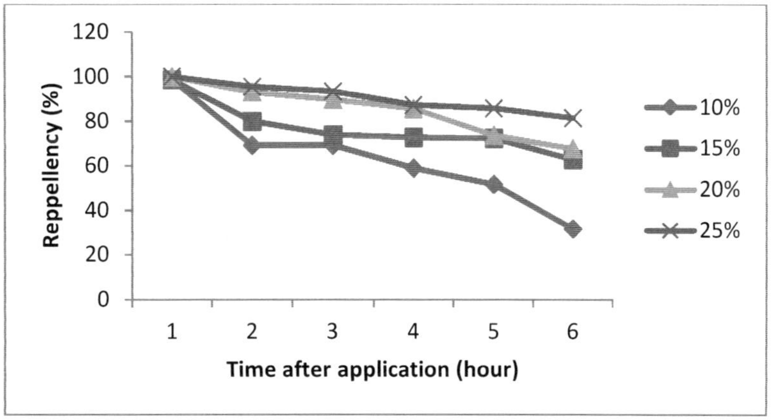

Figure 2. Protection ability of sweet basil oil against imago of *A. aegypti* across application time

The repellent activity of sweet basil oil against A. aegypti obtained in this study was higher than previously reported. The most increased repellency of O. basilicum after one-hour application at a concentration of 20% was 79.7%. At the same time, a previous study reported that the repellency of O. basilicum one hour after application only reaches 22.9% at a concentration of 20% [14]. Sritabutra and Soonwera showed that herbal essential oils dissolved in different oils have another repellent activity [16]. Differences in the chemical content of essential oils in the same plant species can be caused by geographical, genetic, climatic, and seasonal factors. Essential oil from clove oil in olive oil and coconut oil gave the longest of 76.50 min and 96.00 min respectively against A. aegypti [17].

An anti-mosquito lotion can be effective if its protective power is at least 90% and can last for 6 hours [18]. However, the repellency time against mosquitoes should be > 2 h [19] in another country. This study revealed that basil essential oil could be a repellent against A. aegypti mosquitoes. The essential oil of O. basilicum, namely eugenol, thymol, and cineol, is a compound that functions as an insect repellent because of its favors repelling insects, especially mosquitoes [16]. Linalool contains sweet basil oil, potentially controlling insect pests by causing a repellent effect, acute toxicity, non-preference, and a knockdown effect [15].

4. Conclusion

It can be concluded that sweet basil oil mainly consists of linalool (31,78 %), thymol (11,60%), and cineol (0,99%). That volatile oil has high larvacidal activity with LC50 and LC90 were 72,08 ppm and 121,47 ppm against A. aegypti larvae. The repellent activity of sweet basil oil increases with the oil concentration. However, at 25 % concentration, this sweet basil oil could protect from Aedes aegypti bites for up to six hours with repellency levels up to 81,3 %.

Acknowledgments
The authors would like to thank the Institute for Health Research and Development (BP4B2) Donggala, Central Sulawesi, which has provided facilities to extract sweet basil oil and repellency tests.

References

[1] Kementrian Kesehatan RI 2021 *Profil Kesehatan Indonesia Tahun 2020* Jakarta; Depkes RI https://wwwkemkesgoid/downloads/resources/download/pusdatin/profil-kesehatan-indonesia/Profil-Kesehatan-Indonesia-Tahun-2020pdf
[2] Muzinic V and Zeljezic D 2018 Non-target toxicity of novel insecticides Arh Hig Rada Toksikol 69, 86-102
[3] Vontas J, Kioulos E, Pavlidi N, Morou E, Della Torre A, Ranson H 2012 Insecticide resistance in the major dengue vectors Aedes albopictus and Aedes aegypti Pestic Biochem Physiol 104, 126–31
[4] Sunaryo, Ikawati B, Rahmawati, Widiastuti D 2014 Status resistensi vektor demam berdarah dengue Aedes aegypti terhadap malathion J Ekol Kesehat 2,146-52
[5] Astriani Y, Widawati M 2016 Potensi tanaman di indonesia sebagai larvasida alami untuk Aedes aegypti. Spirakel 8 (2),37-46
[6] Govindarajan M, Sivakumar R, Rajeswary M, Yogalakshmi K 2013 Chemical composition and larvicidal activity of essential oil from Ocimum basilicum (L) against Culex tritaeniorhynchus, Aedes albopictus and Anopheles subpictus (Diptera: Culicidae) Experimental Parasitology 134 7–11
[7] Islam T Iqbal J, Abdullah K, Khan EA 2017 Evaluation of some plant extracts against maize weevil, Sitophilus zea mais (Coleoptera: Curculionidae) under laboratory conditions Pak J Agr Sci 54, 737–41
[8] Javanmardi J, Khalighi A, Kashi A, Bais, HP, Vivanco JM 2002 Chemical characterization of basil (Ocimum basilicum L) found in local accessions and used in traditional medicines in Iran. J. Agric. Food Chem. 50, 5878–83
[9] Rodríguez-González A et al. 2019 Insecticidal Properties of Ocimum basilicum and Cymbopogon winterianus against Acanthoscelides obtectus, Insect Pest of the Common Bean (Phaseolus vulgaris, L.). Insects 10, 151; doi:10.3390/insects10050151
[10] Ranjitha, J, Vijiyalakshmi 2014 Facile methods for the extraction of essential oil from the plant species—A review Int J Pharm Sci Res 5(4), 1107–15
[11] Sastrohamidjojo H 2004 Kimia Minyak Atsiri 1 ed Diedit oleh Mudasir Yogyakarta Gadjah Mada University Press
[12] Finney DJ 1971 Probit Analysis Cambridge University Press, London, UK
[13] Word Health Organization (WHO) 2009 Guidelines for efficacy testing of mosquito repellents for human skin, Who/Htm/Ntd/Whopes/20094, bll 1–6
[14] Kardinan A 2007 Potensi Selasih Sebagai Repellent Terhadap Nyamuk Aedes aegypti jurnal Litri, 13(2), 39–43
[15] Silva SM, Cunha JPAR, Carvalho SMD et al 2017 Ocimum basilicum essential oil combined with deltamethrin to improve the management of Spodoptera frugiperda Ciência e Agrotecnologia 41, 665-75
[16] Sritabutra D and Soonwera M 2013 Repellent activity of herbal essential oils against Aedes aegypti (Linn) and Culex quinquefasciatus (Say) Asian Pac J Trop Dis 3(4), 271-76
[17] Nollet LML, en Rathore HS 2017 Green Pesticides Handbook, Green Pesticides Handbook London New York: Taylor dan Francis Group
[18] Komisi Pestisida Departemen Pertanian 1995 Metode Standar Pengujian Efikasi Pestisida Departemen Pertanian, Jakarta 1-HL 4/9-95

[19] Thai Industrial Standards Institute 1986 Mosquito repellents Bangkok: Thai Industrial Standards Institute Ministry of Industry; p 15

Neutrophil lymphocyte ratio (NLR) in smokers

H Harun[1], J G Ismail[2], S Anwar[3,4], I K Rupawan[5]

[1]Departement of Clinical Pathology. Medical Faculty, Tadulako University, Palu, Indonesia
[2]Medical Faculty, Tadulako University, Palu, Indonesia
[3]Departement of Anatomy, Medical Faculty, Tadulako University, Palu, Indonesia
[4]SMF of Pulmonology, Medical Faculty, Tadulako University, Palu, Indonesia
[5]Departement of Anatomy Pathology, Medical faculty, Tadulako University, Palu, Indonesia

haeraniharun.unhas@gmail.com

Abstract. Smoking is a severe health problem. Smoking has been shown to increase the risk of systemic inflammation. Neutophil Lymphocyte Ratio (NLR) is used as a marker of systemic inflammation and monitors prognosis, morbidity, and mortality in many diseases. This study aimed to analyze NLR in smokers. This is an observational analytic study conducted at Tadulako University, Palu. A total of 48 subjects consisted of 24 active smokers and 24 non-smokers. Subjects were given a questionnaire containing a history of the disease, smoking status, and the number of cigarettes consumed per day. Blood samples were also taken for hematological examination. NLR of the subjects was then compared with the appropriate statistical test. The NLR in the smoker's group was 1.54 ± 0.81 and in the non-smoker's group 1.87 ± 1.76. This NLR was not significantly different between smokers and non-smokers, with a p-value of 0.578. Most of the smoking subjects are light and moderate smokers, and it is possible that this condition has not been able to cause a significant increase in systemic inflammation.

Keywords: Smoking, NLR, Tobacco Smoke.

1. Introduction

Smoking is one of the leading causes of morbidity and mortality. One in ten adults worldwide dies from smoking, with a mortality rate of 5,000,000 people per year [1]. In 2013 WHO (World Health Organization) reported that 29% of adults in Indonesia smoke every day. This is exacerbated by the increase in smokers as much as 32.8% in the young adult age range. The increase in the number of smokers in Indonesia is expected to increase every year if there is no intervention in preventing smoking, controlling the production and use of tobacco as a raw material for cigarettes. One of the efforts made by the government is to create a smoke-free area [1-2].

Exposure to cigarette smoke that lasts a long time will trigger inflammation in the respiratory tract and lung parenchyma of smokers. Exposure to existing cigarette smoke can increase the release of pro-inflammatory cytokines by epithelial cells, which will affect the growth, differentiation, and activity of leukocytes [3]. The association between exposure to cigarette smoke and leukocyte parameters indicates examining the inflammatory status through the leukocyte component. Neutrophilia occurs in smokers because continuous exposure to nicotine can stimulate neutrophils to produce IL-8, a chemoattractant and neutrophil activator. Meanwhile, lymphocytosis is associated with chronic cell damage due to constituents of cigarette smoke exposure [1][4].

Neutrophil Lymphocyte Ratio (NLR) serves as a biological marker in assessing inflammatory status, infection, pre-postoperative assessment, determination of the prognosis of cardiovascular disease and malignancy, and in conditions such as tuberculosis. When inflammation occurs, neutrophils in the blood will increase up to four to five times the standard limit. At the same time, lymphocytes tend to be more constant due to the consistent proliferation of lymphoid tissue, lymph, and blood. This difference in the distribution of cells during inflammation is the basis for the use of NLR [1][4][5]. This study aims to analyze the value of NLR as a marker of inflammation in smokers.

2. Method

This is observational analytic research conducted at Tadulako University on lecturers, staff, and students. Subjects who met the inclusion criteria were given a questionnaire containing a history of the disease, smoking history, number of cigarettes consumed, and body mass index (BMI). The subject's venous blood sample will also be taken for hematological examination using a hematology analyzer. Data from questionnaires and hematological tests were collected and analyzed.

The research subjects were divided into two groups, smokers and non-smokers. The smoker's group was then subdivided into light, moderate and heavy smokers depending on the number of cigarettes consumed by the subjects. The NLR value was obtained by dividing the number of neutrophils by the number of lymphocytes from the results of the hematological examination. The NLR values were then compared between the smokers and non-smokers groups using the Mann-Whitney test. The NLR value is significantly different if the p-value is <0.05.

3. Result

The subjects for the smoker's group were 24 people, and the issues for non-smokers were 24 people. The data in Table 1 shows the characteristics of the subjects.

Tabel 1. Sample characteristics

Characteristics	Smoker n=24			Non-Smoker n=24		
	n (%)	**Mean ± SD**	**Median (Min-Max)**	**n (%)**	**Mean ± SD**	**Median (Min-Max)**
Age (Years)						
20-30	9 (37,5)	24,78 ± 2,94	25 (21 - 30)	14 (58,3)	24,00 ± 4,18	22 (20-30)
31-40	7 (29,1)	35,86 ± 2,47	36 (33 - 39)	5 (20,8)	34,80 ± 3.34	35 (31-39)
41-50	4 (16,7)	45,00 ± 3,36	45 (41 - 49)	4 (16,7)	46,75 ± 2,21	47 (44-49)
51-60	4 (16,7)	55,50 ± 1,29	55,5 (54 - 57)	1 (4,16)	-	-
BMI[a] (kg/m²)						
< 18,5	3(12,5)	17,7 ± 0,17	17,8 (17,5 – 17,8)	2 (8,3)	17,65 ± 0,49	17,6 (17,3-18,0)
18,5 – 24	15(62,5)	21,8 ± 1,82	22,0 (18,7 – 24,2)	16 (66,6)	22,04 ± 2,10	22,3 (18,6-24,5)

24 – 30,5	5(20,8)	25,8 ± 1,09	25,7 (24,7 – 27,2)	5 (20,8)	26,80 ± 1,99	26 (24,8-29,4)
>30,5	1(4,16)	-	-	1 (4,16)	-	-
Smoking Status						
Light	9 (37,5)	7,22 ± 2,72	6,00 (4-10)	-	-	-
Medium	12 (50)	13,67 ± 2,06	12,00 (12-16)	-	-	-
Heavy	3 (12,5)	-	-	-	-	-
WBC[b] (10^9/L)						
< 4,2	0	-	-	-	-	-
4,2-10,8	23 (95,8)	7,8 ± 8,0	8,0 (5,3 – 10,0)	23 (95,8)	7,3 ± 1,45	7,3 (4,5-9,5)
>10,8	1 (4,16)	-	-	1 (4,16)	-	-
Neutrophil (10^9/L)						
< 2,0	0	-	-	0	-	-
2,0 – 7,5	20 (83,3)	40,3 ± 6,17	49,6 (37,3 – 57,6)	17 (70,8)	48,5 ± 5,79	50,8 (36,4-57,9)
>7,5	4 (16,7)	64,6 ± 3,94	64,3 (60,9 – 68,8)	7 (29,17)	117,4 ± 144,56	63 (59,5-445,2)
Lymphocyte (10^9/L)						
<1,0	0	-	-	0	-	-
1,0-4,0	5 (20,83)	24,7 ± 6,02	28,3 (15,1 – 29,1)	4 (16,7)	24,52 ± 2,92	24,3 (21,2-28,3)
>4,0	19 (79,17)	40,2 ± 5,80	40,1 (31,1 – 52,4)	20 (83,3)	40,17 ± 6,32	39,0 (30,1-53,4)
NLR[c]	24 (100)	1,54 ± 0,81	0,74 – 4,44	24 (100)	1,87 ± 1,76	0,68 – 9,62

[a]BMI: Body Mass Index
[b]WBC: White Blood Cell
[c]NLR: Neutrophil Lymphocyte Ratio

Most of the subjects were in the smoker's group aged 20-30 years (37.5%) and the non-smoker group (58.3%). Most BMI in the group of smokers and non-smokers was at the value of 18.5 – 24 kg/m2, respectively 62.5% and 66.6%. In the smoker's group, the number of light smokers was 37.5%, moderate smokers were 50%, and heavy smokers were 12.5%. For the NLR value in the smoking group, the average was 1.54 ± 0.81, and in the non-smoker group, 1.87 ± 1.76.

Tabel 2. Comparison of NLR in smokers and nonsmokers

	Smoker Mean ± SD	Non-Smoker Mean ± SD	p[a]
NLR	1,54 ± 0,81	1,87 ± 1,76	0,578

[a]Mann-Whitney test

Comparison of NLR values in the smokers and non-smokers groups by Mann-Whitney test have p-value = 0.578. This indicates that there was no significant difference in NLR in the two groups in this study.

4. Discussion

Tadulako University is a smoke-free environment, but smokers are still found even though most consume cigarettes outside the university environment. Smoking activity in the university environment is unknown, and there is still a possibility that this group of smoker's smokes in the campus area. Even though the non-smoker group is not active smokers, there is a possibility that they are passive smokers. The risk of inflammation increases in the sedentary smoking group with prolonged exposure, affecting the pulmonary inflammatory microenvironment and immunity to infection and vaccination [6].

The results of the Mann-Whitney test showed that there was no significant difference in the NLR value in the smokers and non-smokers groups. Both groups had NLR values within normal limits. The average value of NLR in healthy adults is 0.78 – 3.53 [7]. This shows no difference in NLR values in the smoker's group compared to the non-smoker group. The level of inflammation assessed by NLR in this group can be almost the same as in the non-smoker group. In the group of smokers in this study, most of them were light and moderate smokers. It is possible that the level of inflammation caused by smoking in the light and moderate smoker groups was not sufficient to cause inflammation which resulted in a significant increase in NLR in smokers.

Other markers of inflammation such as leukocyte count, differential count of neutrophil and lymphocyte also had almost the same mean values in both groups. This shows that the level of inflammation in smokers and non-smokers is not much different. However, it is necessary to consider other factors that also affect the level of inflammation, including nutritional factors, unknown diseases, exposure to the environment, infection, and other factors [8-9].

5. Conclusion

There was no significant difference in NLR between the smokers and non-smokers groups. It is possible that the level of inflammation caused by smoking in the light and moderate smoker groups was not sufficient to cause inflammation which resulted in a significant increase in NLR in smokers.

References

[1] Nusa GB, Widyastiti NS Perbedaan Neutrophil Lymphocyte Ratio pada Subjek Bukan Perokok, Perokok ringan dan Perokok Sedang-Berat Diponegoro Med J 2016; 5: 903–10
[2] Sarah DAS, Naftali Z, Yusmawan W The Difference of Neutrophil Lymphocyte Ratio (NLR) in Conventional and Electric Smokers Dipoengoro Med J 2020; 9: 283–7
[3] Lee J, Taneja V, Vassallo R Cigarette smoking and inflammation: cellular and molecular mechanisms J Dent Res 2012; 91: 142–9
[4] Shenwai M, and Aundhakar D Effect of cigarette smoking on various hematological parameters in young male smokers Indian J Basic Appl Med Res IJBAMR 2012; 2: 386–92
[5] Harun H, Muhammad N, and Sumarni S, et al Neutrophil Lymphocyte Ratio and Acid-Fast Bacilli in Tuberculosis Indian J Forensic Med Toxicol 2021; 15: 1091–6
[6] Bhat TA, Kalathil SG, and Bogner PN, et al Secondhand Smoke Induces Inflammation and Impairs Immunity to Respiratory Infections J Immunol 2018; 200: 2927–40
[7] Forget P, Khalifa C, and Defour J-P, et al What is the normal value of the neutrophil-to-lymphocyte ratio? BMC Res Notes 2017; 10: 12

[8] Galland L Diet and inflammation Nutr Clin Pract Off Publ Am Soc Parenter Enter Nutr 2010; 25: 634–40
[9] Liu G, Huang Y, Reis FS, et al Impact of Nutritional and Environmental Factors on Inflammation, Oxidative Stress, and the Microbiome 2019 BioMed Res Int 2019; 2019: e5716241

A comparative study of quality of life between elders living in a temporary shelter (huntara) and elders living at home in Petobo village

F Zulmulatifah[1], N Syamsi[2], A A M Tanra[2]

[1]Medical Student, Faculty of Medicine, Tadulako University
[2]Department of Pharmacology, Faculty of Medicine, Tadulako University

alfiamuthmainnah@gmail.com

Abstract. Natural disasters can result in various changes in a person's life. In addition to physical changes, psychological changes also significantly affect the quality of life of the elders. Quality of life is an important indicator to assess the success of health care interventions, both prevention, and treatment. The environment where older adults live is a fundamental problem that causes elders' good and bad health status. This study aims to determine the difference in the quality of life among the elders who live in a temporary shelter (Huntara) and the elders who live at home after the disaster in the Petobo Village community. The study design is a comparative study. The study was conducted in September-October 2019. The number of respondents was 194 elders consisting of 97 people living in the Huntara and 97 people living at home. The sampling technique is purposive sampling. The result of the Mann- Whitney U test show that there is a significant difference between the quality of life of older adults living in the Huntara and living at home in the Petobo village, namely the physical health domain at QoL ($p=0.000$), psychological domain at QoL ($p=0.001$), the social relations domain at QoL ($p=0.010$) and the environmental domain at QoL ($p=0.000$). The conclusion is there is a significant difference between the quality of life of older adults living in the Huntara and living at the home of Petobo Village in each domain.

Keywords: Elderly, Quality of life, Natural Disaster, Huntara

1. Introduction

Natural disasters in Palu-Sigi-Donggala (PASIGALA) on 28 September 2018 were earthquakes, tsunami, and liquefaction. According to BNPB (2018), the incident caused approximately 2,101 deaths. Besides the fatalities, thousands of houses were damaged or vanished along with liquefaction events. This incident caused residents who lost their houses and other possessions to move to refugee camps or temporary shelters (HUNTARA) built by the government. To this date, there are approximately 221,450 people who have been relocated.

The refugees in the Huntara consist of families of various ages, ranging from infants to older adults. In addition to physical changes, psychological changes also have a significant impact on the quality of life of older adults, such as memory loss, loneliness, fear of losing loved ones, fear of facing death, and depression which will affect the quality of life an elder[1-2].

Quality of life is an essential indicator for determining the effectiveness of health-care interventions, both preventive and treatment. Quality of life encompasses physical dimensions and social role performance, emotional states, intellectual and cognitive functions, and a sense of health and life satisfaction[3].

Health is one element that plays a vital role in the quality of life of older adults. Health shows the degree to which an older adult can enjoy the essential things that happen in life and is a measure of the quality of life of elders[2-4].

The fundamental problem that causes older adults' good and bad health status is the elders' environment. Inequalities in the elderly's places of residence result in differences in access to health treatments. The living environment is an essential factor that affects the quality of life of older adults. The different living environments may change the role of elders in terms of adapting. For elders, changes in family, socio-economic, and social community resulted in setbacks in adapting to the new environment and interacting with their social environment. When older adults move into new places, they will be exposed to new situations and friends, which will force them to adapt positively or negatively. Variances in residency can lead to differences in the elderly's physical, social, economic, psychological, and religious-spiritual environments which can affect the health status of elders[5].

2. Methods

This study is a comparative study, which aims to determine the difference in the quality of life of older adults who live in Huntara and older adults who live at home after the disasters in Petobo village. The study was carried out in the Huntara and the residencies in Petobo village. The number of samples was determined by the sample size calculation for proportion estimation. Based on the analysis, the study requires 97 older adults in each group, meaning we need 194 older adults in total.

The WHOQOL-BREFF questionnaire used in this study contains a list of written statements used as a guide in interviews, consisting of ideas that had been prepared and had their validity and reliability tested. The questionnaire consists of 24 facets, including four domains: i) physical health, consisting of 7 questions; ii) psychological, six questions; iii) social relationship, three questions; and iv) environment, eight questions. The WHOQOL-BREFF also measured two facets of quality of life in general, such as i) overall quality of life; and ii) general health. The results will be categorized as good quality of life when the transformed scores are more than the average value (mean) of each domain and poor quality of life when the transformed scores are less than the mean value for each part.

3. Results

Table 1. Distribution of frequency and percentage based on demographic data of respondents living in huntara and at homes (n=194).

Demographic Data	Huntara		Home	
	f	%	f	%
Age groups				
60 – 69 years	72	74,2	80	82,5
70-79 years	24	24,7	15	15,5
> 80 years	1	1,1	2	2
Occupation				
Retire	9	9,3	27	27,8
Civil servant	0	0	8	8,2
Business and sales	7	7,2	28	28,9
Farmer	12	12,4	2	2,1
Blue-collar workers	2	2	1	1
Unemployment	65	67,1	29	29,9
Others	2	2	2	2,1

According to the age characteristics of respondents, most older adults are in the 60-69 years age group, with 72 people (74.2 %) living in the Huntara and 80 people (82.5 %) living at home. The most significant proportion was in the 70-79 years old age group, with 24 elderly individuals (24.7%) living in the Huntara and 15 people (15.5%) living at home. Then, among those over the age of 80 groups, only one elder (1.1 %) lives in the Huntara, and two elders (2 %) live at home.

According to the characteristics of respondents based on occupation, most older adults living in Huntara do not have a job, with 65 (67.1%) were unemployed and 9 were retirees (9.3 %). Furthermore, as many as 12 older adults (12.4 %) worked as farmers, 7 (7.2 %) worked as a business, and sales employees, two (2%) as blue-collar workers, and two (2%) had other jobs. In addition, for the elderly living at home, there are 29 (29.9%) who are unemployed, and 27 are retirees (27.8%). Twenty-eight people work as business and sales employees (28.9 %). Then, 8 (8.3%) worked as civil servants, two as farmers (2%), one person as a manual worker (1%), and two people in other occupations (2 %).

The univariate analysis describes the quality of life of older adults in four domains: physical, psychological, social, and environmental. The WHOQOL-BREF questionnaire was used to assess the quality of life. The results of the calculation in each domain is a raw score which then needs to be transformed into a score on a range of 0-100.

Table 2. Descriptive statistics of quality of life in each domain.

	Mean	**SD**	**Minimum**	**Maximum**
Domain 1 (Physical Aspect)				
Elderly living in Huntara	59,99	11,690	38	88
Elderly living at home	70,28	14,021	38	88
Domain 2 (Psychological Aspect)				
Elderly living in Huntara	51,13	10,987	31	75
Elderly living at home	56,09	9,406	31	69
Domain 3 (Social Aspect)				
Elderly living in Huntara	59,44	10.987	19	100
Elderly living at home	64,73	11,971	31	81
Domain 4 (Environment Aspect)				
Elderly living in Huntara	49,76	13,114	19	75
Elderly living at home	66,88	7,113	38	81

The mean value is used to classify the excellent quality of life and the poor quality of life. In domain one or the physical aspect, older adults living in Huntara have a mean value of 59.99, while older adults living at home have 70.28. In domain two, or the psychological aspect, the mean value for older adults living in the Huntara is 51.13, and the mean value for older adults living at home is 56.09. In domain three or social aspect, the mean value for older adults living in the Huntara is 59.44, and the mean value for older adults living at home is 64.73. In domain four or environmental aspects, the mean value for older adults living in the Huntara is 49.76, and the mean value for older adults living at home is 66.88. Then, the quality of life is categorized as good if the score in each domain is more than equal to the mean value, and the quality of life is classified as poor if the score in each field is less than equal to the mean value.

Table 3. Frequency distribution by the quality of life in elderly

Variable	Elderly living in Huntara		Elderly living at Homes	
	f	%	f	%
Domain 1 (Physical Aspect)				
Good	45	46,4	48	50,5
Poor	52	53,6	49	49,5
Domain 2 (Psychological Aspect)				
Good	41	42,3	56	57,7
Poor	56	57,7	41	42,3
Domain 3 (Social Aspect)				
Good	46	47,4	62	63,9
Poor	51	52,6	35	36,1
Domain 4 (Environment Aspect)				
Good	58	59,8	58	59,8
Poor	39	40,2	39	40,2

Table 3 shows that the physical quality of life of elders living in Huntara is poor, with 52 people (53.6%). Meanwhile, 49 (49.5%) elders living at home also had poor physical aspects. Quality of life on the psychological part, 56 older adults (57.7%) were categorized as inferior, while 56 older adults (57.7%) were classified as good. In the social aspect, poor quality of life was found in 51 older adults (52.6%) who lived in the Huntara, while 62 elders (63.9%) had a good quality of life. Similar results were found in the environmental quality of life. Older adults have a good quality of life, with 58 people (59.8%) in older adults living in Huntara and living at home.

The bivariate analysis aims to examine the effect between the independent variable and the dependent variable. The independent variables of this study were older adults who lived in the Huntara and older adults who lived at home. The dependent variable was the quality of life. The bivariate test in this study used the Mann-Whitney U non-parametric test; hence no normality test was performed.

Table 4. Results of the mann whitney u. statistical test

	Domain 1	Domain 2	Domain 3	Domain 4
Mean Whitney	2.614E3	33393.000	3715.000	1239.500
Wilcoxon W	7.368E3	8146.000	8468.000	5992.500
Z	-5.398	-3.419	-2.572	-8.974
Asymp. Sig.	.000	.001	.010	.000

Table 4. shows the results of the statistical test of bivariate analysis, indicating a significant difference (p0.05) between the place of residence and the four dimensions of quality of life. The description of Asym. Sig value in domain 1 with values (p=0.00), domain 2 (p = 0.001), domain 3 (p=0.10) and domain 4 (p=0.000).

4. Discussion

Both living in Huntara or at home in Petobo Village, the respondents mainly were aged 60-69 years old (>50%). There was 73 male (75.3%) living in Huntaran and 70 male (72.2%) living at home. Sixty-five of them were unemployed (67.1%). Nine respondents were retired, and the others were blue-collar workers and doing other jobs such as teacher, handyman, and housemaid. As for those living at home, only 29 people (29.9%) did not work, and 27 (27.8%) were retirees. The majority of elders living at home (28 people or 28.9%) were doing personal business or entrepreneurship.

The analysis of the physical domain of the elderly in this study shows 52 respondents (53.6%) living in Huntara had poor quality of life. On the other hand, those living at home slightly differed between those who had a good quality of life (48 people or 49.5%) and poor quality of life (49 people or 50.5%). There is a significant difference between the quality of life between the elderly living in Huntara and those living at home (p = 0.000). These findings are similar to previous research conducted by Putri6 that found that the physical aspects of the quality of life of the elderly living with their families and those living in social institutions differed significantly. The study explains that the population and life expectancy of the elderly is increasing by the year, so their quality of life must be an essential aspect to improve their life expectancy.

A study by Nuryanti[7] established that the elderly aged >60 years old often complained about their chronic illness. The reduction of physical ability in the elderly naturally happens because of the degenerative process, leading to physiological and biochemical changes in their bodies. Moreover, this condition will influence the overall ability and function of the body itself.

Physical well-being is strongly related to the health state. Elders will experience physical, cognitive, and psychosocial life[8]. These conditions have been assessed by the instruments used in this study. The questions related to the physical domain are health status, drug consumption, energy to carry out activities, social skill, and sleep quality of the respondents. This study shows that the respondents living in the Huntara had poor sleep quality. They stated that the noises from activities in the interconnected houses contributed to their poor sleep quality.

The analysis of the psychological domain found 56 respondents (57.7%) living in Huntara had poor quality of life. Nevertheless, 56 respondents (57.7%) living at home had a good quality of life. The statistical analysis of Mann-Whitney shows a significant difference in the quality of life between the respondents residing in Huntara and those living at home. Erickson's psychosocial theory explains that elders are in an integrity state where a person has to get used to many kinds of success and failure in life[9]. Physiological changes in their life include social changes, role changes according to the development task, depression level, and emotional stability[10].

Aspects of psychological quality of life in the elders are their perception of their body and appearance, their feelings, their perception about themselves, and their ability to think, learn, memorize, and concentrate, which describe the cognitive state that enables the elders to focus[11]. The difference in the quality of life for the psychological domain between the respondents living in Huntara and living at home might be due to the traumatic feeling towards the disasters in Huntara a year ago. These lead to difficulty in concentrating. The elderly often compared the condition before and after the disasters, which might influence their physiological state. Research conducted by Ramlan (2011) found that emotional support for the elderly from their family is critical to avoid being neglected in a family[12].

The social domain analysis shows 52 respondents (52.6%) with poor quality of life in Huntara. However, most of the respondents living at home had a good quality of life (56 respondents or 57.7%). There is a significant difference in the quality of life in the social domain for those living in Huntara and those living at home (p=0.010). This study uses only three questions to assess this aspect, so the raw score for this domain is relatively low.

A study by Putri[6] found a significant difference in the quality of elderly residence for a social domain. The study shows that the elders living at a nursing home had poor quality of life (80%) while the elders living with their family had a good quality of life (67.5%). The personal aspect of the elders comprises personal relation (the relation between an individual and other people), social support (help from the surrounding environment), and sexual activity of each person[11].

Support from family and community will influence the life of the elders in a community. The inadequate support from their surroundings will drive a negative change for them, which later affects their quality of life13. Kuntjoro reveals that it is essential to keep the elders doing beneficial activities to maintain their physical and mental health as social support. It is a form of social integrity where elders have a sense of community, where they can share the feeling, concern of anything or do creative activities together[14].

Older adults need social support from their spouses, children, and other elders. These supports include acceptance, attention, awards, or help from others. These supports will increase their quality of life for the social domain[15]. Social and community belonging were the relation between each elder and their surroundings, involving the happy feeling of being together with friends, having an active social relationship, and not facing difficulty in social relation. Spiritual activities will help the elders find the meaning and dignity of their life[8].

The difference in the quality of life for the respondents living in Huntara and at home might be due to nearby support. Those residing in Huntara shared the same feeling. Therefore they tended to be less supportive than those living at home who got help from their family. The respondents living in Huntara lacked activities, especially social activities where they could share feelings, experiences, thought, and attention.

The environment domain consists of financial resources (financial condition), where the elders live and experience to do recreation or fun activities[11]. The analysis of the environment domain reveals that both the respondents residing in Huntara and at home had a good quality of life (58 people or 59.8% in each group). However, the statistical analysis shows a significant difference in both groups ($p=0.000$).

The surrounding environment plays a vital role in the elderly quality of life. The different backgrounds might cause the change of the quality of life, where the elders are demanded to adapt. They should adapt well, especially for their family role, social, and economy, unless they had difficulties building interaction with other people in their social environment. The elders living in a nursing home should get used to the new environment, such as new friends, although some of them might find struggling[7]. Education and economic background of the elders living in a community play essential roles in fulfilling the need for an adequate environment, including the availability of clean and healthy residence, availability of information, transportation, and affordability of health services[9].

A study by Widya[16] establishes the same result that there is no significant difference between the quality of life of the elders living in PTSW and those living with their families. They already resided in PTSW Gan Mabaji for five years, which means that they might be doing good adaption. This study shows that the respondents living in Huntara had good adaptation in the environment domain. This might contribute to no significant difference in the quality of life between those residing in Hutara and at home.

The overall analysis using Man-Whitney U finds no significant differences in quality of life between those living in Huntara and at home ($p>0.05$ for each domain). It shows that we accept the research hypothesis (H1) and reject the H0 of a different quality of life between the respondents living in Huntara and at home in Petobo Village after the earthquake, tsunami, and liquefaction PASIGALA.

5. Conclusions

a. Based on the result and discussion above, we can conclude:
b. The analysis of the physical domain shows that there were 45 respondents (46.4%) who had a good quality of life, and 52 respondents had a poor rate of live (53.6%) living in Huntara. In addition, 48 respondents (49.5%) had a good quality of life, and 49 respondents (50.5%) had poor quality of life living at home.
c. The analysis of the physiological domain shows that 41 respondents (42.3%) had a good quality of life, and 56 respondents had poor quality of life (57.7%) living in Huntara. In addition, 56 respondents (57.7%) had a good quality of life, and 41 respondents (42.3%) had poor quality of life living at home.
d. The social domain analysis shows that 46 respondents (47.4%) had a good quality of life, and 51 respondents had poor quality of life (52.6%) living in Huntara. In addition, 62 respondents (63.9%) had a good quality of life, and 35 respondents (36.1%) had poor quality of life living at home.
e. The analysis of the environment domain shows that there were 58 respondents (59.8%) who had a good quality of life, and 39 respondents had a poor rate of live (40.2%) living in Huntara. In addition, 58 respondents (59.8%) had a good quality of life, and 39 respondents (40.2%) had poor quality of life living at home.

There is a significant difference in the quality of life for each domain between the respondents living in Huntara and at home in Petobo Village.

References

[1] Ebersole P, Hess P, Touhy T, Jett K Gerontological nursing & health aging 2nd ed St Louis, Missouri : Mosby,Inc; 2005
[2] Sari MT, Susanti Gambaran Kualitas Hidup Lansia di Panti Sosial Tresna Werdha Budi Luhur dan Lansia di Keluruhan Paal V-Kota Jambi Jurnal Ilmiah Universitas Batanghari Jambi 2017;17 (2)
[3] Putri, DLNA, Akhmadi, Subekti H Gambaran kualitas hidup lansia yang tinggal di Panti Sosial Tresna Werdha Yogyakarta Unit Budhi Luhur JIK 2008;3 (2)
[4] Tamher S, Noorkasiani Kesehatan Usia Lanjut dengan Pendekatan Asuhan Keperawatan Jakarta : Salemba Humanika; 2012
[5] Tude R Pengaruh Perilaku Hidup Sehat Terhadap Kualitas Hidup Pada Lansia di Panti Werdha Salib Putih Salatiga Jawa Tengah [Skripsi] Universitas Kristen Satya Wacana Salatiga; 2016
[6] Putri ST, Fitriana LA, Ningrmn A, Sulastri A Studi Komparatif : Kualitas Hidup Lansia Yang Tinggal Bersama Keluarga dan Panti JPKI 2015;1 (1)
[7] Nuryanti Hubungan Perubahan Peran Diri dengan Tingkat Depresu pada Lansia yang Tinggal di UPT PSLU Pasuruan, Babat Lamongan Jurnal Unair 2012
[8] Rohmah AI, Purwaningsih, Bariyah K Kualitas Hidup Lanjut Usia Jurnal Keperawatan 2012;3 (2)
[9] Setyoadi Perbedaan tingkat Kualitas Hidup Lansia di Komunitas dan Panti Scientific Journal UMM 2010
[10] Damojo BR Geriatri-Ilmu Kesehatan Lanjut Usia Bagi Petugas Kesehatan Jakarta : DEPKES RI; 2004
[11] Anbasaran SS Gambaran Kualitas Hidup Lansia Penderita Hipertensi di Wilayah Kerja Puskesmas Rendang Periode 27 Februari sampai 14 Maret 2015 Universitas Udayana; 2015
[12] Ramlah Hubungan Pelaksanaan Tugas Kesehatan dan Dukungan Keluara dengan Lansia di Wilayah Kerja Puskesmas Kassi-kassi Makassar; 2011
[13] Potter PA & Perry AG Basic Nursing 6th ed St Louis Mosby Elsevier; 2005
[14] Fitria Interaksi Sosial dan Kualitas Hidup Lansia di Panti Wredha UPT Pelayanan Sosial Lanjut Usia dan Anak Balita Binjai[Skripsi] Fakultas Ilmu Keperawatan : Universitas Padjadjaran; 2011
[15] Hayati S Pengaruh Dukungan Sosial Terhadap Kesepian Pada Lansia [Skripsi] Universitas Sumatera Utara; 2010
[16] Widya Perbedaan Kualitas Hidup Antara Lansia yang Tinggal di Keluarga dengan Lansia di Panti Sosial Tresna Werdha [Skripsi] UIN Alauddin Makassar; 2016

Effectiveness of moringa oleifera on triglyceride levels in diabetic wistar rats (Rattus norvegicus) induced with streptozotocin (STZ)

T Setyawati[1], R Adawiyah[1], R M Walanda[1], Riski[2], R Chandra[2]

[1]Department of Biochemistry, Faculty of Medicine, Tadulako University, Palu Indonesia

[2]Medical Student, Faculty of Medicine, Tadulako University, Palu Indonesia

tridentist@gmail.com

Abstract. Dyslipidemia is a lipid metabolism syndrome in an abnormal situation inside the body marked with increasing or decreasing the fraction of plasma lipid. This condition can be treated with an antioxidant found on Kelor Leaves (Moringa oleifera) that contains high flavonoid and saponin and potency to control the lipid levels inside the body. This study aims to know the effectiveness of Moringa oleifera to decrease the lipid profile of white rats. The design of this study is a true experimental pre-test and post-test with a group of the control. The sample, white rat (Rattus norvegicus) grouped into six groups and inducted by STZ 50 mg/kg BW. Group I is the standard control, group II is a negative control that is not given the treatment, group III is a positive control, group IV, V, VI is the group that given the treatment of Moringa oleifera with the concentration of 50 mg/kg BW, 100 mg/kg BW, and 200 mg/kg BW respectively. The data were analyzed using the parametric statistical test, Two-Way ANOVA. The Two-Way ANOVA test showed no significant difference in triglycerides levels that showed P-value 0,827 ($p > 0,05$) after the treatment. The induction of Moringa oleifera has not significantly decreased the triglycerides levels of the white rat (Rattus norvegicus).

Keywords: Moringa oleifera, fraction, lipid profile, triglyceride, diabetic, streptozotocin

1. Introduction

Diabetes is a common chronic disease and a significant threat to human health. The hallmark of diabetes is blood glucose levels that are higher than usual, caused by defective insulin secretion or impaired biological effects, or both.[1] Diabetes can cause chronic damage and dysfunction of various tissues, especially the eyes, kidneys, heart, blood vessels, and nerves. [2] Diabetes mellitus can cause serious health problems such as hypercholesterolemia. Hypercholesterolemia is a condition where cholesterol concentration in the blood is higher than usual [3].

One of the traditional plants commonly found in Central Sulawesi, especially Palu City, is the Moringa leaf (Moringa Oleifera). Moringa is a plant found in almost all environmental areas in Central Sulawesi. Many people plant these plants in their yards because they are rich in health benefits. These plants have substances that act as anticholesterol. Many people can traditionally use this Moringa plant as anti-cholesterol, anti-bacterial, anti-function, laxative, anti-inflammatory, anti-tumor, anti-pyretic, and treat rheumatism. Other benefits of Moringa leaves are to treat allergies, aches and pains, festering wounds, and prevent hypertension, lower cholesterol levels in the body, lower blood sugar levels, and lower uric acid levels.[4].

According to Tian et al. (2011), flavonoid compounds can inhibit Fatty Acid Synthase (FAS), an enzyme that is very important in fat metabolism. Inhibition of FAS can directly reduce the formation of fatty acids, thereby reducing the appearance of triglycerides. The mechanism of saponins can reduce triglyceride levels, namely by inhibiting the absorption of cholesterol and triglycerides in the intestine and increasing the reaction for the formation of bile acids from cholesterol and then excreted through feces.[6]

The mechanism of STZ in increasing triglyceride levels begins with decreased secretion and decreased insulin sensitivity. Decreased secretion and reduced insulin sensitivity leading to insulin deficiency, increasing fat degradation. As a result, fat accumulation occurs. The liver will take up the accumulated fat for lipoprotein synthesis, and the rest is broken down into Acetyl CoA, which will be used in the tricarboxylic acid cycle. If the accumulation of fat is significant, the amount of fat that enters the liver for lipoprotein synthesis will increase, which will result in hyperlipidemia (increased cholesterol and triglyceride levels).[7]

2. Methods

This research is pure experimental research using pre-test and post-test group designs. This research is used to determine the effectiveness of Moringa leaf extract where researchers will observe the control group and the treatment group after being given an action. In this study, the samples were grouped into six groups. Group I as a standard control was not given any treatment, group II as a positive control, group III a negative control, IV, V, AND VI as a group treated with Moringa leaf quercetin extract at 50mg/kg BW, 100mg/kg BW and 200mg/kg BW.

3. Results

Based on research that has been done on experimental animals, white male rats aged 2-3 months with body weight ranging from 200-300 grams as many as 17 tails. Which were divided into six groups, namely four regular control groups, four negative control groups, four positive control groups, 4 Moringa leaf water fraction groups at 50mg/kg BW, four groups Moringa leaf water fraction dose of 100mg/kg BW, four groups of Moringa leaf water fraction dose 200mg/kg BW. The results of measuring glucose levels and triglyceride levels are as follows:

Table 1. The average measurement of rat blood glucose level

Groups	**Day-0 (mg/dL)**	**Day-14 (mg/dL)**	**Difference (mg/dL)**
Normal Control	119,7±43,7	102,7±21,1	17,0
Negative Control	289,1± 78,5	95,8±21,9	193,3
Positive Control	527,5±71,3	172±65,0	355,5
Moringa Extract 50mg/kgBw	448,9±100,4	68,9±14,5	380,0
Moringa Extract 100mg/kgBW	445±135,4	150,5±40,9	294,5
Moringa Extract 200mg/kgBW	448,2±44,6	125,5±44,5	322,6

Description: Day 0 = Blood glucose levels after streptozotocin induction; Day 14 = Blood glucose levels on day 14

In table 1, it can be seen that the blood glucose levels of rats after being induced by streptozotocin increased, and then triglyceride levels were checked in rats

Table 2. The average measurement of rat triglyceride levels

Groups	Day 0 (mg/dL)	Day 14 (mg/dL)	Difference (mg/dL)
Normal Control	72,78±26,84	101,21±35,17	28,43
Negative Control	115,08±19,84	135,43±37,58	20,35
Positive Control	124,31±12,81	105,32±8,41	18,99
Moringa Extract 50mg/kgBW	131,61±11,28	126,84±38,39	4,76
Moringa Extract 100mg/kgBW	87,15±32,68	105,24±64,26	18,09
Moringa Extract 200mg/kgBW	100,96±40,47	100,66±25,38	0,3

Description: Day 0 = Triglyceride levels after streptozotocin induction; Day 14 = Day 14 triglyceride levels.

In table 2, it can be seen that the average triglyceride levels of rats after streptozotocin induction. After 14 days of treatment, there was no decrease in triglyceride levels in the Moringa leaf water fraction treatment group at a 100 mg/kg BW dose with an average triglyceride level of 105.24 mg/dL. In the water fraction group, Moringa leaves at 50 mg/kgBW and 200 mg/kgBW decreased triglyceride levels with an average triglyceride level of 126.24 mg/dL and 100.66, respectively.

Based on table 2 of the average triglyceride levels of rats in each group after streptozotocin induction on day 0, the mean triglyceride level did not increase significantly. The mean triglyceride levels in the negative control group, positive control, and treatment groups <145 mg/dL were within normal limits of the six groups, the group of rats with the highest increase in triglyceride levels was the 50 mg/kgBW Moringa leaf water fraction group is 131.61 mg/dlBW and the Moringa leaf water fraction group is 200 mg/kgBW which is 100.96mg/dl. On the 14th day, there was a decrease in triglyceride levels in the Moringa leaf fraction group at doses of 50 mg/kgBW and 200 mg/kgBW, but the levels were still within normal limits < 145 mg/dL.

4. Discussion

In this study, selecting white male rats could not affect the menstrual cycle and pregnancy and female rats. It could provide better research results, and rats aged 3-4 months had similarities with adult humans and had not experienced intrinsic aging [8] Before the treatment, the rats were adapted for one week to adapt to their environment; the rats were given adequate food and drink during maintenance.

In this study, metformin was used, which has a blood lipid-lowering effect and is associated with a reduction in free fatty acids and/or a reduction in DM weight loss.[9] The mechanism of metformin lowers triglyceride levels through the activation mechanism of AMPK (activated protein kinase) in the liver. AMPK is a serine/threonine-protein kinase that plays a vital role in regulating catabolic/anabolic pathways by acting as cellular energy and nutrient sensor. AMPK consists of a hetero-trimeric complex containing a catalytic subunit and two regulatory band subunits. Each subunit has several isoforms (a1, a2; b1, b2; g1, g2, g3) encoded by different genes, providing multiple heterotrimeric combinations with tissue-specific distributions.[10]

In this study, Streptozotocin was used as an inducer of diabetes (hyperglycemic) conditions in experimental animals. The blood sugar levels of rats became high, and in diabetes conditions, hypercholesterolemia could occur due to insulin resistance. Insulin resistance in DM causes a decrease in the Lipoprotein lipase (LPL) enzyme activity in muscle and adipose tissue. It increases the transfer of free fatty acids from peripheral adipose tissue to the liver [11]. Streptozotocin also interferes with the antioxidant defense system and increases free radical formation, which contributes to oxidative stress in diabetes. The formation of nitric oxide (NO) plays a vital role in destroying the pancreas by inducing cell death (apoptosis).[12]

STZ induction in experimental animals can cause hyperglycemia, hyperlipidemia, and increased lipid peroxidation.[13] Hyperlipidemia after STZ injection is caused by increased lipid absorption in the small intestine and increased (abnormal) changes in -cholesterol acyltransferase activity in the small intestine. Lack of insulin can cause excessive fatty acid metabolism. Increased free fatty acids will increase the secretion of very-low-density lipoprotein (VLDL) accompanied by triglycerides and cholesterol.[14] According to Putri's research (2017), in diabetes conditions, there is impaired insulin secretion, which is one of the hormones that affects triglyceride levels. The insulin hormone can reduce blood triglyceride levels because insulin can prevent hydrolysis triglycerides.

Based on the results obtained, there was no increase in triglyceride levels after streptozotocin induction after four days; namely, the average triglyceride level was 103.77 mg/dl. This level shows results within normal limits. According to Putri (2015), DM induction with STZ can increase serum triglyceride levels in adult Wistar rats within five weeks after STZ injection. However, in this study, only for four days after STZ induction, triglyceride levels were obtained within normal limits. According to Tsalissavrina (2006), the standard limit for rat blood triglyceride levels is 26-145 mg/dl. According to Ariany (2019), the condition of rats can affect triglyceride results because STZ induction is possible to cause liver damage. The liver is the site of triglyceride synthesis and will be formed in VLDL. In this study, it was suspected that rats had liver damage so that triglyceride synthesis was less than optimal. Low triglyceride production causes triglyceride levels to be within normal limits.

The effectiveness of the water fraction of Moringa leaves on reducing triglyceride levels was measured by the Two Way Anova test and obtained a P-value of 0.827 ($P>0.05$). Based on the test results, the research hypothesis H0 was accepted, and H1 was rejected; there was no effect of the water fraction of Moringa oleifera leaves on decreasing triglyceride levels in male diabetic white rats (Rattus norvegicus). According to Elly's research (2015) that the aqueous fraction of Moringa leaf extract at a dose of 150 mg/KgBW, 300 mg/KgBW, and 600 mg/KgBW can reduce rat triglyceride levels and statistically with a test based on the results of ANOVA analysis ($p<0.05$) showed that the administration of Moringa leaf extract has a treatment effect on reducing triglyceride and blood glucose levels in rats. While the researchers got the results, there was no statistical effect through the ANOVA test with a value ($p>0.05$) with the dose of water fraction given, namely 50mg/kgBW, 100mg/kgBW, and 200mg/kgBW. This study is in line with the research conducted by Darrashcytha et al. (2017) on "The Effect of Giving Moringa Leaf Extract (Moringa Oleifera) on Triglyceride Levels and High-Density Lipoprotein in Dyslipidemic White Rats (Rattus Norvegicus)" that administration of Moringa leaf extract (Moringa oleifera) did not have a significant effect on triglyceride levels ($p = 0.051$) and has a substantial impact on HDL levels ($p = 0.000$) in white rats induced by high-fat diet with PTU.

According to Mohammed et al. (2020), STZ-induced diabetic mice will experience necrosis in the liver characterized by increased ALT, AST, and GGT activity in plasma, which may indicate the hepatotoxic effect of STZ. Type 1 diabetes, with poor glycemic control, has been reported to be associated with hyperlipidemia and hepatic steatosis. A liver disease characterized by fatty liver accumulation with necroinflammation and hepatocellular injury. Diabetes also increases the risk of liver cirrhosis. However, while the mechanism of development and progression of chronic liver disease in diabetes is still poorly understood, the contribution of oxidative stress and inflammatory cytokines is clear. Chronic hyperglycemia induces oxidative stress through various mechanisms, including forming advanced glycation end products, polyol pathway, hexosamine pathway, activation of protein kinase C, glucose auto-oxidation, and mitochondrial superoxide overproduction to various organ and tissue dysfunctions through reactive oxygen production.

The results also showed a decrease in triglycerides in the water fraction of Moringa leaves at a dose of 50 mg/kgBW and a water fraction of Moringa leaves at 200mg/kgBW not statistically significant. This decrease occurred due to bioactive compounds and phytochemicals from Moringa leaf extract (Moringa Oleifera). According to Wardhani (2020), through phytochemical analysis, Moringa Oleifera contains tannins, saponins, flavonoids (quercetin and dodgem alcohol), phenolic acids, alkaloids, -sitosterol, carotenoids, terpenes, and isothiocyanates, has antioxidant activity, can regulate fat and blood sugar metabolism. The content of flavonoids, tannins, saponins, and sitosterol reduces excessive triglycerides in the body. Isothiocyanates, phenolic acids, flavonoids, and terpenoids exhibit hypoglycemic activity. This weight loss effect is due to carotenoids, alkaloids, flavonoids, and isothiocyanates in Moringa (Moringa Oleifera) leaf extract. Rats fed a high-fat diet, an aqueous extract of M.oleifera leaves with an oral dose of 1 g/kgBW per day for 30 days experienced a significant reduction in total cholesterol in serum, liver, and kidney 14.4%, 6.4 %, and 11.1%.[23] According to Chumarket et al. (2008), an oral daily dose of 100 mg/kg aqueous extract of M. oleifera leaves can reduce total serum cholesterol and lipoprotein cholesterol by 50% and 75%, respectively, while carotid plaque reduces 97%.

This study also showed increased triglyceride levels in the moringa leaf water fraction at a dose of 100 mg/kgBW from 87.15 mg/dl to 105.24 mg/dl. These results indicate an increase but are still within normal limits. According to the researcher's assumption, theoretically, this condition may be due to the accumulation of fat in the liver, thereby increasing the synthesis of triglycerides in the liver. Triglyceride levels that are still within normal limits may be caused by a balance between fat accumulation which increases triglyceride levels in the body with a decrease in triglyceride levels due to the phytochemical content present in the water fraction Moringa leaves such as flavonoids, tannins, saponins, and -sitosterol. So, suppose the rate of fat accumulation used for the synthesis of triglycerides is balanced with the rate of reduction in triglyceride levels due to the effects of phytochemical content. In that case, it will produce triglyceride levels that are suspended or within normal limits.

In the positive control group given metformin decreased triglyceride levels. This is in line with research by Golay (2008), that metformin reduces triglyceride levels by 10-20%, reduces VLDL synthesis in the liver, reduces 5-10% total cholesterol, lowers LDL-C, and increases HDL-C. According to Tandon (2007), Metformin can control risk factors related to cardiovascular disease in patients with type 2 diabetes and obesity.

Triglyceride levels in the standard control group with negative control increased. Based on the results of Allo's research (2013), in the negative control group who were only given CP 551 concentrate food and aquades without drug administration and the water fraction of Moringa leaves, it was seen that: triglyceride levels in this group increased although still within normal limits. Things that might affect the results obtained from negative controls are that they have not reached the pure line, there are differences in the ability to adapt to the environment and food, which can then affect the diet and physical activity of rats so that the triglyceride levels of rats increase in normal controls. This condition is possible because of the fat intake in the food given to the rats, namely pig feed CP 551 which contains protein content of 18.50-20.50%, the fat content of 4.00%, fiber 8.00%, and calcium 0.90%. [26]

5. Conclusions

Based on the results of the study on the effectiveness of the water fraction of Moringa leaf (Moringa oleifera) lipid profile of white rats (Rattus norvegicus) hyperglycemia induced by streptozotocin, it can be concluded that Moringa leaf extract (Moringa oleifera) was not practical for reducing triglyceride levels in white male rats (Rattus norvegicus) which were injected with streptozotocin. The concentration of Moringa leaf extract (Moringa oleifera) that the researchers used was not effective in reducing triglyceride levels in white male rats (Rattus norvegicus) which were injected with streptozotocin

References

[1] Lonappan, A, Bindu, G, Thomas, V, Jacob, J, Rajasekaran, C, Mathew, K T Diagnosis of diabetes mellitus using microwaves Journal of Electromagnetic Waves and Applications, 2007; 21(10): 1393-1401

[2] Krasteva, A, Panov, V, Krasteva, A, Kisselova, A, Krastev, Z Oralcavity and systemic diseases—Diabetes Mellitus Biotechnol Equip 25 2011; 2183–2186

[3] Tandi, J, Rizky, M, Mariani, R, Alan, F Uji Efek Ekstrak Etanol Daun Sukun (Artocarpus altilis (Parkinson Ex FA Zorn) Terhadap Penurunan Kadar Glukosa Darah, Kolesterol Total dan Gambaran Histopatologi Pankreas Tikus Putih Jantan (Rattus norvegicus) Hiperkolesterolemia-Diabetes Jurnal Sains dan kesehatan, 2017; 1(8): 384-396

[4] Yanti, E., Vino, R, N Pengaruh pemberian rebusan daun kelor (Moringa oleifera) terhadap penderita hipertensi Jurnal Ilmu Kesehatan. 2018; 3(1): 24-29

[5] Tian, W X, Ma, Z F, Zhang, S Y, Sun, Y H, and Li, B H Fatty Acid Synthase Inhibitor from Plants and Their Potential Application in The Prevention of Metabolic Syndrome Journal of Cancer Research and Clinical Oncologi Cancer 2011; 8:1-9

[6] Arauna, Y, Aulanni'am, dan Oktavianie, D A Studi Kadar Trigliserida dan Gambaran Histopatologi Hepar Hewan Model Tikus (Rattus norvegicus) Hiperkolesterolemia yang Diterapi dengan Ekstrak Air Benalu Mangga (Dendrophthoe petandra) Artikel Ilmiah 2012 Program Studi Pendidikan Dokter Hewan Universitas Brawijaya, Malang

[7] Wahyuni, P T, Syauqy, A Pengaruh pemberian pisang kepok (Musa paradisiaca forma typical) terhadap kadar glukosa darah puasa pada tikus sprague dawley pra sindrom metabolik (Doctoral dissertation, Diponegoro University) 2015; 4(2): 547-556

[8] Kamila Uji Efektivitas Kombinasi Ekstrak Etanol Sambiloto (Andrographis paniculata) dan Daun Afrika (Vernonia amygdalina) Terhadap Penurunan Kadar Gula Darah Pada Tikus Wistar yang Djoiinduksi Aloksan *Skripsi* 2017 Program Studi Kedokteran Universitas Muhammadiyah Palembang

[9] Tandon, V R Metformin therapy: Benefits beyond glycemic control *International Journal of Diabetes in Developing Countries*, 2007; *27*(1): 1-4

[10] Geerling, J J, Boon, M R, van der Zon, G C, *et al* Metformin lowers plasma triglycerides by promoting VLDL-triglyceride clearance by brown adipose tissue in mice *Diabetes*, 2014; *63*(3): 880-891.

[11] Sinaga, U K L, Ismawati, I, Asni, E (2018) Pengaruh Pemberian Asam Alfa Lipoat terhadap Kadar Kolesterol Total pada Tikus Diabetes Melitus Tipe 2 *Jurnal Ilmu Kedokteran*, 2018; *11*(2): 18-23

[12] Ismawati, I, Mukhyarjon, M, Romus, I, Sonia, D Efek asam alfa lipoat terhadap insulitis pada tikus diabetes melitus tipe 2 *Jurnal Gizi Klinik Indonesia*, 2019; *16*(2): 58-63.

[13] Adji, D Hubungan konsentrasi malondialdehida glukosa dan total kolesterol pada tikus putih yang diinjeksi dengan streptozotocin *Jurnal Sain Veteriner*, 2008; *26*(2): 73-77

[14] Ermawati, D, Rachmawati, B, Widyastiti, N S Efek suplementasi β-carotene terhadap kolesterol total, trigliserida dan malondialdehid pada tikus sprague dawley yang diabet *Jurnal Gizi Indonesia (The Indonesian Journal of Nutrition)*, 2014; *2*(2): 47-52

[15] Putri, S, R, Arta, F, Tasmini 2017 Pengaruh ekspresi lipoprotein lipase terhadap kadar trigliserida serum pada tikus model diabetes melitus tipe 2 yang terpapar kuersetin jangka pendek Prosiding seminar ilmiah PBBMI 2015 2017; 131-139

[16] Putri, S R, & Anggraini, D I Obesitas sebagai faktor resiko peningkatan kadar trigliserida. *Jurnal Majority*, 2015; *4*(9): 78-82

[17] Tsalissavrina, I, Wahono, D, Handayani, D Pengaruh pemberian diet tinggi karbohidrat dibandingkan diet tinggi lemak terhadap kadar trigliserida dan HDL darah pada *Rattus novergicus* galur wistar *Jurnal Kedokteran Brawijaya*, 2006; *22*(2): 80-89

[18] Ariany, D Efek ekstrak daun kelor *(Moringa Oleifera*) terhadap kadar trigliserida tikus jantan galur Sprague Dawley diabetes melitus yang diinduksi Streptozotocin *Tesis* 2019; Fakultas Kedokteran Universitas Islam Negeri Syarif Hidayatullah Jakarta
[19] Elly Wardani, E, Hadi Sunaryo, H, Muhammad Zaki Sopiani, Z, & Mulki Fatahillah, M (2015) Aktivitas antihipertrigliserida dan antihiperglikemia ekstrak daun kelor (*Moringa oleifera Lam*) pada tikus hipertrigliserida diabetes *Media Farmasi*, 2015; *12*(2): 199-212
[20] Darrashcytha, P P S, Rizki, M, & Priyambodo, S Pengaruh Pemberian Ekstrak Daun Kelor *(Moringa Oleifera)* terhadap Kadar Trigliserida dan High Density Lipoprotein pada Tikus Putih (Rattus Norvegicus) Dislipidemia *Jurnal Kedokteran*, 2017; *6*(3.1):
[21] Mohammed, R R, Omer, A K, Yener, Z, Uyar, A, Ahmed, A K Biomedical effects of Laurus nobilis L leaf extract on vital organs in streptozotocin-induced diabetic rats: Experimental research *Annals of Medicine and Surgery*, 2020; 6(10): 1-10
[22] Wardhani, T M Pemanfaatan Tanaman Kelor (Moringa oleifera Lam.) sebagai Sumber Terapi Preventif dan Kuratif pada Pasien Perlemakan Hepar dengan Sindrom Metabolik *SCRIPTA SCORE Scientific Medical Journal*, 2020; *1*(2): 12-12
[23] Stohs, S J, & Hartman, M J Review of the safety and efficacy of Moringa oleifera *Phytotherapy Research*, 2015; *29*(6): 796-804
[24] Chumark P, Khunawat P, Sanvarinda Y,et al Thein vitroandex vivoantioxidant properties, hypolipidemic andantiatherosclerotic activities of water extract of MoringaoleiferaLam leave *J Ethnopharmacol* 2008; 116: 439–446
[25] Allo, I G, Wowor, P M, & Awaloei, H Uji efek ekstrak etanol daun jambu biji (Psidium guajava L) terhadap kadar kolesterol total tikus Wistar (Rattus norvegicus) *eBiomedik*, 2013; *1*(1): 371-378
[26] Gas, K J I, Mahardika, I G, Suasta, I M Pengaruh Penggantian Ransum Komersial Dengan Ampastahu Terhadap Penampilan Babi Ras *Jurnal Peternakan Tropika*, 2015; *3*(3): 482-49

Optimum conditions for bioadsorption of lead (Pb) by durian skin waste biomass (*duriobzibethinus*)

S Aminah[1], I Said[1], P Ningsih[1], A Alfiana[1]

[1]Chemistry Education, Department of Mathematics and Natural Science Education, Faculty of Teacher Training and Education, Tadulako University

purnamaningsih@yahoo.com

Abstract. The purpose of this research is to figure out what the optimum conditions are for durian skin biomass to absorb Pb from its solution. The optimum conditions were determined by determining the optimal weight, pH for biosorption, contact time for lead biosorption, and initial concentration. Atomic Absorption Spectrophotometry (AAS) was used to determine the adsorbed analyte. The durian skin used in this study came from local durian skin waste in the Parigi district of Central Sulawesi. The weight of the adsorbent was varied between 40, 60, 80, 100, and 120 mg. The contact time was varied between 20, 40, 60, 80, and 100 minutes. The pH was varied between 2, 3, 4, 5, and 6, and the concentration was varied between 50, 100, 150, 200, and 250 ppm to determine the optimum conditions for Pb adsorption. The results showed that the optimum conditions for biosorption of durian peels were 120 mg with a 92.2 percent absorption capacity, 60 minutes of contact time with a 97.89% absorption capacity, pH 5 with a 99.81 percent absorption capacity, and a concentration of 50 ppm with a 99.59% absorption capacity.

Keywords : optimum, bioadsorption, durian, biomass

1. Introduction

Central Sulawesi is one of Indonesia's regions with many natural resources. Durian fruit is one among them. Durian flesh is the most commonly utilized part; however, durian skin is less widely used and will end up as garbage. Using durian peel waste as biomass is one of the preventative measures taken.

Humans use heavy metals in gasoline, batteries, paint, and various other industries, one of which is lead metal (Pb). Large amounts of lead can be dangerous to living things since it is poisonous and can interfere with the nervous system and kidney function. As a result, to regulate heavy metal pollution in the environment, the maximum metal content in waste to be discharged into the environment must be limited. The adsorption method is one of the techniques for reducing heavy metal waste. The adsorption method is a more practical option than other ways since it has a greater efficiency level and does not produce hazardous compounds as a side effect.

Adsorption is a reversible separation process that occurs when gas or liquid molecules are absorbed by a solid. There are two components in the adsorption process: the adsorbate, which is the substance that is absorbed, and the adsorbent, which is the substance that absorbs. Adsorbents are solids that can absorb liquids into the surface, whereas adsorbates are organic compounds, colors, and moisturizing agents. When a solution comes into contact with a solid adsorbent, the adsorbate molecules transfer

IOP Conf. Series: Earth and Environmental Science 1075 (2022) 012021 doi:10.1088/1755-1315/1075/1/012021

from the solution to the solid until the adsorbate concentration is dissolved and the solid is in equilibrium[1].

Biomass is a material derived from organic compounds that may be replenished by living organisms, including both animals and plants. With the content included in it, cellulose, biomass can be used to adsorb metal ions. Heavy metals in wastewater can be removed using these groups as ion exchangers and adsorbents. There are numerous varieties of biomass, one of which being durian skin. Durian skin contains components that are high in cellulose (50-60%), carboxymethylcellulose, and lignin, according to study (5%). This cellulose can be used to bind metals[2]. It is feasible to use durian skin biomass as an adsorbent to reduce lead (Pb) metal levels based on the features of the durian skin.

Precipitation, coagulation, and ion exchange are just a few of the physical and chemical mechanisms that have been used to reduce Pb levels in water. However, processing these ways is still costly, especially in developing countries. The adsorption process is a popular purification and separation technique in the industry because it is more cost-effective in water and wastewater treatment[3]. It is also a popular technique for reducing metal ions in wastewater.

Based on this, a study was carried out to see how effective durian skin biomass (Durio zibethinus) was at adsorbing lead metal (Pb). Adsorbent concentrations of 40, 60, 80, 100, and 120 mg were utilized. 20, 40, 60, 80, and 100 minutes, respectively contact time, are among the options. pH 2, pH 3, pH 4, pH 5, and pH 6 are the different pH levels. The adsorption capacity is then measured by altering the lead metal solution concentration (Pb).

2. Methods

2.1. Materials and tools

The following items were utilized in this study: 100 mL Erlenmeyer flasks, a 25 mL spatula, 50 mL volumetric flasks, 100 mL beakers, a dropper, an oven, a desiccator, a digital balance, a blender, a stirring rod, filter paper, aluminium foil, a 200-sieve mesh, a shaker, a funnel, a spray bottle, and the AAS. Durian skin (Durio zibenthinus), solid $Pb(NO_3)_2$, filter paper, label paper, aquadest, and HCl were employed in this experiment.

2.2. Procedures

2.2.1. Durian peel biomass adsorbent preparation (durio zibethinus). Parigi, Central Sulawesi, provided the sample. Then, using clean running water, clean the durian skin. Durian skin should be cut into small pieces and then ground until smooth. Smoothing the durian skin with a blender, sifting the durian skin that has been mashed using a 100-mesh sieve until a more delicate durian skin powder is obtained, the coarse powder is smashed and filtered back. The biomass is weighed using a digital balance before being stored in a dry and closed container.

2.2.2. Preparation of a Pb stock solution with a concentration of 1000 ppm $Pb(NO_3)_2$. The stock solution is made by placing 1.6 grams of solid Pb(NO3)2 in a beaker, adding 5 mL of distilled water to the beaker containing solids, stirring until completely dissolved, and then adding 5 mL of distilled water to the beaker containing solids. Into a 1000 mL volumetric flask, pour the solution, add distilled water to the mark and shake until homogenous.

2.2.3. Determination of the optimal weight of durian skin biomass for lead (Pb) metal adsorption. Various adsorbents weighing 20, 40, 60, 80, 100, and 120 mg were combined with 50 mL of 100 ppm Lead solution at pH 5 in a 100 mL Erlenmeyer tube, and the Erlenmeyer tube was covered with aluminium foil to determine the optimum weight. And using rubber bands, shake for 60 minutes using a shaker. Measure the absorbance of the solution using AAS at a wavelength of 217 nm after filtering the mixture using Whatman filter paper.

2.2.4. Determination of the optimum contact time for the adsorption of lead copper (Pb) by durian skin biomass (durio zibethinus). Weigh the durian skin to the desired weight and place it in a 100 mL Erlenmeyer. 50 mL of a Pb solution with a 100-ppm concentration, Aluminium foil is used to close the Erlenmeyer tube, tied with rubber. Shake the mixture using a shaker for 20 minutes, 40 minutes, 60 minutes, 80 minutes, and 100 minutes. Using Whatman filter paper, we filtered the mixture and then measured the absorbance of the filtrate with an AAS at 217 nm.

2.2.5. Determination of the ideal pH for lead (Pb) adsorption in durian skin biomass (durio zibethinus). Place the durian powder in a 100 mL Erlenmeyer and weigh it to the required weight. In 5 Erlenmeyer with pH 2, 3, 4, 5, and 6 versions, pour 50 mL of 100 ppm Pb solution. Shake for the optimum contact time, then let stand, filter with Whatman filter paper, and measure the absorbance of the filtrate with an AAS at 217 nm.

2.2.6. Determination of the optimal copper (Pb) adsorption concentration durian skin biomass (durio zibethinus). Weigh the durian skin powder to the optimum weight, then place it in a 100 mL Erlenmeyer with 50 mL of Pb solution at varying concentrations of 50, 100, 150, 200, and 250 ppm at a pH of 5, then shake the mixture for 60 minutes. Filter the mixture with Whatman filter paper, and then analyze the filtrate with an AAS at 217 nm.

3. Result and discussion

Durian skin (Figure 1) is one part of the durian that has not been appropriately used thus far; durian skin is just wasted and becomes garbage, contributing to pollution in the environment. Despite the fact that the durian skin has not been exploited to its full potential, it has been discovered that the durian skin's content can be used as a bio adsorbent to absorb heavy metals in water. This study aimed to determine the best conditions for the durian peel biomass adsorbent in the adsorption of lead (Pb) metal ions.

Figure 1. Durian skin

3.1. The effect of durian skin biomass weight on durian bark biomass adsorption of lead metal

The weight of the adsorbent was determined in this work using a durian skin biomass adsorbent to determine the maximum amount of adsorbent that may be utilized in the adsorption process to maximize adsorbent efficiency[5].

One of the parameters that can affect the adsorption process and determine the capacity of the adsorbent during the addition of the adsorbate concentration is the weight of the adsorbent. In investigating lead ion adsorption by durian skin biomass, several weights of 40 mg, 60 mg, 80 mg, 100 mg, and 120 mg were used to determine the best weight. A Pb solution with a concentration of 100 ppm at pH 5 was utilized, and the contact time was 60 minutes. Table 1 shows the effects of measuring the adsorbed Pb content on the durian skin biomass weight change.

Table 1. The effects of adsorbed Pb concentration on weight variations in durian skin biomass

Adsorbent Weight (mg)	Abs	pH	Contact Time (Minute)	Ci (mg/L)	Ceq (mg/L)	Cb (mg/L)	$^{x}/_{m}$ (mg/g)	% Adsorbed
40	0,406	5	60	100	23,6441	76,3559	95,4449	~~76,3559~~
60	0,583	5	60	100	38,6441	61,3559	51,1299	61,3559
80	0,534	5	60	100	34,4915	65,5085	40,9428	65,5085
100	0,467	5	60	100	28,8136	71,1864	35,5932	71,1864
120	0,219	5	60	100	7,7966	92,2034	4,1162	92,2034

The concentration of lead absorbed (Cb) by biomass can be calculated using equation below. Pb concentration in %

% Adsorption = ((Ci-Ceq))/Ci x 100

% Pb Absorbed = Cb/Ci x 100

Cb = Ci – Ceq

Where:
Cb = Absorbed Lead Concentration (mg/L)
Ci = Initial concentration of Pb solution (mg/l)
Ceq = Concentration of unabsorbed Pb solution (mg/L)

Based on these results, the measurement curve for the weight variation of durian peel biomass on Pb metal may be displayed in Figure 2.

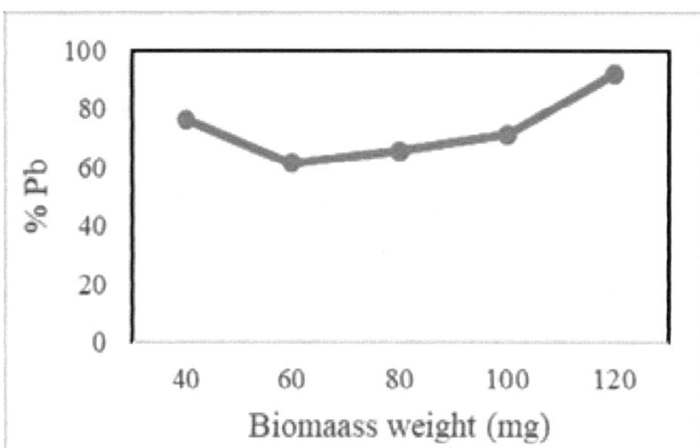

Figure 2. Relationship curve between durian skin biomass weight and Pb absorbed

According to the weight data, the adsorbent 40 mg has a higher absorption capacity of 76.36% than 60, 80, and 100 mg, which only has 61.36 % to 65.51 %. However, the adsorption increased from 61.36 % to 91.64 % at 60, 80, 100, and 120 mg. This illustrates that the adsorption process is affected by the weight of the adsorbent since as the weight of the adsorbent increases, the value of percent Pb adsorbed to ions increases as well until reaching equilibrium. The percentage increased by 92.2% when the adsorbent was 120 mg in weight. This is due to a rise in the number of adsorbents that interact with lead metal, as well as the effect on the cell density contained in the adsorbent, resulting in a very efficient interaction, with the more weight of the adsorbent used, the larger the number of pores used to adsorb metal ions[6].

According to the data, the 40 mg adsorbent weight has a better absorption capacity. Adsorption will increase as the number of adsorbents increases, potentially increasing efficiency and decreasing adsorption. There are charges in the adsorbent that are opposed to the metal
ions. Particles on the surface of the adsorbent contain a positively charged active site that interacts with the charge of negatively charged metal ions. At high weight, a decrease in adsorption reflects an accumulation of adsorption charges, reducing adsorption area[7].

The weight of the adsorbent that interacts with the lead metal causes the rise in absorption that happens. The density of the adsorbent cells in the solution increased adsorption, resulting in relatively effective interaction between the active center of the adsorbent cell wall and Pb ions; with the more substances adsorbed, the more active centers of the biomass reacted. As a result, as the amount of adsorbent increases, the ratio no longer holds, affecting the adsorbent's uptake of Pb metal ions[8]. The results revealed that as the weight of the durian peel biomass adsorbent increased, so did the number of particles and surface area of the durian skin biomass, increasing the active adsorption site and adsorption efficiency. In contrast, the adsorption capacity decreased with adsorbent[9]. This is supported by the perspective of[10], who claims that as the adsorbent weight grows, the proportion of adsorption increases while the adsorption decreases.

3.2. Durian skin biomass variation of optimal contact time against metal lead adsorption

The adsorption rate, defined as a change in concentration over time, is strongly connected to contact time in the adsorption process. The maximum adsorption limit in removing adsorbate can be found by stirring with a shaker by determining the optimum time in the adsorption process. The contact time was calculated based on the optimum weight attained during the prior treatment[11]. The best contact period in investigating lead ion adsorption by durian skin biomass was determined at 20 minutes, 40 minutes, 60 minutes, 80 minutes, and 100 minutes. A Pb solution with a concentration of 100 ppm at pH 5 and an optimal weight of 120 mg durian skin biomass was utilized. Table 2 shows the results of the Pb concentration measurement on the adsorption of Pb metal as a function of contact time variation.

Table 2. Measurement results of Pb concentration adsorbed by durian skin biomass at various contact times.

Adsorbent weight (mg)	Abs	pH	Contact time (Minute)	C_i (mg/L)	C_{eq} (mg/L)	C_b (mg/L)	x/m (mg/g)	% Adsorbed
120	0,171	5	20	100	3,7288	96,2712	40,113	96,2712
120	0,322	5	40	100	16,5254	83,4746	34,7811	83,4746
120	0,168	5	60	100	3,4746	96,5254	40,2189	97,892
120	0,26	5	80	100	11,2712	88,7288	36,9703	88,7288
120	0,274	5	100	100	12,4576	87,5424	36,476	87,5424

The measurement curve of the contact time variation of durian skin biomass on Pb metal can be displayed in Figure 3 based on these data results.

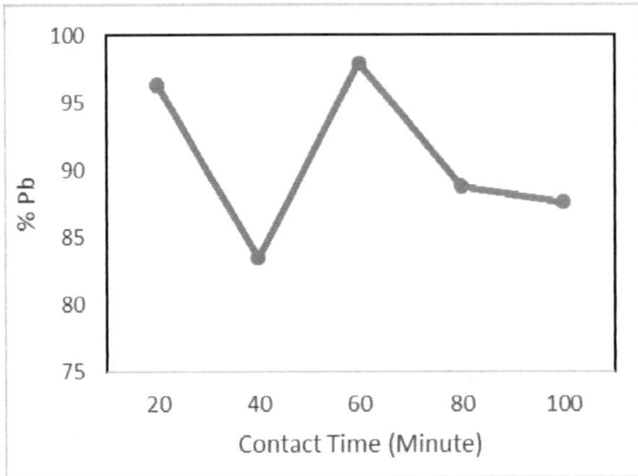

Figure 3. Relationship curve between durian peel biomass contact time and absorbed Pb

The results obtained from determining the optimum contact time indicate that the optimum contact time occurs at 60 minutes with an absorption capacity of 97.89% based on the optimum contact time variation data. Furthermore, while Pb metal ions were absorbed in 96.27% in the 20th minute, they were absorbed in 83.47% in the 40th minute. The adsorption process increased and decreased in the 20th and 40th minutes before reaching the optimum contact time due to reversible adsorption (back and forth), causing the adsorbent's absorption on Pb metal ions to become unstable[12].

The adsorption process experienced a decrease in absorption after reaching the optimum contact time, namely at 80 minutes with an absorption capacity of 88.73 percent and at 100 minutes with an absorption capacity of 87.54%. This is because the adsorbent can be said to have reached equilibrium at 60 minutes. After all, the metal removal efficiency of lead can decrease after passing. According to the analysis above, the efficiency and adsorption capacity declined after reaching the optimum time for removing the adsorbent. This is because desorption might occur when the adsorbent and the adsorbate period exceed the optimal contact time. The process of eliminating adsorbate from the surface of an adsorbent is known as desorption. This occurs because the adsorbent's character has become saturated, allowing the adsorbate molecules that have been adsorbed to return to the solution[13].

3.3. The effects of pH on lead adsorption by durian skin biomass

In studies on lead ion adsorption by durian skin biomass, pH variations of 2,3,4,5 and 6 were used to determine the optimal pH. Table 3 shows the measurement results of pH changes on lead adsorption by durian skin biomass.

Table 3. The results of adsorbed Pb concentration measurements at various pH levels

Adsorbent weight (mg)	Abs	pH	Contact time (Minute)	C_i (mg/L)	C_{eq} (mg/L)	C_b (mg/L)	x/m (mg/g)	% Adsorbed
120	0,008	2	60	100	0,3861	99,6139	41,5264	99,6139
120	0,007	3	60	100	0,3366	99,6634	41,5264	99,6634
120	0,005	4	60	100	0,2376	99,7624	41,5677	99,7624
120	0,004	5	60	100	0,1881	99,8119	41,5883	99,8119
120	0,006	6	60	100	0,2871	99,7129	41,547	99,7129

Based on these data results, it can be seen that the measurement curve of the pH variation of Pb metal by durian skin biomass can be seen in Figure 4.

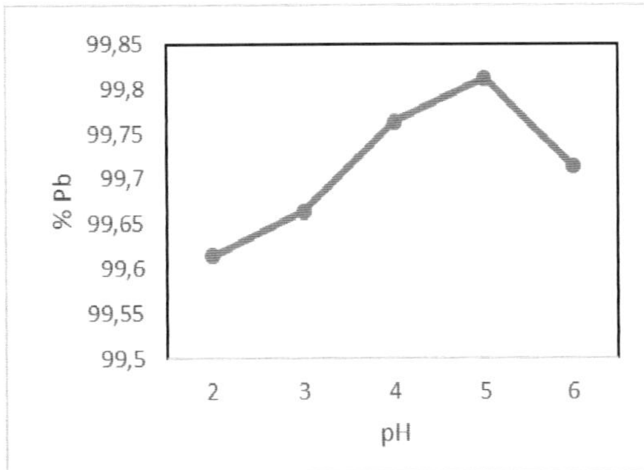

Figure 4. The curve of the relationship between pH and Pb absorbed by durian skin biomass

The degree of acidity (pH) is one of the parameters that influence the adsorption process of metals in solution because the presence of H+ ions in solution will compete with cations for binding to the active group. The interaction between ions and the active group of the adsorbent is also affected by pH[15].

The higher the pH, which means a decrease in acidity, the higher the adsorption power; however, the rise is insignificant, according to the data acquired in calculating the optimum pH. At pH 2, 3, 4, and 5, absorption capacity increased, reaching 99.61 % at pH 2, 99.66 % at pH 3, 99.76 % at pH 4, and 99.81 % at pH 5, whereas absorption capacity at pH 5 6 fell to 99.71%. This suggests that the adsorption of lead metal by durian skin biomass is affected by pH. The decrease in absorption caused by H+ negative functional groups will react with H+, preventing metal ions from attaching [16]. Based on the findings above, the best pH for adsorption of lead metal using durian skin biomass is pH 5, with a percentage of 99.81 % absorption.

3.4. From the data obtained in determining the optimum pH variation of optimum concentration of Pb metal ions by durian skin biomass

The adsorption capacity was determined using different concentrations of 50, 100, 150, 200, and 250 ppm at an ideal pH of 5 and a biomass weight of 120 mg with a 60-minute contact period. The goal of determining the optimum concentration variation of lead metal adsorption by durian skin biomass is to determine the optimum concentration of lead solution adsorption by the adsorbent durian peel biomass. Table 4 shows the findings of establishing the optimum concentration variation for lead metal adsorption by durian skin biomass.

Table 4. The results of the measurement of the adsorption capacity of lead metal by durian skin biomass

Weight Adsorbent (mg)	Abs	pH	Contact time (Minute)	C_i (mg/L)	C_{eq} (mg/L)	C_b (mg/L)	$^x/_m$ (mg/g)	% Adsorbed
120	0,014	5	60	50	0,256	49,744	41,4533	99,488
120	0,024	5	60	100	0,5536	99,4464	41,436	99,4464
120	0,056	5	60	150	1,506	149,4464	41,2333	98,96
120	0,124	5	60	200	3,5298	196,4702	40,9313	98,2351
120	0,183	5	60	250	5,2857	244,7134	40,7857	97,8857

Based on these data results, it can be seen that the curve of the relationship between adsorption capacity and % Pb absorbed by durian skin biomass can be seen in Figure 5.

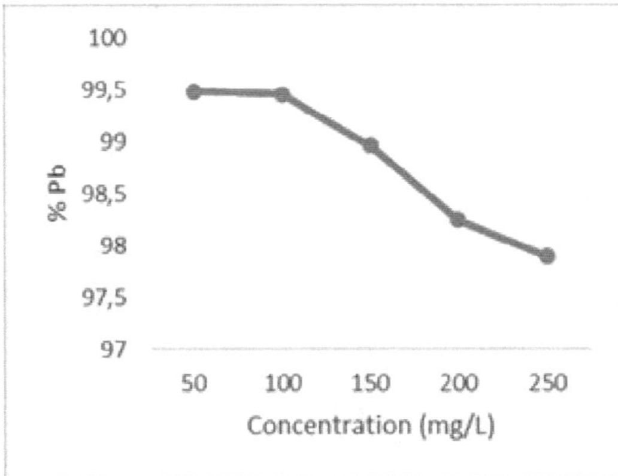

Figure 5. The curve of the relationship between adsorption capacity and %Pb absorbed

The efficacy of adsorption at various concentrations revealed that the optimal condition occurred at a concentration of 50 ppm with a 99.49 % absorption capacity, indicating that the adsorbent had reached equilibrium at this concentration. While at a concentration of 100 ppm, 99.45 %, 150 ppm, 98.96 %, 200 ppm, 98.24 %, and 250 ppm, 97.89 %, the absorption capacity continues to diminish. This is because, at a concentration of 50 ppm, the amount of adsorbent injected is in equilibrium.
The stronger the adsorption on the metal, the more adsorbent is applied. This leads to greater metal ion adsorption values, which are proportional to the number and surface area of the adsorbent.

4. Conclusion

According to studies based on trials and estimates, Durian skin biomass can act as a biosorbent. The weight of the adsorbent, contact time, pH, and concentration all have a significant impact on absorption. The ideal weight was found to be 120 mg, with a Cu metal absorption percentage of 99.08 %. The best contact time was 100 minutes, with a percentage of Cu metal absorption of 98.31%. The ideal pH was pH 5, with a Cu metal absorption percentage of 99.72%. At a concentration of 50 ppm, with a Cu metal absorption percentage of 98.92%, the adsorption capacity was in equilibrium with varied concentrations.

Acknowledgments

The authors acknowledge the Faculty of Teacher Training and Education-Tadulako University for financial funding in this study.

References

[1] Ashbahani (2013) Pemanfaatan Limbah Ampas Tebu sebagai Karbon Aktif untuk Menurunkan Kadar Besi pada Air Sumur *Jurnal Teknik Sipil Untan* vol 13, no 1 (2013): h, 106

[2] Apriani 2013 Pengaruh Konsentrasi Aktivator Kalium Hidroksida (KOH) terhadap Kualitas Karbon Aktif Kulit Durian sebagai Adsorben Logam Pb pada Air Gambut. *Prisma Fisika* Vol I, No 2 (2013), Hal 82 – 6

[3] *Selvi, K., Pattabi S and Kardivelu K (2001) Removal of Cr (Vl) from aqueos solution by adsorption onto activated carbon Bioresour technol Vol 80 : 87-9*

[4] Irwanto (2014) Studi Pemanfaatan Kalsium Karbonat (CaCO3) dari Serbuk Cangkang Telur Sebagai Adsorben Terhadap Ion Raksa (Hg2+) Skripsi : tidak diterbitkan

[5] Satriani, D, Ningsih, P, & Ratman (2016) Serbuk dari Limbah cangkang Telur Ayam Ras Sebagai Adsorben Terhadap Logam Timbal (Pb) Jurnal Akademika Kimia, 5(3), 103- 8

[6] Podala, K., Walanda, D K., & Napitupulu, M (2016) Biocharcoal Dari kulit kakao (Theobrema cacao L) untuk Mengadsorpsi Ion Logam Timbal Jurnal Akademika Kimia 4(3) 136- 42

[7] Setiawan, I K. A, Napitupulu, M, &Walanda, D K. (2018) Biocharcoal dari Kulit Rambutan (Nephelium Lappaceum L) sebagai Adsorben Zink dan Tembaga Jurnal Akademika Kimia 7(4), 193 - 9

[8] Radyawati, (2011) Pembuatan Biocharcoal dari Kulit Pisang Kepok untuk Penyerapan Logam Timbal (Pb) dan Logam Seng (Zn) Skripsi Sarjana pada FKIP Universitas Tadulako Palu : tidak diterbitkan

[9] Apriliani, A (2010) Pemanfaatan Arang dari Ampas Tebu Untuk Menyerap Ion Logam Cd, Cr, Cu, dan Pb dalam Air Limbah Skripsi pada UIN Syarif Hidayatullah Jakarta : Diterbitkan

[10] Baros, J L, Maedo G R, Duarte M M, Silvia E P & lobato (2003) Biosorpsi Cadmium Using The Fungus Asprgillus Niger *Breaz Journal chemistry*, 20 : 1-17

[11] Syauqiah, I, Amalia, M, & Kartini, H A (2011) Analisis Variasi Waktu dan Kecepatan Pengaduk pada Proses Adsorpsi Limbah Logam Berat dengan Arang Aktif. *Info Teknik*, 12(1), 11-20

[12] Martina, D, Hastuti, R, & widodo, D S (2016) Peran Adsorben Selulosa Tongkol Jagung (*Zea Mays)* dengan Polivinil Alcohol (PVA) untuk Penyerapan Logam Timbal (Pb^{2+}) *Jurnal Kimia Sains dan Aplikasi* 19(3), 77-82

[13] Windiastuti, R, T, & Santoso, T (2020) Determination of Optimum pH and Contact Time from the Adsorption Process of Cu(II) Ions by Corn Cob (*Zea Mays*) Biomass *Jurnal Akademika Kimia,* 9(4), 224 – 9

[14] Ginting, F D (2008) Pengujuan Alat Pendingin Sistem Adsorpsi Dua Adsorber Dengan Menggunakan Methanol 100 mL Sebagai Refrigeran Jurnal Universitas Indonesia : Jakarta

[15] Nurmalasari, D, Hastuti, R, & widodo, D S (2015) Pengaruh Penambahan Polivinil Alcohol pada Biomassa Tongkol Jagung-Bulu Ayam Sebagai Adsorben Campuran Ion Logam Tembaga dan Kromium *Jurnal Kimia Sains dan Aplikasi* Vol 18 No 1 : 18-23

[16] Alfiany, H , S Bahri, & Nurakhirawati (2013) Kajian penggunaan arang aktif tongkol jagung sebagai adsorben logam Pb dengan beberapa Aktivator asam *Jurnal Natural Science* 2(3) : 75-86

Optimum conditions for absorption of copper (Cu) from its solution using durian skin waste biomass (Duriobzibethinus)

I Said[1], S Aminah[1], P Ningsih[1], S Chairunnisa[1]

[1]Chemistry Education, Department of Mathematics and Natural Science Education, Faculty of Teacher Training and Education, Tadulako University

purnamaningsih@yahoo.com

Abstract. This study aimed to determine the optimum conditions for durian skin biomass to absorb copper metal (Cu) from its solution. Determination of optimum conditions includes optimum weight, optimum pH for biosorption, biosorption contact time, and initial concentration of copper metal (Cu). Determination of the adsorbed analyte used Atomic Absorption Spectrophotometry (AAS). In this study, the durian skin used was obtained from local durian skin waste from the Parigi Regency, Central Sulawesi Province. The parameters tested were the weight of the adsorbent or the weight of the powder with variations of 50 mg, 70 mg, 90 mg, 110 mg, and 120 mg, variations in contact time 20 minutes, 40 minutes, 60 minutes, 80 minutes, and 110 minutes, variations in pH 2, 3, 4, 5, 6 and concentration variations of 50 ppm, 100 ppm, 150 ppm, 200 ppm, and 250 ppm to determine the optimum conditions for adsorption of copper (Cu). The results of this study indicate that the optimum conditions for absorption of copper metal by durian skin biomass respectively, the optimum weight for biosorption was 120 mg with an absorption capacity of 99.08%, the optimum contact time was 100 minutes with an absorption capacity of 98.31%, the optimum pH was 5 with the absorption capacity of 99.72% and the optimum concentration of 50 ppm with an absorption capacity of 98.92%.

Keywords: Absorption, Durian, Biomass

1. Introduction

Heavy metals can be found in natural and artificial combinations and garbage (waste). Humans, animals, plants, and the ecosystem are all at risk from heavy metal pollution in the environment. Heavy metals can enter and accumulate in the human body via the food chain, beverages, or the air. Heavy metals are classed as B3 waste in specific quantities, which can harm the ecosystem since they are hazardous to animals and people [1]. Adsorption, precipitation processes, filtration, electrochemistry, reverse osmosis, oxidation/reduction, ion exchange, evaporation, and membrane technology are some techniques that can be used [2].

An adsorption process is a common approach for reducing heavy metal ions in wastewater and is an effective purification and separation process utilized in the industry since it is regarded as more economical in wastewater treatment [3]. Biosorption refers to biomass's ability to bind heavy metals from solutions via metabolic or chemical-physical processes. The presence of active groups in biological materials adsorbents, according to [4], plays a function in the adsorption process of heavy metals with these materials.

Durian peel (Durio zibethinus) is a good material to utilize as a copper adsorbent since it contains 50-60% carboxymethylcellulose and 5% lignin [5]. Because durian skin has a high cellulose content, it can be used as a heavy metal bio adsorbent [6]. Based on the preceding description, a study utilized durian skin waste biomass to determine the best conditions for absorbing Cu from its solution (Durio zibethinus). This research aims to determine the best conditions for durian skin biomass to absorb Cu from its solution. Optimal circumstances were measured, including optimal weight, ideal pH for biosorption, biosorption contact time, and initial Cu concentration.

2. Method

2.1. Materials and instruments

100 mL Erlenmeyer, 25 mL spatula, 50 mL volumetric flasks, 100 mL beakers, dropper, oven, desiccator, analytical balance, blender, stirring rod, filter paper, aluminum foil, 100 sieves, mesh, shaker, funnel, spray bottle, and Atomic Absorption Spectrophotometry (SSA) GBC 932 AA were the instruments used in this study. Durian skin (Durio zibenthinus), CuSO4.5H2O, aqua dest, HCl, label paper, and filter paper were utilized in this experiment.

2.2. Procedures

2.2.1. Durian skin biomass adsorbent preparation (Durio zibethinus). The sample was taken at Parigi, in the province of Central Sulawesi. The first step is to clean the durian skin with clean running water, cut the durian skin into small pieces, and then dry the durian skin sample in an oven at 105°C for 3 hours to reduce the water content in the durian skin. The durian skin sample was then chilled at room temperature, smoothed with a blender, sieved the mashed durian skin with a 100-mesh sieve to obtain a more delicate durian skin powder. The coarse powder was mashed and sifted again, and the biomass was weighed with a digital balance. Then store it in a dry, closed container.

2.2.2. Preparation of copper (Cu) stock solution 1000 ppm $CuSO_4.5H_2O$. To make the stock solution, start by putting 4.024 grams of solid $CuSO_4.5H_2O$ into a beaker, then adding 5 mL of distilled water into a beaker containing solids, stirring until completely dissolved, then pouring the solution into a 1000 mL volumetric flask and topping it off with distilled water to the limit mark. Then shook until the mixture was completely homogenous.

2.2.3. Determination of the optimum weight of copper (Cu) metal adsorption by durian skin biomass. Various adsorbents weighing 50, 70, 90, 110, and 120 mg were mixed with 50 mL of 100 ppm copper solution at pH 5 in a 100 mL Erlenmeyer tube, then closed the Erlenmeyer tube with aluminium foil and tied with rubber, shaken for 60 minutes, filtered the mixture using Whatman filter paper, and measured the absorbance of the solution using AAS at a wavelength of 324.8 nm.

*2.2.4. Determination of optimum contact time for copper (Cu) metal adsorption by durian skin biomass (Durio zibethinus).*Weigh the durian rind to the optimum weight and place it in a 100 mL Erlenmeyer tube, add 50 mL of Cu solution with a concentration of 100 ppm, close the Erlenmeyer tube with aluminium foil and rubber, and shake the mixture with a shaker for 20, 40, 60, 80, and 100 minutes. Using Whatman filter paper, the mixture was filtered, and the absorbance of the filtrate was measured using an AAS at a wavelength of 324.8 nm.

*2.2.5. Determination of optimum pH for copper (Cu) metal adsorption by durian skin biomass (Durio zibethinus).*Weigh the durian rind to the optimum weight, then place it in a 100 mL Erlenmeyer, pour 50 mL of 100 ppm Cu solution into 5 100 mL Erlenmeyer fruits with pH variations of 2, 3, 4, 5, and 6, shake with a shaker for the optimum contact time previously obtained, then allow to stand, filter the mixture using Whatman filter paper, and analyze the absorbance of the filtrate with an AAS.

2.2.6. Determination of the optimum concentration of copper (Cu) adsorption by durian skin biomass (Durio zibethinus). Weigh the durian peel powder to the desired weight, then place it in a 100 mL Erlenmeyer with 50 mL Cu solution at varying concentrations of 50, 100, 150, 200, and 250 ppm with pH 5 agitate the mixture for 60 minutes. Filter the mixture with Whatman filter paper and examine the filtrate with an AAS at 324.8 nm.

3. Result and discussion

The findings of this study include weight, contact time, pH, and concentration as optimum conditions for absorption of copper metal (Cu) from its solution using durian skin waste biomass (*Durio zibethinus*).

3.1. Durian skin biomass weight variation on lead metal adsorption by durian skin biomass. durian skin biomass adsorbent preparation.

The AAS test was used to determine the optimum copper (Cu) metal biosorption weight by durian skin biomass, with five sample variations of 50,70,90,110, and 120 mg. The goal of obtaining the optimum adsorbent weight is to identify the highest quantity of adsorbent that may be utilized for the adsorption process to maximize adsorbent efficiency [7]. Table 1 shows the results of the Cu concentration measurement at the optimum weight.

Table 1. Data on the results of determining the optimum weight on the Cu metal adsorption process by durian skin biomass

Adsorbent weight (mg)	C_i (mg/L)	C_{eq} (mg/L)	C_b (mg/L)	% Adsorbed
50	100	1,6070	98,393	98,39
70	100	1,0988	99,9012	98,90
90	100	1,0987	98,9013	98,65
110	100	1,1059	98,8941	98,89
120	100	0,9198	99,0802	99,08

Based on these data results, it can be seen that the measurement curve for the weight variation of durian skin biomass on copper metal (Cu) can be seen in Figure 1.

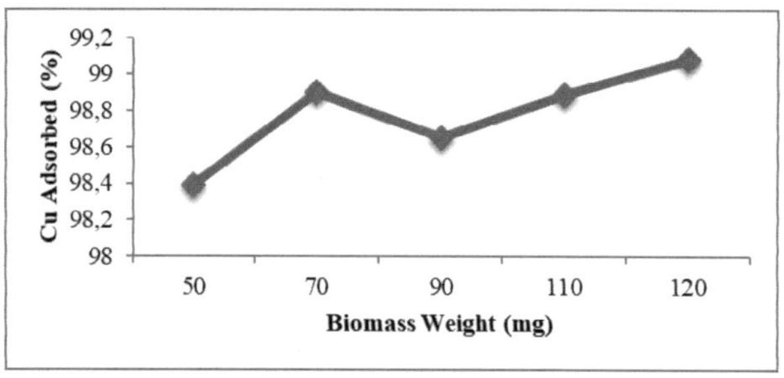

Figure 1. Relationship curve between biomass weight of durian skin and absorbed Cu

The amount of Cu absorbed is affected by the weight of the adsorbent employed, as shown in Figure1. The adsorbent 50 mg has a low absorption capacity compared to other weight variants of the adsorbent, 98.39%. With an adsorption capacity of 98.90%, 110 mg with an adsorption capacity of 98.89 percent, and 120 mg with an adsorption capacity of 99.08%, there was an increase in absorption.

As the concentration of biomass rises, so does the amount of solute biosorbent, whose surface area expands and the amount of metal-bound increases. On the other hand, as the biosorbent dose increases, the amount of biosorbent solute per unit weight at a given location decreases. This is assumed to be owing to the saturation of the sorbent's surface pores, which have been filled with heavy metals, preventing the biosorbent from reabsorbing. The increase in adsorption capacity becomes minor at a certain point. This is because an increase in the active site is not accompanied by an increase in the volume of the aqueous medium used as a reaction site [8].

With the addition of the adsorbent, the weight of the durian peel biomass adsorbent increased proportionally to the number of particles and surface area of the durian skin biomass, resulting in an increase in the active adsorption site and an increase in adsorption efficiency during the adsorption capacity decreased [3].

According to the statistics above, the optimum weight for adsorption of copper metal utilizing durian skin biomass is 120 mg, with a 99.08 percent absorption rate.

3.2. Variation of optimum contact time against copper (Cu) metal adsorption by durian skin biomass

Because contact time is a quantity that might represent the kinetics of an adsorbent in interacting with the adsorbate, adsorption time is one of the parameters in the process of adsorption [9]. The goal of determining the optimum contact time is to determine how long the metal must be exposed to the adsorbent before being absorbed.

In investigating copper metal adsorption by durian skin biomass, the optimum contact time was determined at various durations of 20, 40, 60, 80, and 100 minutes. A Cu solution with a concentration of 100 ppm and a pH of 5 was utilized, with an optimal weight of 120 mg durian skin biomass. Table 2 shows the results of the Cu concentration measurement at the optimum contact time.

Table 2. Measurement results of Cu concentration adsorbed by durian skin biomass at various contact times.

Contact Time	C_i (mg/L)	C_{eq} (mg/L)	C_b (mg/L)	% Adsorbed
20	100	2,994	97,006	97,01
40	100	3,060	96,939	96,94
60	100	2,874	97,126	97,13
80	100	2,959	97,040	97,04
100	100	1,692	98,307	98,31

Based on these data results, it can be seen that the measurement curve for the variation of the contact time of durian skin biomass on Cu metal can be seen in Figure 2.

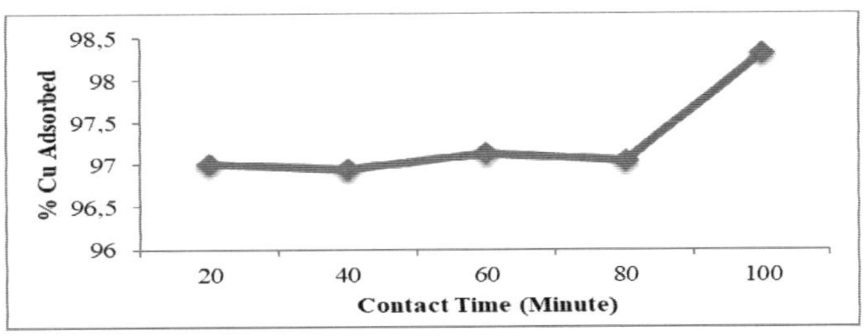

Figure 2. Relationship curve between durian peel biomass contact time and absorbed Cu

Figure 2 determination of the optimum contact time reveals that the difference in contact time from 20 to 80 minutes is insignificant. Furthermore, compared to other contact times, the absorption of Cu metal, which is 98.31%, increased dramatically at the 100th minute. The adsorption process increased and decreased in the 20th and 80th minutes before reaching the optimum contact time. The adsorption that happened was reversible (back and forth), causing the adsorbent absorption on Cu metal to become unstable [10].

According to the research, the time necessary to absorb Cu metal optimally is 100 minutes. The graph also shows that the Cu adsorption process runs steadily after the optimum contact period attained equilibrium. The adsorption capacity increases at the optimum time, that is, when the equilibrium time has been reached.

3.3. Variation of pH on lead adsorption by durian skin biomass

Because the process of heavy metal ion absorption by biomass is significantly controlled by pH, the optimum pH in the study of Cu metal adsorption by durian skin biomass was determined at pH variations of 2,3,4,5 and 6. The goal of determining the optimum pH is to figure out what pH is required for optimal metal adsorption by the adsorbent [11]. Table 3 shows the assessment results of pH changes on Cu metal adsorption by durian skin biomass.

Table 3. Measurement results of adsorbed Pb concentration at various pH

pH	C_i (mg/L)	C_{eq} (mg/L)	C_b (mg/L)	% Adsorbed
2	100	1,436	98,564	98,56
3	100	1,1769	98,8231	98,82
4	100	0,6108	99,3892	99,39
5	100	0,2807	99,7193	99,72
6	100	0,9175	99,0825	99,08

Based on these data results, it can be seen that the measurement curve for the pH variation of Cu metal by durian skin biomass can be seen in Figure 3.

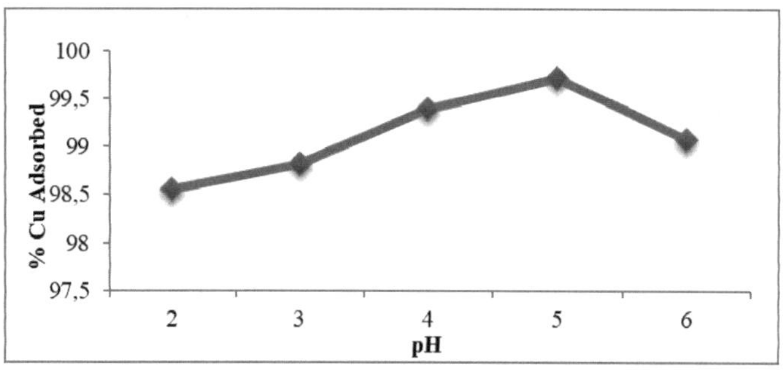

Figure 3. The curve of the relationship between pH and Cu absorbed by durian skin biomass

The curve for obtaining the optimum pH in Figure 3 indicates that it increased from pH 2 to pH. At pH 5, the most adsorption occurs, but at pH 6, the capacity decreases again. This means that the higher the pH, the lower the acidity, and hence the more significant the metal ion absorption. Because pH affects the surface charge of the adsorbent and any species in the adsorption process, the magnitude of the pH value of the solution impacts the adsorption process [12].

Based on the data above, the optimum pH using durian skin biomass in adsorption of copper metal is at pH 5 with a percentage of 99.72% absorption

3.4. Variation of optimum concentration of Cu metal ions by durian bark biomass

The absorption ability of an adsorbent is highly influenced by the concentration of a metal ion solution. The goal of varying the concentration of a solution is to figure out the relationship between metal ion adsorption power in varied ideal concentrations of durian skin biomass. The adsorption capacity was determined by increasing concentrations of 50, 100, 150, 200, and 250 ppm at an optimal pH of 5 and an optimum biomass weight of 120 mg over a 100-minute contact time. Table 4 shows the findings of establishing the optimum concentration variation for lead adsorption by biomass Durian skin.

Table 4. The results of measuring the adsorption capacity of copper metal by concentration variations

Concentration	C_i (mg/L)	Check	C_b (mg/L)	%Cu Adsorbed
50 ppm	50	0,5369	49,4631	98,92
100 ppm	100	4,6206	95,379	95,37
150 ppm	150	8,6722	141,327	94,22
200 ppm	200	11,2133	188,787	94,39
250 ppm	250	12,4796	237,520	95,01

Based on these data results, it can be seen that the curve of the relationship between adsorption capacity and % Cu absorbed by durian skin biomass can be seen in Figure 4.

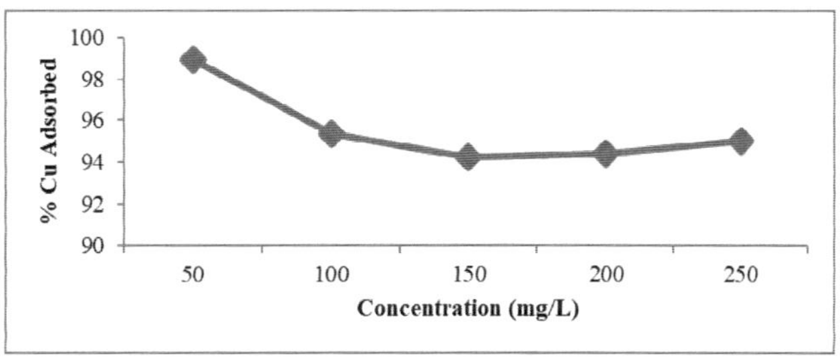

Figure 4. The curve of the relationship between adsorption capacity and % Cu absorbed

The collision theory helps explain the effect of concentration on the adsorption process. Because there are more molecules in each unit area of space at higher concentrations, collisions between molecules are more likely. The larger the number of crashes, the greater the chance of producing effective collisions, resulting in a faster reaction [13].

The concentration of copper ions surged at 50 ppm and then declined, as seen in the graph. Because there are many adsorption groups with active groups proportional to the area of the adsorbent, this is achievable. When the active group is saturated, the increase in absorbate concentration decreases; nevertheless, if the adsorbate does not saturate the adsorbate, the rise in concentration increases the amount of adsorbate adsorbed [14].

The heavy metal's initial concentration determines the mechanism of heavy metal absorption. Specific sites absorb heavy metals at low concentrations, but as metal concentrations rise, specific areas become saturated, and exchange sites become filled [15].

According to the preceding description, the optimum concentration of copper metal in durian skin biomass is 50 ppm, with a 98.92% absorption rate.

4. Conclusion

According to a study based on trials and estimates, Durian skin biomass can act as a biosorbent. The weight of the adsorbent, contact time, pH, and concentration all have a significant impact on absorption. The ideal weight was found to be 120 mg, with a Cu metal absorption percentage of 99.08 percent. The best contact time was 100 minutes, with a rate of Cu metal absorption of 98.31 percent. The ideal pH was pH 5, with a Cu metal absorption percentage of 99.72 percent. At a concentration of 50 ppm, with a Cu metal absorption percentage of 98.92 percent, the adsorption capacity was in equilibrium with varied concentrations.

Acknowledgments

The authors gratefully acknowledge the financial support of the Faculty of Teacher Training and Education-Tadulako University for this study.

References

[1] Ariyani, dan Sukma Budi (2019) Karakteristik bioadsorben dari limbah kulit durian untuk menyerapan logam berat Fe san Zn pada air sumur Jurnal teknologi proses dan inovasi industri (Vol 4 hal 23)

[2] Deng, S, Wanga,DP, Zhangb, G, and DoucaState, Y (2016)Polyacrylonitrile-based Fiber Modified with Thiosemicarbazide by Microwave Irradiation and its Adsorption Behavior for Cd(II) and Pb(II) Journal of Hazardous Materials, 307:64-72

[3] Apriliani, A (2010) Pemanfaatan Arang Ampas Tebu sebagai Adsorben Ion Logam Cd, Cr, Cu, dan Pb dalam Limbah Air Limbah Jurnal Program Studi Kimia Fakultas Sains dan Teknologi Jakarta

[4] Ahalya, N, Ramachandra, TV, and Kanamadi, RD (2003) Biosorption of Heavy Metal Research Journal Of Chemical And Environment Vol 7(4), 71-9

[5] Apriani, R, Faryuni, ID, dan Wahyuni, D (2013): Pengaruh Konsentrasi Aktivator Kalium Hidroksida (KOH) terhadap Kualitas Karbon Aktif Kulit Durian sebagai Adsorben Logam Fe pada Air Gambut Jurnal Prisma Fisika, Vol1 No2 Hal 82-6 Fakultas Matematika dan Ilmu Pengetahuan Alam Universitas Tanjungpura Pontianak

[6] Ariyani, dan Sukma Budi (2019) Karakteristik bioadsorben dari limbah kulit durian untuk menyerapan logam berat Fe san Zn pada air sumur Jurnal teknologi proses dan inovasi industri (Vol 4 hal 23)

[7] Satriani, D; Ningsih, P; dan Ratman, Serbuk dari Limbah Cangkang Telur Ayam Ras sebagai Adsorben terhadap Logam Timbal (Pb), 2016, Jurnal Akademika Kimia, 5(3), 103-8

[8] Mulyawan, R, Saefumillah, A, dan Foliatini, F (2015) Biosorpsi Timbal Oleh Biomassa Daun Ketapang Molekul, 10(1), 45-56

[9] Safrianti, I, Wahyuni, N, dan Zaharah, T A (2012) Adsorpsi timbal (II) oleh selulosa limbah jerami padi teraktivasi asam nitrat: pengaruh pH dan waktu kontak Jurnal Kimia Khatulistiwa, 1(1)

[10] Syauqiah, I, Amalia, M, dan Kartini, H A (2011) Analisis Variasi Waktu dan Kecepatan Pengaduk pada Proses Adsorpsi Limbah Logam Berat dengan Arang Aktif Info Teknik, 12(1), 11-20

[11] Windiastuti, R, T, and Santoso, T (2020) Determination of Optimum pH and Contact Time from the Adsorption Process of Cu(II) Ions by Corn Cob (Zea Mays) Biomass Jurnal Akademika Kimia, 9(4), 224-9 A reference

[12] Febriani, H, Puspitasari, D J, dan Sosidi, H (2021) Adsorbsi Ion Logam Cu (II) Menggunakan Biomassa Daun Genjer (Limnocharis flava) KOVALEN: Jurnal Riset Kimia, 7(2), 131-6

[13] Widhianti, W D (2010) Pembuatan Arang Aktif dari Biji Kapuk (Cieba pentandra L) Skripsi

[14] Husna, N U, Mohadi, R, dan Said, M (2019) Studi Adsorpsi Ion Cr (Vi) Menggunakan Asam Humat Dari Tanah Gambut Dengan Bantuan Radiasi Sinar Uv (Doctoral dissertation, Sriwijaya University)

[15] Saeed, A, Waheed-Akhter, M, and Iqbal, M (2005) Removal and recovery of heavy metals from aqueous solution using papaya wood as a new biosorbent Sep Purif Technol 45:

Risk factors of chronic rhinitis in furniture workers of West Palu, Central Sulawesi.

C R Nayoan[1] A H Q Thalib[2]

[1]ENT – HNS Department, Faculty of Medicine, Tadulako University, Palu, Central Sulawesi, Indonesia
[2]Faculty of Medicine, Tadulako University, Palu, Central Sulawesi, Indonesia

Ch.lapadji@gmail.com[1], qalby.thalib@yahoo.com[2]

Abstract. Chronic Rhinitis is an inflammatory disease of the nose characterized by sneezing, snotty and nasal congestion, intermittently or persistently. The prevalence of rhinitis in the United Kingdom reaches 29%, in Indonesia is not known clearly. The types of work that cause rhinitis are chronic like farmers, laboratory workers, carpenters or painters, food industry workers and, health workers. Chronic Rhinitis can affect the quality of the worker's life, and it can reduce work productivity Objective: To determine the risk factors for chronic rhinitis on the workers in wood furniture which was contacted by sawdust in the West Palu area Method: This study uses analytical research methods with a cross-sectional approach. The sampling technique is total sampling, and the total sample in this study is 27 workers. Results: The study showed a significant relationship with the incidence of rhinitis to the carpenters who work on wooden furniture in the western Palu on age ($p = 0.027$), duration of exposure ($p = 0.041$. There was no significant relationship between the use of PPE ($p = 0.100$) Conclusion: The factors that can influence the incidence of chronic rhinitis to furniture workers are age and duration of exposure

Keywords: Chronic Rhinitis, Furniture Worker, Sawdust, Palu

1. Introduction

Rhinitis originated from two Greek words; "rhin / rhino," which means nose, and "it is," which means inflammation. So, rhinitis means nose inflammation or an inflammatory condition affecting the nasal mucosa[1]. The common rhinitis symptom includes repeated sneezing, accompanied by other symptoms such as watery rhinorrhoea, nasal congestion, itchy nose, and eyes with profuse lacrimation[2]. In the field of otorhinolaryngology (ENT), occupational diseases are commonly found in occupational rhinitis, which is the same with chronic rhinitis, hearing loss, corrosive esophagitis, and head injuries[3]. Occupational rhinitis is chronic rhinitis caused by exposure to materials from the workplace, including dust, fumes, vapors, and gases. These symptoms may not appear until several months or even years after the first exposure. Therefore, these symptoms can be suspected as occupational rhinitis if there is an association between time and openness at work, and the condition improves if they are not at work[4].

The prevalence of rhinitis in the UK reaches 29%, while Denmark is 31.5%, and in the USA around 33.6%. The exact prevalence of rhinitis in Indonesia is poorly known, but data from various hospitals shows that rhinitis has a frequency ranging from 10-26%[5]. In occupational rhinitis or chronic rhinitis, it is estimated that 15% of industrial workers are the most likely to suffer from occupational rhinitis (48%), followed by administrative workers (29%) and material processing workers (13%). Several occupations considering high risk for chronic rhinitis include farmers, laboratory workers, carpenters or painters, food industry workers, and health workers[6].

One occupation that needs attention to the risk of rhinitis is wooden furniture workers because exposure to wood dust as a raw material for producing furniture is harmful[6]. Working in an environment with wood dust causes inhalation of particles through the respiratory tract, including the nose. Wood dust is the second most common cause of occupational rhinitis. Exposure to wood dust causes inflammatory diseases in the upper and lower respiratory tract. Because there is still a lack of literature explaining the mechanism of the inflammatory process caused by wood dust in the respiratory tract, especially the nose, we urge to have a better scientific understanding through this study since furniture workers are commonly found in Palu, especially West Palu.

2. Method

This study was an analytical observational study using a cross-sectional design. This study was conducted in Palu City, West Palu District, Central Sulawesi, in July 2019. The sampling technique used was total sampling, where the entire population–all wooden furniture workers in Palu City, West Palu District, Central Sulawesi, act as a study sample. The inclusion criteria in this study included: Wooden furniture workers in West Palu District, Palu City, Central Sulawesi, who were directly exposed to sawdust, used PPE while working, were willing to sign informed consent, and were ready to complete the study questionnaires. Meanwhile, the exclusion criteria for this study included: wooden furniture workers who are absent or sick and wooden furniture workers who have a history of illness.

3. Results

The population in this study is wooden furniture workers in the West Palu district. 32 participants were eligible for inclusion criteria and then recruited for this study, while the rest 27 people met the exclusion criteria. The sampling technique used was total sampling with predetermined inclusion and exclusion criteria. The data collected in this study are primary data through questionnaires completion to determine factors that influence the incidence of chronic rhinitis in wooden furniture workers exposed to sawdust in the district of West Palu, Central Sulawesi.

Table 1. Distribution of time of exposure, use of PPE, age, and incidence of chronic rhinitis in wooden furniture workers with sawdust exposure in West Palu District, Central Sulawesi

Variable	Frequency	%
Length of exposure (hours)		
< 8	3	11,1%
> 8	24	88,9%
Use of PPE		
Yes	13	48,1%
No	14	51,9%
Age (years)		
< 20	0	0%
20-30	13	48,1%
30-40	5	18,5%

>40	9	33,3%
Chronic Rhinitis		
Sufferer	19	70,4%
Non-sufferer	8	29,6%

Based on Table 1., it is known that the majority of the length of exposure is > 8 hours (88.9%, n=24). For the use of the PPE variable, it is known that most wooden furniture workers do not wear PPE (51.9%, n=14) while working, compared to workers who wear PPE. The age group of respondents in this study was mostly 20-30 years old (48.1%, n=13), while the minority age group was 30-40 years old (18.5%, n=5). This study also found that 19 people (70.4%) of 27 respondents suffered from chronic rhinitis.

Table 2. The correlation test results between the length of exposure with chronic rhinitis in wooden furniture workers exposed to sawdust in West Palu District

			Chronic Rhinitis		***P*-value**
			Non-sufferer	**Sufferer**	
Length of exposure	< 8 hours	Frequency	1	2	0.041
		%	12.5	10.5	
	> 8 hours	Frequency	7	17	
		%	87.5	89.5	

Table 2. shows that the correlation between length of exposure with chronic rhinitis in wooden furniture workers exposed to sawdust in West Palu District, Central Sulawesi, is statistically correlated with the p-value = 0.041 ($p<0.05$). The correlation between PPE use and chronic rhinitis in wooden furniture workers exposed to sawdust in West Palu District, Central Sulawesi, is not statistically correlated with the p-value = 0.100 ($p>0.05$), as shown in table 3.

Table 3. The correlation test results between the use of PPE with chronic rhinitis in wooden furniture workers exposed to sawdust in West Palu District

			Chronic Rhinitis		***P*-value**
			Non-sufferer	**Sufferer**	
Use of PPE	Yes	Frequency	4	9	0.100
		%	50.0	47.4	
	No	Frequency	4	10	
		%	50.0	52.6	

Table 4. The correlation test results between age with chronic rhinitis in wooden furniture workers exposed to sawdust in West Palu District

			Chronic Rhinitis		***P*-value**
			Non- Sufferer	**Sufferer**	
Age (years)	20-30	Frequency	7	6	0,025
		%	87.5	31.6	
	30-40	Frequency	0	5	
		%	0.0	26.3	
	> 40	Frequency	1	8	
		%	12.5	42.1	

Table 4. shows that the correlation between age and chronic rhinitis in wooden furniture workers exposed to sawdust in West Palu District, Central Sulawesi, is statistically correlated with the p-value = 0.025 ($p<0.05$).

4. Discussion

This study found that the majority of the length of exposure is > 8 hours (88.9%, n=24). For the use of the PPE variable, it is known that most wooden furniture workers do not wear PPE (51.9%, n=14) while working, compared to workers who wear PPE. Additionally, the age group of respondents in this study was mostly 20-30 years old (48.1%, n=13), while the minority age group was 30-40 years old (18.5%, n=5). The statistical analysis performed in this study was chi-square test, with obtained p-value = 0.041 ($p <0.05$), meaning that there is a significant correlation between the length of exposure to the risk factors of chronic rhinitis in wooden furniture workers exposed to sawdust in the West Palu district, Central Sulawesi.

According to a study by Solikhah[7], the risk of workers experiencing health problems due to wood dust can be prevented. Prevention efforts that can be done are 1) reducing dust and fume in the air, either by using exhaust ventilation, installing dust control devices, and cleaning the workspace regularly, 2) using PPE when working, 3) applying the rules of no working hours which exceed 8 hours/day, and 4) set a threshold value. The threshold value is a common hazard factor in the workplace as a time-weighted average level/intensity that workers can accept without causing illness or health problems for workers in carrying out their work every day, which is not more than 8 hours/day or 40 hours/week[7].

The results of this study are not in line with Damayanti's[6] study in Semarang, which found that the study findings were considered not statistically significant. Although the length of exposure per day is not statistically significant, the incidence of occupational rhinitis / chronic rhinitis is frequent in workers who work 8 hours or more per day compared to workers who work less than 8 hours per day [6]. According to the existing theory, if workers are frequently exposed, and the length of exposure is accumulated into years of working, then there will be continuous exposure for years. If there is constant exposure over time, it can lead to irreversible tissue damage, so that the sensitivity of the airway will increase both to allergens and non-allergens[6].

This study found no significant correlation between the use of PPE and chronic rhinitis. These results are in line with research conducted by Quadrusman[4] in Makassar, which states that there is no correlation between PPE use and the incidence of occupational rhinitis / chronic rhinitis. Therefore, it can be concluded that although there is no correlation between the use of PPE and the incidence of rhinitis, the risk of developing rhinitis will increase in people who do not use masks properly. Furthermore, this statement is in line with the results of a study conducted by Fachrudin[8], which was found that people who do not wear masks properly are at twice the risk of having rhinitis compared to people who wear masks. This condition is due to the higher exposure to sawdust for people who do not wear masks properly[8].

The statistical analysis performed in this study was a chi-square test, with obtained p-value = 0.025 ($p <0.05$), meaning that there is a significant correlation between respondents age with the risk factors of chronic rhinitis in wooden furniture workers exposed to sawdust in the West Palu district, Central Sulawesi. According to the available scientific evidence, more than 30 years is a productive age at work. Therefore, chronic rhinitis is commonly found in this age group compared to those aged 30 and under. This study finding is in line with other studies which state that rhinitis is frequently found in the 3rd to 4th decade of life. However, the results of this study are not in line with research conducted by Damayanti [6], which found that there was no significant correlation between age and the incidence of occupational rhinitis / chronic rhinitis [6]. In this study, the risk factors for chronic rhinitis in wooden furniture workers exposed to sawdust in West Palu District, Central Sulawesi, are the length of exposure and age. However, in this study, although PPE use was not related to the chronic rhinitis in wooden furniture workers in West Palu District, Central Sulawesi, the workers who did not use PPE were more likely to suffer from chronic rhinitis.

IOP Conf. Series: Earth and Environmental Science 1075 (2022) 012023 doi:10.1088/1755-1315/1075/1/012023

5. Conclusion

Based on the study results and discussion that have been described previously, several conclusions can be drawn as follows:

a. There is a correlation between age and the incidence of chronic rhinitis in wooden furniture workers in West Palu with a p-value of 0.025 ($p<0.05$)
b. There is a correlation between length of exposure and the incidence of chronic rhinitis in wooden furniture workers in West Palu with a p-value of 0.041 ($p<0.05$)
c. There is no significant correlation between PPE use and the incidence of chronic rhinitis in wooden furniture workers in West Palu.

References

[1] Sabilarussdi, and Roestiniadi, DS Diagnosis, and management of occupational rhinitis 2010; 6(1)
[2] Rafi, M, Adnan, A, and Masdar, H Overview of allergic rhinitis in students of the medical faculty of the University of Riau in 2013-2014 2015; 2(2)
[3] Manuputty, AC, Rahardjo, SP, Djamin, R, and Perkasa, F The relationship between exposure to wheat dust and the quality of life of patients with occupational rhinitis 2010; 18(1)
[4] Quadarusman, E, Rahardjo, SP, Punagi, AQ, and Djamin, R The risk of occupational rhinitis in workers exposed to wheat dust 2011; 41(1)
[5] Widuri, A IgE antibody therapy in allergic rhinitis 2009; 9(1)
[6] Damayanti, A R, Yusmawan, W, Naftali, Z Risk factors for occupational rhinitis in car painting workers using spray paint (study on car painting workshops in Semarang city) 2016; 5(4)
[7] Sholikhah, A, M and Sumarmaji The relationship between worker characteristics and total dust content with respiratory complaints in wood industry workers x in Lumajang district 2015; 1(1)
[8] Fahrudin I 2006 Occupational rhinitis and related factors Thesis of Graduate Program in Occupational Medicine Jakarta: FKUI

Monitoring tools for liquefaction potential in a region based on the internet of things (IoT) with photovoltaic (PV) sources

R Fauzi[1], A Mustari[1], R Santoso[1], Alamsyah[1], Y A Rahman[1]

[1]Department of Electrical Engineering, Engineering Faculty, Tadulako University

rfauzi@gmail.com

Abstract. From several records of liquefaction events that have occurred, it is known that liquefaction events during an earthquake are more likely to occur in coarse-grained soil conditions. In this case, the soil will quickly lose shear strength due to increased pore water stress due to cyclic loads (earthquake loads) very rapidly and in a moment. Due to the loss of load and shear stress, the soil properties, initially solid, turn into liquid (Mase, L. Z., & Somantri, A. K. 2016). This study will design a liquefaction potential monitoring tool in an internet of things (IoT) based area with Photovoltaic (PV) sources. This tool provides information to the public to identify areas that can cause liquefaction. The design of the liquefaction potential monitoring tool in a room will also offer data detected by vibration sensors, temperature sensors, and humidity sensors below the ground surface. When the vibration sensor notices a vibration in sequence, the temperature and humidity sensors will activate and read the changes that occur in the soil.

Keywords: Liquefaction, data logger, IoT

1. Introduction

Liquefaction is an event where the soil loses shear strength due to increased pore water stress due to the occurrence of cyclic loads (earthquake loads) very quickly and in a moment. Due to the loss of load and shear stress, the soil properties, initially solid, turn into liquid [2],[4]. From several records of liquefaction events that have occurred, it is known that liquefaction events during an earthquake are more likely to be in coarse-grained soil conditions (granular soil), this type of soil includes gravel, fine sand, muddy sand, or former swamp soil with little or no clay. Grounds of this type do not have significant attractive forces between particles in the earth. Some opinions that liquefaction occurs only in granular soils based on observations in most of the history of liquefaction in the world [3]. Although there are also findings that indicate it occurs in fine-grained soils [6].

Seeing the potential for very damaging liquefaction has a terrible impact on infrastructure. It results in casualties, so it is essential to design a tool to determine the potential for liquefaction in an area, especially in Palu, Central Sulawesi, and its surroundings. This tool will be used to record changes that occur below the ground surface shortly after a minor earthquake occurs in an area. The value captured by the sensor is then sent to the webserver to analyze the potential for liquefaction in the area. This study will design a liquefaction potential monitoring tool in an internet of things (IoT) based space with Photovoltaic (PV) sources. This tool provides information to the public to identify areas that can cause liquefaction. The design of the liquefaction potential monitoring tool in a room will also offer data detected by vibration sensors, temperature sensors, and humidity sensors below the ground surface. When the vibration sensor notices vibration in sequence, the temperature and humidity sensors will activate and read the changes that occur in the soil [1].

2. Material and methods

2.1. Research road map

In achieving the objectives of this research, several previous studies have been developed that have supported this research. Preliminary studies and research to be conducted are described in the form of a road map, which can be seen in Figure 1:

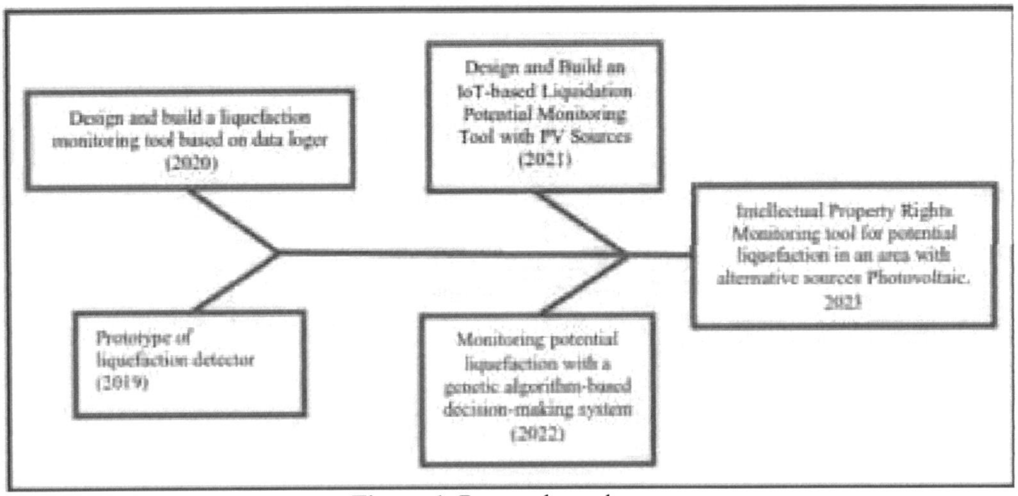

Figure 1. Research road map

2.2. Description of the tool design stage

During this stage, several activities are carried out, including:

1. Assemble the minimum system consisting of the Arduino nano atmega 328 board and the port that connects the sensor and arduino. The sensors used are temperature, humidity, and vibration in this case.
2. Create a database connected to the webserver for storing sensor data that has been read and sent by the microcontroller.
3. Make a panel where the sensor circuit holder, minimum system and solar charge control (SCC), and the battery, in this case, the 12V/10A battery, are stored and protected from weather changes.
4. Make the photovoltaic panel holder and the circuit panel as a place to attach the board and the connecting cable between the photovoltaic and SCC and the sensor circuit and cable installed below the ground surface, in this case, the temperature and humidity sensor.

2.3. Description of the tool design stage

During this stage, several activities are carried out, including:

1. Test the input voltage value from the photovoltaic (PV) to the 12V/10A battery and the control circuit.
2. Measure the value of the Analog To Digital Converter (ADC) from the temperature sensor, humidity sensor, and vibration sensor.
3. Conduct lab tests on the performance of the liquefaction potential monitoring tool by taking samples of environmental models that have liquefaction potential.

2.4. System design

The data logger system used consists of a sensor as a reader of the output value of the PV, where the sensors used are temperature sensors and voltage sensors. Sensor readings are carried out with an analog to digital converter (ADC) device integrated into the minimum microcontroller system.

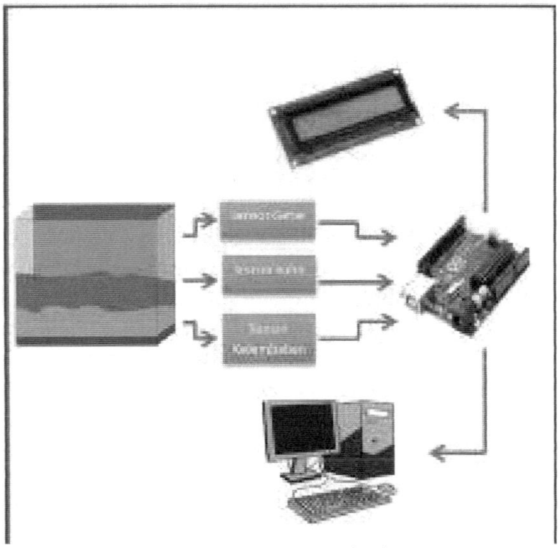

Figure 2. System design

Figure 3. Connection of hardware implementation

Figure 4. A panel of monitoring tools

Figure 5. Wiring and connection

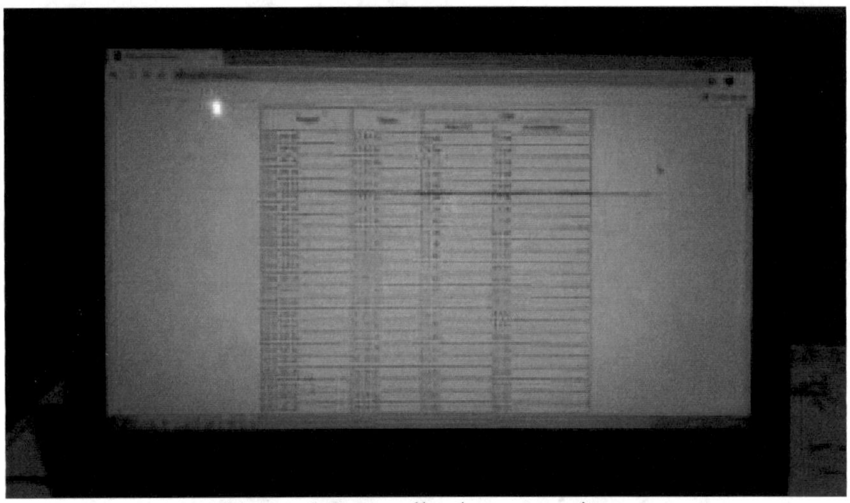

Figure 6. Data collection on monitors

3. Results and discussion

The ultimate goal of this research is to monitor the process of reading sender data in the form of vibration, temperature, and humidity sensors based on the Internet of Things (IoT) which in this case, the alternative energy source used in this tool is sourced from Photovoltaic by utilizing the intensity of sunlight. This tool can monitor soil conditions from several parameters: vibration, temperature, and humidity. These three parameters are variables that often appear in areas with liquefaction potential. In this way, the monitoring process is expected to inform the potential for liquefaction in a room. However, many supporting data are needed to strengthen this liquefaction potential, one of which is a significant change in ground level. The data read by the three sensors will then be sent to the webserver for later data analysis and processing into parameter data for potential liquefaction markers. The display is as follows:

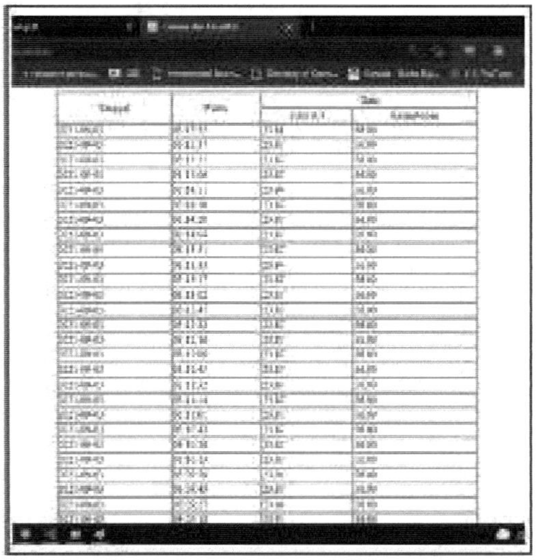

Figure 7. Data of sensors in webserver

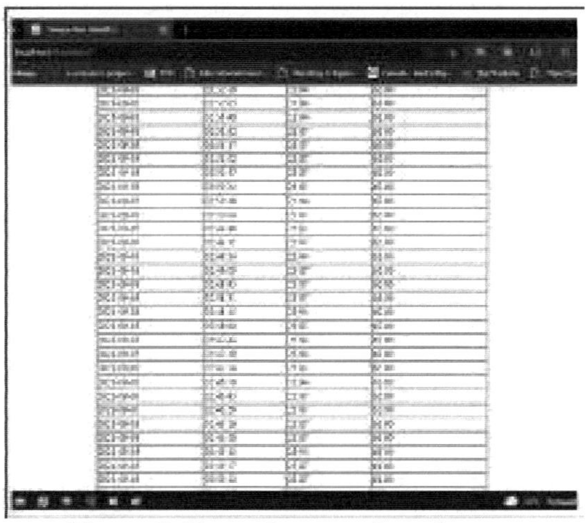

Figure 8. Data of sensors in webserver

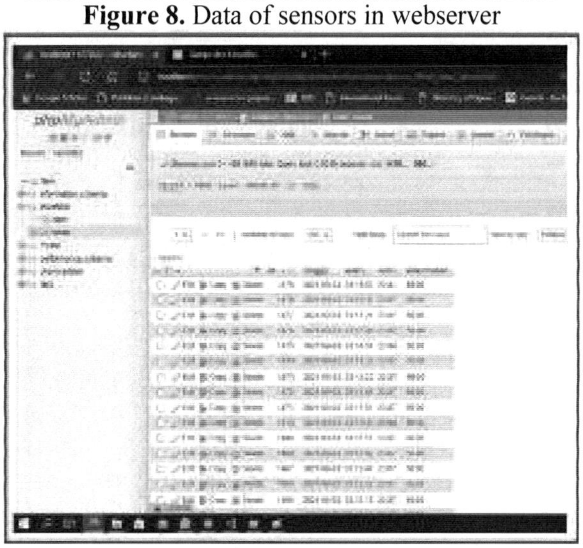

Figure 9. Database input sensors

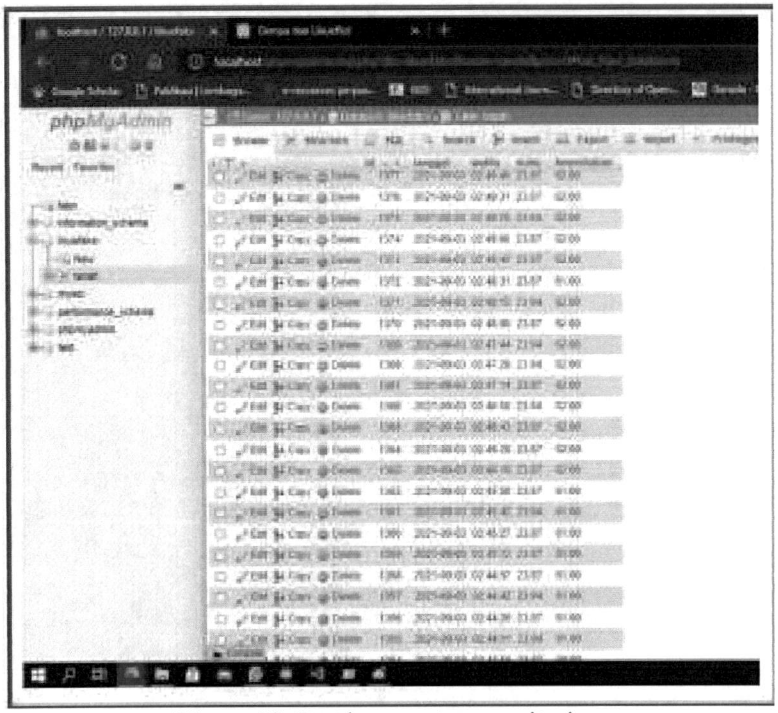

Figure 10. Database system monitoring

In the table above, it can be seen that the data has been received on the webserver. The data is in data read by temperature and humidity sensors. Temperature and humidity sensors read values from the environment in analog data. This analog data is then processed into digital data by the microcontroller, which is already incorporated in the minimum system board of the MCU esp the 32 node system. This MCU node then sends the data obtained to the webserver, which is displayed in the form of a sequential table on the webserver that has been formed.

4. Conclusions

For the measurement of temperature, humidity, and vibration data, it can be read, but sometimes there are errors due to the susceptible condition of the vibration sensor. However, environmental conditions in general, including temperature and humidity, can be adequately measured. Photovoltaic installed as an alternative power source has functioned well and is relatively stable in providing an electricity supply to the liquefaction potential recording device.

Acknowledgment
The authors would like to thank all those who have supported this research: Electrical Engineering of Tadulako University and the office Regional Disaster Management Agency of the Province of Central Sulawesi, Indonesia.

References

[1] Burange, A W, & Misalkar, H D (2015) Review of Internet of Things in Development of Smart Cities with Data Management & Privacy
[2] Enden Mina, Rama Indera, Sudirman (2018) "Analisa Potensi Likuifaksi Berdasarkan Data Spt (Studi Kasusproyek Pembangunan Gedung Baru Untirta Sindang Sari)" JURNAL FONDASI 71
[3] Firda Zettyara, Harimurti, Yulvi Zaika (2018) "Analisis Potensi Likuifaksi Akibat Gempa Bumi

Menggunakan Data Cpt (Cone Penetration Test) Di Kabupaten Tulungagung" Jurnal Teknik Sipil 12

[4] Kusuma, Rama Indera, Enden Mina, and Nuri Kurniawan (2017) "Analisa Potensi Likuifaksi Dari Data Cpt (Studi Kasus Sinter & Coke Plant Area Krakatau Posco)" JURNAL FONDASI 31

[5] Mase, L Z, & Somantri, A K (2016) Analisis Potensi Likuifaksi di Kelurahan Lempuing Kota Bengkulu Menggunakan Percepatan Maksimum Kritis Potensi : Jurnal Sipil Politeknik, 18(1)

[6] Miftachul Janah, Supriyadi, Ian Yulianti, Yuni Dwi Trisnowati (2016) Analisis Seismotektonik Papua Bagian Utara Melalui Pemetaan Sebaran Mekanisme Fokus Unnes Physics Journal 5 (2)

Effect of partial fly ash as cement substitute on the fracture toughness

A P N Siregar[1], A Rivani[1], H Mellisa[1], A Aswin[2], Karlina[3]
[1]Jurusan Teknik Sipil Universitas Tadulako (Untad) Palu
[2]Graduated student Teknik Sipil Universitas Tadulako (Untad) Palu
[3]Proram Studi Diluar Kampus Utama (PSKDU) Untad-Morowali

** Corresponding author email: atur_pns@yahoo.com*

Abstract. *The development of construction has an impact on recent innovations. An alternative material is needed to replace some cement by considering the fracture toughness to reduce the impact of failure on construction. One of them is fly ash. The fracture toughness (KIC) is the fracture of a material to evaluate the ability of constituents containing defects to resist fracture. Test specimens with dimensions of 50 × 100 × 350, 50 × 150 × 500, 50 × 300 × 650 and 50 × 400 × 1250. The addition of fly ash with levels of 0%, 5%, 10%, and 15% has an effect on the compressive strength, namely in the cement binding process so that at a certain percentage level it will increase the strength of the concrete and the condition of concrete collapse is affected by binding cement with reduced aggregate due to the addition of fly ash is evidenced by the type A, B, C and D specimens that have decreased in the addition of fly ash from 0% condition.*
Keywords: *Fracture Toughness, Fly Ash, Compressive Strength Test, Three Point Bend Test*

1. Introduction

The development of construction has impact on the innovations recently both in the field of meodelling dan materials. Concrete is one of the materials often used in the field constructions and infrastructure [1].

SNI 03-2834-2002 defines that concrete is a mixture consisting of Portland cement or other hydraulic cement, gravel, sand and water with and or without additives to form a solid mass. From the above definition, it is found that the concrete constituents are cement, gravel, sand, water and other necessary additives

In accordance with the needs, the types of cement are developed and made according to their requests that are tailored to the needs of the construction, but Portland cement is still the type of cement commonly used in construction in Indonesia. Based on data from the Indonesian Cement Association (ASI) on its website, cement production in Indonesia has increased until 2018 produced 30,047,831 tons of cement.

Based on the above background, an alternative material is needed as partial replacement cement by considering on the fracture concrete in order to reduce the impact of failure on construction, as we know that the collapse behavior of concrete is constituents materials influenced.

One of the alternative binders that can be used is fly ash, which is the residue from burning coal at the Mpanau Palu Steam Power Plant (PLTU). With the addition of fly ash in concrete, it is hoped that it will have an influence on aspects of the collapse behavior. While the most dominant basic material for making concrete is aggregate which is 60% - 70% of the total volume of concrete [2] which has a good influence on the fracture toughness of concrete with the Three Point Bend test is the level of aggregate hardness [3]. By using the same aggregate material, it can be seen the effect of fly ash on the fracture toughness of concrete.

2. Literature Review

2.1. Concrete

Concrete is a composite material that has been widely used in the construction sector. [4] defines concrete as a material composed of a mixture of Portland cement or other hydraulic cement, fine aggregate, coarse aggregate and water and with or without other additives. Thus concrete is a mixture of several aggregates which are chemically bonded together by the hydration product of Portland cement with or without other alternative materials added. It can be concluded that the basic ingredients for making concrete namely cement, aggregate (fine and coarse), water and if needed can be added with additional materials, so that the mechanical properties of concrete depend on these materials.

According to [5] there are several parameters that affect the mechanical properties of concrete including the cement content of the mixture, the strength and cleanliness of the gravel, good mixing method, the position of the concrete placement, the finishing and compaction of the concrete, the chloride content of not more than 0.15% in the concrete for exposed and 1% of unexposed concrete and its performance quality.

2.2. Concrete Composition Material

The basic ingredients for making concrete are:

a. Portland Cement

Portland cement is the most commonly used construction material, besides being easy to apply and to obtain so it is widely used in construction. ASTM C-150-1985 defines Portland cement as hydraulic cement produced by pulverizing clinker consisting of calcium and hydraulic silicates, essentially containing one or more calcium sulfate as an additive which is ground together with the main ingredients.

b. Fine Aggregate

Material/particle with a diameter between 0 - 5 mm or fine aggregate that can pass through a 5 mm test sieve is called sand. For the gradation limits, the fine aggregates are grouped into 4 zones as shown below, Figure 1.

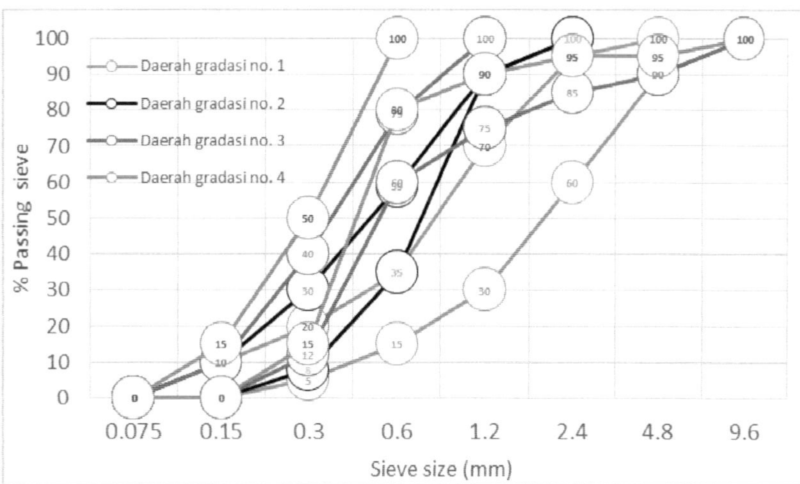

Figure 1. Sand gradation limit [5]

c. Coarse Aggregate

Gravel or crushed stone is coarse aggregate as a mixture of concrete. In general, what is meant by coarse aggregate is particle with a grain size of more than 5 mm, having a hard shape and without being porous. For the gradation limits, coarse aggregate is grouped as shown in Figure 2 below.

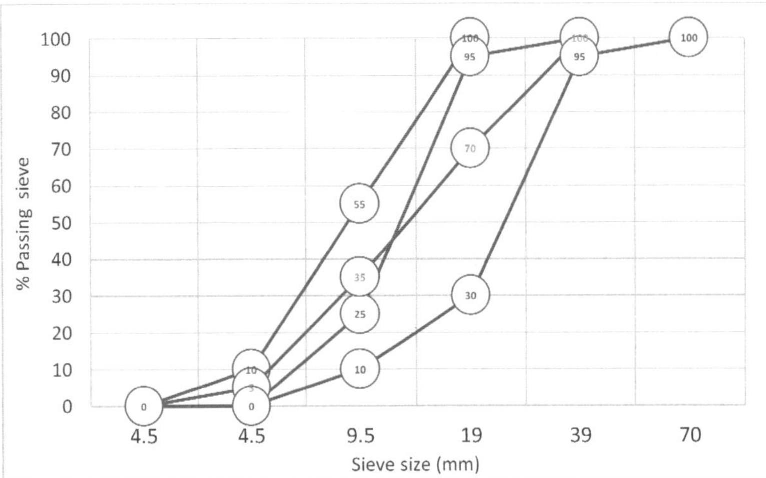

Figure 2. Gradation limit of gravel [5]

d. Water

Water is one of the most essential materials in the manufacture of concrete which serves to hydrate cement, mortar or paste. It takes about 25% of the weight of cement for hydration process water between cement and water, but in reality the value of water per weight of cement will be difficult to mix if it is less than 0.35 [6].

e. Fly Ash

Fly Ash is grouped into two types, namely class F and class C according to ASTM C.618. The results of burning anthracite or bituminous coal get fly ash for class F and while burning lignite or sub-bituminous coal produces class C. Meanwhile in the [7], the specification of fly ash as an additional mixed material for concrete mentioned there are 3 kinds of fly ash namely:

1. Class F fly ash is fly ash produced from the combustion of coal with anthracite type with a temperature of 1560°C
2. Class N fly ash is the product of natural pozzolans for example from diatone soil, shale, tuft, and pumice.
3. Class C fly ash is the result of combustion with a carbon content of 60%; Fly ash has characteristics like cement with lime content above 10%.

Based on the journal [8] Abu Batu Ex. PLTU Mpanau shows that Fly Ash is in the class F category with the chemical composition in the table below:

Table 1. Chemical Content of Fly Ash Ex. PLTU

Senyawa Kimia	Kandungan (%)
SiO_2	51,50
Al_2O_3	17,26
Fe_2O_3	25,54
CaO	2,09
K_2O	1,23
TiO_2	0,95
MnO	0,48
P_2O_5	0,25
Senyawa Lain	0,70
Total	100

2.3. *Fracture Mechanics*

In crack mechanics, the result of loading that exceeds the design will have an effect on the cracks so that the material collapses. Cracks that occur in the material can be classified into 3 modes, namely opening mode (mode 1), in-plane shear mode (mode 2), and out of plane shear mode (mode 3).

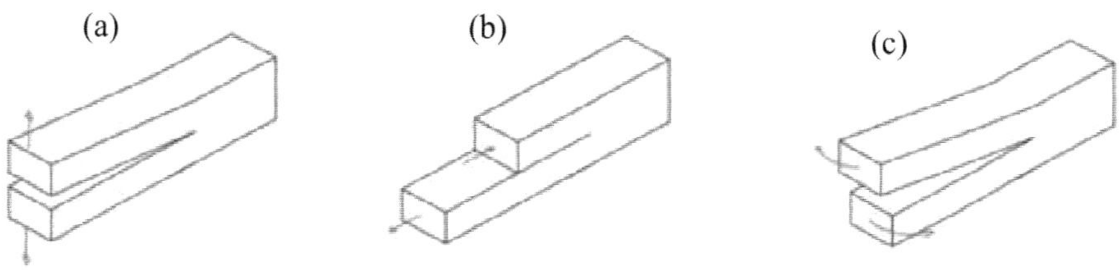

Figure 3. Fault Mode: (a) Opening Mode – Mode I, (b) In-Plane Shear – Mode II, (c) Out-of-Plane Shear – Mode III [9]

In planning with the mechanical fracture method, according to the fracture toughness the mode that will be used in this test is the opening mode (mode 1)

Mode 1 cracks are often referred to as exposure mode or pull mode. Because the crack tip is exposed to the normal force so that the midline of the crack faces away evenly and the displacement of the crack face is perpendicular to the plane of the crack [9].

2.4. *Fracture Toughness*

Cracks in the material will always exist during the fabrication process or the manufacture of test objects. When the material is stressed, the stress is concentrated around the crack. The result of crack propagation when experiencing stress will exceed the stress value of the concrete material. The amount of energy required for crack propagation is a way to indicate the fracture toughness of a material [10]. The fracture parameter commonly used is called the stress intensity factor (SIF or KIC) in Linear Elastic Fracture Mechanic (LEFM).

The relationship between stress intensity factor (SIF), applied load speed and specimen geometry can be derived to obtain a calculation called fracture toughness. Fracture toughness was calculated by means of a three-point bend beams test according to the recommendation of [11] for determining the collapse toughness of concrete (1990). The equations used are:

$$K_{Ic} = \sigma\sqrt{\pi a}\, g_1\left(\frac{a}{D}\right) \qquad \sigma = \frac{3PS}{2D^2 b} \tag{1}$$

Equation (1) above can be written as follows:

$$K_{Ic} = \frac{3PS\,\sqrt{\pi a}\, g_1\left(\frac{a}{D}\right)}{2D^2 b}$$

With,

$$g_1\left(\frac{a}{D}\right) = \frac{1{,}99 - \left(\frac{a}{D}\right)\left(1-\frac{a}{D}\right)\left[2{,}15 - 3{,}93\left(\frac{a}{D}\right) + 2{,}70\left(\frac{a}{D}\right)^2\right]}{\sqrt{\pi}\left(1+2\left(\frac{a}{D}\right)\right)\left(1-\left(\frac{a}{D}\right)\right)^{3/2}} \tag{2}$$

Collapse toughness of concrete or commonly known as fracture toughness (KIC) is the collapse toughness of a material to evaluate the ability of constituents containing defects to resist fracture. Collapse strength of concrete can be calculated by the following formula:

$$K_{Ic} = 3\,(Pc + 0{,}5W)\,\frac{s(\pi a_0)^{0{,}5}\, g\left(\frac{a_0}{D}\right)}{2D^2 b} \tag{3}$$

3. Research Methods

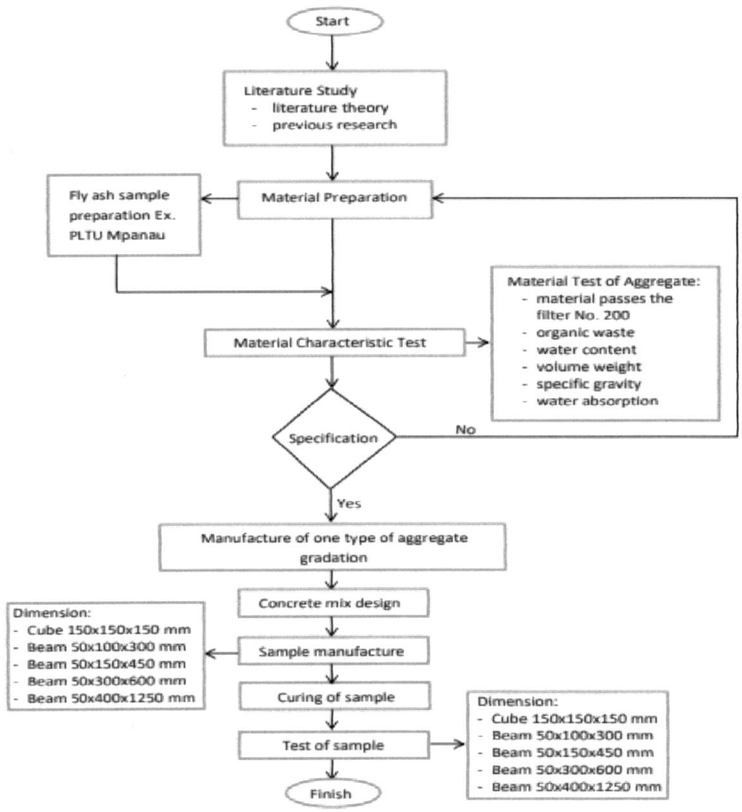

Figure 4. Research Flowchart

4. Results and Discussion

4.1. Material Inspection Results

A concrete plan requires an aggregate inspection so that the material used meets the specifications and treatment that we will provide and obtains data to make a concrete mix design. To obtain material that meets the requirements, inspections are carried out including testing the content of mud through the No. 200 filter, testing organic impurities, testing water content, testing density, testing specific gravity and testing absorption for gravel and sand. The following is summarized in Table 2.

Table 2. Material Test Results

Testing		Value	Specification
Bulk Density		2.24 %	Maks 1 %
SSD Density		0.27 %	Maks 2 %
Apparent Density		No.1	
Absorption		0.77 %	
Sand Moisture Testing		0.24 %	
Fill weight of Coarse Agg.	weight loss	1.333 gr/cm^3	Min 1.2 gr/cm^3
	solid content weight	1.533 gr/cm^3	
Fill weight of Fine Agg.	Weight Loss	1.397 gr/cm^3	
	Solid Fill Weight	1.533 gr/cm^3	
Gravity and Absorption of Coarse Agg.	Bulk Density	2.57	2.5 – 2.7
	SSD Density	2.59	
	Apparent Density	2.67	
	Absorption	0.87 %	
Gravity and Absorption of Fine Agg.	Bulk Density	2.57	
	SSD Density	2.57	
	Apparent Density	2.59	
	Absorption	0.28 %	

4.2. Aggregate Sieve Analysis

To find out what zone the gravel enters, a sieve analysis is carried out which refers to [12]. The results of the sieve analysis test can be seen in Figure 5 for the combined gradation below

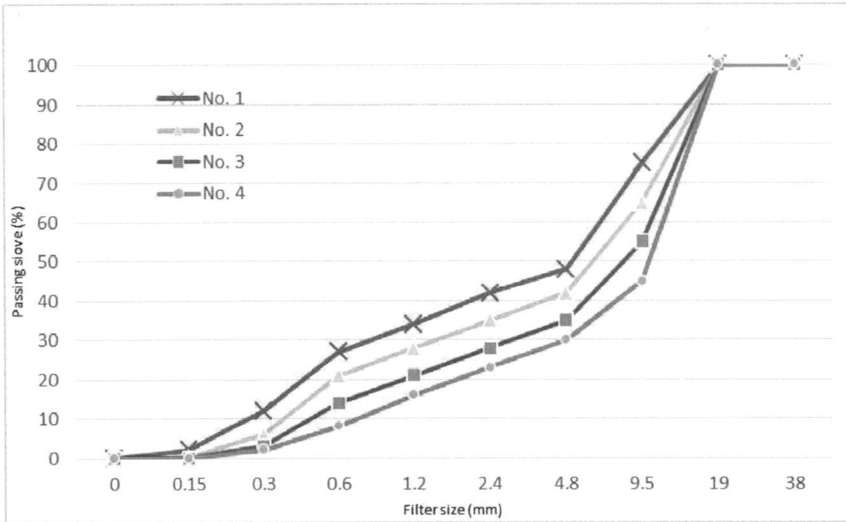

Figure 5. Combined aggregate grading graph

4.3.Design of Concrete Mixture Composition

The design of the concrete mix composition can be calculated based on the material inspection data. With reference to [4], the calculation of the concrete mixture with the composition of the mixture per m^3 is presented in Table 3 below.

Tabel 3. Mixed composition result

Mix	Mixed Proportion (kg/m³)				
	Coarse Agg.	Fine Agg.	Cement	fly ash	Water
FA 0 %	89.806	48.795	32.288	-	16.142
FA 5 %	89.806	48.795	30.674	1.614	16.142
FA 10 %	89.806	48.795	29.059	3.229	16.142
FA 15 %	89.806	48.795	27.445	4.843	16.142

4.4. Compressive Strength Test Results

The results of the compressive strength test with the percentage of fly ash at the age of 28 days are presented in Table 4 and Figure 6 below.

Tabel 4. Concrete compressive strength test results

Fly ash percentage	No. Test Object	Compressive Strength 28 days (MPa)	Average
	1	31,1	
0 %	2	31,1	31,9
	3	33.4	
	1	26,9	
5 %	2	28,7	28,4
	3	29,8	
	1	26,9	
10 %	2	26,4	26,5
	3	26,2	
	1	26,0	
15 %	2	29,1	27,1
	3	26,2	

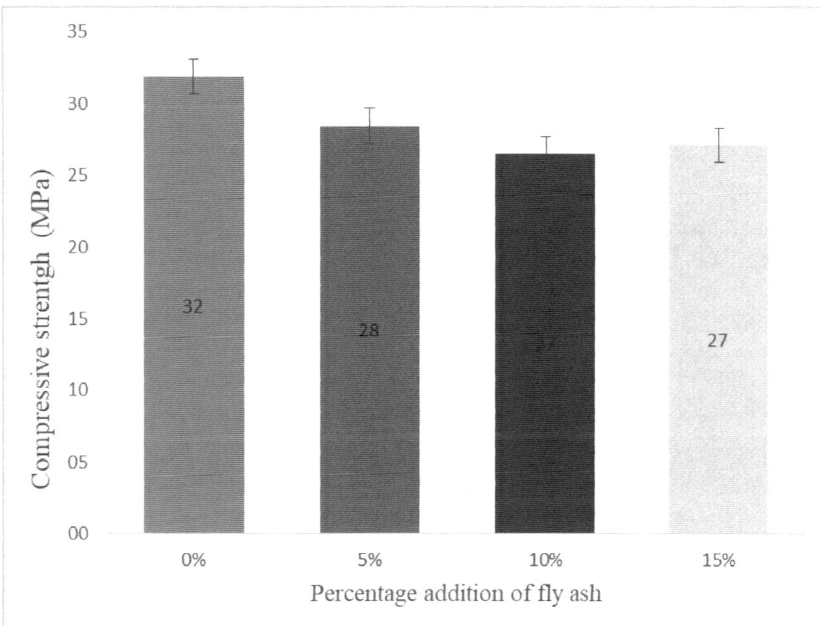

Figure 6. Concrete Compressive Strength Diagram

4.5. *Three Point Bend Test Results*

The results of the three-point bend test with the percentage of fly ash at the age of 28 days are presented in Table 5 below.

Tabel 5. Three-point bend test results

Fly ash Percentage	Test Object		Load	Average
	Dimension	Type		
		A-1	8,5	
	50 × 400 × 1250	A-2	9,0	9,2
		A-3	10,0	
		B-1	11,0	
	50 × 300 × 650	B-2	13,0	12,0
		B-3	12,0	
0 %		C-1	4,5	
	50 × 150 × 500	C-2	5,5	5,0
		C-3	5,0	
		D-1	4,0	3,5
	50 × 100 × 350	D-2	2,5	
		D-3	4,0	
		A-1	8,0	
	50 × 400 × 1250	A-2	9,0	9,0
		A-3	10,0	
		B-1	10,5	
	50 × 300 × 650	B-2	11,5	11,0
		B-3	11,0	
5 %		C-1	5,0	
	50 × 150 × 500	C-2	4,0	4,3
		C-3	4,0	
		D-1	3,5	
	50 × 100 × 350	D-2	3,5	3,3
		D-3	3,0	
		A-1	9,5	
	50 × 400 × 1250	A-2	11,5	9,5
		A-3	7,5	
		B-1	15,0	
	50 × 300 × 650	B-2	12,0	14,2
		B-3	15,5	
10 %		C-1	4,6	
	50 × 150 × 500	C-2	5,0	4,7
		C-3	4,5	
		D-1	4,0	
	50 × 100 × 350	D-2	4,0	3,8
		D-3	3,5	
		A-1	12,5	
	50 × 400 × 1250	A-2	11,5	10,8
		A-3	8,5	
		B-1	8,8	
	50 × 300 × 650	B-2	9,0	10,1
		B-3	12,5	
15 %		C-1	4,0	
	50 × 150 × 500	C-2	4,5	4,3
		C-3	4,5	
		D-1	3,5	
	50 × 100 × 350	D-2	3,5	3,3
		D-3	3,0	

IOP Conf. Series: Earth and Environmental Science 1075 (2022) 012025 doi:10.1088/1755-1315/1075/1/012025

4.6 Effect of Fly Ash on Compressive Strength

Effect of the fly ash addition on the compressive strength is intended to determine the quality of concrete on the percentage of fly ash addition. The results are as referred to in Table 4.14 which is summarized as follows:

1) Percentage of fly ash 0 % = 31.9 MPa
2) Percentage of fly ash 5% = 28.4 MPa
3) Percentage of fly ash 10% = 26.5 MPa
4) Percentage of fly ash 15 % = 27.1 MPa

The results obtained with the existing percentages namely 0%, 5%, 10% and 15%, it shows that the increase in the addition of fly ash will reduce the compressive strength or quality of the test object but in the condition of adding 15% fly ash there is an increase in the compressive strength of the percentage addition of 10 % where the results are not in line with the research of [13] which shows a decrease in compressive strength along with the addition of fly ash levels. Although in this study is using the same fly ash material source.

From these differences, the author again refers to [14] on the effect of the level of aggregate hardness on the concrete compressive strength, showing the influence of the abrasion value on the concrete compressive strength. So the author concludes that the difference is influenced by the concrete constituent materials other than fly ash, thus the effect of fly ash in this case the author concludes that fly ash has an influence on the compressive strength namely the cement binding process so that at a certain percentage level it will increase the strength of the concrete or even increase the strength of the concrete. Reducing the concrete strength itself, so it is necessary to conduct separate research on the cement binding process due to the addition of fly ash in order to determine the time required for the specimen to reach its optimum strength.

4.7. Influence of Fly Ash on Collapse Strength of Concrete

For type A specimens, it can be seen from the results of the KIC analysis (Table 5) that the addition of fly ash as much as 15% of the total cement showed the highest value of 239.08 MPa.mm$^{1/2}$, followed by a percentage of 10% with an average KIC value. 209.97 MPa.mm$^{1/2}$, then the percentage of 0% is 202.79 MPa.mm$^{1/2}$ and finally at 5% has decreased with an average KIC value of 199.52 MPa.mm$^{1/2}$, can be seen in Figure 7 below.

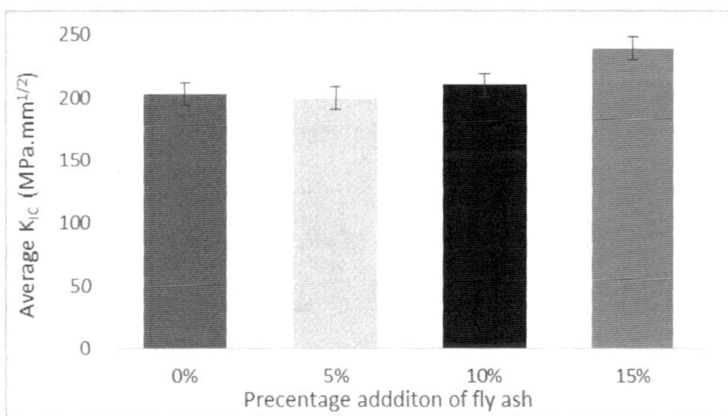

Figure 7. Comparison Diagram of Average KIC Value and FA Percentage of Type A Test Objects

For type B specimens, see Figure 8 shows the highest average KIC value, namely the percentage of 10% addition of fly ash is 323.46 MPa.mm$^{1/2}$, followed by concrete without the addition of fly ash (0% FA content) with a value of 274.48 MPa.mm$^{1/2}$, then at 10% fly ash content, namely 251.80 MPa.mm$^{1/2}$ and finally with the lowest average KIC value, namely 15% FA content with an average value of 231.05 MPa.mm$^{1/2}$.

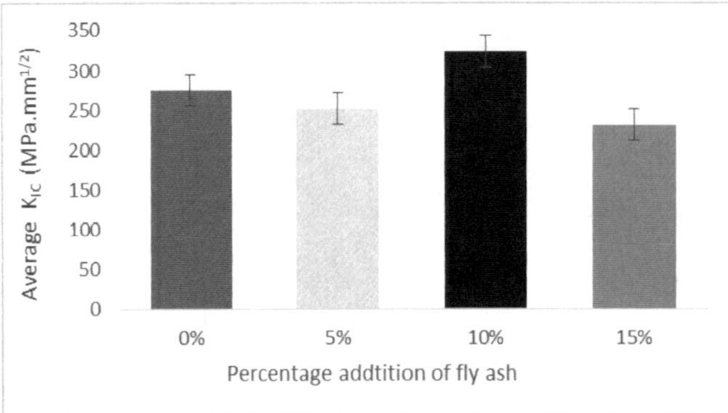

Figure 8. Comparison Diagram of Average KIC Value and FA Percentage of Type B Test Objects

For the results of the type C test object in Figure 9, it shows the KIC value at a percentage of 0%, namely 131.25 MPa.mm$^{1/2}$, then the percentage is 10% with a KIC value of 123.42 MPa.mm$^{1/2}$, then after that for the percentage 5% and 15% are 113.90 MPa.mm$^{1/2}$ and 113.88 MPa.mm$^{1/2}$ respectively.

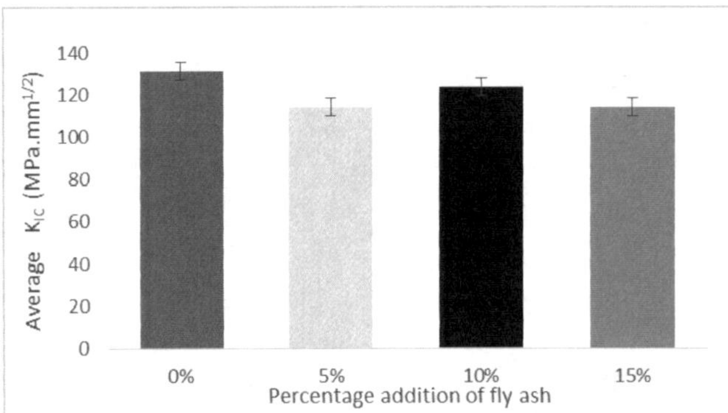

Figure 9. Comparison Diagram of Average KIC Value and FA Percentage of Type C Test Objects

In the test object type D (Figure 10), the KIC value was obtained with the highest value of 10%, namely 108.74 MPa.mm$^{1/2}$, the second percentage of FA was 0% with an average value of 99.33 MPa.mm$^{1/2}$, the third was FA 5% = 94,63 MPa.mm$^{1/2}$ and the last is FA 15% = 94.61 MPa.mm$^{1/2}$.

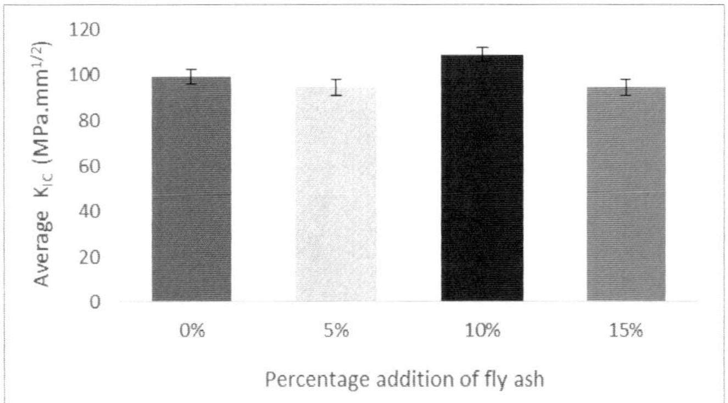

Figure 10. Comparison Diagram of Average KIC Value and FA Percentage of Type D Test Objects

From Table 5 above, it is known that the higher the addition of FA to concrete will affect the bonding between cement and aggregate so that the percentage of cracks which in the type A sample for FA 0% and 5% decreased from 31.25% to 10.94% then experienced the increase in the FA 10% sample and decreased again in the FA 15% namely from 30.26% to 10.94%.

In the type B sample, it shows the same conditions as the type A sample where the percentage of aggregate cracking is in the FA sample of 0%, then 10%, followed by the last 5% sample In the type C sample, for the 0% FA sample with the percentage of cracking that is 27.59% then it decreases at the addition of 5% FA after that it has increased successively up to the 15% FA sample.

In the type C sample, for the 0% FA sample with the percentage of cracking that is 27.59% then it decreases at the addition of 5% FA after that it has increased successively up to the 15% FA sample.

Different from the previous conditions, sample D showed different conditions where the largest aggregate crack was in the 15% FA sample then followed by the 5% sample, the third was in the 0% FA sample, the last in the 10% FA sample.

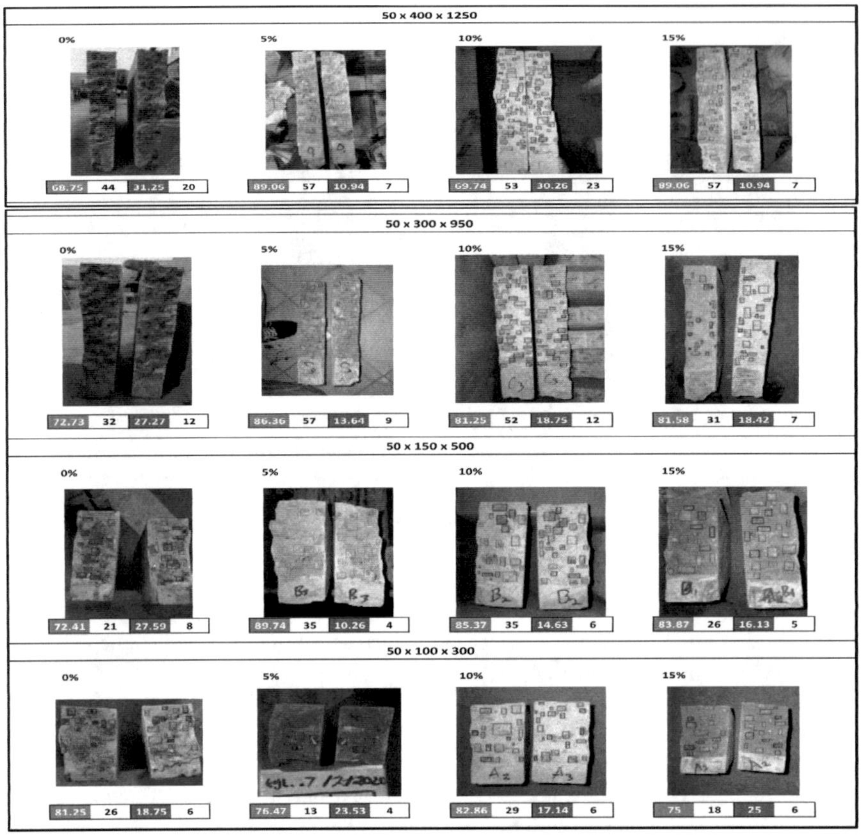

Figure 11. Effect of Addition of Fly Ash Collapsed on Three Point Bend Test Results

5. Conclusion
   1. From the results of the compressive strength test, the author concludes that fly ash has an influence on the compressive strength, namely the cement binding process so that at a certain percentage level it will increase the strength of the concrete or even decrease the strength of the concrete itself.
   2. The results of the three-point bend test show the value of the KIC parameter for each type of test object with the percentage of addition of fly ash which includes 0%, 5%, 10% and 15% showing varying values. In type A test specimens, 15% FA showed the highest value, namely 239.08 MPa.mm$^{1/2}$ then 10% with a value of 209.97 MPa.mm$^{1/2}$ and 0%, namely 202.79 MPa.mm$^{1/2}$ last at 5 % decreased with a KIC value of 199.52 MPa.mm$^{1/2}$. In type B specimens with the highest value of 10% is 323.46 MPa.mm$^{1/2}$, 0% with a value of 274.48 MPa.mm$^{1/2}$, 10% is 251.80 MPa.mm$^{1/2}$ and 15% with a value of 231.05 MPa.mm$^{1/2}$. For type C specimens, the KIC value from the largest is 0%, namely 131.25 MPa.mm$^{1/2}$, 10% is 123.42 MPa.mm$^{1/2}$, 5% with 113.90 MPa.mm$^{1/2}$ and 15%, namely 113.88 MPa.mm$^{1/2}$. Finally, for type D specimens, values of the largest order were obtained, namely 10% = 108.74 MPa.mm$^{1/2}$, 0% = 99.33 MPa.mm$^{1/2}$, 5% = 94.63 MPa.mm$^{1/2}$ and 15% = 94.61 MPa.mm$^{1/2}$.
   3. From the visual observation, it shows that the percentage of damage to the aggregate is lower than the aggregate that is released or in this case that is not cracked, so the author concludes that this behavior is influenced by the binding of cement to the aggregate which is reduced due to the addition of fly ash as

evidenced on the object test type A, type B and type C which experienced a decrease in the addition of fly ash from the condition of 0% or without fly ash.

REFERENCES

[1] Anonymous. 2018. Cement Industry in Indonesia. [Online] Available at: https://asi.or.id/cement-industry-in-indonesia/ [Accessed 24 November 2020].
[2] Siregar, A. P. N. 2016. The Effect of Aggregate Size Distribution on Fracture Energy of Concrete, Guildford: University of Surrey.
[3] Siregar, A. P. N., Pasaribu, E. L. & Suarnita, I. W. 2019. The effect of coarse aggregate hardness on the fracture toughness and compressive strength of concrete. MATEC Web of Conferences, 258(04011), p. 5.
[4] SNI 03-2834-2002. Procedures for Making Normal Concrete Mixture Plans. Pustran, Balitbang, DPU.
[5] Mulyono, T. 2003. Concrete Technology. 1 ed. Yogyakarta: ANDI.
[6] Tjokrodimuljo, K. 1996. Concrete Technology. 1 ed. Yogyakarta: Nafiri.
[7] SK SNI S-15-1990-F. Fly ash as a concrete mix specification. Public Works Department Publishing Foundation.
[8] Adam, A. A., Ramadhan, B. R. & Maricar, S., 2019. The Effects of Water to Solid Ratio, Activator to Binder Ratio, and Lime Proportion on the Compressive Strength of Ambient-Cured Geopolymer Concrete. Journal of the Civil Engineering Forum, May, 5(2), p. 162.
[9] Brook, D. 1984. Elementary Engineering Fracture Mechanics. Martinus Nijhoff Publisher, Hague, Netherlands
[10] Elfgren, L. & Shah, S. P. 1989. Analysis of Concrete Structures by Fracture Mechanics. s.l.: RILEM Workshop.
[11] RILEM TC89-FMT (1990) Fracture mechanic of concrete test method. Materials and Structures 23: 247-252
[12] SNI 03-1968-1990. Test method of fine and coarse aggregate sieve analysis. Indonesian National Standard.
[13] Hadar, S.Z.A. 2009. Analysis of Compressive Strength of Concrete with Fly Ash Additive ex. PLTU Mpanau. Tadulako University, Palu.
[14] Pasaribu, Emma L. 2018. Final Project: The Effect of Hardness of Coarse Aggregate on Collapse Strength of Concrete. Tadulako University, Palu.

The Study of Liquefied Soil by Using Swedish Weight Sounding Test in Jono Oge Village, Sigi Regency

Hendra Setiawan*,1, Ihzam Ilaiya1, Ida Sri Oktaviana1, Irdhiani1, Iffah Fadliah1, Yusfi Hasanah1

1Jurusan Teknik sipil, Fakultas Teknik, Universitas Tadulako, Jalan Soekarno Hatta Km. 9, Palu, Sulawesi Tengah, Indonesia (94118)

**E-mail: hendra3909@gmail.com*

Abstract:. The liquefaction that occurred on September 28, 2018, caused a massive lateral spreading in certain areas. This study aims to determine and compare the properties of land affected and not affected by liquefaction based on the Swedish Weight Sounding test, as well as the potential for liquefaction at Jono Oge village. The results of this study indicate that the laboratory test results show that the dominant soil classification on the affected and the unaffected soil is poorly graded sand silt (SP-SM).Whereas the results of field testing showed that the Nsw value on the affected soil is in the range 3,33 n/m – 263,16 n/m, the Nsw value on the unaffected soil is in the range 10,00n/m – 357,14 n/m, the qa value on the affected land is in the range up to 240,53 kN/m^2, the qa value on the unaffected land is in the range of up to 315,71 kN/m^2, the qu value on the affected land is in the range of 2.25 kN/m^2 – 247,37 kN/m^2, the qu value of unaffected land is in the rangc of 2.25 kN/m^2 – 312,86 kN/m^2.

1. Introduction

Liquefaction is a phenomenon that occurs when the soil loses its strength and stiffness due to cyclic loading, and a result, the soil mass undergoes a transition. The increase in pore water pressure due to changes in stress that occur causes the soil properties to change from solid to liquid. This phenomenon can be characterized by the presence of ground movement in the horizontal direction, water seepage of coming out of soil fractures, the movement of sloping buildings, and the occurrence of subsidence. From previous earthquake records, it is known that the liquefaction phenomenon has the potential to occur in loose saturated granular soils. There were some cases of earthquakes that triggered by liquefaction, namely the Niigata Earthquake (1964), the Alaska Earthquake (1964), the Flores Earthquake (1992), and the Central Sulawesi Earthquake (2018).

The Central Sulawesi earthquake that occurred on September 28, 2018, with a magnitude of 7.4 Mw triggered several areas to experience disasters. The tsunami in the outskirts of Palu City and Donggala Regency, fires in Balaroa Village, and the occurrence of liquefaction phenomena in Palu City and Sigi Regency

The liquefaction phenomenon that occurred in Sigi Regency was in Sibalaya Village, Jono Oge Village, while in Palu City it occurred in Balaroa and Petobo Villages. Based on the research of Day (2002) and Tsuchida (1970) it is known that several soil properties have the potential for liquefaction, including sandy soil, groundwater level, and uniform distribution of soil grains. Based on this and the phenomena that have occurred, the researchers chose the village of Jono Oge to research soil properties in the area, the method used to obtain soil properties was using field testing, namely Swedish Weight Sounding.

2. Literature Review

2.1. The Earthquake

According to [1], an earthquake is a movement or vibration that occurs in the earth's crust resulting from endogenous energy. The definition of endogenous energy itself is the energy or strength of the bowels of the earth that occurs due to changes in the earth's crust. An earthquake is a vibration or a series of vibrations that occur from the earth's crust which has a temporary nature (not eternal) which then spreads in all directions [2].

Based on the definition of experts, it can be concluded that an earthquake is a vibration that occurs in the skin of the earth that originates from the bowels of the earth and spreads in all directions. The spread of these vibrations causes a cyclic (repeated) load on the soil, causing the pressure experienced by the soil to increase, triggering liquefaction.

2.2. Liquefaction

[3] defines liquefaction as the change in the properties of granular soils, from solid to liquid. The meaning of the word liquid is not the change of soil into the water but is interpreted as a change like the soil which tends to be like water. This is due to an increase in pore water pressure in the soil. [4] states that liquefaction is the process of reducing the shear strength of the soil due to seismic loads when an earthquake occurs. [5] stated that liquefaction occurs in loose sandy soil (not dense) and saturated with water. Based on this description, liquefaction is the transformation of loose sandy soil and water-saturated into liquid due to an increase in pore water pressure and a decrease in soil shear strength triggered by seismic loads during an earthquake.

Based on forementioned information, it can be concluded that liquefaction events are the phenomenon of soil loss of strength and stiffness due to earthquake loads so that the nature of the soil changes from solid to liquid. This event occurs in the consistency of loose saturated granular soil with drainage properties in the soil. Fine sand (sand), silty sand, and loose sand are soils that have the potential to liquefy due to cyclic (repetitive) loads. The granular soil layer has good porosity, allowing this soil layer to drain, store and absorb water which results in this soil layer being water-saturated.

The process of changing the condition of the saturated soil to become liquid due to the increase in pore water pressure until it reaches a point equal to the total stress due to the cyclic load from the earthquake so that the effective soil stress is equal to 0 (zero). This shows that liquefaction is a phenomenon that occurs when the soil loses strength and stiffness in a short time.

2.2.1. Liquefaction Process

The structure of the soil layer that is susceptible to liquefaction is granular soil that has a groundwater table below the granular soil layer. During an earthquake, the granular soil reacts to the cyclic loading caused by the earthquake, resulting in increased pore water pressure. Increasing soil pore water pressure causes water to flow up the surface in the form of mud or sand and changes the soil phase from a solid to a liquid phase [6].

Granular soils with moderate to lose density and high water table elevations, when there is no vibration, the soil will be stable because the soil grains are interlocking. When an earthquake occurs, the soil tends to decrease in volume and result in an increase in pore water pressure so that the effective shear strength of the soil will decrease.

2.2.2. Factors Causing Liquefaction

The factors that cause liquefaction according to [6] include:

- Intensity, duration, and characteristics of earthquakes
- Groundwater level
- Soil type
- Relative density (Initial relative density)
- Soil grain distribution

2.3. Soil Definition

Soil is a collection of minerals, organic matter, and relatively loose deposits, which are located on top of bedrock. The relatively weak bonds between the grains can be caused by carbonates, organic matter, or oxides deposited between the particles. The space between the particles can contain water, air, or both [7].

According to [8] soil is defined as mineral particles that are cemented or loose as a result of weathering of rocks, where the pores between the particles are filled with air and/or water. Due to the influence of weather and other influences, the soil undergoes weathering resulting in changes in the size and shape of the grains. Weathering of rocks can be caused by mechanical, chemical, and organic weathering.

2.3.1. Soil Classification

The soil classification system was created basically to provide information about the characteristics and physical properties of the soil. Due to the wide variety of properties and behavior of soils, classification systems generally classify soils where the soils have similar physical properties. Soil classification is also useful for a more detailed study of the state of the soil and the need for testing to determine the technical properties of the soil such as compaction characteristics, soil strength, density, etc [9]. The soil classification method used is the Unified Soil Classification System (USCS) [10].

The USCS classification system is the most widely used in experimental civil engineering work that is used to classify soils based on this system is sieve analysis, grouping this system into two groups, namely Coarse-grained-soil and fine-grained soils. The USCS classification system can be seen in Table 1.

2.3.2. Sieve Analysis

Sieve analysis is a widely used method for determining soil classification. Soil grain size is generally depicted on a graph with a relationship between percent pass and grain size, making it possible to know the distribution. Soil with large and small grain sizes evenly distributed is well graded, while soil with almost uniform grain size means poorly graded soil [7].

Types of soil that are prone to liquefaction are loose sandy soil to saturated medium. [11] summarized the results of sieve analysis on many soils known to have liquefied and which were not liquefied during an earthquake. The grain size curve that separates liquefied and non-liquefied soil can be seen in Figure 1.

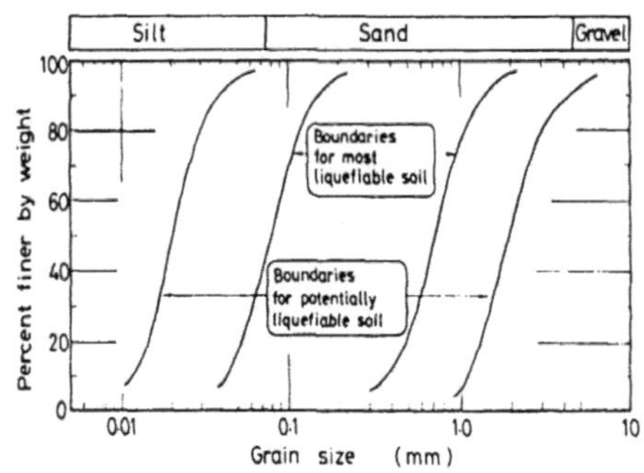

Figure 1 Grain size distribution curve vulnerable land liquefaction [11]

Table 1. Unified Soil Classification System (USCS)

<table>
<tr><th colspan="3">Major Division</th><th>Symbol</th><th>Typical Name</th><th colspan="3">Classification Criteria</th></tr>
<tr><td rowspan="8">Coarse-grained soils (> 50% retainer on No. 200 ASTM Sieve)</td><td rowspan="4">Gravels, 50% or more of coarse fraction retained on No.4 ASTM Sieve</td><td rowspan="2">Clean gravel</td><td>GW</td><td>Well-graded gravels and gravel-sand mixtures, little or no fines</td><td rowspan="8">Classification on the basis of percentage of fines. < 5% passing No. 200 ASTM Sieve-GW, GP, SW, SP. > 12 passing No. 200 ASTM Sieve-GM, GC, SM, SC.</td><td colspan="2">C greate thanr 4 :
Cc between 1 and 3</td></tr>
<tr><td>GP</td><td>Poorly-graded gravels and gravel-sand mixtures, little or no fines</td><td colspan="2">Not meeting both criteria for GW</td></tr>
<tr><td rowspan="2">Gravel with fines</td><td>GM</td><td>Silty gravels, gravel-sand-silt mixtures</td><td>Atterberg limits plot below A-line or PI < 4</td><td rowspan="2">Shaded in chart, double symbol</td></tr>
<tr><td>GC</td><td>Clayey gravels, gravel-sand-clay mixtures</td><td>Atterberg limits plot above A-line or PI > 7</td></tr>
<tr><td rowspan="4">Sands, > 50% of coarse fraction passes No.4 ASTM Sieve</td><td rowspan="2">Clean sands</td><td>SW</td><td>Well graded sands and graverly sands, little or no fines</td><td colspan="2">Cu greater than 6 :
Cc between 1 and 3</td></tr>
<tr><td>SP</td><td>Poorly-graded sands and graverly sands, little or no fines</td><td colspan="2">Not meeting both criteria for SW</td></tr>
<tr><td rowspan="2">Sands with fines</td><td>SM</td><td>Silty-sands, sand-silt mixtures</td><td>Atterberg limits plot below A-line or PI < 4</td><td rowspan="2">Shaded in chart, double symbol</td></tr>
<tr><td>SC</td><td>Clayey sands, sand-clay mixtures</td><td>Atterberg limits plot above A-line or PI > 7</td></tr>
<tr><td rowspan="6">Fine Grained soils, (50% or more passes NO. 200 ASTM Sieve)</td><td colspan="2" rowspan="3">Silts and Clays (LL 50% or less)</td><td>ML</td><td>Inorganic silts, very fine sands, rock flour, silty or clayey fine sands</td><td colspan="3" rowspan="6">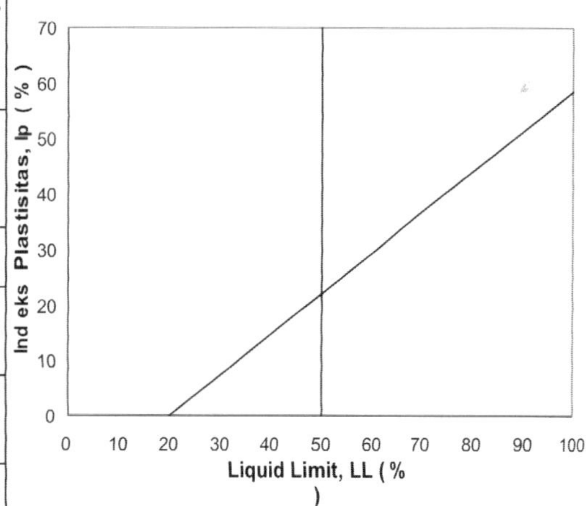</td></tr>
<tr><td>CL</td><td>Inorganic clays or low to medium plasticity, graverly clays, sandy clays, silty clays, lean clays.</td></tr>
<tr><td>OL</td><td>Organic silts andorganic silty clays of low plasticity</td></tr>
<tr><td colspan="2" rowspan="3">Silts and Clays (LL >50%)</td><td>MH</td><td>Inorganic silts, micaceous or diatomacecous fine sands or silts, elastic silts</td></tr>
<tr><td>CH</td><td>Inorganic clays of high plasticity, fat clays</td></tr>
<tr><td>OH</td><td>Organic clays of medium to high plasticity</td></tr>
<tr><td colspan="3">Highly organic clays</td><td>PT</td><td>Peat, muck and other highly organic soils</td><td colspan="3">Visually distinguishable according to ASTM D-2488</td></tr>
</table>

Source : [12]

2.4. Swedish Weight Sounding

The Swedish Weight Sounding is a simple penetration tester that is manually operated under a dead load of 100 kg where a number of half turns is required for penetration of 25 cm from the rod [13]. In 1914 SWS was standardized by the Geotechnical Commission of the Swedish State Railway.

According to JIS A 1221, Swedish Weight Sounding is a tool for measuring the static penetration resistance of native soils and for assessing soil hardness or subsoil composition. Swedish Weight Sounding is very simple and has relatively good penetration ability and is used for soil investigations to a depth of 10 m.

It is estimated that about 20,000 Swedish Weight Sounding tests are performed annually in Sweden [14]. This test has also been used in several countries in Eastern Europe as well as countries such as Singapore and Algeria [15]. [16] also reported the use of Swedish Weight Sounding to estimate the foundation depth and bearing capacity for several buildings in the southern region of Tehran, Iran.

The Swedish Weight Sounding test can also be used to obtain soil profiles that are affected and not affected by liquefaction or landslides, and this tool can also identify continuous and discontinuous soil profiles. Due to the relative ease of use of this tool, the Swedish Weight Sounding test is often performed for soil investigations in earthquake-devastated areas. The picture of the Swedish Weight Sounding tool can be seen in Figure 2.

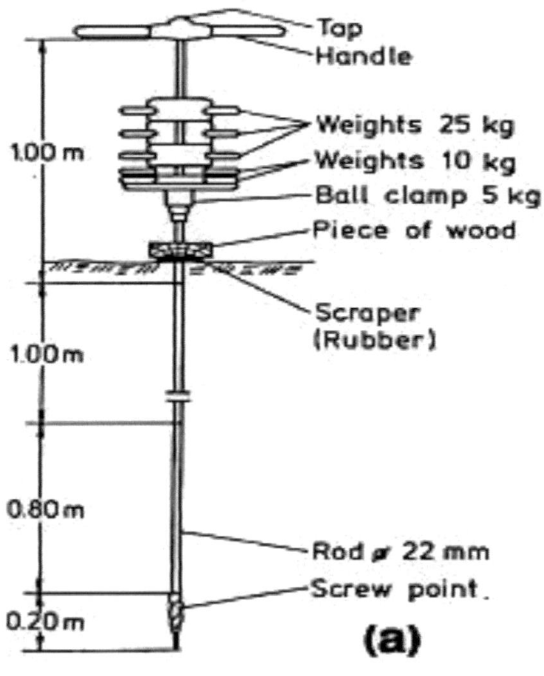

Figure 2 Swedish Weight Sounding Tool [13]

3. Research sites

3.1. Research Flow Chart

The research is planned to be conducted at the location of Jono Oge Village, South Dolo District, Sigi Regency, Central Sulawesi. The research flow chart and location are shown in Figure 3 and 4, respectively.

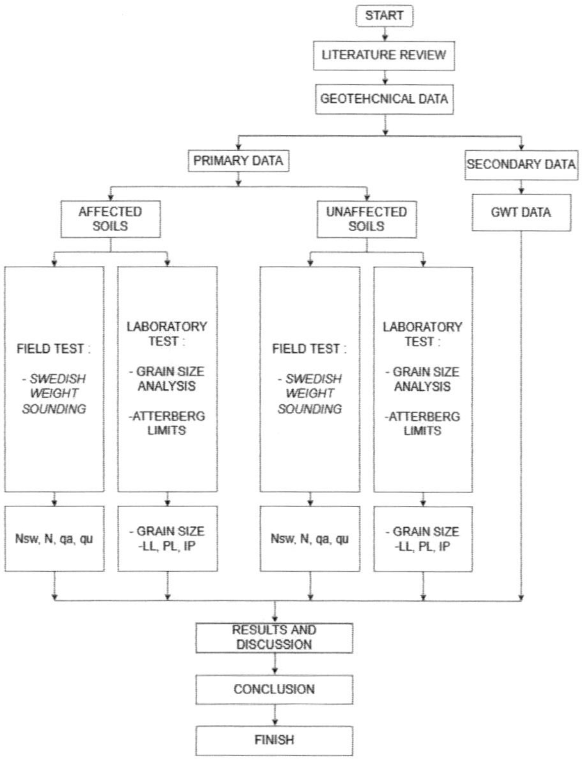

Figure 3 Research Flow Chart

3.2. Sampling Location

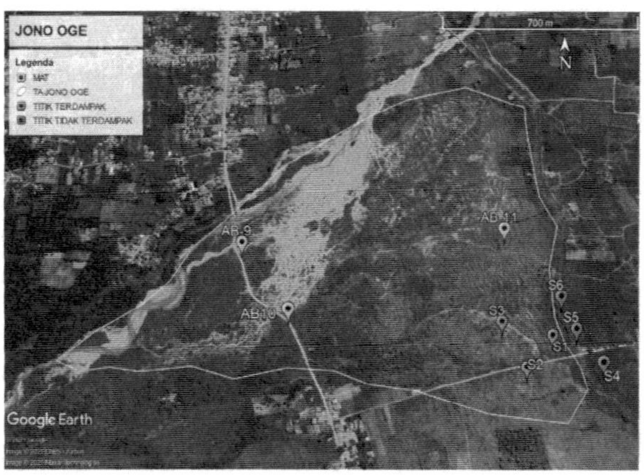

Figure 4 Location Research

According to [17], the sample is part of the number and characteristics possessed by the population. Based on this thought, the sampling that will be carried out must be able to represent and provide an overview of the characteristics of the location under study. Before taking the sample, the researcher surveyed the research location. The samples to be taken are samples that represent areas affected by liquefaction and areas not affected by liquefaction, as shown in Figure 4 the test points representing the affected areas are in the inner area of the yellow line while the outer area of the yellow line represents areas not affected by liquefaction. In other words, areas affected or not affected by liquefaction can be separated by the location of the Gumbasa Irrigation Channel which on the bottom or west side of the irrigation channel is the affected area while the top or east side of the irrigation channel is an area that is not affected by liquefaction. The number of samples to be taken is 6 (six) points, of which 3 (three) points represent the affected area where the distance between points varies, including S1 – S2 is 121 m away, S2 – S3 is 138 m away, and S3 – S4 is 140 meters away. m and 3 (three) points represent unaffected areas where the distance between points varies, including S4 – S5 is 112 m, S5 – S6 is 98 m, and S6 – S4 is 206 m.

4. Results and Discussion

4.1. Primary Data

4.1.1. Sieve Analysis

The filter analysis test was carried out by first taking samples from each affected area and not affected by liquefaction, with a total of 6 soil samples. Each sample was in the oven for 24 hours, then weighed as much as 1000 grams for each sample and continued with the Atterberg limits test. Then the following results are obtained.

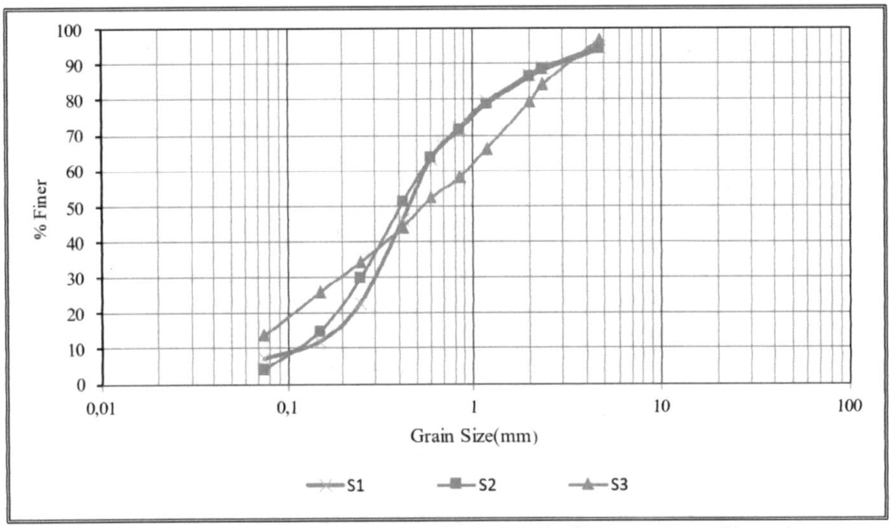

Figure 5 Gradation of grain size for each sample in liquefaction affected area

Figure 5 shows that of the 3 soil samples in the area affected by liquefaction, they are categorized as non-plastic coarse-grained soils, with sample classification S1 in the form of SP-SM, namely sand with poor grades and mixed with silt, sample S2 in the form of SP, namely poorly graded sand and sample S3. in the form of BC, namely silty sand.

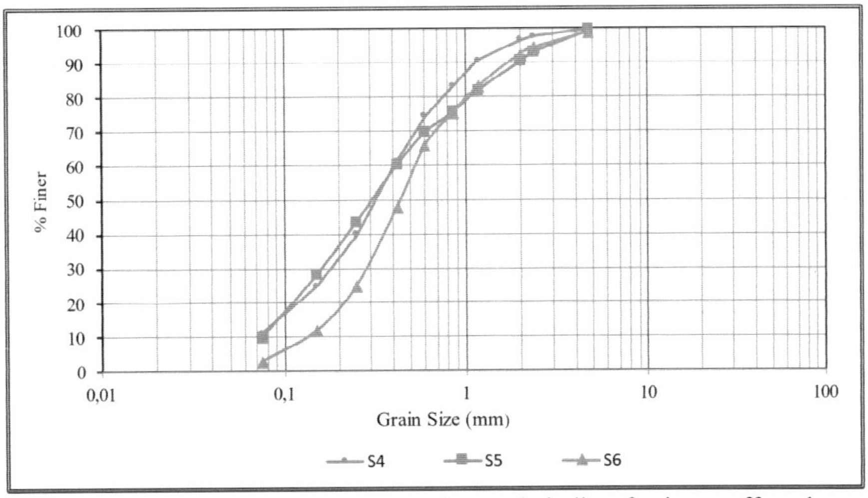

Figure 6 Gradation of grain size for each sample in liquefaction unaffected area

Figure 6 shows that from 3 soil samples in areas that are not affected by liquefaction, they are categorized as non-plastic coarse-grained soils, with sample classifications S4 and S5 in the form of SP-SM, namely poorly graded sand and mixed with silt, and sample S6 in the form of SP, which is poorly graded sand.

4.1.2. Atterberg's Limits

Atterberg limits testing was carried out on all soil samples, affected and not affected soils by liquefaction, while the test results showed that the affected and unaffected soils were non-plastic soils.

4.1.3. Swedish Weight Sounding

Tests were carried out on each affected area and not affected by liquefaction as much as 3 points, using the Swedish Weight Sounding tool. The location of the test point can be seen in Figure 7.

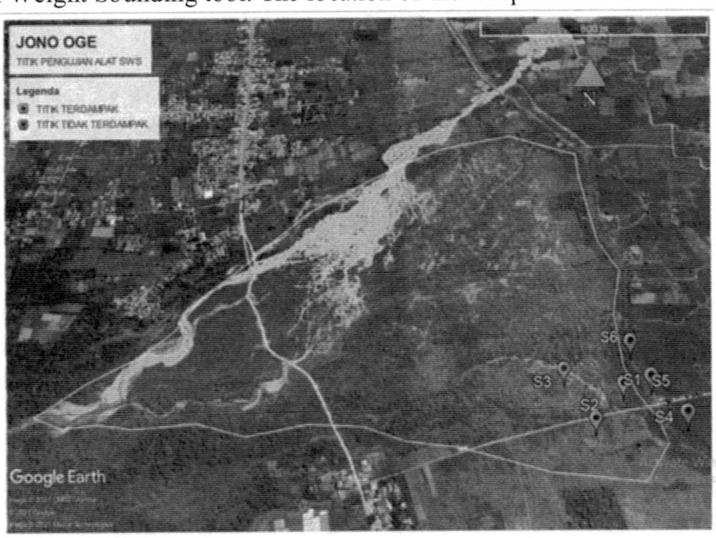

Figure 7 Location of testing points
Source : *https://earth.google.com, 2021*

It can be seen from Figure 8 below that if the Nsw value is divided into each 1 m depth range, then the Nsw value at a depth of 0 - 1 m is in the range of 4.00 n/m – 270.27 n/m, the Nsw value at a depth of 1-2 m is in the range of 8.00 n/m – 60.00 n/m, the value of Nsw at a depth of 2-3 m is in the range of 4.67 n/m – 357.14 n/m and the value of Nsw at a depth of 3-4 m is in the range of 3.33 n/m – 322.58 n/m.

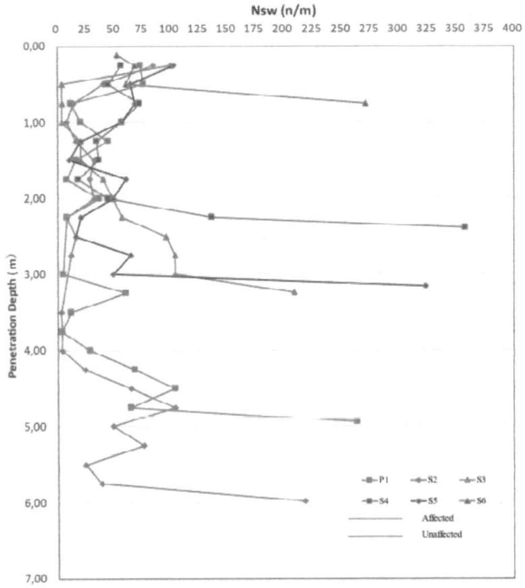

Figure 8 Graph of Nsw Value Relationship to Penetration Depth

Figure 9 below presents that if the qa value is divided into each 1 m depth range, then the qa value at a depth of 0 - 1 m is in the range of 33.20 kN/m2 – 246.22 kN/m2, the qa value at a depth of 1-2 m is at the range of 36.40 kN/m2 – 78.00 kN/m2, the value of qa at a depth of 2-3 m is in the range of 33.73 kN/m2 – 315.71 kN/m2, and the value of qa at a depth of 3-4 m is in the range of 32.67 kN/m2 – 288.06 kN/m2.

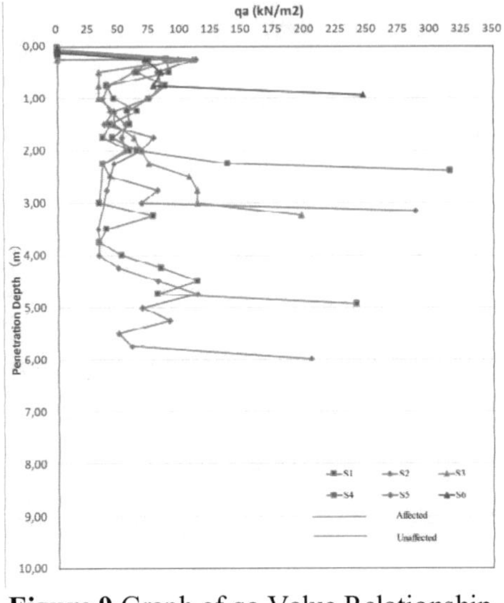

Figure 9 Graph of qa Value Relationship to Penetration Depth

Figure 10 below reveals that if the value of qu is divided into every 1 m depth range, then the qa value at a depth of 0 - 1 m is in the range of 2.25 kN/m2 – 247.70 kN/m2, the value of qu at a depth of 1-2 m is in the range 51.00 kN/m2 – 90.00 kN/m2, the value of qu at a depth of 2-3 m is in the range of 48.00 kN/m2 – 312.86 kN/m2, and the value of qu at a depth of 3-4 m is at range 47.50 kN/m2 – 286.94 kN/m2.

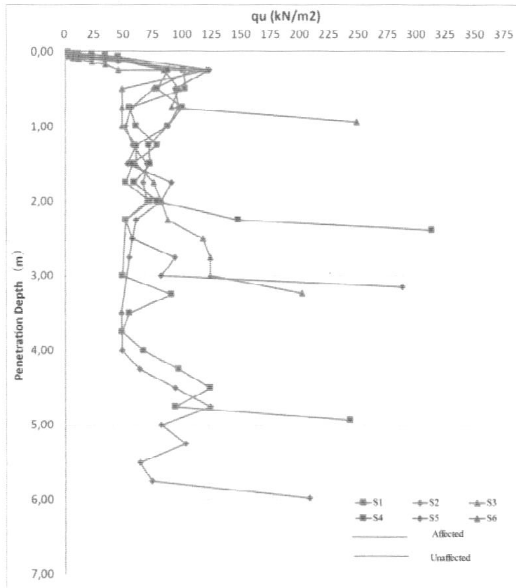

Figure 10 Graph of qu Value Relationship to Penetration Depth

4.2. Secondary Data

Groundwater level (GWL) data obtained from the JICA Survey For Disaster Information Collection In Indonesia 2020, which is located in Jono Oge village, at point AB-9 the GWLvalue is 3.93 meters, AB-10 is the GWLvalue is 2,31 meters, and AB-11 the GWL value is 9.03 meters. The location of the Ground Water Level probe is shown in Figure 11.

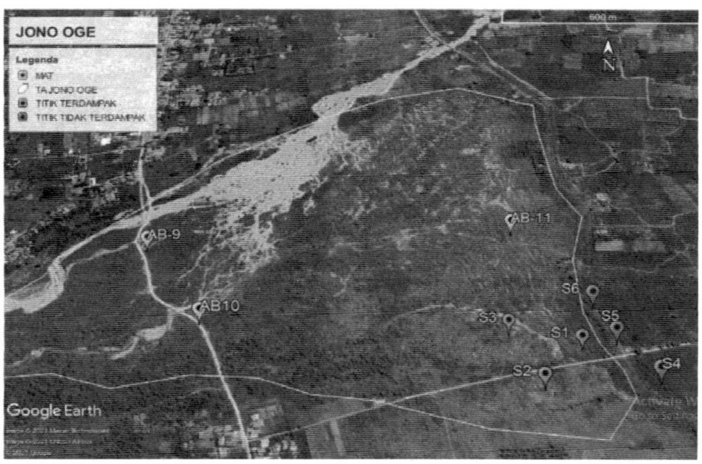

Figure 11 Probling point GWL (Ground Water Level)

Source : *https://earth.google.com, 2021*

4.3. Discussion

4.3.1. Sieve Analysis and Atterberg Limits

Based on the sieve analysis results, the liquefaction potential based on the grain size can be determined, then connected to the Tsuchida liquefaction potential curve (1970) and the results are shown in Figures 12 – 13.

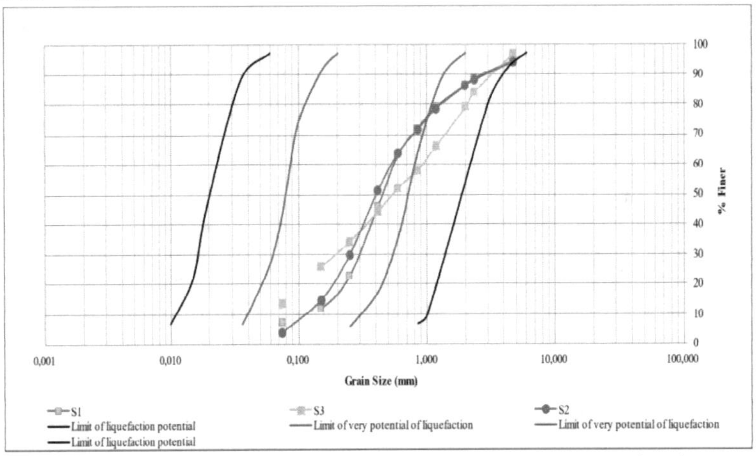

Figure 12 Graph of Liquefaction Potential based on Grain Size Distribution in Liquefaction Affected Areas

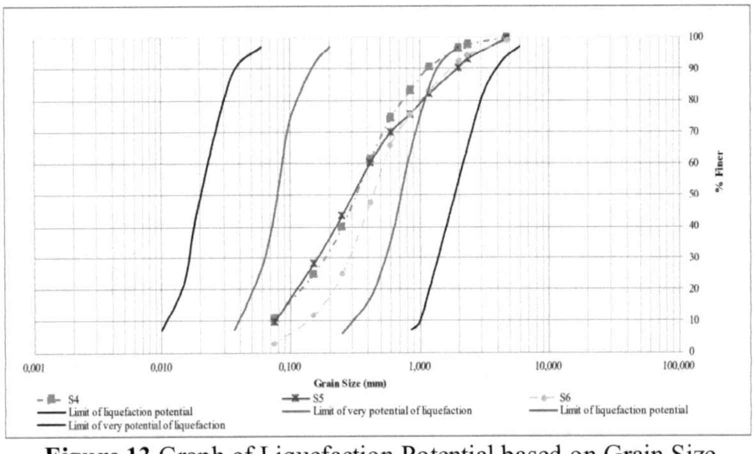

Figure 13 Graph of Liquefaction Potential based on Grain Size Distribution in Liquefaction Unaffected Area

Based on the Liquefaction Potential Curve by [11] which is shown in Figures 12 and 13, it show that the locations affected or not affected by liquefaction have the potential for liquefaction from moderate to very potential. In addition, from laboratory testing, namely sieve analysis and Atterberg limits, soil types obtained using the USCS (Unified Soil Classification System) method in the affected area, point S1 is SP-SM (poorly graded sand and silty sand), point S2 is SP (Sand poorly graded, contains little or no fines) and at point, S3 is SM (silty sand). Meanwhile, in areas not affected by S4 SP-SM (poorly graded sand and silty sand), S5 point SP-SM (poorly graded sand and silty sand), and point S6 SP (Poorly graded sand, little or no fine grains). Soil affected and not affected by liquefaction is a non-plastic soil, this is based on the Atterberg limits test where the LL and PL values in each sample owned produce LL values: 0 and PL: 0 and IP: 0.

Poorly graded soils have particles that do not fill voids and have the potential for soil contraction, resulting in higher pore water pressure during earthquakes, which can cause liquefaction [7].

4.3.2. Swedish Weight Sounding

Through the Swedish Weight Sounding test in the affected and unaffected areas, the results shown in Figures 8 - 10, it is known that the values of Nsw, qa, and qu in the affected areas for a depth of 0.25 m obtained a large value, the value of Nsw in S1 is 73 n/m, S2 is 84.75 n/m, and S3 is 100 n/m, then this value decreases as the penetration depth increases until it reaches the hard soil layer, from the data obtained it can be seen that the surface soil in the area affected by liquefaction has a good carrying capacity, this assumption is supported by the liquefaction incident in Jono Oge village on September 28, 2018, where the surface soil at the time of the study was the boundary between soil that experienced lateral spreading and did not experience lateral spreading.
Based on these data, a comparison can be made between affected and unaffected soils at the same depth of penetration, the comparisons are in Tables 2 – 4.

Table 2 Nsw Value Comparison

Depth	Nsw Value					
	Affected Area			Unaffected Area		
	S1	S2	S3	S4	S5	S6
(m)	(n/m)	(n/m)	(n/m)	(n/m)	(n/m)	(n/m)
0,00-1,00	12,00-73,33	8,00-84,75	4,00-100,00	44,00-72,00	56,00-102,86	51,85-270,27
1,00-2,00	8,00-44,00	16,00-32,00	20,00,48,00	18,00-44,00	10,00-60,00	
2,00-3,00	4,67-8,00	8,00-16,00	56,00-104,00	136,00-357,14	16,00-64,00	
3,00-4,00	4,00-60,00	3,33-4,00	208,33		332,56	
4,00-5,00	64,00-253,16	23,00-104,00				
5,00-6,00		24,00-217,39				

From table 2 it is known that at each penetration depth, the value of Nsw in the unaffected area is the highest. At a depth of 0.00 – 1.00 m the highest Nsw value is at point S6 representing the unaffected area with a value of 270.27 n/m, at a depth of 1.00 – 2.00 m the highest Nsw value is 60.00 n/m obtained from the unaffected area, namely point S5, at a depth of 2.00 – 3.00 m the highest Nsw value is found in the unaffected area, namely point S4 with a value of 357.14 n/m, and at a depth of 3.00 - 4.00 m The highest Nsw value is obtained from point S5 representing the unaffected area with a value of 322.56. This shows that unaffected soil has better resistance than affected land to liquefaction.
Based on the test data, it was found that soils with Nsw values < 70 n/m had more potential for liquefaction than soils with Nsw values > 70 n/m. This value is obtained by taking the average of the Nsw values in each region, both affected and non-affected areas.

Table 3 qa Value Comparison

Depth	qa Value					
	Affected Area			Unaffected Area		
	S1	S2	S3	S4	S5	S6
(m)	(kN/m²)	(kN/m²)	(kN/m²)	(kN/m²)	(kN/m²)	(kN/m²)
0,00-1,00	39,60-90,80	36,40-97,80	33,20-110,00	65,20-87,60	74,80-112,29	71,48-246,22
1,00-2,00	36,40-65,20	43,80-55,60	46,00-69,40	33,40-62,20	38,00-78,00	
2,00-3,00	33,73-36,40	36,40-43,80	74,80-113,20	138,80-315,71	42,80-81,20	
3,00-4,00	33,20-78,00	32,67-33,20	196,67		288,06	
4,00-5,00	81,20-240,53	49,20-113,20				
5,00-6,00		49,20-203,91				

From table 3 it is known that at each penetration depth, the value of qa in the unaffected area is the highest value. The highest qa value at a depth of 0.00 – 1.00 m is 246.22 kN/m2, namely at point S6, a depth of 1.00 – 2.00 m is 78.00 kN/m2, namely at point S5, depth is 2.00 – 3.00 is 315.71 kN/m2 at point S4, and at a depth of 3.00 – 4.00 is 288.06 kN/m2 at point S5.

Table 4 qa Value Comparison

Depth	qa Value					
	Affected Area			Unaffected Area		
	S1	S2	S3	S4	S5	S6
(m)	(kN/m²)	(kN/m²)	(kN/m²)	(kN/m²)	(kN/m²)	(kN/m²)
0,00-1,00	2,25-102,00	2,25-108,56	2,25-120,00	2,25-99,00	2,25-122,14	2,25,247,70
1,00-2,00	51,00-78,00	57,00-69,00	60,00-81,00	70,50-78,00	52,50-90,00	
2,00-3,00	48,50-51,00	51,00-57,00	87,00-123,00	147,00-312,86	57,00-93,00	
3,00-4,00	48,00-90,00	47,50-48,00	201,25		286,94	
4,00-5,00	93,00-242,37	63,00-123,00				
5,00-6,00		63,00-208,04				

From table 4 it is known that at each penetration depth, the value of qu in the unaffected area is the highest. The highest qa value at a depth of 0.00 – 1.00 m is 247.70 kN/m2, namely, at point S6, a depth of 1.00 – 2.00 m is 90.00 kN/m2 at point S5, depth is 2.00 – 3.00 is 312.86 kN/m2 at point S4, and at a depth of 3.00 – 4.00 is 288.94 kN/m2 at point S5.

4.3.3. Groundwater Level

Groundwater level (GWL) investigation points arenot far from Jono Oge Village as shown in Figure 11, sourced from the JICA Survey For Disaster Information Collection In Indonesia, 2020, it was found that at point AB-11 which is not far from the test point Swedish Weight Sounding is a point, the GWL is at 9.03 m, while the AB-9 GWL is at 3.93 m and the AB-10 GWL is at 2.31 m. The researcher assumes that the MAT value at point AB-11 at the time before liquefaction occurs is at a range that is not much different from points AB-9 and AB-10, this is based on the location of point AB-11 which is right next to the irrigation channel which results in seepage. of the channel is very influential on the GWL at that point. This shallow GWLvalue is predicted to increase the potential for liquefaction in liquefaction-affected areas.

Conclusion

Based on the results of laboratory and field tests carried out, it can be concluded as follows:

- In areas affected by liquefaction, the soil classifications in the form of SP-SM, SP, and SM are non-plastic soils, and the Nsw values are in the range of 3.33 n/m – 263.16 n/m, qa is in the range of up to 240.53 kN/m2, and qu is in the range of 2.25 kN/m2 – 242.37 kN/m2. Meanwhile, in areas not affected by liquefaction SP-SM and SP are non-plastic soils, and the obtained Nsw values are in the range of 10.00 n/m – 357.14 n/m, qa is in the range up to315.71 kN/ m2, and qu are in the range of 2.25 kN/m2 – 312.86 kN/m2
- Jono Oge Village indeed has a high liquefaction potential. This is based on the graph of the relationship between the Nsw value and the depth of penetration in the affected soil, which has an average value of less than 70 n/m, more potential to experience liquefaction compared to an average value of more than 70 n/m. In addition, Tsuchida's liquefaction potential curve on soil samples from Jono Oge Village shows that all soil samples have high liquefaction potential.

References

[1] Bayong Tjasyono H.K. (2006). *Ilmu Kebumian dan Antariksa*. Bandung: PT. Remaja Rosdakarya . UPI.

[2] Mulyo, A. (2004). *Pengantar Ilmu Kebumian*. Pusaka Setia:Bandung

[3] Marcuson, W.F., III. (1978). Definition of Term Related to Liquefaction. Journal of Geotechnical Engineering (104), pp 119712000.

[4] Muntohar, A.,S. (2010). Mikro-zonasi Likuifaksi dan Penurunan Tanah Akibat Gempa Bumi, Penelitian Dosen. Yogyakarta: UMY.

[5] Towhata, I. (2008). Geotechnical Earthquake Engineering, Spinger-Verlag Berlin Heidelberg.

[6] Day, Robert W. (2002). *Geotechnical Earthquake Engineering Handbook*. McGraw-Hill Company. New York, NY

[7] Hardiyatmo, H. C. (2002). *Mekanika Tanah 1, edisi ke-2*. Beta Offset, Yogjakarta.

[8] Sosrodarsono, Suyono. (1984). *Mekanika Tanah & Teknik Pondasi*, Pradnya Paramita, Jakarta.

[9] Bowles, J.E., (1984). *Physical and Geotechnical Properties of Soils,* McGraw-Hill Book Company, USA.

[10] Unified Soil Classification System (USCS). (2006). *Standard Pracitice for Classification of Soil for Engineering Purpose.* ASTM Standard D2487, West Conshohocken, PA.

[11] Tsuchida, H. (1970). *Prediction and countermeasure against the liquefaction in sand deposits.* In *Abstract of the seminar in the Port and Harbor Research Institute* (pp. 31-333).

[12] Das, Braja M., 1995. *Mekanika Tanah (Prinsip-prinsip Rekayasa Geoteknis) Jilid 1,* Erlangga, Jakarta.

[13] JIS. (1995). JIS A 1221 Japanese Industrial Standard. Method of Swedish Weight Sounding Rev. 4: Japanese Standards Association.

[14] Broms, B.B., and Flodin, N. (1988). History of Soil Penetration Testing, Penetration Testing, A.Balkema Inc.

[15] Bergdahl U, Broms BB and Muromachi T. (1988)"Swedish Weight Sounding Test: international reference test procedure," Proceedings of 1st International Conference on Penetration Testing, Orlando, USA, Vol.1, 71-90.

[16] Habibi M, Cheshomi A and Fakher A. (2006). "A case study of liquefaction assessment using Swedish Weight Sounding," Proceedings of 4th International Conference on Earthquake Engineering, Taipei, Taiwan.

[17] Sugiyono. (2011). Metode Penelitian Kuantitatif, Kualitatif dan R&D. Bandung: Afabeta.

Slope stability using S*implified Bishop* method in Kebun Kopi area Donggala regency, central Sulawesi province

S Ramadhani[1], Martini[2,3], K A Chauf[4], A Dwijaka[5], M Z Bierhofa[6] and Y Gagaramusu[7]

[1]Department of Civil Engineering, Tadulako University, Jl. Soekarno-Hatta Palu, Central Sulawesi, Indonesia, Email: sriyatiramadhani@gmail.com
[2] Department of Civil Engineering, Tadulako University, Jl. Soekarno-Hatta Palu, Central Sulawesi, Indonesia, Email: martinisudihardjo@gmail.com
[3]Corresponding Author
[4] Department of Civil Engineering, Tadulako University, Jl. Soekarno-Hatta Palu, Central Sulawesi, Indonesia, Email: kus_untad@yahoo.co.id
[5] Department of Civil Engineering, Tadulako University, Jl. Soekarno-Hatta Palu, Central Sulawesi, Indonesia, Email: agusdwidjaka@gmail.com
[6] Department of Civil Engineering, Tadulako University, Jl. Soekarno-Hatta Palu, Central Sulawesi, Indonesia, Email: moh.zicobierhofa@gmail.com
[7]Departement of Education, Tadulako University Jl. Soekarno-Hatta Palu, Central Sulawesi, Indonesia Email: yusdingagaramusu@gmail.com

Abstract. Kebun Kopi is prone to landslides, especially during the long rainy season. On September 16, 2020, there was a landslide at KM 42, which caused the closure of the road. Slope handling in the that area has been carried out by providing reinforcement and erosion control, however, it was not sufficient for handling landslide cases. This study aimed to determine the slope stability at KM 42 based on simplified Bishop. The research was conducted in Nupa Bomba Village, Central Sulawesi Province. The data used in this research are secondary data which is the result of previous research. Secondary data were used in the form of slope geometry and laboratory test data. The collected data were analyzed using calculations according to the standards. Slope stability analysis was carried out by manual calculations using the simplified Bishop method and compared with numerical calculations using the Slide program. The results indicate that the value of the minimum safety factor has an average difference 0.22% when compared to the numerical analysis.The disadvantage of manual calculations is that the analysis process is quite long and less accurate, this can be minimized by using software analyzing slope stability problems, especially in the Kebun Kopi area.

1. Introduction

Kebun Kopi is a trans-Sulawesi route that connects provinces on the island of Sulawesi, including North Sulawesi, Gorontalo, South Sulawesi and Southeast Sulawesi. Kebun Kopi has an uphill and winding road and is located on a hilly contour with a slope of 30-65° [1]. The slopes of Kebun Kopi are formed naturally or artificially because of the construction or widening of the Trans Sulawesi road.
Kebun Kopi is prone to landslides, especially during the long rainy season. If a landslide occurs in this area, it will cause extraordinary traffic jams. According to [2], Kebun Kopi is prone to soil movement because the area has soil that is not dense, there are ground vibrations caused by vehicles, erosion and high rainfall.

Kebun Kopi area is included in the Tanantovea District which has two seasons: the dry season and the rainy season [3]. On September 16, 2020, a landslide occurred in Kebun Kopi, specifically at KM 42 which caused the closure of the Trans Sulawesi road. Continuous rain or rain with high rainfall is considered to be one of the triggering factors that play a significant role in the process of landslides, because it can affect the process of saturation of the soil to change in the dimensions of space and time. Slope handling in Kebun Kopi area has been carried out by providing reinforcement and erosion control, but this has not been maximized for handling landslide cases, on the slopes of the KM 42 Kebun Kopi, reinforcement has also been installed, but landslides still occur. Based on the above background, it is necessary to conduct research on the characteristics of landslides and how the slope stability in Kebun Kopi area, especially KM 42.

This study aimed to determine the slope stability of KM 42 Kebun Kopi based on the simplified Bishop method. This research was conducted on the existing slope that experienced a landslide on September 16, 2020 with coordinates 0°43'43,671" S latitude and 119°59'214" East longitude (Figure 1). The research site is Nupa Bomba Village, Tanantovea District, Donggala Regency, Central Sulawesi Province.

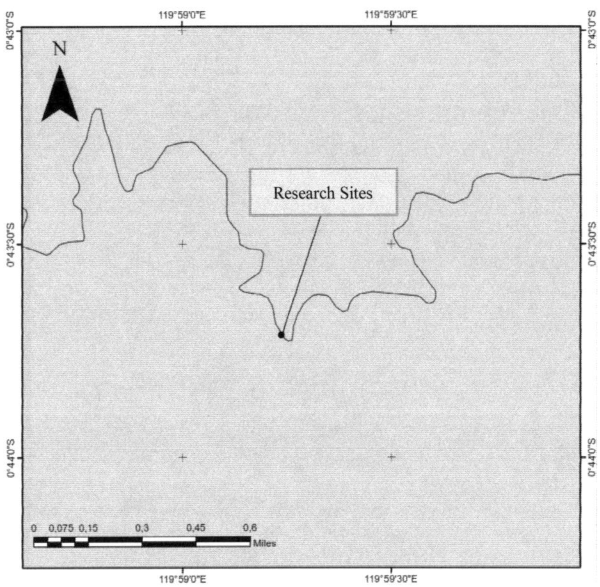

Figure 1. Research Sites

2. Literature Review

2.1. Slope Movement

Varnes [4] prefers to use the term slope movement than landslide. The term of landslide is widely used and undoubtly, it is used as a term for almost all types of slope movement. Cruden [5] suggests a simple definition of landslide as the movement of a mass of rock, soil or debris or soil down a slope. Cornforth [6] states that the term of landslide includes all slope movements that occur naturally or artificially. If the gravitational force acting on the slope exceeds the shear strength of the slope-forming material, the slope will collapse and movement will occur. External factors, such as weather conditions, can result in a reduction or increase in slope strength.

Varnes [4] explains that the types of landslides are grouped into several classifications as follows:

- Fall is a fall or rock mass moving through the air, including free fall motion, jumping and rolling of boulders and debris without much contact with each other. This type of movement includes debris lawina, avalanche of stone, debris and soil.

- Slides are movements caused by failure through one or more observable or predictable planes. Slides are divided into two types, namely slides and slides. If the ground motion is affected by translational motion and the composition of the material changes a lot, it is called a slide. But if the avalanche with the composition of the material does not change much and is generally influenced by rotational motion, it is called a slump.
- Flow is a movement that is influenced by the amount of content or soil moisture content, occurs in moving materials that generally cannot be recognized. Included in the type of dry flow movement is sandrun, rock fragment flow, loess flow. While the types of wet flow movement are sand-silt flow, fast soil flow, slow soil flow, mud flow, and debris flow.
- A complex landslide is a combination of two or three types of movement above. In general, multiple landslide occur in nature, but usually there is one type of movement that stands out or is more dominant.
- Horizontal motion/lateral spread, is a type of landslide that is influenced by the horizontal movement of rock material.
- In both translational and rotational types of landslides, there is a boundary between the moving and stationary masses (called the slip plane). The depth of the boundary from the ground surface is very important for the description of the landslide.

2.2. Limit Equilibrium Method
Limit equilibrium analysis is a method of analyzing the balance of a potentially moving mass by comparing the driving force and the resisting force along the sliding plane of the landslide, an illustration of the forces acting on the slip plane can be seen in Figure 2. The comparison of the two forces will produce a slope safety factor (FS) value, with the equilibrium limit condition being reached when the FS value = 1 [7].

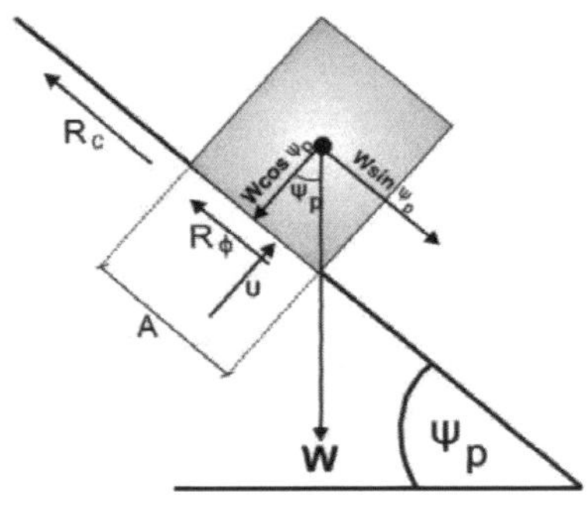

Figure 2. Illustration of the forces acting on the slip plane of a slope [7].

$$Fs = \frac{resisting\ force}{driving\ force} = \frac{R_C + R_\phi}{S} = \frac{c_A + W \cos\psi\, P \tan\phi}{W \sin\psi p} \quad (1)$$

with
FS : slope safety factor,
c : cohesion on the slip plane (kN/m2),
A : area of rock mass that has the potential to move (m2),
W : weight of rock mass above the slip plane (kN),
Ψp : the angle of inclination of the slip plane (°),
Φ : internal friction angle (°).

The FS value on a slope is the smallest value calculated on all slip planes that have the potential to produce landslides on that slope. The limit equilibrium method can be distinguished as follows:

2.2.1.The analysis method uses the total block or total mass

Limit equilibrium analysis with this method will be explained based on the illustrations in Figure 3 (a) for tension cracks on the upper slope and Figure 3 (b) for tension cracks on the slope face.

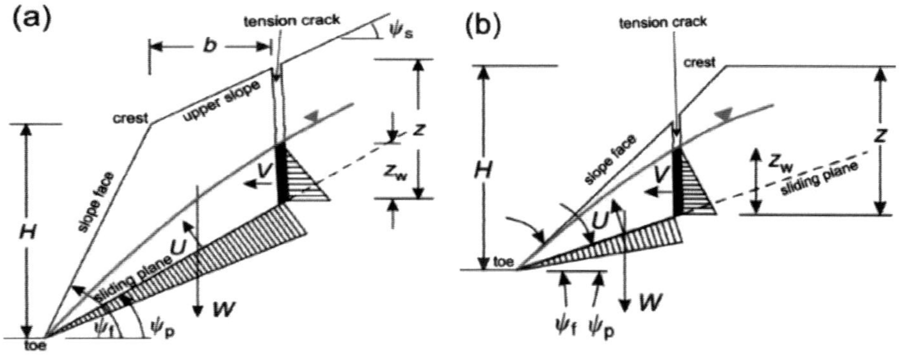

Figure 3. Geometry of the plane failure slope and the forces acting on it. a) tension crack and upper slope, b) tension crack on the slope face [8].

Based on the illustration in figure 3, without involving the influence of external forces such as seismic ground motion and rock reinforcement, the FS value can be calculated using the following formula [8]:

$$SF = \frac{RF}{DF}\, cA + (W cos\psi p - u - v sin\psi p) tan\phi \quad (2)$$

$$A = (H + b\ tan\ \psi s - z) cosec\psi p \quad (3)$$

$$u = \frac{1}{2}\, \Upsilon_w z_w((H + b\ tan\ \psi s - z) cos\psi p \quad (4)$$

$$v = \frac{1}{2}\Upsilon_w {z_w}^2 \quad (5)$$

with

DF : driving force (kN/m),
RF : resisting force (kN/m),
c : cohesion on the slip plane (kN/m2),
A : area of sliding area (m2),
W : weight of rock mass above the slip plane (kN),
Ψ_P : the angle of inclination of the slip plane (°),
φ : internal friction angle (°),

2.2.2. Analytical method using mass divided into several slices

This method is used for inhomogeneous material analysis. The calculation of the slope safety factor value is based on the integration of the forces acting on each slice. Several methods that use this principle and are most often used include the modified Bishop which is suitable for circular landslide and the Janbu method which is suitable for circular, non-circular, polygonal, and planar landslide [7]. In this study, the General Limit Equilibrium (GLE) method was used.

According to Abramson, et al. [9] the analysis of slope stability based on the limit equilibrium method is carried out by dividing the sliding soil mass into several slices which are considered as a unique shear block as shown in Figure 4.

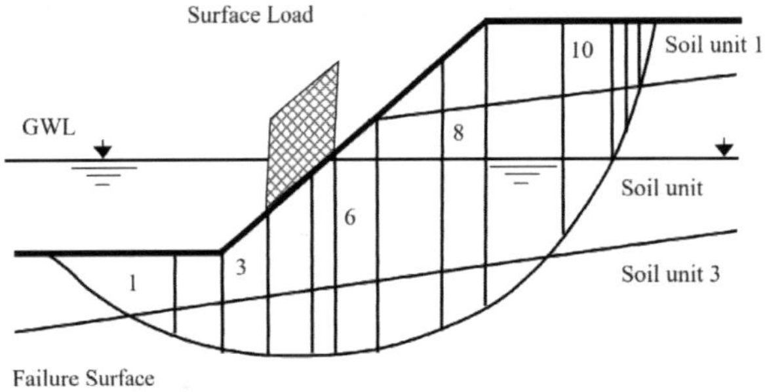

Figure 4. Distribution of sliding soil mass [9]

The numerical model is more appropriate for boundary equilibrium analysis because of the following benefits [10]:

- complex phenomena can be modeled (water effects, dynamics, discontinuities, etc.),
- capacity to introduce constitutive models,
- the safety factor is not constant along the sliding surface,
- The slope safety factor can be calculated by the Itasca program by reducing the shear strength of the soil gradually until the slope collapses, but this does not represent a surface defined by failure, the numerical program can display a two-dimensional picture of the shear.

2.3. OMS Method

The Ordinary/Fellinius method assumes that all forces acting on the wedge have a zero resultant in the direction perpendicular to the slip plane. The safety factor (F) is defined as the ratio between the number of moments of shear resistance along the landslide plane (Mr) and the number of moments of the mass of the landslide mass (Md), namely:

$$FS = \frac{\sum M_r}{\sum M_d} = \frac{R\sum_{i=1}^{i=n}(ca_i + N_i \tan\varphi)}{R\sum_{i=1}^{i=n} W_i \sin\theta_i} \tag{6}$$

If there is water on the slope, the pore water pressure in the landslide area does not increase the moment due to the soil that will slide. If there are forces other than self-weight, such as building loads on a slope, the moment due to the load is calculated as Md.

2.4. Bishop Simplified Method

This method assumes that the forces acting on the sides of the wedge have zero resultant in the vertical direction. The safety factor (F) is defined as:

$$F = \frac{R\sum_{i=1}^{i=n}(c'a_i + (N_i - u_i a_i)\tan\varphi')}{\sum_{i=1}^{i=n} W_i X_i} \tag{7}$$

with:

c' : effective soil cohesion (kN/m2),
ui : pore water pressure at slice i (kN/m2),
j' : effective friction angle in the soil (o),
xi : R Sin q

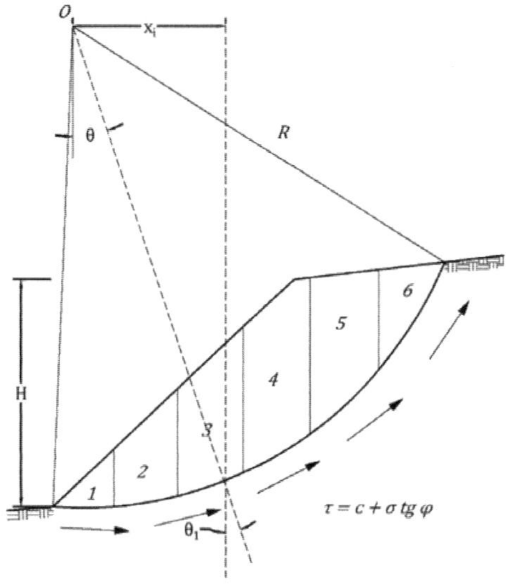

Figure 5. Slice Method [11]

With:

R	: radius of landslide circle (m),
c	: cohesion (kN/m2),
ai	: the length of the arc of the circle at the ith slice (m),
j	: internal friction angle (°),
n	: number of slices,
Wi	: weight of the ith slice soil mass (kN),
qi	: angle of inclination as in Figure 6 (°).

2.5. Spencer Method

Spencer's method assumes that the forces acting in the wedge are parallel in direction and both satisfy the balance of forces and the balance of moments. If the slope angle is determined, the safety factor can be calculated by trial and error.

2.6. Janbu Method

The Janbu method considers the vertical balance force of each slice as the horizontal balance force of all slices.

2.7. Morgenstern method

This method similar with the Spencer method.

2.8. Safety Factor Criteria

In one group of the structure design has been based on the factor of safety against shear. These structures include gravity and dam infill as well as rock and soil slopes, all involving potential slippage along a well-defined failure surface [12]. The safety factor is defined as the factor with which the shear strength parameter can be reduced to bring the slope or dam foundation into a state of equilibrium. The safety factors used in conventional geotechnical practice are based on logical experience [13]. The safety factor for a design of various types of rock in engineering problems ranging from a safety factor value greater than 1 for extreme loading, seismic and gravity analysis in dam design. The safety factor value is greater than 1.5 for permanent rock slopes and the safety factor value is greater than 2 for block fall-out in the tunnel. safe factor criteria can be seen in table 1.

Table 1. Safety Factor Criteria [12]

Safety Factor (F)	Incident
F < 1	A collapse occured
1 ≤ F < 1,5	Critical condition (not significant in design)
F ≥ 1,5	Stable conditionl (design above critical value)

3. Research Method

The data used in this study is secondary data which is the result of previous research. Secondary data used in the form of slope geometry data and laboratory test data.

The data collected is processed with calculations according to standards. Calculation of slope stability analysis was carried out by manual calculations and compared with numerical calculations with the simplified Bishop method with the help of Slide 7 from previous studies. The method used in slope stability analysis is the simplified Bishop. The steps for analyzing data manually using the simplified Bishop method are as follows:

- Draw a circular slip plane divided into several vertical slices, usually for the width to be the same.
- Measure the width, height and angle of inclination of each slice
- Calculate the area (A), soil weight (W) and arc length (L) of the landslide area from each slice.
- Calculating dead load and live load due to additional load on the slope
- Calculation of the area, soil weight and arc length of the landslide circle from each slice.
- Area A (i)=B(i)*H(i), 6. Weight of soil, W (i) = γ * A(i) W (i)= ϒ*A(i)
- Lenght of the landslide circle $L(i) = \frac{\alpha\,(i)}{360} * 2\pi R$
- Calculation of the safety factor (F) with the equation :

$$F = \frac{\sum_{n-1}^{n=0}(c.\Delta L_n + W_n.\ Cos\ \alpha_n.\tan\emptyset}{\sum_{n=1}^{n=0} W_n.Sin\ \alpha_n} \tag{8}$$

- The calculation of the safety factor F is repeated to obtain a qualified F value (FS ≥ 1.5)
- Comparing the value of the safety factor calculated manually with the value of the safe factor calculated by the numerical calculation from previous studies using the silde7 software

4. Findings and discussion

4.1. Topography Data

Topography is determined by taking measurements in the field using a GPS device (Figure 6a). The height of the existing slope at the location of the landslide is 26 meters with a slope angle of 60° (Figure 6b)

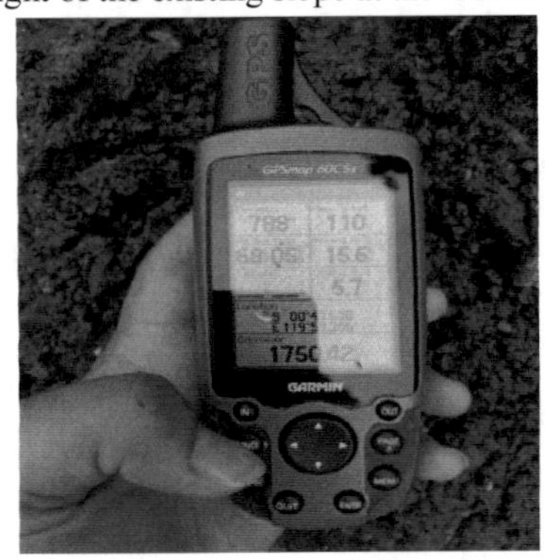

Figure 6. Measurements at the research location a) Slope measurement using GPS and b) Slope

4.2. Soil Characteristics

Soil characteristics data were obtained from previous research conducted by [14]. The taken Soil data that was obtained during the rainy season at the landslide location which can be seen in Table 2.

Table 2. Soil Characteristics in landslides area [14]

No	Description	Symbol	Unit	Landslides								
				Toe			Midle			Peak		
				1	2	3	1	2	3	1	2	3
					Gravel Sand			Gravel Clay			Gravel Clay	
1	Water content	*w*	%		27.05			26.23			29.33	
2	Specific gravity	*Gs*	-		2.64			2.65			2.67	
3	unit wieght	γ_b	gr/cm³		2.3			2.3			2.2	
4	dried unit weight	γ_d	gr/cm³		1.8			1.8			1.7	
5	Porosity	*n*	-		0.595			0.596			0.611	
6	Saturation degree	*S*	%		48.690			47.214			49.861	
7	Pore number	*e*	-		1.467			1.472			1.571	
8	Liquid limit	*LL*	%		23.33			24.16			22.84	
9	Plastic Limit	*PL*	%		-			14.09			11.69	
10	Plastic Index	*PI*	%		NP			10.08			11.15	
11	Gravel	*G*	%		2.70			11.10			17.50	
12	Sand	*S*	%		87.48			67.60			64.90	
13	Silt/Clay	*M/C*	%		9.82			21.30			17.60	
14	Coefficienr of Uniformity	*Cu*	-		-			-			-	
15	Gradation Coefficiei	*Cc*	-		-			-			-	
16	Cohesion	*c*	kPa	19.32	22.46	19.42	42.66	19.22	19.24	16.47	16.28	13.24
17	Sliding Angle	ϕ	°	42.36	40.74	35.83	39.03	36.30	37.23	39.46	35.35	38.14
18	Color	-	-		Gray			Brown			Brown	

4.3. The Slope Stability Analysis using the Simplified Bishop Method

The Slope stability analysis using manual calculations applying the Simplified Bishop method. In this study, the method uses the concept of the safety factor value (F) in analyzing the existing slope modeling that slides in conditions during the rainy season. The results of the analysis will be compared to the results of numerical analysis using the Simplified Bishop method with the help of the Slide Program. The analysis uses the existing formula with trial and error calculations to determine the slip plane. To get the minimum safety factor value, several alternative slide planes are drawn with radii 46, 47,389, 48, and 49 m. The results of calculations from several alternative slip planes, obtained the value of the safety factor as in table 3, the minimum safety factor value was obtained at the alternative radius (R = 49 m) with a safety factor value (F) of 1.345 For the calculation of the safety factor (F) with an alternative radius (R) 49 m can be seen in Figure 7.

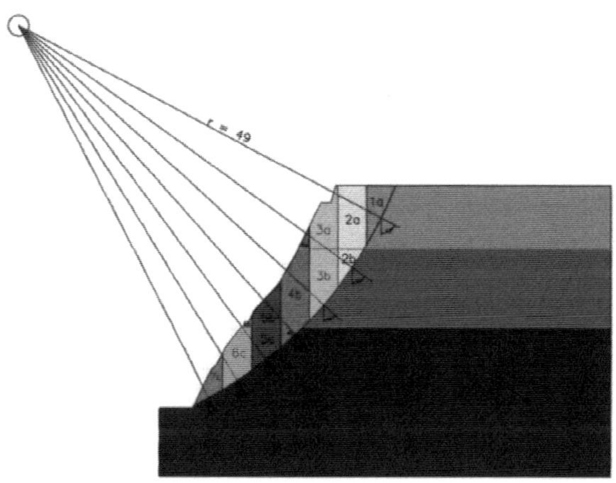

Figure 7. The slip plane at the alternate radius (R) is 49 m with a minimum safety factor of F=1.345

Table 3. Calculation of safety factor for slip plane with R = 49 m

slice no	Large m2	γ' Kn/m³	Wn (Kn/m)	αn	Sinαn	Cosαn	b_n	φ °	c kn,m²	tan (φ) (Kn/m)	Wn Sinαn	A=c*bn + Wn tan(φ)	Fs Right	B=cos(αn)+(sin (αn)tan(φ)/Fs)	A/B	Fs Left
1a	12,689	16,67	211,526	63	0,891	0,454	3,473	38,14	13,24	0,785	188,471	212,078	1,345	0,974	217,701	
2a	25,565	16,67	426,169	55	0,819	0,574	3,473	38,14	13,24	0,785	349,097	380,622	1,345	1,052	361,874	
2b	7,289	16,67	121,508	55	0,819	0,574	3,473	36,3	19,22	0,735	99,533	156,007	1,345	1,021	152,805	
3a	17,914	16,67	298,626	48	0,743	0,669	3,473	38,14	13,24	0,785	221,923	280,472	1,345	1,103	254,284	
3b	22,527	16,67	375,525	48	0,743	0,669	3,473	36,3	19,22	0,735	279,070	342,602	1,345	1,075	318,699	
4a	1,768	16,67	29,473	43	0,682	0,731	3,473	38,14	13,24	0,785	20,100	69,125	1,345	1,130	61,199	
4b	27,505	16,67	458,508	43	0,682	0,731	3,473	36,3	19,22	0,735	312,702	403,559	1,345	1,104	365,600	1,345
4c	3,377	16,67	56,295	43	0,682	0,731	3,473	42,36	19,32	0,912	38,393	118,430	1,345	1,194	99,211	
5b	11,908	16,67	198,506	37	0,602	0,799	3,473	36,3	19,22	0,735	119,464	212,568	1,345	1,127	188,561	
5c	13,207	17,65	233,104	37	0,602	0,799	3,473	42,36	19,32	0,912	140,285	279,653	1,345	1,207	231,762	
6b	0,510	16,67	8,502	32	0,530	0,848	3,473	36,3	19,22	0,735	4,505	72,996	1,345	1,137	64,174	
6c	18,322	17,65	323,383	32	0,530	0,848	3,473	42,36	19,32	0,912	171,367	361,974	1,345	1,207	299,819	
7c	8,068	17,65	142,400	28	0,469	0,883	3,473	42,36	19,32	0,912	66,853	196,945	1,345	1,201	163,954	
				Total							1944,91				2615,691	-

The above table 3 shows a description of the calculation of the slice method based on the simplified Bishop's equation. The results of slope stability analysis using the simplified Bishop method based on modeling of existing landslides under rainy season conditions when compared numerically with slide7 software are shown in Table 4.

Table 4. Comparison of safety factors with silde7 software and manual

Alternative Radius (R), m	**Safety Factor, FS**		**difference**	**% difference**
	Slide7	**Manual**		
46	1.248	1.597	0.349	0.279
47.398	1.596	1.491	0.295	0.247
48	1.188	1.391	0.203	0.171
49	1.151	1.345	0.194	0.168

The value of the safety factor obtained based on numerical calculations using slide 7.0 software and manually shows a value that is not much different, the average percentage difference for the two methods is 0.22%.

5. Conclusion

The results of the analysis show that for the existing slopes the landslide location, is in critical condition both in the dry season and the rainy season, where the value of 1≤ FK <1.5 while in locations that do not landslide in a stable condition in the dry season and rainy season FK 1 ,5. On the existing slope of the landslide, it is necessary to make sloping or tiered, so that the slope in the study area becomes stable.

6. References

[1] Dona I R and Sudiar N Y 2015 Identifikasi bidang gelincir menggunakan metode geolistrik tahanan jenis konfigurasi Schlumberger di Bukit Lantiak Kecamatan Padang Selatan (The identification of the slip surface using the geoelectric resistivity method of the Schlumberger configuration in Bukit Lantiak, Padang Selatan District). *Pillar Of Physics*, vol 5(1).

[2] Ani S 2017 Aktivitas Masyarakat Dalam Kegiatan Ekonomi Di Daerah Rawan Longsor (Kasus: Jalur Poros Tawaeli-Toboli *Jurnal Pendidikan Geografi*, vol 5(2)

[3] BPS K D 2017 *Kecamatan Tanantovea dalam Angka 2017*, Badan Pusat Statistik Kabupaten Donggala.
[4] Varnes D J 1978 Slope movement types and processes (*Special report*) pp 176:11-33.
[5] Cruden D M 1991 A simple definition of a landslide. *Bulletin of the International Association of Engineering Geology-Bulletin de l'Association Internationale de Géologie de l'Ingénieur*, vol 43(1), pp.27-29.
[6] Cornforth D H 2005 *Landslides in practice. Investigation, Analysis and Remedial/Preventive Options in Soil*, pp.14-15.
[7] de Vallejo L G and Ferrer M 2011 *Geological engineering* CRC Press.
[8] Hoek E and Bray J 1981 *Rock Slope Engineering* (revised third edition): The Institute of Mining and Metallurgy. London, England.
[9] Abramson L W, Lee T S, Sharma S and Boyce G M 2001. *Slope stability and stabilization methods*. John Wiley & Sons.
[10] Kourdey A, Alheib M and Piguet J P 2001 Evaluatio of slope stability numerical methods *17th International Mining Congress and Exhibition of Turkey*, pp. 705-710.
[11] Hardiyatmo H C 2012 *Tanah Longsor dan Erosi: Kejadian dan Penanganan*. Edisi Pertama. Gadjah Mada University Press, Yogyakarta.
[12] Hoek E 1991 September When is a design in rock engineering acceptable?. *In 7th ISRM Congress*. OnePetro.
[13] Duncan J M 2000 Factors of safety and reliability in geotechnical engineering. *Journal of geotechnical and geoenvironmental engineering*, vol 126(4), pp.307-316.
[14] Bierhofa M Z, Ramadhani S, Martini & Chauf K A 2021 Kajian Longsor Kebun Kopi KM42 dengan Menggunakan Metode Kesetimbangan Batas. *Konferensi Nasional Teknik Sipil 17*

Potential of Liquifaction at Nasanapura Hospital Petobo Village Palu City

A Rahayu[1], I Uno I[2], N Hidayat[3], A Dwijaka[4], M Yusuf[5]
[1,2,3,4]Department of Civil Engineering, Engineering Faculty, Tadulako University
[5] Under graduated student at Department of Civil Engineering, Tadulako University
*Corresponding Author: astri2018.1440@gmail.com

Abstract. The liquefaction event that occurred in Petobo Village, Palu City, Central Sulawesi, was caused by the Palu earthquake on September 28th, 2018 with a magnitude of 7.4 and a depth of 10 km. This study aims to determine the characteristics of the soil and also whether the area still has the potential for liquefaction based on the Test *Swedish Weight Sounding (SWS),* and *Cone Penetration Test (CPT).* The test is carried out at a location that is the limit of the cessation of liquefaction, which has a variety of soil types. Determination of soil characteristics is carried out by laboratory tests, Sieve Analysis and Atterberg Limits. The results of laboratory testing, the soil types at point A are poorly graded sand and silty sand (SP-SM). As for the results of SWS at point D the soil with an Nsw value of more than 180.00 n/m is at a depth of 3.68 m with a Nsw value of 274.73 n/m. From the results of the CPT calculation it was found that the soil at Point E at a depth of 0.80 m – 6.40 m has potential for liquefaction and soil at a depth of 6.60. m – 8.40 m no liquefaction potential.

Keywords: ***Liquefaction, Petobo, Swedish Weight Sounding, Cone Penetration Test***

1. Introduction

The liquefaction event that occurred in Petobo Village, Palu City, Central Sulawesi, was caused by the Palu earthquake on September 28th, 2018 with a magnitude of 7.4 Mw and a depth of [1]. The earthquake occurred due to land fault activity, namely the Palu Koro Fault. The Palu Koro Fault is a fault that extends approximately 240 km north (Palu City). The geological condition of Palu City which is dominated by Alluvium material with quaternary age and shallow water table depth is a factor in the occurrence of liquefaction in Petobo Village [2].

Nasanapura hospital is a public facility located in the Petobo area and is still being used today. The location of this hospital is right beside an area prone to liquefaction. Because of the liquefaction phenomenon, it is very important for designers to consider the potential for liquefaction when planning a building. Liquefaction is defined as the transformation of granular material from a solid to a liquid state as a result of an increase in pore water pressure and loss of effective stress [3]. Liquefaction is a phenomenon where the soil loses a lot of strength (strength) and stiffness (stiffness) for a short time. There are 3 main factors that cause liquefaction, namely:

1. Coarse-grained soil
2. The position of the groundwater level not far from the ground surface
3. Earthquake

2. Material and Methods

2.1. Location of Research

The research site is located in the Nasanapura hospital, Petobo Village, South Palu District, Palu City, Central Sulawesi Province. The research location can be seen in the image below.

Figure 1. Research location

2.2. Site Test Equipment, SWS and CPT.

SWS was first introduced by the Geotechnical Committee of State Railways of Sweden, and is a common penetration test in the Nordic countries as well as in Japan. Habibi et al. [4], reported the use of SWS to estimate foundation depth and bearing capacity for several buildings in the southern region of Tehran, Iran.

The Swedish Weight Sounding test equipment consists of a screw point (diameter 25 mm), a set of loads (1 x 5 kg, 2 x 10 kg, and 3 x 25 kg, a total of 100 kg), a number of rods with a diameter of 22 mm, and a handle for turning the rod. Figure 2 shows a schematic view of the SWS test equipment.

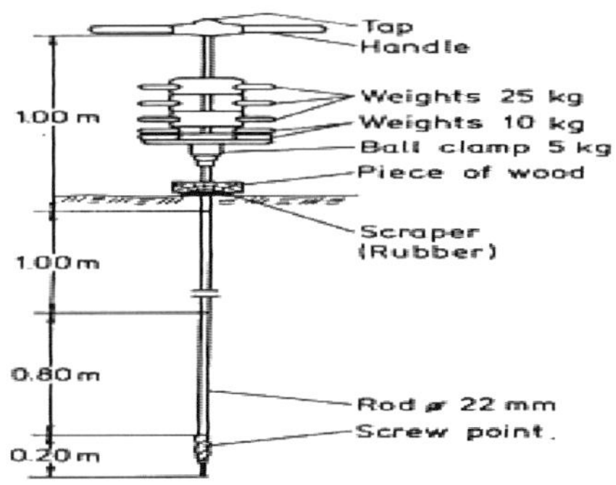

Figure 2. The SWS test equipment [5].

The CPT test or more often called the sondir test is a static penetrometer that is widely used in Indonesia. This tool comes from the Netherlands and is known as the Dutch-cone Penetrometer Test. The working principle of this tool is to press the tip of the penetrometer (conus) down with a pressing machine anchored to the ground. There are two kinds of penetrometer tips that are commonly used, conus (standard type) and biconus (friction sleeve or adhesion jacket type). With this test soil type can be determined (Figure 3) and soil hardness is obtained from the value of qc.

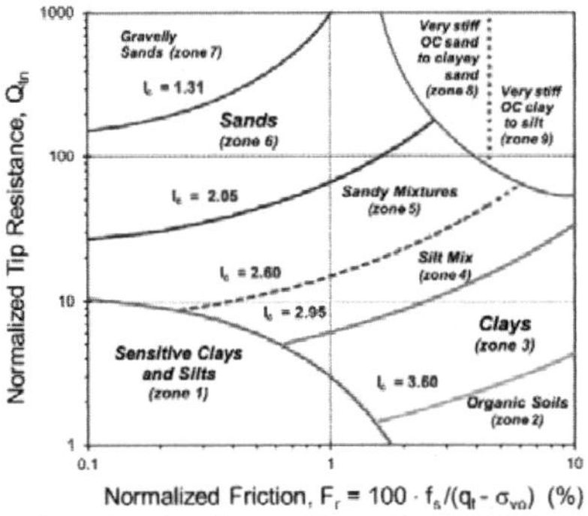

Figure 3 Graph of soil type based on CPT test results
Source: Mayne [6].

2.3. Methods

The complete research method is presented in the following figure:

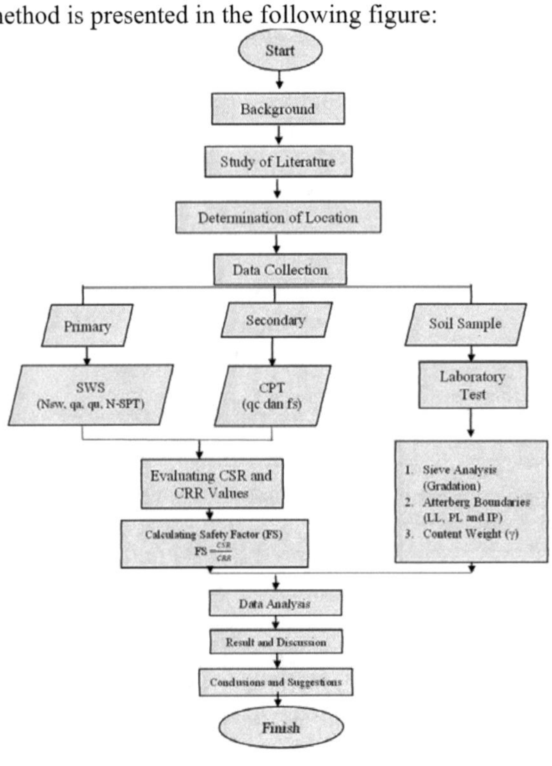

Figure 4. Flowchart of research

3. Results and Discussion

3.1 Physical Properties of Soil

Soil physical properties are properties that can be seen visually without being given special treatment on the soil. Soil properties are highly dependent on grain size. The grain size is used as the basis for classifying soils [7], [8].and percentage of liquifaction potential [9].

3.2 Grain size Analysis

Point A is the location which is the limit of the cessation of liquefaction. The grain size analysis results are presented in the following tables and graphs:

Table 1. The results of grain size analysis at point A

Point	Depth of Sample (m)	Percentage of pass # No.4 (%)	Percentage of pass # No.200 (%)	Classification (*Unified*)
	0.20	99.05	14.87	SP - SM
	0.40	98.88	8.87	SP - SC
A	0.60	96.35	13.99	SP - SM
	0.80	94.15	12.62	SP - SM
	1.00	95.2	13.42	SP - SM

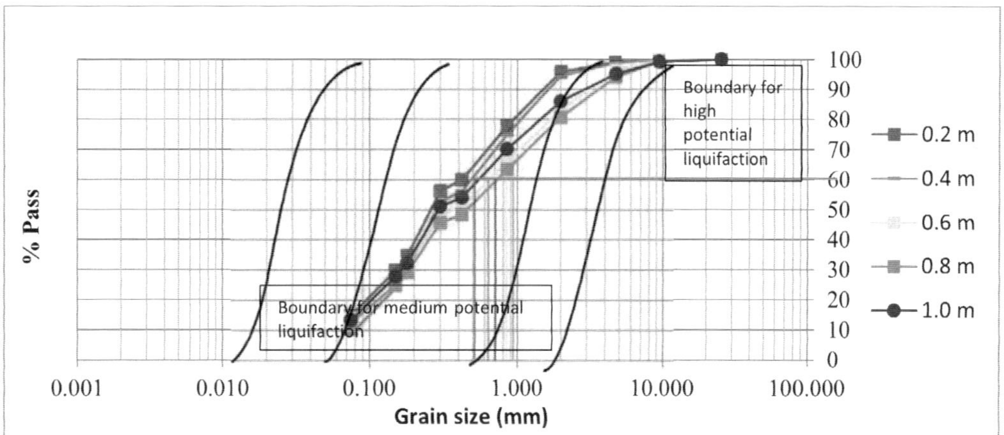

Figure 5. Graph of grain size distribution

Table 2. Atterberg Limit at point A

Point	Depth of Sample	LL	PL	PI	Classified (Unified)
	(m)	(%)	(%)	(%)	
	0.20	22.15	19.90	2.25	CL - ML
	0.40	25.30	23.64	1.66	CL - ML
A	0.60	24.25	21.87	2.38	CL - ML
	0.80	22.80	-	-	ML
	1.00	22.60	16.24	6.36	CL

Table 3. Percentage of Liquifaction Potential based on grain size distribution at point A

Liquifaction Potential	Value (%)
High	39,06
Medium	29,28
Low	31,67

Based on the table and figure above, it can be seen that the soil sample at point A at a depth of 0.2 – 1.0 m is categorized as coarse-grained soil because more than 50% of the fraction passes filter no. 4. and includes soils with high liquefaction potential. This may occur because the sample used is a sample whose location is right at the limit of the cessation of liquefaction.

From Table 2 it can be seen that the sample at point A at a depth of 0.2 m – 1.0 m consists of clay and inorganic silt with low to medium plasticity (CL-ML), inorganic clay (CL), very fine sand (ML).

3.3 Swedish weight sounding

This test is carried out at points in areas not affected by liquefaction close to areas affected by liquefaction using the Swedish Weight Sounding test kit, with the following data:

Table 4. Swedish Weight Sounding Test Results Point D

Load	Sum of half cycle /25 cm	Penetration depth		Sum of penetration	Sum of half cycle /1m	Allowable Bearing Capacity	Unconfined Compressive Strength	N	
Wsw (N)	Na	D (cm)	D (m)	L (cm)	Nsw	qa (kN/m2)	qu (kN/m2)	coarse material	fine material
a	b	c	d	e	f	g	h	i	j
0.05		6.50	0.07			0.00	0.00	0.10	0.15
0.15		11.50	0.12			0.00	0.01	0.30	0.45
0.25		14.00	0.14			0.00	0.01	0.50	0.75
0.50		18.50	0.19			0.00	0.02	1.00	1.50
0.75		23.00	0.23			0.00	0.03	1.50	2.25
1.00		30.50	0.31			0.00	0.05	2.00	3.00
1.00	9.00	50.00	0.50	19.50	46.15	66.92	34.66	5.09	5.31
1.00	0.00	52.00	0.52	2.00	0.00	30.00	0.05	2.00	3.00
1.00	10.00	100.00	1.00	48.00	20.83	46.67	15.67	3.40	4.04
1.00	14.00	125.00	1.25	25.00	56.00	74.80	42.05	5.75	5.80
1.00	12.00	200.00	2.00	75.00	16.00	42.80	12.05	3.07	3.80
1.00	6.00	250.00	2.50	50.00	12.00	39.60	9.05	2.80	3.60
1.00	0.00	257.50	2.58	7.50	0.00	30.00	0.05	2.00	3.00
1.00	7.00	275.00	2.75	17.50	40.00	62.00	30.05	4.68	5.00
1.00	5.00	300.00	3.00	25.00	20.00	46.00	15.05	3.34	4.00
1.00	7.00	325.00	3.25	25.00	28.00	52.40	21.05	3.88	4.40
1.00	10.00	350.00	3.50	25.00	40.00	62.00	30.05	4.68	5.00
1.00	50.00	368.20	3.68	18.20	274.73	249.78	206.09	20.41	16.74

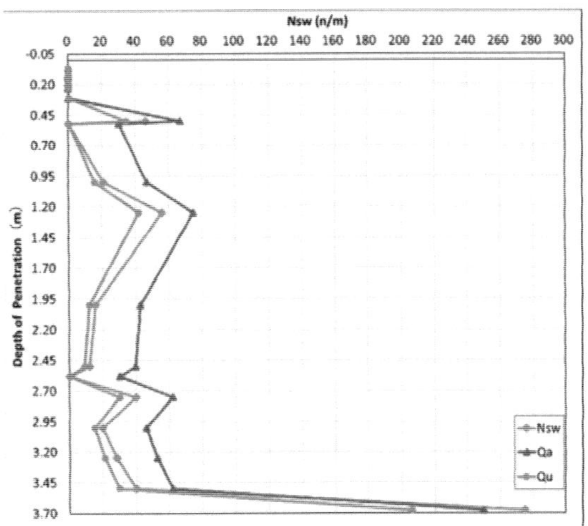

Figure 6. Graph of the relationship between Nsw, qa and quSWS with the penetration depth at point D

From Figure 6, it can be seen that the results of the Swedish Weight Sounding test on the sample point D contained soil with an Nsw value of more than 180.00 n/m at a depth of 3.68 m with an Nsw value of 274.73 n/m. For soils with Nsw values between 120.00 – 180.00 n/m liquefaction can occur, although it has a smaller probability when compared to soils with Nsw values below 120.00 n/m.

3.3.1. Calculation of Liquefaction Potential Based on SWS Data

The method used in conducting the liquefaction potential analysis is the procedure used by Seed and Idriss [10], with N data taken from the SWS value converted to the N-SPT value:
Depth = 3.00 m, Earthquake magnitude (M) = 7.4, Calculation of liquefaction potential for point D
Safety factor (FS), FS = CRR/CSR = 0.12/0.39 = 0.36 (liquefied)

Table 5. Liquifaction Potential based on Swedish Weight Sounding Test at Point D

Depth (m)	N	CSR	CRR For M=7,5	CRR	FS	Description
0.07	0.10	0.398	0.094	0.127	0.32	Liquified
0.12	0.30	0.398	0.102	0.137	0.34	Liquified
0.14	0.50	0.398	0.110	0.149	0.37	Liquified
0.19	1.00	0.398	0.128	0.174	0.44	Liquified
0.23	1.50	0.398	0.144	0.199	0.50	Liquified
0.31	2.00	0.398	0.154	0.211	0.53	Liquified
0.50	5.09	0.397	0.272	0.394	0.99	Liquified
0.52	2.00	0.397	0.137	0.178	0.45	Liquified
1.00	3.40	0.395	0.150	0.188	0.48	Liquified
1.25	5.75	0.394	0.192	0.245	0.62	Liquified
2.00	3.07	0.391	0.125	0.147	0.38	Liquified
2.50	2.80	0.388	0.118	0.135	0.35	Liquified
2.58	2.00	0.388	0.109	0.124	0.32	Liquified
2.75	4.68	0.387	0.138	0.159	0.41	Liquified
3.00	3.34	0.386	0.121	0.137	0.36	Liquified
3.25	3.88	0.385	0.125	0.142	0.37	Liquified
3.50	4.68	0.384	0.132	0.149	0.39	Liquified
3.68	20.41	0.383	0.788	0.988	2.58	Non Liquified

From the calculation of liquefaction potential at SWS point D, soil at a depth of 0.07 m – 3.50 m was identified as having liquefaction potential. While at a depth of 3.68 m identified no potential for liquefaction.

3.4. Cone Penetration Test (CPT)

CPT or more commonly called the sondir test is a static penetrometer that is widely used in Indonesia. With this test soil type can be determined and soil hardness is obtained from the value of qc and also find the liquefaction potential.

3.4.1 Determining Soil Type Based on CPT Data

Determination of soil type is based on the formula suggested by Mayne [6], which uses the WBT index (Ic) and the Qtn value which can be seen in Figure 3. The following is an example of determining soil density based on sondir data point E, at a depth = 1 m as follows:

qc = 15.075 kg/cm2

qt = 15.075 x 98.1 = 1478.86 kPa

fs = 0.906 kg/cm2 = 0.906 x 98.1 = 88.88 kPa

1. Calculating the bulk weight of the soil (γ)
2. Calculate the overburden pressure (σvo)
3. Calculate the friction ratio (Fr)
4. Calculating the value of Ic

For that, iterate the value of n with the help of Microsoft Excel. From the iteration results, the value of n = 1.00 is obtained. Then the value of Ic = 2.99 and Qtn = 27.59 m. Furthermore, it is checked if the soil type meets the zone (Figure 3).

From the results of the calculation and plot of the Qtn – Fr value in the graph above, it can be concluded that the soil is included in zone 4 which is a type of silty soil – loamy silt to silty clay.

The calculation above is an example of calculating soil type determination for point E with a depth of 1 m. Furthermore, the determination of soil types with varying depths at point E can be seen in Table 6.

3.4.2 Calculation of Liquefaction Potential Based on CPT Data

The example calculation below based on [10],[11],[12] uses data at point E with a soil layer depth of 2.4 m and has a groundwater level of ± 8.20 m and Layer depth h_l = 0.2 m

σ_{vo} vertical total stress		= 39.33 kN/m^2
σ'_{vo} vertical effective stress		= 17.75 kN/m^2
Unit weight of density(γ)		= 17.55 kN/m^2
Unit weight of water (γ_w)		= 9.81 kN/m^3
Effective Unit weight (γ')	= γ - γ_w	= 7.74 kN/m^3
Magnitude (Mw)		= 7.4 Mw
a_{maks}	= 0.33 g	= 2.943 m/s^2
Acceleration of gravity (g)		= 9.81 m/s^2
qc	= 20.1 Pa	= 2010 kN/m^2
Pa		= 100 kN/m^2
Fs	= 0.453 kg/cm^2	= 44.44 kN/m^2

a) Calculating the value σ_{vo}, σ'_{vo}, the value (rd)

b) Calculating *Cyclic Stress Ratio* (SCR)

$$CSR = \frac{amaks}{g} \times \frac{\sigma vo}{\sigma' vo} . rd$$
$$= \frac{2{,}943}{9{,}81} \times \frac{42{,}843}{19{,}299} . 0{,}981$$
$$= 0{,}654$$

c) Calculating the value of Q, ratio friction, Ic, Kσ , qc1N

d) Calculating *Cyclic Resistance Ratio* (CRR)

- $CRR_{7,5} = 93 \left(\frac{qc1N}{1000}\right)^3 + 0{,}08 = 0{,}0889$
- Magnitude scaling factor, MSF= $6{,}9 \exp \frac{-M}{4} - 0{,}058 = 1{,}03$
- CRR = $CRR_{7,5}$. MSF . Kσ= 0,0889 x 1,03 x 1 = 0,092

e) *Factor of Safety* (FS)

$$FS = \frac{CRR}{CSR} = \frac{0{,}092}{0{,}654} = 0{,}14 \text{ (liquified)}$$

Table 6. Classification of soil types and Liquefaction Potential at point E

Depth (m)	Qtn	Fr (%)	Type of soil	qc (kg/cm²)	CSR	CRR	FS	Liquifaction
0.00	0.00	0.00	-	0.00	0.00	0.00	0.00	-
0.20	55.62	3.61	Silty Sand	25.13	0.63	0.74	1.17	No.
0.40	50.23	2.26	Silty Sand	40.20	0.63	1.04	1.63	No.
0.60	79.48	1.29	Sand	70.35	0.63	2.86	4.52	No.
0.80	33.32	2.26	Silty Sand	40.20	0.63	0.42	0.66	liquified
1.00	27.59	6.09	Clayey Silt	15.08	0.63	0.10	0.15	liquified
1.20	51.06	4.61	Silty Sand	10.05	0.64	0.09	0.13	liquified
1.40	34.14	9.49	Clayey Silt	5.03	0.64	0.08	0.13	liquified
1.60	30.72	5.72	Sandy Silt	5.03	0.65	0.08	0.13	liquified
1.80	16.06	10.04	Clay	3.02	0.65	0.08	0.13	liquified
2.00	14.45	9.71	Clay	5.03	0.65	0.08	0.13	liquified
2.20	32.90	4.69	Sandy Silt	10.05	0.65	0.08	0.13	liquified
2.40	41.40	2.30	Silty Sand	20.10	0.65	0.09	0.14	liquified
2.60	103.42	1.84	Sand	25.13	0.65	0.10	0.15	liquified
2.80	272.32	1.31	Sand	35.18	0.65	0.12	0.19	liquified
3.00	272.28	2.03	Sand	45.23	0.65	0.16	0.24	liquified
3.20	267.93	2.74	Clayey Sand	50.25	0.65	0.17	0.27	liquified
3.40	48.68	2.74	Silty Sand	50.25	0.64	0.17	0.26	liquified
3.60	48.64	2.74	Silty Sand	50.25	0.64	0.16	0.25	liquified
3.80	68.33	1.30	Silty Sand	70.35	0.64	0.27	0.43	liquified
4.00	78.15	1.14	Sand	80.40	0.63	0.34	0.54	liquified
4.20	87.97	2.02	Silty Sand	90.45	0.63	0.42	0.67	liquified
4.40	87.93	2.02	Silty Sand	90.45	0.63	0.40	0.63	liquified
4.60	102.68	1.30	Sand	105.53	0.62	0.55	0.88	liquified
4.80	77.99	1.14	Silty Sand	80.40	0.62	0.27	0.44	liquified
5.00	63.17	1.41	Silty Sand	65.33	0.62	0.18	0.29	liquified
5.20	72.99	1.83	Silty Sand	75.38	0.62	0.22	0.36	liquified
5.40	87.74	1.52	Silty Sand	90.45	0.62	0.31	0.50	liquified
5.60	92.63	1.44	Sand	95.48	0.61	0.33	0.54	liquified
5.80	92.59	1.92	Sand	95.48	0.61	0.32	0.52	liquified
6.00	97.48	1.82	Sand	100.50	0.61	0.34	0.56	liquified
6.20	112.23	1.58	Sand	115.58	0.61	0.46	0.75	liquified
6.40	126.98	1.40	Sand	130.65	0.61	0.59	0.98	liquified

6.60	142.73	1.25	Sand	145.73	0.60	0.76	1.26	No.
6.80	131.83	3.38	Clayey Sand	135.68	0.60	0.60	1.00	No.
7.00	136.72	0.98	Sand	140.70	0.60	0.63	1.06	No.
7.20	141.61	0.94	Sand	145.73	0.60	0.67	1.12	No.
7.40	146.50	1.21	Sand	150.75	0.59	0.70	1.18	No.
7.60	156.32	1.42	Sand	160.80	0.59	0.80	1.35	No.
7.80	185.86	1.20	Sand	190.95	0.59	1.23	2.08	No.
8.00	195.68	1.36	Sand	201.00	0.59	1.36	2.31	No.
8.20	195.64	1.82	Sand	201.00	0.59	1.31	2.23	No.
8.40	195.59	1.82	Sand	201.00	0.58	1.25	2.15	No.

From the results of the Cone Penetration Test (CPT) data processing for point E, soil at a depth of 0.20 m – 0.60 m has no potential for liquefaction, at a depth of 0.80 m – 6.40 m has potential for liquefaction and soil at a depth of 6, 60 m – 8.40 m no liquefaction potential. For point F, the soil at a depth of 0.2 m has no potential for liquefaction, at a depth of 0.40 m – 10.20 m has potential for liquefaction and soil at a depth of 10.40 m – 11.60 m has no potential for liquefaction.

From Figure 7, the graph of the relationship between soil depth and the factor of safety (FS) at point E, it is found that the FS values vary. That is, at a depth of 0.2 m – 0.6 m, the FS value is more than 1, then it goes down again at a depth of 0.8 m – 6.4 m with an FS value less than 1 and rises again at a depth of 6.6 m – 8, 4 m with an FS value of more than 1. The highest FS value is at a depth of 0.6 m with an FS value of 4.52.

From the results of field and laboratory tests, it can be concluded that the type of soil in the Nasanapura Hospital varies greatly in the depth of hard soil. And also at one point to another the type of soil will be different even though the distance between the points is very close. It is proven by the Swedish Weight Sounding (SWS) test with a depth of 3.68 m, the Cone Penetration Test (CPT) test with a groundwater level of 8.20 m and a depth of 8.4 m. This is because the land in the Nasanapura Hospital (where testing and sampling) is the boundary of the area affected by liquefaction on September 28th, 2018.

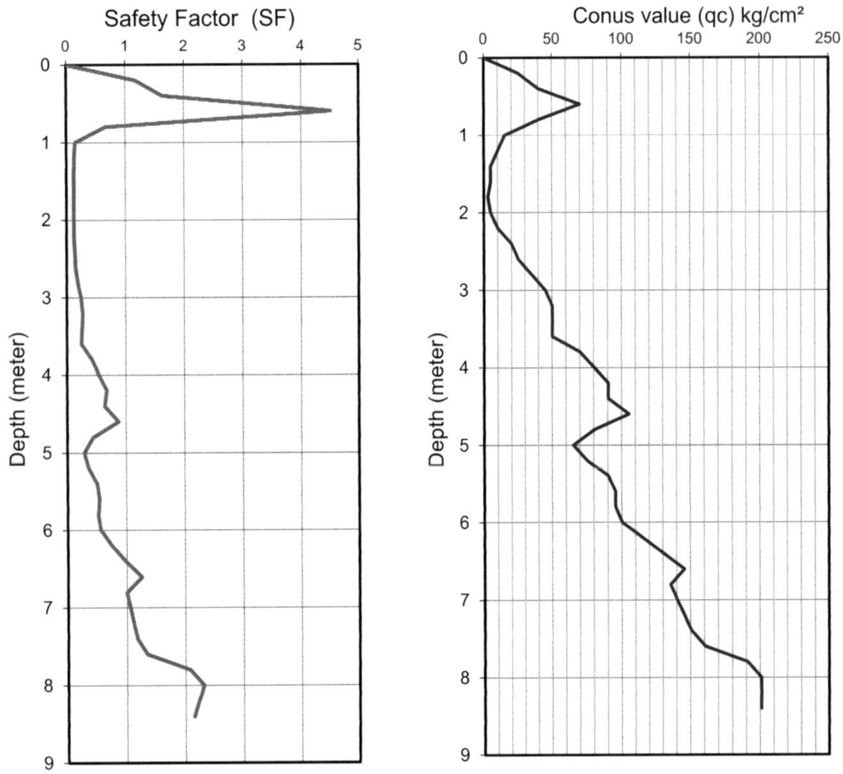

Figure 7. Graph of the relationship between soil depth and cone value (qc) at CPT point E and Graph of the relationship between soil depth and factor of safety (FS) at CPT point E

4. Conclusions

Based on the results of field and laboratory tests conducted on several soil samples, it can be concluded as follows:

a. From the results of the grain size analysis test and following the soil classification based on USCS, it was found that the type of soil at Point A with a depth of 0.2 m, 0.6 m, 0.8 m and 1.0 m is SP-SM, namely poorly graded sand and silty sand. At a depth of 0.4 m, the soil type was SP-SC, namely poorly graded sand and loamy sand. From the results of the sieve analysis test on the potential for liquefaction using the Tsuchida chart, it is explained that the test location is an area that has the potential and has the potential for liquefaction. Then it is supported by the Atterberg Limits test, that the sample at point A at a depth of 0.2 m – 1.0 m consists of clay and inorganic silt with low to medium plasticity (CL-ML), inorganic clay (CL), sand very smooth (ML). Furthermore, in the Fill Weight test carried out on 2 samples with a depth of 0.6 m and 0.8 m, the test results obtained an average dry volume weight of 1.5 g/cm3 and an average wet volume weight of 1.8 g/cm3. cm3.

b. From the results of the calculation of the SWS Point D which is converted to the N-SPT value and following the liquefaction requirements if FS < 1 then the results obtained at point D a depth of 0.07 m - 3.50 m obtained that the tested soil includes soil that has the potential for liquefaction except at a depth of 3.68 m with no potential for liquefaction.

c. From the calculation results of the Cone Penetration Test (CPT) (Point E) with a groundwater level of ± 8.20 m, the soil at Point E has a depth of 0.20 m – 0.60 m with no potential for liquefaction, at a depth of 0.80 m – 6.40 m has potential for liquefaction and soil at a depth of 6.60 m – 8.40 m has no potential for liquefaction.

Acknowledgment
The authors would like to thank all those who have supported this research: DIPA Engineering Faculty of Tadulako University

References:

[1] Irsyam M et. al. (2018) *Kajian Gempa Palu Provinsi Sulawesi Tengah 28 September 2018 (M7.4)* PusGen Bandung
[2] Widyaningrum (2012) *Penyelidikan Geologi Teknik Potensi Likuifaksi Daerah Palu Provinsi Sulawesi Tengah*
[3] Marcuson W F III (1978) *Definition of Term Related to Liquefaction. Journal of Geotechnical Engineering (104)* pp 1197-200
[4] Habibi *et al* (2006) *Manufacturing Using and Developing Swedish Weight Sounding Test for Geotechnical Investigations* MS Dissertation School of Civil Engineering Tehran University Tehran
[5] Bergdahl U Broms B B and Muromachi T (1998) *Swedish Weight Sounding Test International Reference Test Procedure* Penetration Testing ISOPT-I De Ruiter (ed.) A A Balkema Inc 71-90
[6] Mayne PW (2014) *Interpretation of Geotechnical Parameters from Seismic Piezocone Tests* Georgia Institute of Technology Atlanta
[7] Hardiyatmo H C (1992) *Mekanika Tanah I Jakarta PT. Gramedia Pustaka Utama* Jakarta
[8] Hardiyatmo H C (2002) *Mekanika Tanah I, Edisi ke-3,* Gadjah Mada University Press Yogyakarta
[9] Tsuchida H (1970) *Prediction and Countermeasure Against Liquefaction in Sand Deposits* Proceedings of Abstract of the Seminar of the Port and Harbour Research Institute Ministry of Transport Yokosuka Japan
[10] Seed H B and Idriss I M (1971) *Simplified Procedure for Evaluating Soil Liquefaction Potential. Journal of the Soil Mechanics and Foundations Division* ASCE
[11] Youd T L dan Idriss I M (2001) Liquefaction Resistance of Soils *Summary Report from the 1996 NCEER and 1998 NCEER/NSF Workshop on Evaluation of Liquefaction Resistance of Soils* Journal of Geotechnical and Geoenvironmental Engineering April 2001 297-31
[12] Boulanger R W and Idriss I M (2014) *CPT and SPT Based Liquefaction Triggering Procedures* University of California At Davis

Beach protection model in Palu Bay coastal area post tsunami 2018

Setiyawan[1], Saparuddin[2], Vera WA[3], Fildawati Y[4], Wisnu W[5]

[1,2,3]Department of Civil Engineering, Engineering Faculty, Tadulako University
[4,5]Bachelor Civil Engineering Undergraduate's Program at Tadulako University

setiyawanvip@yahoo.co.id

Abstract. Talise Beach is a coastal area that has the potential of marine products and also as a means of tourism for the local community. Recently, Talise Beach and several areas along Palu Bay have experienced a lot of erosion by the impact of waves that threaten Country Road (Trans Sulawesi), settlements and other public places as a result of the earthquake and tsunami that occurred in 2018 with a wave height of 3- 4 m which resulted in erosion and retreating of the coastline and even damaged several main road infrastructure and several other public facilities. The erosion problem is handled by constructing a Revetment-type coastal protection along the Talise Beach. The purpose of this study was to determine the height of the waves at Talise Beach, then to plan the technical design of the coastal protection structures at Talise Beach and to calculate the shear stability of the planned buildings. To test the stability of the Revetment, secondary data are needed in the form of wind data for 10 years (2011-2020), land contour data and tidal data for sea water and waves that are observed directly in the field. The results obtained from the calculation are the Hs is 4.466 meters, Ts is 4.5 seconds, the elevation of the lighthouse is 4.8 meters, the design water level elevation is 2.494 m. The revetment that is built consists of 3 layers using the main material is crushed stone.

Keywords: Stability, Revetment, Beach, Wave.

1. Introduction

According to [1], coastal protective buildings are buildings used to protect the beach from damage due to attacks of sea waves, currents, reduce the energy of waves that reach the beach and change the rate of sediment transport along the coast. Coastal protective buildings can be masonry, concrete, turap, and wood. The surface of the building can be upright, slanted, curved or neighboring. One example of a beach protector building is revetment.

Talise Beach is a beach located not far from Palu City, Central Sulawesi. Or rather along Palu Bay. Talise Beach is a tourist attraction of Sulawesi that has a very beautiful natural panorama [2]. On this beach there is a stretch of white sand with calm waves and hills that are so enchanting that visitors are spoiled by the beauty of the panorama but now the condition of Talise Beach is very worrying because of the shift in the coastline caused by the Tsunami and Earthquake that occurred in 2018. Palu is located on the Palu-Koro fault line, which runs north-south along the edge of Palu Bay. Geologists estimate the fault segment has the highest derailment movement in Indonesia at 4 cm (1.6 inches) per year, indicating the area is at higher risk of earthquakes. In this earthquake, two sides of the fault line scratched each other horizontally [3].

Because of the damage that occurred, talise beach needs to be handled so that the damage can be reduced or overcome. The treatment is to create a beach protection building on the beach Talise namely revetment, to protect the coast from the continuous blow of the ocean waves talise beach and prevent abrasion that occurs so as to prevent the shifting of the coastline that is getting closer to thehighway. Based on the background described above, the author wants to do research and raise the title " Planning Revetment in Talise Beach Palu City, Central Sulawesi".

Based on the formulation of the problem that has been stated above, the objectives of this study are to find out the high tides of the sea at Talise Beach, to technically plan a beach protector building on Talise Beach and to know the stability of the coastal protective buildings in Talise Beach against the bolster.

2. Material and methods

2.1. Description of study

The planned research location is located in Talise Beach, East Palu District, Central Sulawesi Province, approximately 2 km north of Palu City,this beach is one of the tourist attractions in Palu City [4]. This beach stretches in Palu City so that Donggala Regency, the location of the research site can be seen in Figure 1.

Figure 1. Research site on talise beach [4]

The location of revetment installation is along Talise Beach starting from Cumi-Cumi street to Kampung Nelayan street (along 7,2 km). The location is an area of former tsunami and earthquake in 2018, and is also a tourist attraction, residential area, schools and highway infrastructure. The beach material consists of fine to rough sand, gravel, coral shards and the remains of shell fragments.

The area that will be the location of the research is East Palu District, Palu City, precisely at the location of Talise Beach with a review of research on revetment buildings along 7,2 km. East Palu district is right in the middle of the city located in a position between 0 °44'50" and 0 ° 49'50 " South Latitude and 119 ° 50'00 " and 119 ° 56'10 " East Longitude [5]. Based on its geographical position, the area of East Palu Subdistrict is mostly flanked by land boundaries between two sub-districts, half of which are limited by one sub-district separated by the Palu River, and the rest is directly adjacent to Palu Bay.

2.2. Literature review

Wave transformation analysis is often done with the concept of ocean waves in equivalent, namely the height of waves in the deep sea when the waves do not undergo refractive. The use of these waves aims to establish the height of waves undergoing refractive, diffraction, and other transformations so that the approximate transformation and deformation of the waves can be done more easily. The height of the ocean waves in the equivalent is given by the form of [7,8]:

$$H'_0 = K_r H_0 \tag{1}$$

A force that works vertically will suppress the structure. If the retaining force is a horizontal force derived from the waves come small and cannot withstand the force pressing will result in a shift in the structure [9]. Security stability against sliding must be greater than the security number (Fs) = 1,5. The stability of buildings to sliding forces can be calculated by equations [10,11]:

$$Fs_{(sliding)} = \frac{\Sigma V}{\Sigma H} > 1{,}5 \tag{2}$$

2.3. Methods

In this study the initial data required are wind, tidal and soil data. Wind, tidal and soil data are obtained from secondary data and also by direct retrieval from the researchsite.

The analysis is carried out in the following stages:

1. Data obtained directly from the research site is wind data. Wind data taken is used to determine the Talise Beach Effective Fetch. Gather hourly wind data for 10 years. With the wind data of these hours it will be possible to known wind with daily mean wind speed. Then the direction and speed data wind can be processed using the WRPLOT View software whose results can be seen in the windrose image where the average maximum wave height at Talise Beach. The wind data in this paper were obtained from ECMWF data for 10 the year of observation is 2011-2020.
2. In this study, the tidal data taken was used to determine the determine the tidal type at the research site and obtain the HHWL (Highest High Water Level), MHWL, MSL, MLWL (Mean Low Water) and LLWL (Lowest Low Water Level) at that location using the Least Square method.
3. In this study, soil data was taken to determine stability against shear by using SPT (Standard penetration test) data which is secondary data.
4. As for in this study bathymetric data is used to create a condition map research locations around Talise Beach. Tools used when measuring is a depthometer, this is an instrument for measuring the depth of the sea which is centered from above sea level to below sea floor. then use surfer software to create a bathymetric map image.
5. Analysis of results and discussion

The complete research method is presented in the following figure:

IOP Conf. Series: Earth and Environmental Science 1075 (2022) 012029 doi:10.1088/1755-1315/1075/1/012029

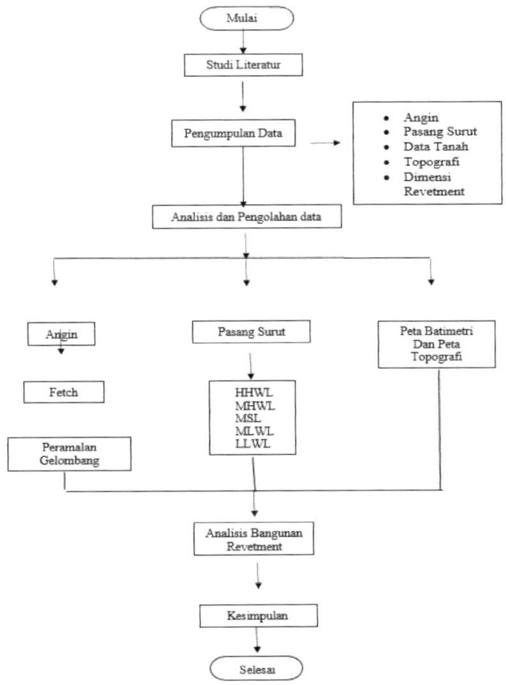

Figure 2. Flowchart research

3. Results and discussion

3.1. Wind speed data

The results of the calculation of rain erosivity are presented in the following tables and graphs:
Wind data is needed to determine the distribution of wind direction and wind speed occurring at the observation site. Wind data used is wind data Year 2011 - 2020 derived from data ECMWF (European Centre for Medium-range Weather Forecast) during 10 years of observation. Below are the calculations from 10 years of wind data along with windrose images. To get windrosedigunakan software WRPlot. Viewable in Figure 3.

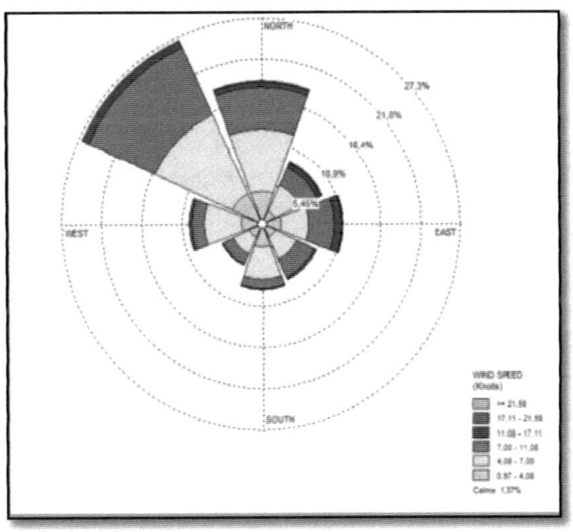

Figure 3. Windrose from wind data for 10 years (2011-2020)

IOP Conf. Series: Earth and Environmental Science 1075 (2022) 012029 doi:10.1088/1755-1315/1075/1/012029

For the maximum wind direction is the Northwest direction, with a maximum speed of 6 knots and a minimum speed of 3 knots.

3.2. Fetch of talise beach

Fetch is effectively used in wave forecasting charts to determine the height, period and duration of waves. Fetch Length Calculation can be seen in figure 4 below.

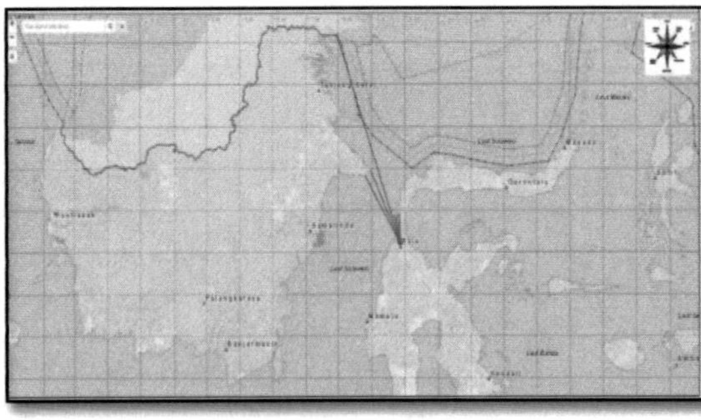

Figure 4. Fetch

3.3. Water level elevation plan

The estimated sea level increase in 2020 is 18 cm = 0.18 m planned for the building age of 10 years [1]. So that the water level elevation plan is as follows:

DWL = HHWL + Δh + SLR

DWL = 2,300 + 0,0144 + 0,18

DWL = +2,4944 m

Water Level Elevation can be seen in figure 5 below.

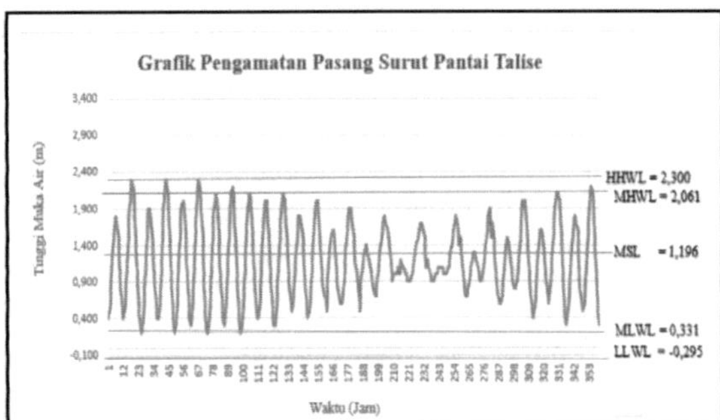

Figure 5. Water level elevation chart plan on talise beach tidal data

3.4. Mercu revetnent elevation calculation

Wave run-ups are searched by using the Run-up graph in figure 6 with the following plot result:

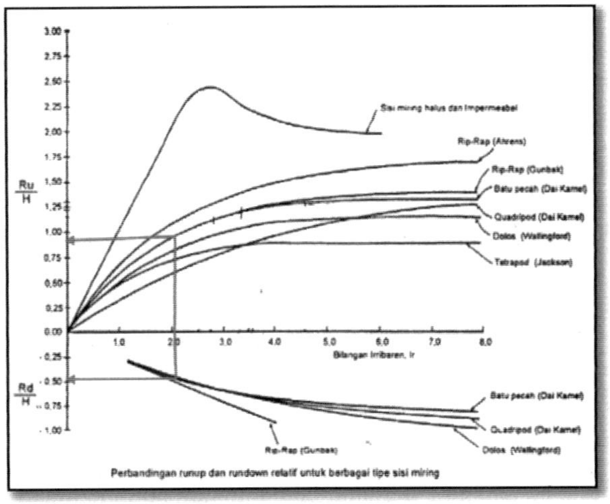

Figure 6. Wave run-up chart

3.5. Toe protection

Toe protection in Talise Beach is shown in the following table and figure:

Table 1. Comparison of dimensions of beach safety buildings

Design Section	Design Existing (m)	Calculation Design (m)
Building height	3	4,8
Peak width	3	3

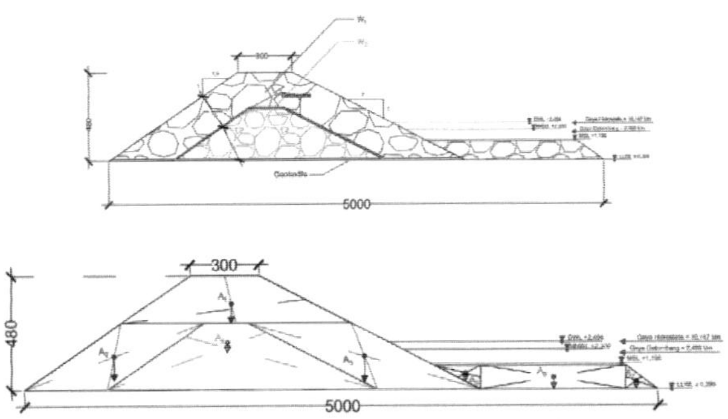

Figure 7. Revetment dimension planning (in cm)

3.6. Revetment building stability analysis

Revetment building stability analysis in Talise Beach is shown in the following table:

Table 2. Calculation of vertical style that is applied

Gaya	Luas (m^2)	Y (t/m3) (Berat Jenis)	W (t/m) (Luas x ybatu)	H (t/m)
1	5,872	1,919	11,269	
2	13,000	2,65	34,450	

3	10,096	26,753	
4	12,522	33,183	
5	3,672	9,732	
6	3,245	8,599	
7	4,025	10,666	
8	8,160	21,625	
9	1,000	2,650	
10	0,147	0,390	
11	0,709	1,879	
Rs			10,147
Rm			2,489
Ph			3,780
Σ		161,196	16,4416

Based on the table above, it can be seen that hydrostatic forces (Rs) 10,15 t/m Rm = 2,49 t/m and Ph = 3,78 t/m. W = 161,20 t/m and H = 16, 42 t/m.

Table 3. Dimensional rccapitulation with B = 50 m (planning design)

Sample	Real Land Data			B (m)	H (m)	Stability Control	
	°	γ (t/m³)	c (t/m²)			Shear>1,5	Control
1	45,0	2,4	0	50	4,8	16,1	Yes
2	38,2	2,1	0	50	4,8	13,7	Yes
3	40,0	2,2	0	50	4,8	12,0	Yes
4	45,0	2,4	0	50	4,8	5,5	Yes

From the table above, it can be seen that changes in with B = 50 m, sample 1 °= 38,2-45,0 t/m3, g = 2,1-2,4 t/m3 and shear 5,5-16,07 > 1,5 and next.

4. Conclusions

The result of this research is significant wave height (Hs) is as high as 4,47 meters and the wave period (Ts) is 4,5 seconds. The result of stability calculation for the planned revetment obtained for samples 1,2,3 and 4 with dimensions B = 27,72 meters meets the requirements of shear stability with Fs ranging from 4,07 – 11,67.

Acknowledgment

The authors would like to thank all those who have supported this research: Enggineering Faculty of Tadulako University and the office Public Works and Water Resources of the Province of Central Sulawesi, Indonesia.

References

[1] Triatmodjo, B (2008) Beach Engineering Beta Offset Yogyakarta
[2] Setiyawan, (2008) Study of Talise Coastline Changes as a "Transit City" Concept of Palu City, Central Sulawesi Province, ITB, Bandung
[3] ECMWF (2021) datasets
[4] Google Earth (2020) "Map of Talise Coast Region" East Palu District
[5] Central Bureau of Statistics of Palu City, Central Sulawesi (2016) "East Palu in Numbers" Publisher BPS Palu City
[6] Regional Development and Investment Planning Agency of Palu City (2014) "Profile of East

Palu District" Publisher of the Regional Development and Investment Planning Agency of Palu City
[7] Abidin, Zainal (2007), "Planning of Offshore Breakwaters and Coastal Protective Walls as Alternatives for Abrasion Prevention in Kalibuntu Beach, Probolinggo Regency", Civil Engineering Thesis, Islamic University of Malang Malang
[8] Burhan Utomo, Basrindu and Dwi Prasetyo Wibowo (2008), "Planning of Protective Buildings for Tambak Mulyo Beach, Semarang", Final Project of Civil Engineering, Diponegoro University Semarang
[9] Puspawati Achjar, Dewi (2007), "Study of Planning for Coastal Protection Buildings to Prevent Beach Erosion in Ampenan, Mataram City, West Nusa Tenggara", Civil Engineering Thesis, Islamic University of Malang Malang
[10] Robydiansah (2012), “Review of the Stability of Shearing and Rolling Parafets in the Grindulu River, Pacitan Regency”, Final Project of Civil Engineering, Yogyakarta State University Yogyakarta
[11] Pratikno et al, (2014) Coast Guard Structure PT Mediatama Saptakarya Jakarta

Climate change impacts on the Palu watershed's precipitation characteristics

Y Arafat[1], I W Sutapa[1], Y Hasanah[1]

[1]Department of Civil Engineering, Engineering Faculty, Tadulako University

iazzyr@gmail.com

Abstract. Climate change occurs naturally and due to accelerated elements due to human activities, which result in deviations from the average climate trend (increasing or decreasing). This study was conducted to identify the effects of climate change on the precipitation characteristics in the Palu watershed (DAS). The climate change analysis method used is Mann-Kendall-Scns. The analysis of the nature of rain uses the standard deviation, which shifts the wet month from the wet month to the dry month using the Oldeman method. This study uses secondary data, namely rainfall data from 1985–2020, from five rainfall stations. The results of this study indicate climate change in the Palu watershed, which is characterised by a negative Z value (decreasing) in January and a positive Z value (increasing) in June. The results of the analysis of the nature of rain in January for the 1985-1995 period were normal (N), the 1996-2006 period was below normal (BN) and the 2007-2020 period was below normal (BN). In June, the 1985–1995 period was below normal (BN), the 1996–2006 period was normal (N), and the 2007–2020 period was normal (N).

Keywords: Climate change, Precipitation characteristics, Palu watershed

1. Applying paragraph styles

Indonesia is one of the countries located in the equatorial region, so the characteristics of meteorological elements, especially rainfall, are strongly influenced by monsoon climate conditions, namely there is a clear difference between the wet season and the dry season [1]. According to the IPCC [2], climate change is defined as the average change in climate that occurs over a long period of time, typically decades or more. Climate change is statistically defined as a change in the trend either up or down of climate elements accompanied by daily, seasonal, and cyclical variations that remain valid for a long period. Climate change is measured based on changes in the main components of the climate, namely temperature, seasons (both rainy and dry seasons), humidity, and wind. Of these variables, the most frequently cited variables are temperature and rainfall [3].

Globally, climate change is a change in climate with reference to the earth's area as a whole on the variability of climate conditions in a place that is statistically real for a long period of time, generally a decade or more [4]. Climate change, it will affect the hydrological processes that occur in the watershed, it affects the flow of the river.

Climate change has brought changes to the rainfall characteristics. In general, the duration of the rainy season is getting shorter, on the contrary the duration of the dry season is getting longer. The number of rainy days tends to decrease, while the maximum daily rain and rainfall intensity tend to increase [6]. A watershed is an area bounded by a topographical divider that stores, accommodates and drains rainwater that falls on it to the main river that empties into the sea or lake [5]. Watershed is a representation of the hydrological processes that occur in the area. The process starts from rainwater that falls in the watershed area then flows into the sea and then evaporates again. This natural process will be disturbed if the rainfall that occurs in a watershed fluctuates [5].

The purpose of this study is to examine how climate change has changed the characteristics of rain in the Palu watershed.

2. Material and methods

2.1. The study area

The location of this research is located in the Palu watershed, which is at the coordinates of 0°55'12,6'' - 0°58'26,38''S and 119°52'23,82'' - 119°52'18,24''E. The Palu River is formed by two main tributaries, namely the Miu and Gumbasa rivers, and flows down to Palu Bay. The Palu River is 90 kilometers long and flows from southeast to northwest along a long valley channel on the Palu Koro Fault. The Palu River originates in the Kulawi Mountains, Sigi Regency. The total area of Palu watershed is 3,043 square kilometers and includes many tributaries such as the Wuno River ,Gumbasa River, Saluki River, Larono River, Sangkulara River, Lewara River, and Saluri River. The following map depicts the research location:

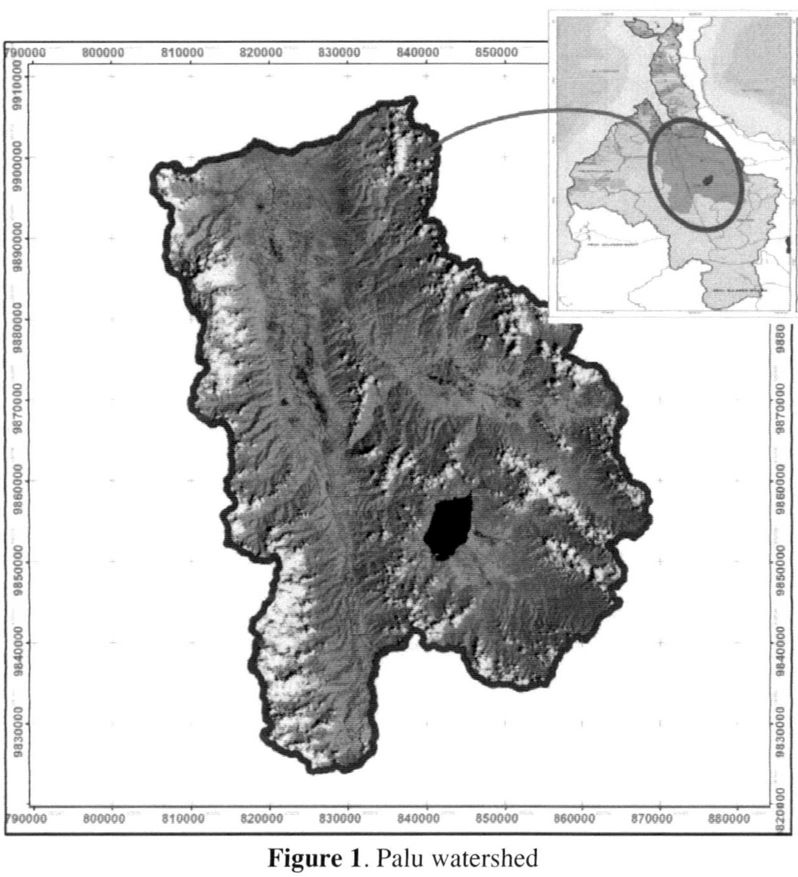

Figure 1. Palu watershed

2.2. *Rainfal data*

The rainfall data used in this study was obtained from five rainfall stations in the Palu watershed. The five rainfall stations have had rainfall data for over 30 years. The rainfall stations are:

1. The Bora Rainfall Station was established in 1976 in the Gumbasa River Basin, 01°01'39''S and 119°55'35'' E.
2. The Kalawara Rainfall Station was established in 1976 in the Gumbasa River Basin, 01°01'0''S and 119°53'31'' E.
3. The Kulawi Rainfall Station was established in 1984 in the Gumbasa River Basin, 01°28'11''S and 120°00'43''E.
4. The Palolo Station was established in 1976 in the Palu-Lariang River Basin, 01°04'07''S and 120°04'44''E.
5. The Wuasa Station was established in 1976 in the Palu-Lariang River Basin, 01°25'53"S and 120°18'55,7"E.

2.3. *Method*

2.3.1. The Mann–Kendall methode. Climate change identification using the Mann–Kendall equation and the MAKESENS 1.0 software. The Mann-Kendall-Sens method is used for trend analysis and climate projections (maximum daily, monthly, and yearly). This yields a "Z" value indicating whether or not climate change has occurred, as well as a rainfall projection for the next few years. The daily minimum, average, and maximum values of climate data are determined in order to determine the trend. Then, the Mann–Kendall analysis was performed [7][8][9]. The non-parametric Sens method is used to determine the magnitude of the trend under the assumption that it is linear [10][11].

$$Z = \begin{cases} (S-1)/\sigma s ....jika......S > ...0 \\ 0.....................jika.....S = ...0 \\ (S+1)/\sigma s ....jika....S < ....0 \end{cases} \tag{1}$$

Where:

$$S = \sum_{k=1}^{n-1} \sum_{j=k+1}^{n} \operatorname{sgn}(X_j - X_k) \tag{2}$$

$$\sigma s = \sqrt{n(n-1)(2n+5)/18} \tag{3}$$

X_j and X_k is the corresponding data value of "j" and "k", j > k

2.3.2. Rainfall characteristics

The standard deviation formula is used to classify rainfall into five different rain characteristics [12]. The five classifications are:

1. Far Below Normal (FBN)

$$FBN = x \leq X - 1,5\ SD \tag{4}$$

2. Below Normal (BN)

$$BN = X - 1,5\ SD < x \leq X - 0,5\ SD \tag{5}$$

3. Normal (N)

$$N = X - 0,5\ SD < x \leq X + 0,5\ SD \tag{6}$$

4. Above Normal (AN)

$$AN = X - 0,5\ SD < x \leq X + 1,5\ SD \tag{7}$$

5. Far Above normal (FAN)

$$FAN = x > X + 1,5\ SD \tag{8}$$

where:
X = watershed average rainfall (mm)
x = monthly precipitation - i (mm)
Standard Deviation (SD)

$$SD = \sqrt{\frac{\sum_{i=1}^{n} xi^2 - \frac{(\sum xi)^2}{n}}{n-1}} \quad (9)$$

Xi = i-month rainfall
n = number of years of observation

2.4. The shift of wet month, moist month and dry month
The analysis of the wet month, moist month and the dry month was carried out by classifying the amount of monthly rainfall with Oldeman's provisions.

Table 1. Category Oldeman rain properties

No	Rainfall	Properties
1	R > 200 mm	Wet month
2	100 < R < 200mm	Moist Month
3	R < 100 mm	Dry Month

3. Results and discussion

3.1. Regional rainfall
Observations of rainfall in the Palu watershed include 5 (five) rainfall stations. Bora Rain Station (1985-2017), Wuasa Rain Station (1985-2017), Palolo Rain Station (1985-2017), Kulawi Rain Station (1985-2020), and the Kalawara Rain Station (1985-2017) so that the calculations use the Thiessen Polygon method for 33 (thirty-three) years of observation. The annual and monthly rainfall patterns in the Palu Basin fluctuate. Based on the average rainfall in the Palu watershed, the highest extreme rain conditions occurred in 2013 of 1770.98 mm/year, and the lowest extreme rain conditions occurred in 2016 of 943.95 mm/year. The results are shown in Figure 2.

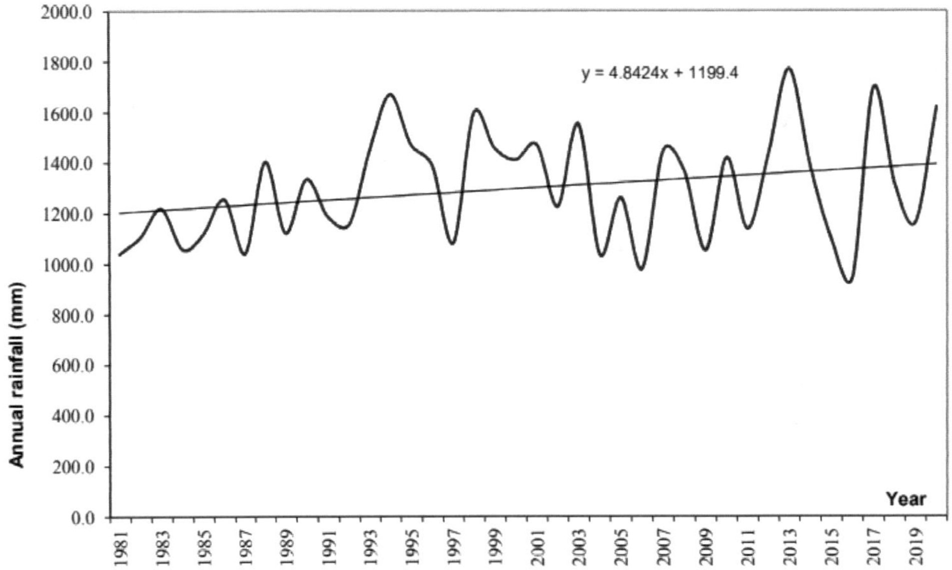

Figure 2. Annual rainfall

3.2. Climate change detection

Calculation of climate change is done using MAKESENS 1.0 software to determine whether there is climate change in the research location. It is indicated by the value of "Z", which is positive (increasing trend) and negative (declining trend). Significant climate change in a positive direction is shown in June. Effective climate change in a negative direction is shown in January, while there is a trend in other months but not significant. The calculation results are presented in Table 2.

Table 2. Recapitulation of climate change calculations

Uraian	From	To	Value	Trend											
				Jan	Feb	Mar	Apr	May	Jun	Jul	Aug	Sept	Oct	Nov	Dec
Daily Max. (mm/day)	1980	2020	36						0.67 Pos, NS						
Monthly (mm/month)	1980	2020	36	-2.15 Neg, YS	0.34 Pos, NS	-0.12 Pos, NS	1.06 Pos, NS	0.12 Pos, NS	1.59 Pos, NS	-0.04 Neg, YS	1.68 Pos, NS	0.94 Pos, NS	-0.41 Neg, YS	-0.04 Pos, NS	-0.20 Neg, YS
Yearly (mm/year)	1980	2020	36						0.45 Pos, NS						

Where

Pos	: Positive or increasing	Z > Zα	.... Significant
Neg	: Negative or increasing	Z< Zα	.... insignificant
YS	: Significant	Z = 0	.... No Trend
NS	: No Significant		
NT	: No Trend		

3.3. Rain characteristics distribution

A relatively significant shift in the type of rain can be seen in the distribution of rain quality over the three periods. These changes occur in January, March, July, November and December. In January, rainfall grew, as evidenced by an increase in the number of normal rain characteristics from the first to the second period and an increase in rainfall above normal in the third period. In March, there was a decrease in rainfall in the second period, marked by the rise in the amount of subnormal rain. An increase in rainfall in the third period was marked by an increase in the nature of normal precipitation. In July, there was an increase in rainfall. This was indicated by a decrease in the amount of rain below normal in the second and third periods and an increase in the nature of normal rain from the first period to the next. The increase also occurred in November and December. In other months there are changes but not so significant.

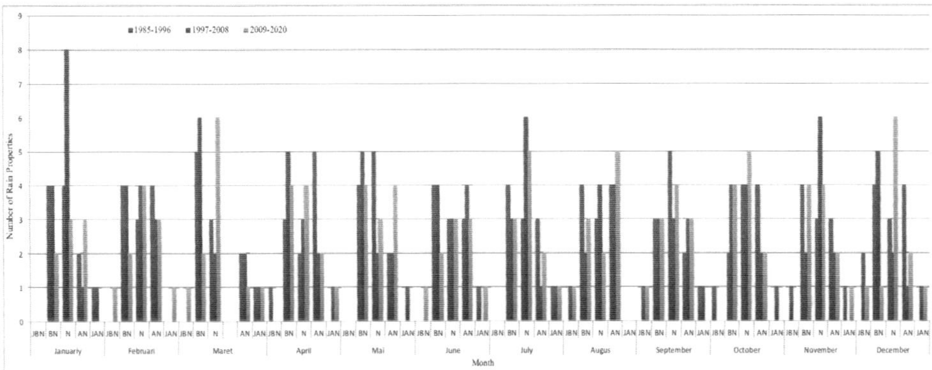

Figure 3. Distribution of rain properties

3.4. Shifting of wet, moist, and dry months.

There are only two types of months in the Palu watershed: humid and dry, with shifts in the humid and dry months occurring in January, June, August, and September. The three months from 1985 to 1995 were humid, whereas the months from 1996 to 2006 and 2007 to 2020 were dry.

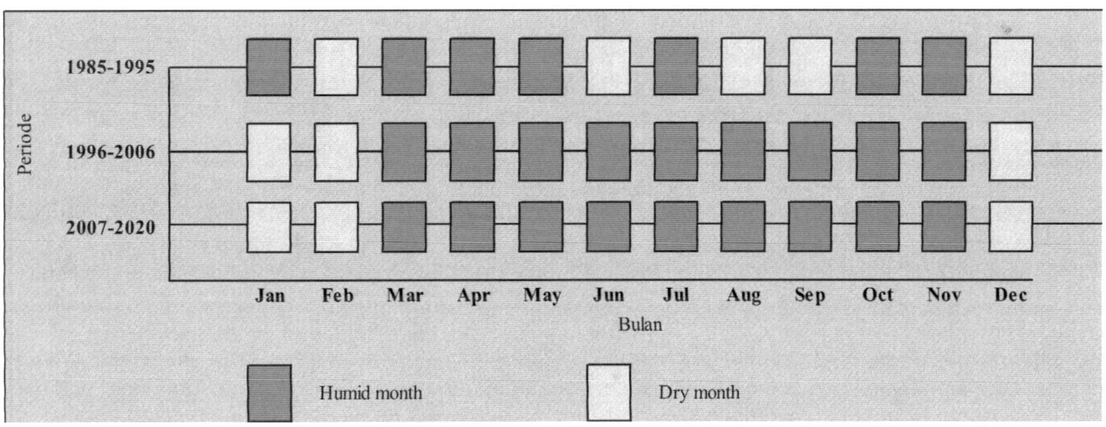

Figure 4. Comparison of the Palu watershed's three seasons of rainy and dry months

4. Conclusion

Significant climate change in a positive direction is shown in June. Effective climate change in a negative direction is shown in January, while there is a trend in other months but not significant. The analysis of the nature of rain uses the standard deviation, which shifts the wet month from the wet month to the dry month using the Oldeman method. This study uses secondary data, namely rainfall data from 1985–2020, from five rainfall stations. The results of this study indicate climate change in the Palu watershed, which is characterised by a negative Z value (decreasing) in January and a positive

Z value (increasing) in June. In the Palu watershed, there are only two sorts of months: humid and dry, with transitions between humid and dry months happening in January, June, August, and September. From 1985 to 1995, the three months were humid, however from 1996 to 2006 and 2007 to 2020, the months were dry.

References

[1] D Susilokarti, S S Arif, S Susanto, and L Sutiarso, "Identifikasi perubahan iklim berdasarkan data curah hujan di wilayah selatan Jatiluhur Kabupaten Subang, Jawa Barat," Agritech, vol 35, no 1, pp 98–105, 2015

[2] O N C CHANGE, "Intergovernmental panel on climate change," World Meteorol Organ, vol 52, 2007

[3] R I Bau, N Musa, and W Pembengo, "Evaluasi Sifat Curah Hujan dan Neraca Air Lahan Pada Pertanaman Cabai Rawit (Capsicum frutescens L) Di Kabupaten Gorontalo," Skripsi, vol 1, no 613413008, 2018

[4] D Murdiyarso, Sepuluh tahun perjalanan negosiasi konvensi perubahan iklim Penerbit Buku Kompas, 2003

[5] I W Sutapa, M Bisri, and L Montarcih, "Modeling Discharge of the Bangga Watershed under Climate Change," in Applied Mechanics and Materials, 2015, vol 776, pp 133–138

[6] H Muhammad, "The Lost of Semarang Coastal Areas due to Climate Change and Land Subsidence," in International Conference on Coastal and Delta Areas, 2015, vol 1, no 1, pp 98–108

[7] B Önöz and M Bayazit, "The power of statistical tests for trend detection," Turkish J Eng Environ Sci, vol 27, no 4, pp 247–251, 2003

[8] S Wang et al, "Long Term Streamflow Response to Climatic Variability in the Loess Plateau, China 1," JAWRA J Am Water Resour Assoc, vol 44, no 5, pp 1098–1107, 2008

[9] A Hakan, K Savaş, and Ş Osman, "Trend analysis of hydrometeorological parameters in climate regions of Turkey," in Conference pre release, BALWOIS, 2010

[10] R C Deo, C A McAlpine, J Syktus, H A McGowan, and S Phinn, "On Australian heat waves: time series analysis of extreme temperature events in Australia, 1950-2005," in Proceedings of the International Congress on Modelling and Simulation (MODSIM07), 2007, pp 626–635

[11] T Salmi, A Määttä, P Anttila, T Ruoho-Airola, and T Amnell, "Detecting trends of annual values of atmospheric pollutants by the Mann-Kendall test and Sen's slope estimates MAKESENS–The excel template application," Finish Meteorol Institute, Helsinki, 2002

[12] Suaydhi, "Analysis of Climate Variability of Indonesia Using a Regional Climate Model" Monash University, 2001

Multi-hazard assessment in Palu Barat District

Sri Mulyati[1], **Rizkhi**[2], **Salhuddin**[3]
[1,2,3]*Department of Architecture, Engineering Faculty, Tadulako University*
*Corresponding Author: physicmulya@gmail.com

Abstract. West Palu District is one of the areas severely affected by the earthquake that occurred on September 28, 2018. The purpose of this study was to assess the level of multi-hazard disasters in the West Palu District. The study was conducted by assessing each disaster hazard, namely the hazard of earthquakes, tsunamis, and liquefaction presented in a multi-hazard map. The earthquake hazard assessment parameter is based on PGA data. The tsunami hazard parameter is based on data from the highest peak of a tsunami wave that has ever occurred. Meanwhile, the liquefaction hazard parameter is based on the ground shear strain (GSS) value. Then the weighting and overlapping methods are carried out to obtain the final result in a multi-hazard assessment map. The results showed that the level of multi-hazard in West Palu District was in the medium and high class. The middle multi-hazard level is in Baru Village, Kamonji Village, and Siranindi Village. In contrast, the high multi-hazard level is in Balaroa Village, Ujuna Village, and Lere Village.

Keywords: Multi-hazard, earthquakes, tsunamis, liquefaction

1. Intruduction

Indonesia is a country that is prone to multiple disasters, such as earthquakes, tsunamis and volcanic eruptions because it is located between the confluence of three tectonic plates (Eurasian, Indian-Australian, and Indian). Meanwhile, multiple effects of physical processes and human activities also contribute to the disasters hazard every year. Based on the Indonesian Disaster Risk Index, Central Sulawesi Province has a high risk of earthquake threats.[1]. One of the leading causes is the Palu Koro Fault which extends from the Makassar Strait to the north coast of Bone Bay with about 500 km faulth length . In Palu city, the fault passes from Palu Bay into the mainland, cuts through the city center, continues to the Lariang River in the Pipikoro Valley, Donggala (south of Palu).[2].

In Palu City, especially West Palu District, the earthquake, tsunami, and liquefaction disasters coincided on September 28, 2018. Based on the same complex disaster conditions in West Palu District, it is necessary to assess the disaster that is focused on making multi-hazard maps. The purpose of this study is to assess the level of multi-hazard disaster in West Palu District. The research focused on earthquake, tsunami, and liquefaction disasters that occurred in West Palu District.

The purpose of this study is to assess the multi-hazard disaster level in West Palu District. The research focused on earthquake, tsunami, and liquefaction disasters that occurred in West Palu District.

2. Literature Review

Similar studies related to hazard studies and multi-hazard studies in Palu City is: The research about multi-hazard risk assessment of Kulon Progo Regency using landslide, flood and tsunami hazard parameters. A Multi-hazard map is obtained from overlying hazard maps using the GIS tool. The result is a risk index in the range of 0.4 to 0.79, which indicates Wates, Kalibawang, and Kokap Sub-District as the highest on multi-hazard risk. In contrast, a low-risk hazard is in Lendah and Nanggulan Sub-District.[3]. Examined the phenomenon of liquefaction (nalodo) in Palu City and Sigi Regency due to the Donggala earthquake on September 28, 2018. The study said that at least three types of soil failure due to liquefaction occurred in the coastal area of Donggala Regency, the mainland area of Palu City and Sigi Regency, namely lateral spread, soil oscillation movement, and soil mass flow.[4]. Examines the vulnerability of the tsunami disaster in Palu City using the BNPB Perka No. 2 of 2012 From the results of the study, obtained two things, namely the distribution of the level of vulnerability to the tsunami disaster, which is divided into three classes, namely low (9 villages), medium (11 villages) and high (15 villages), so that from the high vulnerability class, recommendations for mitigation policies such as activate family planning programs, conduct mitigation socialization, plant mangrove trees, do not clear land for protected forests and natural forests, and do not build buildings in densely built areas. [5].

3. Metods

3.1 Research Site

The study was conducted in West Palu District of Palu City based on a map of the administrative area of the West Palu District. Based on its geographical location, West Palu District is located in the western hemisphere of Palu City at a position between 0°44'50" and 0°49'00" south latitude and 119°51'00" and 119°55'10" east longitude. Most of it is flanked by a land boundary between two sub-districts, half is bordered by one sub-district, separated by the Palu River, and the rest is directly adjacent to Palu Bay. The characteristics of the West Palu District area according to elevation (altitude above sea level (DPL), which is between 0-15 m, the height is measured based on the location of the village office. The entire area is soil and the morphology is relatively flat. The site is directly adjacent to the sea or coastal areas. The beach is Lere Village, while the other areas arc not coastal arcas.

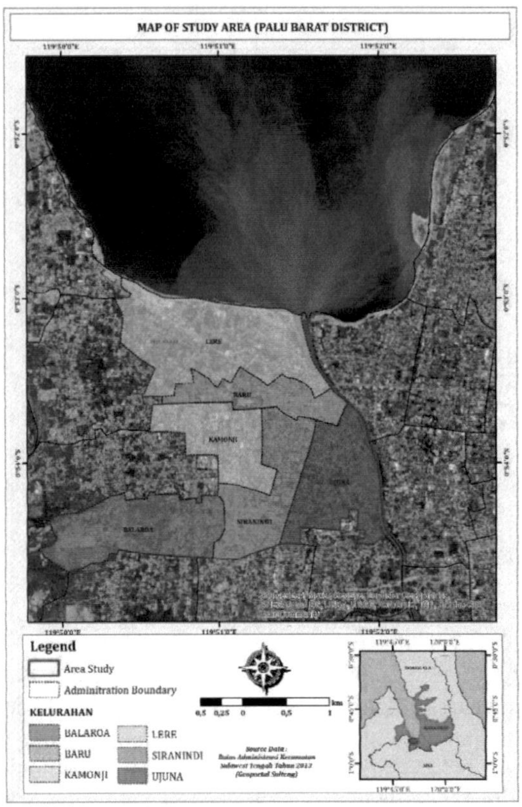

Figure 1. Research location

3.2 Hazard Potential Assessment

3.2.1 Earthquake Hazard

Earthquake hazard assessment uses earthquake data based on PGA (Peak Ground Acceleration) values.).

Tabel 1. Scoring for each earthquake hazard class

PGA Value	Class	Value	Weight (%)	Score
< 0.26	Low	1		0.333333
0.26 – 0.70	Medium	2	100%	0.666667
> 0.70	High	3		1.000000

Source: Perka BNPB 1/2012

3.2.2 Tsunami Hazard

The tsunami hazard assessment is based on the estimated tsunami height.

Tabel 2. Scoring for each tsunami hazard class

Tsunami height	Class	Value	Weight (%)	Score
< 1 m	Low	1		0.333333
1 – 3 m	Medium	2	100%	0.666667
> 3 m	High	3		1.000000

Source: Perka BNPB 1/2012

3.2.3 liquefaction hazard

The liquefaction hazard assessment is based on the value of the Ground Shear Strain (GSS).

Tabel 3. Scoring for each liquefaction hazard class

GSS Score	Class	Value	Weight (%)	Score
10E-6.0 – 10E-4.0	Low	1		0.333333
10E-3.8 – 10E-2.2	Medium	2	100%	0.666667
10E-2.0 – 10E-1.0	High	3		1.000000

Source : modified from Palu City BPBD final report

3.2.4 Multi Hazard Assessment

The multi-hazard potential assessment is obtained from the overlapping of earthquake, tsunami and liquefaction hazard maps.

4. Result and Discussion

4.1 Earthquake Hazard Assessment

Earthquake hazard assessment data based on the peak ground acceleration (PGA) value, are presented in the following table:

Tabel 4. Earthquake Hazard Data and Scoring

Location	PGA Value (gal)	Intensity	Class	Value
Ujuna Village	1.34	X MMI	High	3
Lere Village	1.26	IX-X MMI	High	3
Siranindi Village	1.24	IX-X MMI	High	3
Kamonji Village	1.29	IX-X MMI	High	3
Baru Village	1.24	IX-X MMI	High	3
Balaroa Village	1.24	IX-X MMI	High	3

Source: BPBD Kota Palu, 2018

The results of the earthquake hazard assessment show that the entire West Palu District is in a high earthquake hazard class. This result is strongly influenced by the area adjacent to the earthquake's epicenter, namely the Palu-Koro fault line and the new fault, which is thought to have arisen due to the earthquake on 28 September 2018. The highest PGA value is in Ujuna Village of 1.34 gal, with an X MMI earthquake intensity value. The strength of the X MMI earthquake intensity can cause solid wooden buildings to be damaged, house frames to fall off their foundations, split ground, curved rails, landslides in every river, and on steep ground.

Figure 2. Earthquake Hazard potential map

4.2 Tsunami Hazard Assessment

The tsunami hazard level determination is carried out using the height of tsunami wave crest data presented in the following table:

Tabel 5 Tsunami Hazard Data and Scoring

Location	Tsunami height (m)	Tidal Correction (m)	Inudation (m)	Class	Value
Lere Shop	5.6	5.7	320	High	3
Grandmall Palu	5.6	6.1	320	High	3
Mercure Hotel Palu	9.2	10	468.8	High	3

Source: BMKG, 2018

The hazard assessment results show a high level of tsunami hazard in Lere Village because it is directly adjacent to the coast of Palu Bay. The tsunami that occurred was triggered by a seafloor landslide caused by the September 28, 2018 earthquake. This incident was reinforced by discovering a lot of land subsidence and natural changes on the coast that were very different from before. In Lere Village, the peak of the tsunami wave occurred around the Mercure Palu Hotel as high as 9.2 meters with a tidal correction to 10 meters. Inundation blow as far as 468.8 meters caused by the position of the Mercure Hotel, which is on the offshore edge.

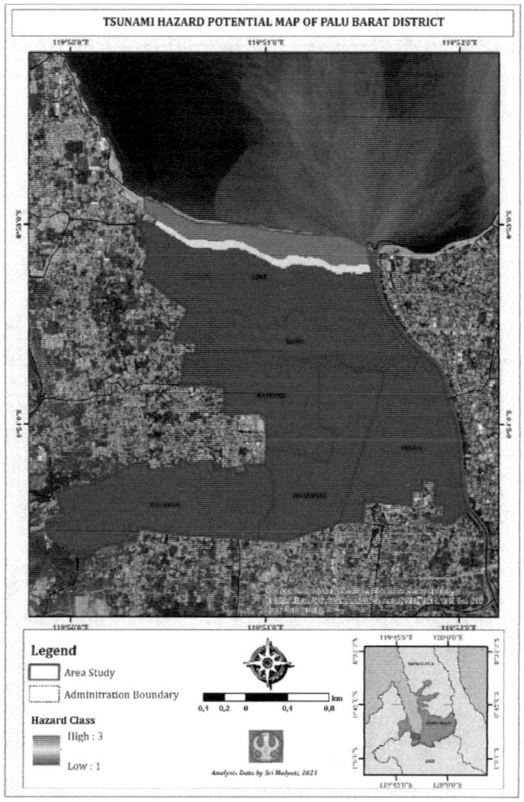

Figure 3. Tsunami Hazard potential map

4.3 Liquefaction Hazard Assessment

The liquefaction hazard assessment is based on the Ground Shear Strain (GSS) value which is presented in the following table:

Tabel 6. Liquefaction Hazard Data and Scoring

Lokasi	Nilai GSS	Class	Value
Ujuna Village	10-E2.1	High	3
Lere Village	10-E2.0	High	3
Siranindi Village	10-E2.2	Medium	2
Kamonji Village	10-E2.2	Medium	2
Baru Village	10-E2.2	Medium	2
Balaroa Village	10-E2.0	High	3

Source: BPBD Kota Palu, 2018

The value of the Ground Shear Strain (GSS) in the soil layer describes the ability of the soil layer material to shift during an earthquake. Due to the level of shock strength, the dynamic soil nature is classified from soil that is still elastic (no effect), plastic-elastic (partially affected) to collapse and melt. The results showed the value of the soil shear strain in the medium and high classes.

The areas with soil shear strain values in the medium class are Siranindi Village, Kamonji Village, and Baru Village. The dynamic properties of soil classification to the level of shock strength in the area are included in the collapsed category. Soil phenomena have a high potential for soil fractures and land subsidence; it is potential for small-scale liquefaction.

Areas that have high ground shear strain values are Balaroa Village, Lere Village and Ujuna Village. The dynamic properties classification of the soil to the level of shock strength in the region is included in the melting category. Soil phenomena have a high potential for soil fractures and land subsidence. It also has liquefaction potential on a medium to high scale. Soil conditions influence the high liquefaction potential in each region. The potential for liquefaction occurs in areas where alluvial fan deposits meet the alluvial weakness of the Palu river. The areas that have a high liquefaction hazard level a located in areas bordering the Palu river.

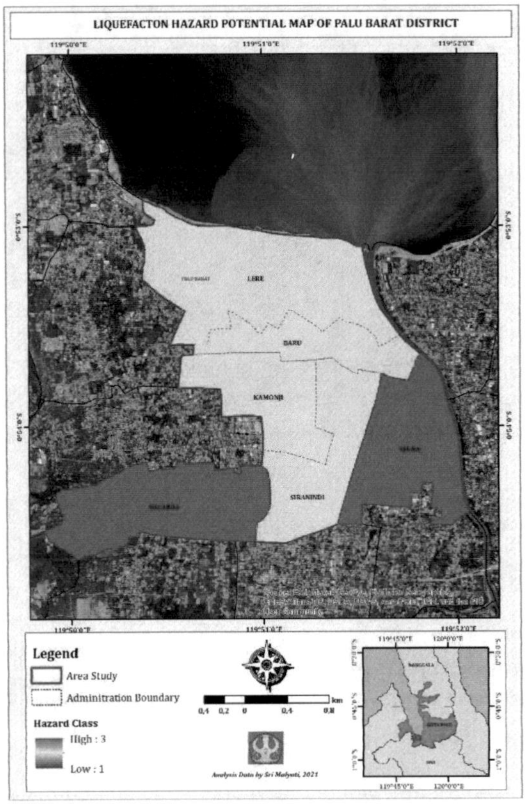

Figure 4. Liquefaction Hazard potential map

4.4 Multi-Hazard Assessment

The results showed that the multi-hazard potential level in Palu Barat District was in the medium and high categories. Areas that have moderate multi-hazard potential are Baru Village, Kamonji Village, and Siranindi Village. Meanwhile, areas that have high multi-hazard potential are Balaro Village, Lere Village and Ujuna Village. This result is affiliated with multiple earthquakes and liquefaction disasters that coincided in Balaroa Village, multiple earthquakes and tsunami disasters in Lere Village and parts of Ujuna Village.

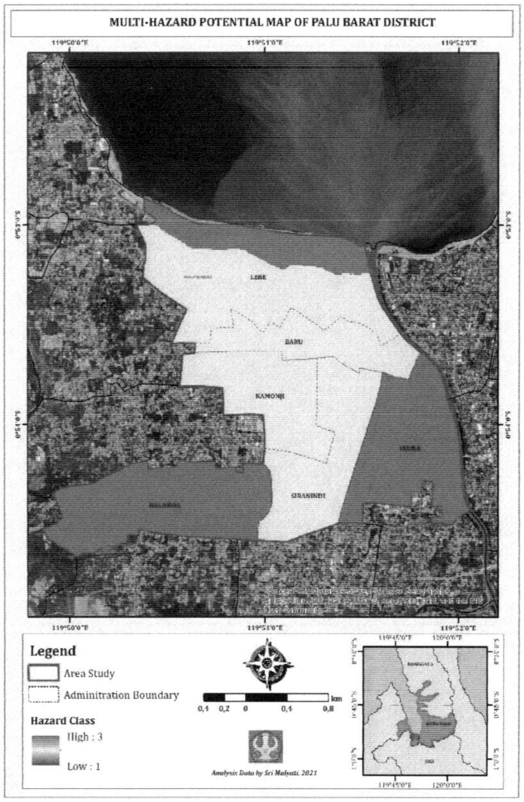

Figure 5. Multi-Hazard potential map

5. Conclusion

The multi-hazard assessment of disasters produces three maps: earthquake hazard maps, tsunami hazard maps, and liquefaction hazard maps. Based on the multi-hazard map of West Palu District, we can generally describe that the level of multi-hazard disasters is categorized as medium and high. The multi-hazard level is in Baru Village, Kamonji Village and Siranindi Village. In contrast, the high multi-hazard level is in Balaroa Village, Ujuna Village and Lere Village.

Acknowledgment

The author would like to thank all parties who have supported this research. This research was funded by DIPA Faculty of Engineering in 2021.

References:

[1] Badan Nasional Penanggulangan Bencana (BNPB), 2016. Indek Risiko Bencana Indonesia, Jakarta
[2] Amri M.R., dkk., 2016. Risiko Bencana Indonesia. Badan Nasional Penanggulangan Bencana (BNPB). Jakarta.
[3] A Arifati, R M Nisaa, A I Tinasar 2017 *Prosiding Seminar NasionalGeografi UMS* 978-602-361-072-3
[4] A Tohari 2018 *Kajian Gempa Palu Provisi Sulawesi Tengah* 978-602-5489-14-3
[5] H T Sarapang, O H A Rogi, P Hanny 2019 *J. Spasial Vol. 6 No. 2* 2442-3262

[6] Peraturan Kepala Badan Nasional Penanggulangan Bencana Nomor 02 Tahun 2012.Pedoman Umum Pengkajian Risiko Becana.12 Januari 2012.Badan Nasional Penanggulangan Bencana (BNPB) Tahun 2012.Jakarta
[7] S Pribadi, I Gunawan, J Nugraha, T Haryono, E Susanto, Irawan, Romando, C Basri, A Mustarang, Heriyanto 2018 *Laporan Survey Tsunami* BMKG

Architectural adaptation of vernacular stilt houses in Palu city

NR Burhany[1,*], **A Jiba RB**[2], **Puteri F**[3], **A Marwah**[4]
[1,2,3,4]*Department of Architecture Engineering, Engineering Faculty, Tadulako University*
*Corresponding Author: ninaburhany@gmail.com.com

Abstract. This study intends to identify architectural adaptations that occur in vernacular stilt houses in Palu City. The research method used is a qualitative approach and field research methods. Typology is used as an analytical tool and will analyze the basic structures, functions, and the process of development of these basic structures to their current form. Then examine the possible influence of local conditions on architectural adaptations that occur. The results of this study will enrich the repertoire of knowledge of Archipelago Architecture about the development of vernacular stilt house architecture. In addition, the results of this research will later be used as reference material in conserving Kaili tribal architecture and developing stilt house architecture in urban community life.
Keywords: Architectural Adaptation, Raised floor construction, Traditional House, Building Conservation

1. Introduction

Indonesian people, in general, know the house on stilts as one type of vernacular architecture that exists in various regions in Indonesia. A stilt house is a form of vernacular architecture that is widespread throughout the world, from the coasts of America, Europe, Africa, Asia, to Australia [1]. Most of the iconic traditional houses of the provinces and regions in Indonesia are in the form of stilts, including in Palu City, Central Sulawesi.

The original inhabitants of Palu City, namely the Kaili tribe, inherited several types of architectural houses on stilts, namely those known as Banua Mbaso, Kataba, and Tinja Kanjai, the houses were based on the social stratification of the Kaili population at that time. Banua Mbaso/Souraja is a special home for the nobles. The Kataba house is a residence used by middle-class aristocrats, meaning "Kataba" means a boarding house consisting of all board materials. The house "Tinja Kanjai" is a house for ordinary people, "Tinja Kanjai" means the house of ties [2].

Culture is something dynamic. Culture is changing and shifting according to the conditions and developments of the times. Likewise, with cultural products, there is a possibility that some will remain unchanged, some will change, and some will even disappear. As a cultural product, stilt houses are also inseparable from these dynamics, so that architectural adaptation can occur in the development of Kaili traditional stilt houses, especially the Banua Mbaso and Kataba type houses which are houses belonging to the Kaili tribe, nobility, and middle-class society.

Compared to the original traditional stilt houses, vernacular stilt houses in Palu city also experienced architectural changes and adaptations. Currently, the development of the city and culture is starting to shift the existence of stilt houses, if we do not identify and document the typology of stilt houses, we may lose track of its development and even lose track of it altogether if it is not conserved.

For conservation, it is necessary to document research results as reference material. Research on architectural adaptations of vernacular stilt houses will show adaptations that occur in vernacular stilt houses so that an overview of their sustainability can be obtained. In addition, the results of this study will enrich the repertoire of knowledge of Archipelago Architecture regarding the development of vernacular stilt house architecture.

2. Methods and Samples

2.1. Methods

This research is qualitative. Qualitative research is multimethod in focus, involving an interpretative, naturalistic approach to its subject matter. This means that qualitative researchers study things in their natural settings, attempting to make sense of, or interpret, phenomena in terms of the meanings people bring to them. Qualitative research involves the studied use and collection of a variety of empirical materials – case study, personal experience, introspective, life story, interview, observational, historical, interactional, and visual texts – that describe routine and problematic moments and meanings in individuals' lives [3]

This research also conducts field research to examine problems in natural settings to interpret. Interpret the observed phenomena by making in-depth observations of objects in the field, considering that this research requires direct measurements and field observations.

The typology is used as an analysis tool, with typology, an architectural object analyzed for changes, namely those concerning the basic structure, basic properties, and the process of development of the building to its current form and function of the object. From the results of the typology analysis, it will be able to determine the type of object and place it correctly in the existing type classification. Typology can explain the changes of a type, in which a type has characteristics that can distinguish it from other types.

According to Rafael Moneo [4], typological analysis can divided into 3 phases:

1) Analyzing typology by digging from history to find out the initial idea of a composition, or in other words knowing the origin or occurrence of an architectural object.
2) Analyzing the typology by knowing the function of an object.
3) Analyzing the typology by looking for a simple form of a building through searching for basic structures and their basic properties.

The typological analysis uses parameters that describe groups of objects based on their basic properties. Habraken [5] offers three ways of distinguishing the types of architectural forms:

1) Spatial System; identify the type and form of space and how the relationship between these spaces, hierarchies, patterns, and orientations
2) Physical Systems; identify the characteristics of its components, namely the material and structure of the space-forming elements.
3) Stylistic System; related to the appearance of the building, the shape and appearance of the facade of the building include elements of ornament, decorative or tectonic.

Then it also examines the possible influence of local conditions on architectural adaptation that occurs, regarding (1) climatic factors, (2) technical and material factors, (3) socio-cultural factors, and (4) community economic factors.

2.2. Samples

A purposive sampling method was used to select sample houses. The selected sample is representative of the population of the banua mbaso and kataba stilt houses.

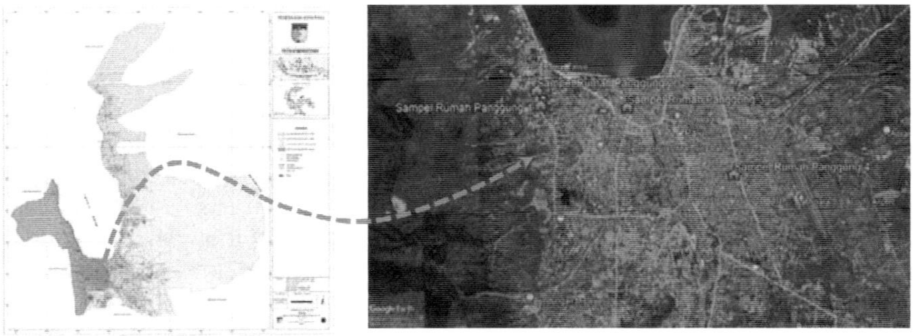

Figure 1. The 4 points of the location of stilt houses in Palu city which are used as research samples.

(a) Stilt House Sample 1

Home Owner : Pari
Tribe: Kaili Ledo
Building Age : ± 100 years
Address : Jl Munifrahman No.120
Kabonena Village
Resource person: Haris (45 years) 4th Generation
Coordinate point: 0°53'36.82"S 119°50'7.74"E

(b) Stilt House Sample 2

Home Owner : Pue Njidi
Tribe: Kaili Ledo
Building Age : ± 100 years
Address: Jl. Munifrahman No.57
Kabonena Village
Resource person: Andi Fitri (34 years) granddaughter of the owner
Coordinate point:0°53'22.36"S 119°50'4.00"E

(c) Stilt House Sample 1

Home Owner : Nurhayati Thayeb
Tribe: Kaili Ledo
Building Age : ± 81 years
Address : Jl. Hos Cokro Aminoto lrg. Baru No. 10, Baru Village
Resource person: Shafa Lisyahada (22 years) granddaughter of the owner
Coordinate point: 0°53'39.96"S 119°51'34.69"E

(d) Stilt House Sample 1

Home Owner Name : Marten Palar
Tribe: Kaili-Manado
Building Age : ± 76 years
Address : Jl. Anoa No.175 North Tatura Village
Resource person: Mardin M. Palar (46) grandson of the owner
Coordinate Point: 0°54'47.47"S 119°53'19.94"E

Figure 2. Reseach Sample, (a) Sample 1, (b) Sample 2, (c) Sample 3, (d) Sample 4

3. Literatur Review

3.1. Stilt House

The stilt house is a practical, comfortable, and adaptable building type. It is one of the oldest forms of human housing, with ancient roots in tropical regions of both southeastern Asia and the Americas. Its history extends back to at least the origin of agriculture. The raised design, verandahs, ventilated roofs, and large apertures for windows and doors combine to enhance ventilation, a highly effective method of natural cooling for houses in the humid tropics [6]. The construction of a stilt house is that the floor plan is lifted from the ground (or water) and supported by pillars. Based on location, stilt houses can be divided into three types, stilt houses in watery areas, stilt houses in semi-watery areas, and stilt houses on land [7].

Based on observations and preliminary surveys, most of the vernacular stilt houses that are still found in Palu City are included in the Banua Mbaso and Kataba categories, although some of them still look like stilt houses there have changed. This research will focus on these two types of buildings.

Fuad Zubaidi's article [8] discusses the Kaili architecture as a vernacular process and product, describes *Banua Mbaso* and *Kataba*;

Banua Mbaso/Souraja or Banua Magau is a stilt house built on rectangular wooden beams made of hardwood. The roof is generally gable. The front and back are covered with a wide board decorated with carvings called Panapiri, on top of the Panapiri on the front and back end a crown or bangko-bangko is carved. The floor and walls are made of planks, while other parts such as rafter beams, girders, and support beams use spinach wood and lime wood. The room is divided into three, namely: "Lonta Karavana" or the front room, "Lonta Tatagana" or the living room, and "Lonta Rarana" or the back room.

Kataba is a residential house for the middle class, Kataba means a plank or plank house (all parts of the house are made of boards). The Kataba type is the same as the Banua Mbaso type, which is a stilt house supported by stone pillars. the roof cover uses rumbia roof. Kataba size is smaller than Banua Mbaso which is 17 x 8 m. The main house measures 10 x 8 m and the kitchen measures 7 x 8 m. The location of the kataba house is generally around the village of Lere, because the beginning of the development and civilization of the city of Palu started from the coast which was also marked by the entry of Islam from the coast called "Karampe". With the development and population of the Kaili people, Kataba's house is not only near the beach but also in the Palu valley and the edge of the hills.

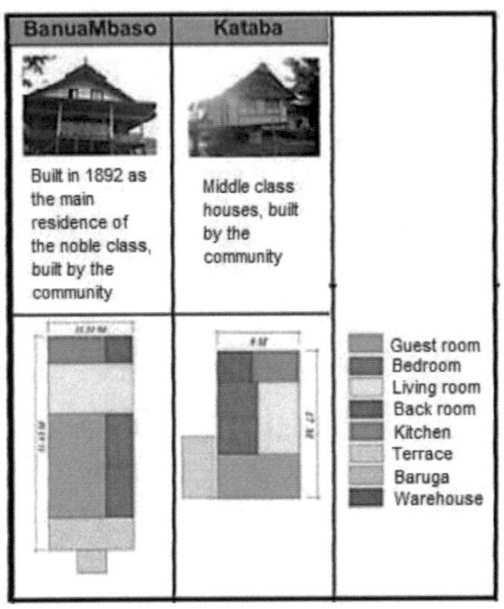

Figure 3. Banua Mbaso and Kataba (Zubaidi, 2009)

- The placement of the pedestal is called Palantara in the Kaili Ledo dialect (Palu city) and Palongo in the Kaili Ija dialect.The number of steps is odd.
- Thc guest room (Lonta Karavana) is a place to receive guests.
- The living room (Lonta Tatangana), is generally limited to the bedroom and dining room, sometimes the dining room is placed in the Lonta Rorana (back room).
- The bedroom (kamara), the room directly adjacent to the kitchen (avu) is intended for women, while the room at the front is designated for men.
- The backroom (Lonta Rarana), generally adjacent to the bedroom and dining room, is equipped with "Avu" (kitchen).
- At the back of the house, there is usually a staircase for women, and sometimes a "Pakuntu" (open space) is added, which is a kind of breezy place for family members.

3.2. Arsitektur Vernakular

According to Yulianto Sumalyo [9], vernacular meaning is the local language, in architecture, this term refers to forms that apply elements of culture, the environment includes the local climate, expressed in architectural forms (layout plans, structures, parts details, ornaments). , etc). Vernacular architecture in particular, is a product of people, place, and culture; it is one facets of identity. Symbolism of architecture can be related to the realization of identity personally and socially. This accretion has reached the level where "architecture as identity" became the equivalent to "architecture as space" and "architecture as a language" [10]

According to Oliver [11], in vernacular architecture, there is a mutual influence between natural/environmental elements and the culture of the people. In the formation of environmental settings, there are several elements that can be used as approaches, including: 1) Climate: polar and semi-polar, relating to continents, deserts, marine, Mediterranean, tropical, sub-tropical, 2) Location and Site: fields, beaches, deserts, forests, grasslands, lowlands, marine, slopes, highlands, valleys, 3) Natural Disasters: earthquakes, floods, landslides, snow, tropical cyclones, 4) Population: from the original place, the impact of density, growth, migration, urbanization, 5) Settlement: clustered, unified, closed area, random, grid, linear, point, organic, intermediate/periphery.

In Mentayani [12], according to Anselm (2006), vernacular architecture emphasizes more on traditions and socio-cultural communities as a measure of human comfort. Therefore, vernacular architecture has the same form or style in one place but is different from other places by adjusting the traditions and socio-cultural conditions of the people. According to Mitchel and Bevan (1992), the vernacular architecture contains four key components originating from local conditions, namely (1) climatic factors, (2) technical and material factors, (3) social and cultural factors, and (4) community economic factors.

Architecture cannot be separated from the environment in which it is located, because the environment is closely related to the humans who inhabit it. Humans take advantage of their environment in order to meet their needs by developing their creativity which in the end is very influential on the shape and style of their artificial environment (architecture) which has implications for their culture [13].

From the definitions and theories regarding Vernacular architecture mentioned above, it can be formulated aspects that will be reviewed and the possibility of influencing the architectural adaptation of existing stilt houses in Palu City, namely climatic factors, technical and material factors, social and cultural factors, and community economic factors.

4. Results and Discussion

4.1 Architectural Adaptation, Stilt House 1

4.1.1 Spatial System Stilt House 1

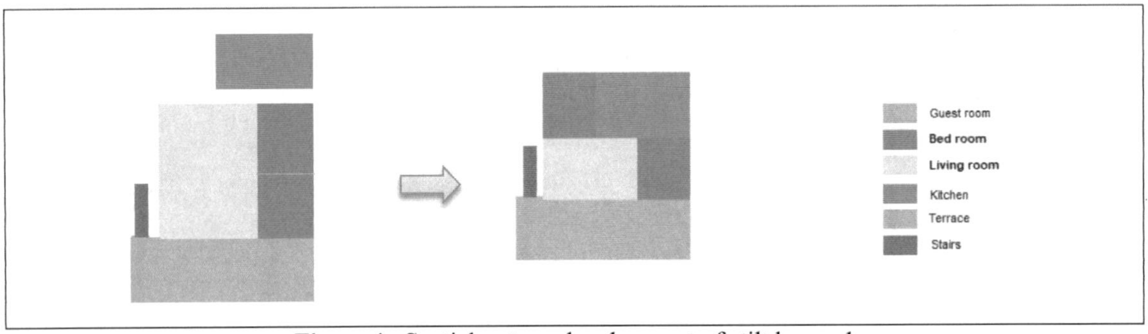

Figure 4. Spatial pattern development of stilt house 1

The Stilt house 1 faces the street. The living room is at the front with access to the left side, where there is a terrace with stairs that go up from behind.

Initially, the kitchen was made separate and lower than the main house. In 2000, because the condition of the kitchen building was damaged, it was removed and moved into the main house building so that the spatial pattern which was similar to the Kataba room pattern then underwent a change.

4.1.2 Physical System Stilt House 1

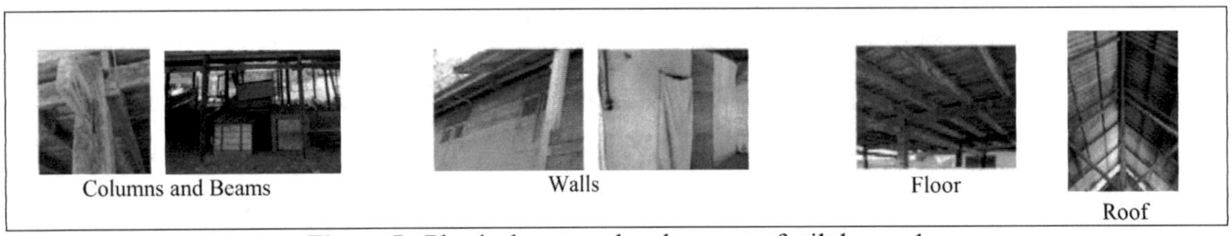

Figure 5. Physical system development of stilt house 1

This stilt house height is 210 cm from ground level, the main structure resting on a pedestal, with columns measuring 12x13 cm and beams 6x10 cm.

The columns and beams are made of Sapo wood using a pen and hole linkage technique. Some of the damaged parts have been refurbished while retaining the original construction system.

The house walls were original using wooden planks, outer walls, and inner walls. After experiencing damage to the interior wall of the house was replaced by using plywood material, while the exterior walls continued to use plank walls.

The building floors maintain using board material by replacing the weathered parts.

From the beginning, the roof was a saddle form of using wooden truss construction and zinc material as a roof covering.

4.1.3 Stylistic System Stilt House 1

The form of the stilted building is still clearly visible because the downstairs room has not changed to massive/closed.

The windows in the front of the house are modified using glass material, while the windows on the left and right sides of the building still use swing windows made of wood.

Figure 6. Stylistic system development of stilt house 1

There are no ornamental/decorative elements at all, and from a tectonic aspect, there is no special processing of construction elements.

Although simple, the use of construction materials and systems produces aesthetic beauty through the horizontal line grooves from the use of wallboards, column and beam connections, and pedestal points under each column.

4.2 Architectural Adaptation, Stilt House 2

4.2.1 Spatial System Stilt House 2

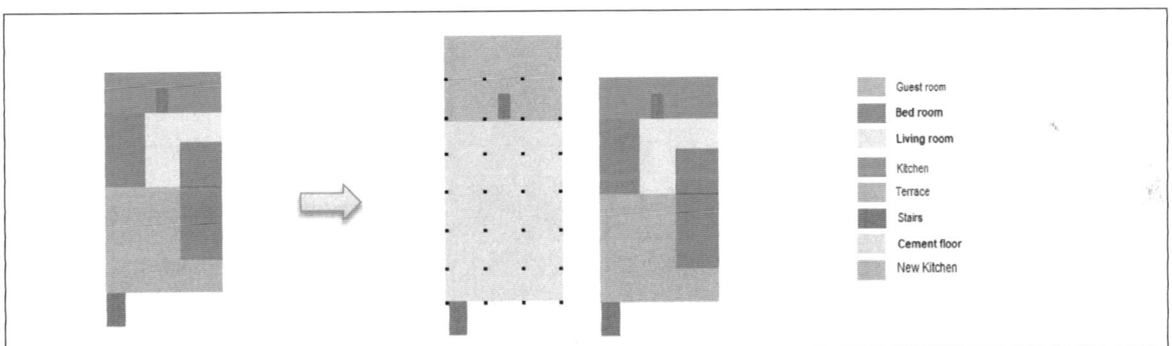

Figure 7. Spatial pattern development of stilt house 2

The stilt house 2 faces the street with access using stairs from the front to a large terrace in front, which functions as a place to receive guests from the general public. There is another staircase at the back in the kitchen area.

At the beginning of construction, the kitchen was in a stilt house, with walls and floors made of boards/wood. When the building materials had begun to rot and were not conducive to being loaded with heavy objects, in 1994, the kitchen moved to the under the house, which was previously a duck coop.

In 1994, the kitchen still had wooden walls and dirt floors. In 2013, the kitchen wall material changed to brick plastered, and the entire floor was hardened using cement and served as a warehouse and garage.

In 1999, the stairs were wooden stairs that used Sapo wood, but due to weathering and starting to tilt, they replaced them with concrete stairs, but the position and size remained similar to the original stairs.

4.2.2 Physical System Stilt House 2

Columns and beams using a pen and hole technique, then locked with dowels in the form of iron/wooden bars. The bottom of the column has made a foundation line, and the entire lower floor is filled with cement mixture.

Figure 8. Physical system development of stilt house 2

The house walls were made entirely of boards/wood, after being weathered/damaged, they were replaced with different materials. At the front, we still use plank walls, plywood walls in the interior, and zinc material on the right-left and rear outer walls. While the additional kitchen space below uses plaster bricks.

The floor structure consists of wooden beams to support the wooden plank floor. In some parts, there are small beams installed on top of support beams on parts of the floor that have begun to rot.

The saddle-shaped roof of the main building resembles the roof of Banua Mbaso, and the upper kitchen uses a pyramid roof whose top is made of terraced/layered, while the lower kitchen roof is a shield roof. The entire roof covering uses zinc material.

4.2.3 Stylistic System Stilt House 2

Figure 9. Stylistic system development of stilt house 2

The form of the stilted building is still quite clearly visible, even there is the use of the downstairs space in the back (kitchen area) is closed using a brick wall, but the front of the main house is still a stage.

The doors and windows still retain their original shape in a two-leaf swing window made of wood.

There are quite a lot of ornamental/decorative elements seen in this building. That existed from the start in the form of carvings on the top of the roof, the plank trim, terrace railings, and additional ornaments created by the homeowner in the form of tendrils on the terrace ceiling..

The tectonic aspect does not have special processing on the construction elements but the use of materials and construction systems produces aesthetic beauty even though it is simple. Such as horizontal and vertical line grooves from the use of board and zinc walls, as well as the form of column and beam connections.

4.3 Architectural Adaptation, Stilt House 3
4.3.1 Spatial System Stilt House 3

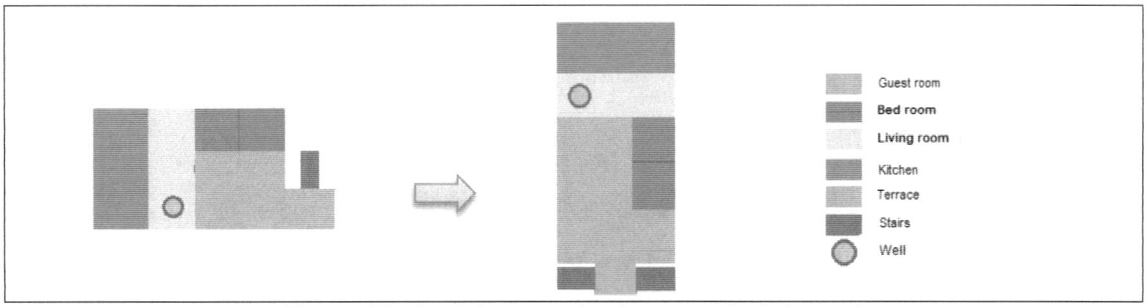

Figure 10. Spatial pattern development of stilt house 3

At the beginning of construction (the 1940s), the stilt house was made to extend to the side, but in the 1970s it was renovated and built extending backward with a similar spatial pattern.

In addition to changing the shape of the building to extend backward, the terrace with 2-way stairs on the left and right is made to resemble the stairs on the Gandaria/Souraja terrace.

The kitchen is no longer a stage and has been replaced with plastered brick walls and ceramic tile floors. In the middle, there is a well with a brick wall as a separator between the house on stilts and the kitchen (stone house).

4.3.2 Physical System Stilt House 3

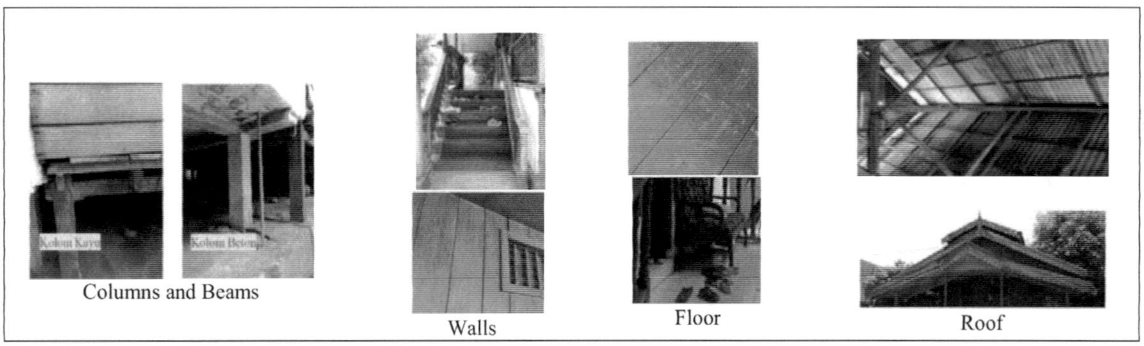

Figure 11. Physical system development of stilt house 3

This building mostly uses wood construction, in some parts such as the stairs and in the living room where there is a well and the kitchen uses brick walls and tiled floors.

The column uses 12/12 cm of wood on stilts with a pin and hole system. The living room uses a concrete column measuring 20/20 cm. The bottom of the column is lined with the foundation, and the entire bottom floor is cemented.

The main house walls use wooden planks, while the back is brick walls. In the front or main part of the house use wood material and the middle of the house on stilts there is a well with a brick wall as a separator between the house on stilts and the kitchen (stone house).

The height stage floor is at 1.40 meters from the ground, using a wooden beam construction with a distance of 60 cm that supports the plank floor. The floors on the terraces and stairs are covered with ceramic tiles.

The gable roof on the main building resembles a Souraja roof. Separate roof covering the terrace and stairs. The entire roofs use zinc material with wooden truss construction.

4.3.3 Stylistic System Stilt House 3

Figure 12. Stylistic system development of stilt house 3

The stilts house form in front is a bit obscured because the front side under the terrace, is covered with a massive wall, but the impression of the stage is still quite clear from the left and right sides. Although the space at the back (kitchen area) is not a stage, the main house is still a stilt house.

There are no ornamental/decorative elements at all, and from a tectonic aspect, there is no special processing of the construction elements.

The front window was modified using an upper ventilation window in the same form to the trellis shape as the staircase railing. Meanwhile, the windows on the side of the building still retain the typical window shape of the Banua Mbaso/Kataba window.

There are several ornaments and decorative elements, in the form of carvings on the top of the roof, the plank trim, and terrace railings that match the trellis on the ventilation.

The pattern of the boards on the ceiling and the vertical lines of the wallboards have an aesthetic value. The tectonic aspect is not treated for construction elements, but it tends to be hidden.

4.4 Architectural Adaptation, Stilt House 4

4.4.1 Spatial System Stilt House 4

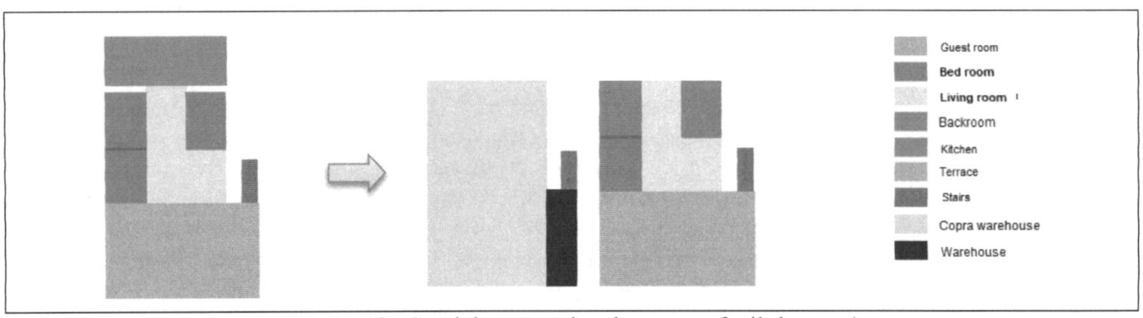

Figure 13. Spatial pattern development of stilt house 4

The Stilt House 4 faces the street. An L-shaped terrace on the right, with stairs going up from behind. Initially, this house was a stilt house similar to Banua Mbaso, with a separate kitchen at the back (now it has collapsed due to weathered materials and due to the earthquake). Then because the owner is a copra entrepreneur, the lower part of the stilt house is covered with a brick wall to function as a copra storage area and has access from the front and both sides.

The upper house is similar to the Souraja building, has three entrances, each of which has two doors for different purposes. The middle door was the entrance for the nobles or respected people and the side door for the ordinary people.

4.4.2 Physical System Stilt House 4

Figure 14. Physical system development of stilt house 4

The column of the house uses palapi/iron wood with a size 15 x 15 cm. The lower column is covered by cement plaster which is connected to the brick wall. The bottom of the column has been made a line foundation and the entire bottom floor is filled with a cement mixture.

The walls on the lower floor are brick walls, to support the security of the room that functions as a copra storage area. While on the top floor using wooden plank walls.

This house uses a cement floor on the lower floor and a plank floor on the upper floor. The structure of the upper floor arrangement of wooden beams mounted on the house pillars, then supporting the wooden plank base beams at a distance of 60 cm.

The house uses a gable roof and a skillion roof for the side terrace, all roofs covered with zinc.

4.4.3 Stylistic System Stilt House 4

Figure 15. Stylistic system development of stilt house 4

The stilted building shape is a bit obscured at the front because the bottom of the terrace is covered by a massive wall, but the impression of the stage is still quite clear from the left and right sides.

There are several ornaments and decorative elements in the form of carvings on the top of the roof, plank trim, terrace railings, and windows and ventilation holes in the form of carvings on the second floor of the building.

The window design in the additional space below adjusts to the window type on the stilt house by simplifying the shape.

The tectonic aspect is not treated for construction elements and even tends to hide because the first floor is massive so that the columns are not visible from the outside, but there is a grid pattern formed from horizontal truss beams/walls and vertical lines from the plank walls that make interesting patterns.

5. Conclusions

The results of this study indicate that based on a sample of houses on vernacular stilts in the city of Palu, an overview of architectural adaptations in buildings that were initially similar to Banua Mbaso and Kataba in the form of:

a. Adaptation to climatic factors occurs in consideration of maintaining the shape of the roof even though it has made improvements or added space. The window shape of even though there is a change in material and design, still uses the same type of window (a swing window with two leaves) because it allows for good air circulation. In addition, there is an adaptation by adding openings/gaps by imitating the Souraja roof model in a space that produces heat more like a kitchen so that it allows the circulation of hot air in the house to get out.
b. Adaptation to engineering and material factors can occur in the addition of columns, beams, and walls, use of different alternative materials, and construction when carrying out repairs or renovations.
c. Adaptation to socio-cultural factors can be seen in changes in spatial patterns both vertically and horizontally, on the exterior and interior that adapt to the needs and changes in lifestyle.
d. Adaptation to community economic factors can be seen in consideration of material selection when carrying out repairs or renovations that are adjusted to economic capacity so that they are not stuck on the same materials as before.

Even though there are adaptation efforts by the owner/occupant of the house, in general, the homeowners find it quite hard to maintain the existence of their stilt houses. Because it needs high cost requires routine maintenance compared to houses with brick and reinforced concrete construction. But on the other hand, they also want to maintain the house on stilts which is the historical heritage of their parents and grandparents.

To prevent the loss of the architectural character of the vernacular stilt houses, from the face of Palu city, it is necessary to reintroduce to the community through building maintenance assistance so that our children and grandchildren can still see firsthand the existence of the stilt houses. Architecture. In addition, it is necessary to encourage the application of the distinctive character of the architecture of stilt houses through innovation in forms and technology in buildings in Palu city so that at least a common thread can be seen from the development of the architecture of stilt houses in the Palu city.

Acknowledgment
The authors would like to thank the Faculty of Engineering, Tadulako University, which has funded this research, and all parties who have assisted in the research process.

References:

[1] Zuber A 2017 *J. Arsir* **1(2)** 175
[2] Fuad Z 2009 *J. Ruang* **1(1)** 30
[3] Norman D K and Yvonna S L 2005. Introduction. The discipline and practice of qualitative research. In *The Sage handbook of qualitative research*, ed. Norman K. Denzin and Yvonna S. Lincoln Thousand Oaks: SAGE Publications. 1–32
[4] Moneo R 1978 *On Typology Oppositins J for Idea and Criticism in Architecture*, Publish for The Institute for Architecture and Urban Studies By The MiT Press 23- 44
[5] N J Habraken 1988 *Type as Social Agreement*, Asian Congress of Architect, Korea: Seoul p 5
[6] Heyward P J 2001 *Up on Stilts: The Stilt House in World History* N.p.: Tufts University p 32
[7] Marwati M 2014 *J. Teknosains* **8(1)** 95-108
[8] Fuad Z 2009 J. Ruang **1(1)** 31-33
[9] Yulianto S 1993 *Arsitektur Kolonial Belanda di Indonesia* Yogyakarta: Gadjah Mada University Press p 35
[10] Abel C 2000 Architecture and Identity: Responses to Cultural and Technological Change. London. New York: Routledge p 137
[11] Paul O 1983 Perfected for Needs: Vernacular Architecture in Education. Habitat International, 7(5-6):337-383.
[12] Ira M 2012 *J. Lanting* **1(2)** 78
[13] Amos R 1969 *House Form and Culture* Prentice Hall Inc New York in p 12

Potential of confine aquifer based on characteristics of the aquifer in phase IA Duyu permanent occupancy at Palu city Central Sulawesi Province

Ninasafitri[1] Zeffitni[2]

[1)]Environmental Engineering Department, Tadulako University, Palu, Indonesia
[2)]Geology Engineering Department, Tadulako University, Palu, Indonesia

zefitni_04@yahoo.com

Abstract. One of the government's efforts towards post-disaster reconstruction in 2018 at Central Sulawesi, especially in Palu City, is to build a Phase IA Duyu Permanent Occupancy. This study aims to determine the potential for confine aquifer based on the aquifer characteristics for freshwater—analysis of aquifer characteristics based on survey results, pumping test data, and hydrochemical parameters. The results showed that the quantitative potential of confine aquifer based on SD-DTW-DY well data was in the medium to high potential class in the Pz3.D0.T0.Qs0.Q0.M1 units, with the distribution of aquifer characteristic values: permeability (K) 53.09 m/day, the piezometric depth is 12.15 m from ground level, the aquifer thickness is 76.32 m, and debit (Q) is 9.50 l/second. Spatially the distribution of aquifers in denudational hills and alluvial plains at the Pakuli Formation with SND-GRV and GRA-DIO lithology. The results of the hydrochemical quality are good quality classes, include: TDS 182.0 mg/l, pH 8.3, Fe 0.00 mg/l, Mn 0.00 mg/l, Cl 11.70 mg /l, SO_4 12.90 mg/l, NO_3 0.00 mg/l, and NO_2 0.18 mg/l. Based on the results of the data analysis, the characteristics of the aquifer in the Duyu permanent occupancy area are classified as medium productive aquifers.

1. Introduction

Groundwater is one of the most important sources of water, especially to meet the needs of people's lives. In line with civilization's development, the increasing demand for freshwater causes groundwater to become an economic resource [1]. However, the presence of groundwater is limited both spatially and temporally [2]. In Central Sulawesi Province, the problem of groundwater is also very complex. The problem of groundwater as a source of basic needs is increasing, especially after the disaster. One of the government's efforts towards post-disaster reconstruction in 2018 at Central Sulawesi, especially in Palu City, is to build a Phase IA Duyu permanent occupancy. In that area, the government made + 230 satellite permanent occupancy units, one of which is freshwater facilities and infrastructure. However, the availability of fresh water in the Permanent Occupancy area is not sufficient because it only uses one well. This study aims to determine the potential for confine aquifer based on the characteristics of the aquifer for freshwater needs in the Duyu Phase IA permanent occupancy at Palu City. The location is part of the administrative area of Duyu Village, Tatanga District, Palu City, Central Sulawesi. The size of Duyu Village is approximately 616 ha. This study aims to identify groundwater potential for freshwater needs in Duyu's permanent occupancy based on the aquifer characteristics of compressed groundwater sources. In figure 1. research location in Duyu Village.

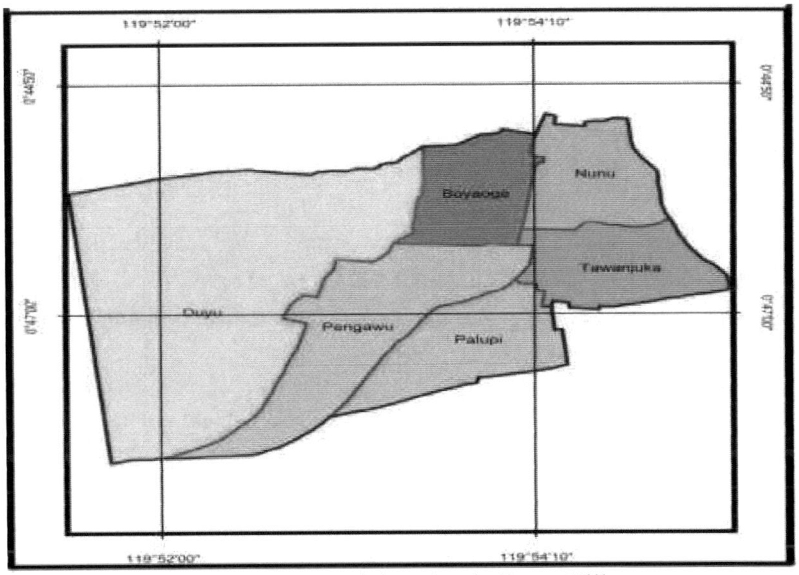

Figure 1. Research location in Duyu Village

2. Materials and Methods

The research location is Duyu Village, Tatanga District, Palu City, Central Sulawesi Province. The location of the permanent regional occupancy is in the administrative area of Duyu Village, Tatanga District, Palu City. The size of the temporary shelter area is 38.3 ha according to the measurement results of the Ministry of ATR/BPN or 6.21% of the administrative area of Duyu Village, which is 616 ha. The research method determines the aquifer characteristics of the SD-DTW-DY compressed groundwater sources based on the survey results, pumping test data analysis, and laboratory tests on hydrochemical parameters [3]. Determination of quantitative potential to determine area Qopt using characteristic aquifer parameters, which include: thickness (D), permeability (K), and transmissivity (T), and well characteristics (piezometric face depth, type discharge, and optimum discharge) [4,5]. In this case, the type of discharge is a discharge that can lower the groundwater level per one meter (l/sec/m) [6]. Another parameter used to determine the Qopt area in each aquifer system is the thickness (D) of the aquifer [7,8].

The groundwater potential is quantitatively determined by the aquifer system model, both from the permeability value, rock porosity value, and fluid properties [9,10]. The potential for groundwater reserves stored in aquifers is primarily determined by the physical properties of the rocks that make up the aquifer [2]. This study determined the potential quantitative level of depressed groundwater for drinking water needs at the study site based on the optimum discharge (Qopt) of SD-DTW-DY wells. The optimum combustion is determined based on pumping test data and quantitative estimation through the determination stages: aquifer thickness, permeability, and transmissivity [11,12]. Some of these parameter values are determined by secondary data, namely the SD-DTW-DY pumping test data. Figure 2. the location of Duyu's permanent occupancy.

Figure 2. Research locations in Duyu permanent occupancy

3. Results and Discussion

Duyu Village is a sub-district in the southern part of the Palu City area, a hilly area with an average height of about 158 m above sea level. Based on data from Tatanga District in 2018 figures, the shape of the land surface in the Duyu Village area consists of 35% of the plain area, 35% of the hills, and 30% of the mountains. These data indicate that Duyu Village has a reasonably high slope level. From the slope map below, it can be seen that a small part of Duyu Village has a slope of 0 to 8%, and most of it has a slope of 8 to 25%. With these relatively hilly conditions, large-scale development requires careful land development work.

The temporary housing development in the Duyu area was developed as a low-density residential area so that spatially it is by its designation. The construction of shelters in the Duyu area has been determined based on the Governor of Central Sulawesi Province Number: 369/516/DIS.BMPR-G.ST/2018 concerning the Determination of Land Locations for Disaster Recovery Relocation in Central Sulawesi Province. In this decree, among other things, it is stated that the location of the shelter for Palu City is 560.90 hectares spread across Tondo – Talise, and Duyu. In the Decree for determining the site, the planned land for constructing the Duyu temporary shelter area is set at 79.30 ha located in Tatanga Village.

The service for drinking water sources by Palu City, Central Sulawesi Province in permanent satellite occupancy built by the Ministry of Public Works and Public Housing for disaster victims located in Duyu Village, Tatanga District, is still minimal. Water flows from the main reservoir in only two hours a day. In the Duyu shelter area, the government built approximately 230 satellite huntap units equipped with facilities and infrastructure. Starting from infrastructure, lighting, green open space, including freshwater [13]. However, until the last few months, the fresh water in the shelter area has not fully flowed normally because it only uses one deep well as a central source to pour into the reservoir before being channeled to the house connection [10,14]. In figure 3. Duyu permanent occupancy conditions and drinking water services.

Figure 3. Duyu permanent occupancy conditions and drinking water services

The results showed that the quantitative potential of confine aquifer based on SD-DTW-DY well data was in the medium to high potential class in the Pz3.D0.T0.Qs0.Q0.M1 units, with the distribution of aquifer characteristic values: permeability (K) 53.09 m/day, the piezometric depth is 12.15 m from ground level, the aquifer thickness is 76.32 m, and debit (Q) is 9.50 l/second [15,16]. The distribution of aquifers is in the hydromorphological units of denudational hills and alluvial plains. Furthermore, the hydrogeological unit is located in the Pakuli Formation with SND-GRV and GRA-DIO lithology. The results of the hydrochemical quality test for drinking water, which are good quality classes, include: TDS 182.0 mg/l, pH 8.3, Fe 0.00 mg/l, Mn 0.00 mg/l, Cl 11.70 mg /l, SO_4 12.90 mg/l, NO_3 0.00 mg/l, and NO_2 0.18 mg/l. Based on the results of the data analysis, the characteristics of the aquifer in the Duyu Permanent Occupancy area are classified as medium productive aquifers [17,18]. Thus, it is necessary to develop sustainable groundwater based on the concept of the Groundwater Basin, which is integrated into appropriate policies and patterns of groundwater resource management [19,20]. In figure 4. distribution of aquifers' characteristic parameters at Duyu Palu Groundwater Basin).

4. Conclusion

Based on the calculation and analysis of aquifer parameters in Duyu's permanent occupancy, it shows that the quantitative potential of groundwater based on SD-DTW-DY well data is in the productive aquifer layer. The distribution of the productive aquifer is in the form of denudational hills in the Pakuli formation with loamy sand lithology and good permeability values. Furthermore, based on the hydrochemical quality test for drinking water, the quality of SD-DTW-DY well water is in the excellent quality class. It can be concluded that the characteristics of the aquifer in Duyu's permanent occupancy area include productive aquifers. Efforts for groundwater conservation need to be carried out at that location so that the groundwater potential can meet the drinking water needs of Duyu's permanent occupancy location.

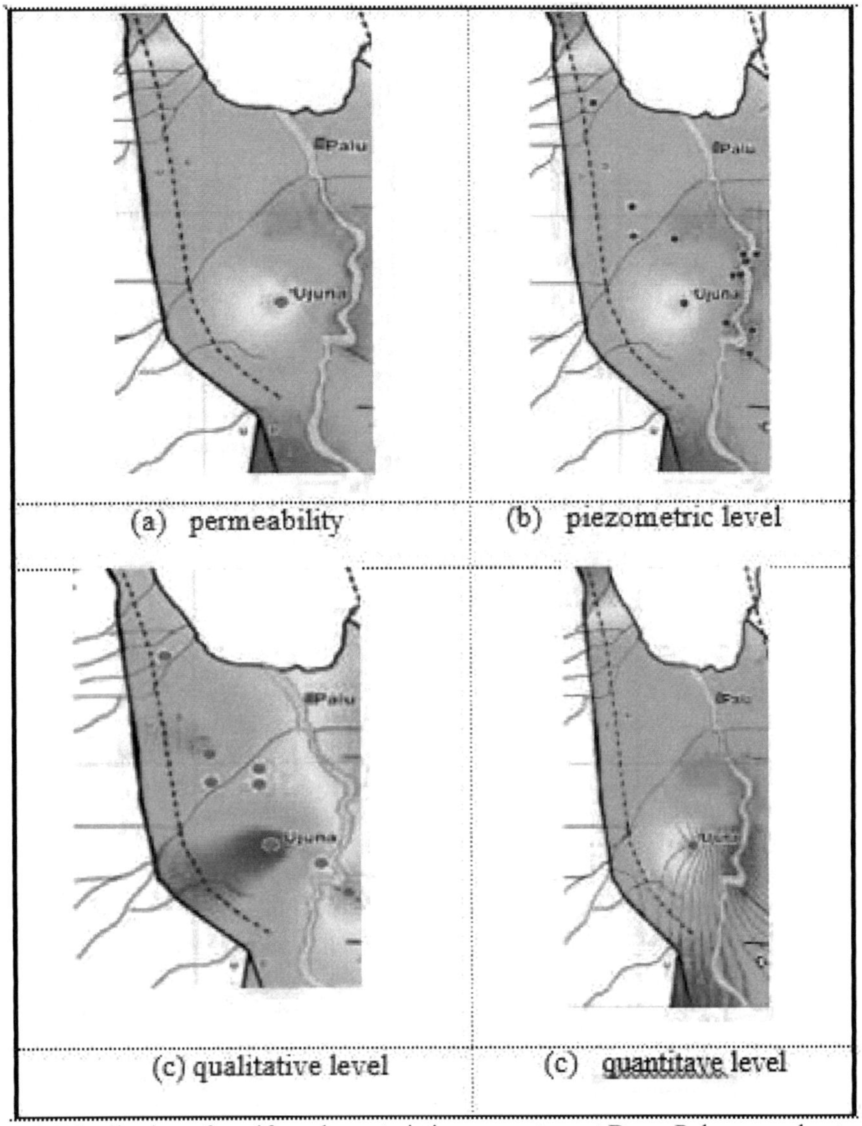

Figure 4. Distribution of aquifers characteristic parameters at Duyu Palu groundwater basin

Acknowledgment
Thanks to the Ministry of Education, Culture, Research and Technology through the Directorate General of Higher Education and Geospatial Information Agency to doing this research smoothly.

References

[1] Jang C-S, Kuo Y-M and Chen S-K 2019 Assessment of shallow groundwater use for irrigating park trees in the metropolitan Taipei Basin according to variability conditions of water quality *Journal of Hydrology X* **2** 100013

[2] Cheng Z, Su C, Zheng Z, Chen Z and Wei W 2021 Characterize groundwater vulnerability to intensive groundwater exploitation using tritium time-series and hydrochemical data in Shijiazhuang, North China Plain *Journal of Hydrology* **603** 126953

[3] Zeffitni -, Basir-Cyio M and Napitupulu M 2019 Zone of Groundwater Quantitative Potential Based on the Characteristics of a Confined Aquifer at the Palu Groundwater Basin Central Sulawesi Province, Indonesia *International Journal on Advanced Science, Engineering and*

Information Technology **9** 1725
[4] Chen X, Zhang K, Chao L, Liu Z, Du Y and Xu Q 2021 Quantifying natural recharge characteristics of shallow aquifers in groundwater overexploitation zone of North China *Water Science and Engineering* **14** 184–92
[5] Chen H, Yong B, Gourley J J, Liu J, Ren L, Wang W, Hong Y and Zhang J 2019 Impact of the crucial geographic and climatic factors on the input source errors of GPM-based global satellite precipitation estimates *Journal of Hydrology* **575** 1–16
[6] Collin M L and Melloul A J 2001 Combined land-use and environmental factors for sustainable groundwater management *Urban Water* **3** 229–37
[7] Zeffitni 2021 Geospatial approach to determining the potential of springs at the Palu Groundwater Basin for post disaster water supply in Central Sulawesi Province *J Phys: Conf Ser* **1763** 012087
[8] Zeffitni, Basir-Cyio M, Napitupulu M and Worosuprojo S 2020 Spatial analysis of the liquefaction vulnerability zone based on the phreatic level at the Palu groundwater basin, Central Sulawesi Province *J Phys: Conf Ser* **1434** 012019
[9] Verstappen H Th 1978 Remote sensing applications for mineral exploration *Earth-Science Reviews* **14** 71
[10] Chao H-C, Pi J-L, You C-F, Shieh Y-T, Lu H-Y, Huang K-F, Liu H-C, and Chung C-H 2021 Hydrogeology constrained by multi-isotopes and volatiles geochemistry of hot springs in Tatun Volcanic Group, Taiwan *Journal of Hydrology* **600** 126515
[11] Mukherjee I and Singh U K 2022 Hydrogeochemical characterizations and quality evaluation of groundwater in the major river basins of a geologically and anthropogenically driven semi-arid tract of India *Science of The Total Environment* **805** 150323
[12] Cavallin A, Marchetti M, Panizza M and Soldati M 1994 The role of geomorphology in environmental impact assessment *Geomorphology* **9** 143–53
[13] Jarzyna J A, Puskarczyk E and Motyka J 2019 Estimating porosity and hydraulic conductivity for hydrogeology on the basis of reservoir and elastic petrophysical parameters *Journal of Applied Geophysics* **167** 11–8
[14] Mutua S, Ghysels G, Anibas C, Obando J, Verbeiren B, Van Griensven A, Vaessens A and Huysmans M 2020 Understanding and conceptualization of the hydrogeology and groundwater flow dynamics of the Nyando River Basin in Western Kenya *Journal of Hydrology: Regional Studies* **32** 100766
[15] Ninasafitri and Zeffitni 2021 Determination risk of liquifaction disaster based on spatial distribution unconfined aquifer at the Tindaki Groundwater Basin, Parigi Moutong District *J Phys: Conf Ser* **1763** 012088
[16] Kresic Neven 2010 Sustainability and management of springs *Groundwater Hydrology of Springs* (Elsevier) pp 1–29
[17] Sedghi M M and Zhan H 2021 Discharge variation of multiple springs associated with a fractured aquifer *Journal of Hydrology* 127030
[18] Bolaane B, Tema N and Phuthologo B 2021 Barriers and coping strategies of households with no access to drinking water and waterborne sanitation in two low-income neighbourhoods in Botswana *Habitat International* **115** 102372
[19] Basir-Cyio M, Mahfudz, Isrun, and Zeffitni 2021 The impact of liquefaction disaster on farming systems at agriculture land based on technical and psychosocial perspectives ed S Clegg *PLoS ONE* **16** e0245591
[20] Zeffitni 2010 *Spatial distribution of groundwater potetial at Palu Groundwater Basin, Central Sulawesi Province* (Gadjah Mada University)

Effect climate change on the potential erosion of the Singkoyo watershed

I W Sutapa[1], Yassir [1], Nina B R[1], Sance L[1], Frederichsen F P[2]

[1]Department of Civil Engineering, Engineering Faculty, Tadulako University
[2]Postgraduate Civil Engineering Master's Program at Tadulako University

wsutapa@yahoo.com

Abstract. Global climate change results in an increase/decrease in rainfall, resulting in frequent occurrence of extreme rains that cause floods and landslides as well as prolonged droughts.. This study aims to determine the effect of climate change on the erosion potential of the Singkoyo watershed. The data used in the form of rain data obtained from Singkoyo Station (2008-2020), geological maps, maps of the earth's appearance and land use maps. The method used in analyzing the erosion potential is USLE. The climate change scenario is carried out by multiplying the average monthly rainfall by -20%, -10%, 10% and 20%. The result of this research is the reduction of rain by 10% and 20% resulted in decreased erosion potential by 3.7 tons/ha/year (13.3%) and 7.3 tons/ha/year (26.2%) . An increase in rainfall of 10% and 20% resulted in the potential for erosion to increase by 3.9 tons/ha/year (13.9%) and 7.9 tons/ha/year (28.1%). The conclusion of this study is that the relationship between rainfall and erosion potential is linear, meaning that an increase in rainfall causes an increase in erosion potential and vice versa.

Keywords: Singkoyo Watershed, Climate Change, USLE Method, Erosion Potential

1. Introduction

Global climate change results in an increase/decrease in rain, an increase in air temperature, changes in seasonal patterns, wind patterns, air humidity, and solar radiation. The most felt impact due to climate change in Indonesia is the climate anomaly. This phenomenon causes the rainy season to last shorter with increased intensity and the dry season to last longer than usual conditions. The increase in human needs as a result of population growth has resulted in an increase in the need for land for agriculture. Meanwhile, land suitable for agriculture is still lacking. As a result, residents use land that is less suitable for agriculture, such as steep slopes. This causes the soil to be easily eroded and carried away by rainwater which causes erosion. Soil damage due to erosion reduces the ability of the soil to support productivity in the watershed.

Several methods have been developed to analyze erosion. Empirical analysis is based on observation and is statistical. The most common method for estimating erosion potential is the Universal Soil Loss Equation (USLE). The USLE method can be used to predict the erosion rate over a long period of time from a farming area with a particular cropping system and land management.

There are several studies on the potential for erosion, including: Research conducted in East Java shows that the USLE method can be used to predict erosion quickly over a large area [1]. The dominant factors that cause the level of erosion hazard in the Amandit Watershed included in the heavy-to-very heavy classification are the slope, high erodibility value, and land cover [2]. The results of the research in the Panasen sub-watershed of Minahasa Regency showed that not all soil that was lifted from the ground surface through the process of erosion, entered the lake and became sediment [3]. The results of erosion research in the Rawapening watershed showed that the decrease in standing area was inversely proportional to the rate of erosion, the amount of erosion and sedimentation. Meanwhile, the built up area is directly proportional to the rate of erosion, the amount of erosion, and sedimentation. Changes in land use that cause the greatest increase in erosion are changes in land use from standing to build up land and changes in agricultural land use to build up land [4]. The results of the study on the erosion hazard index that occurred in the Kawatuna watershed were classified as low to moderate levels. Low erosion occurs in paddy fields and forests, while moderate erosion occurs in mixed gardens, scrublands, and fields. The dominant factor influencing the occurrence of erosion in this area is soil erosion, followed by crop management, and conservation measures [5].

This study aims to determine the effect of climate change on erosion potential in the Singkoyo watershed with a scenario simulating the monthly average rain, decreased by 10% and 20% and increased by 10% and 20%.

2. Material and Methods

2.1. Description of study

The research site is located in the Singkoyo River Basin with an area of 361.96 km^2 and geographically it is located between 122°3'3.3" - 122°22'13" East Longitude and 1°18'14.4" -1°28' 8.8" South Latitude.

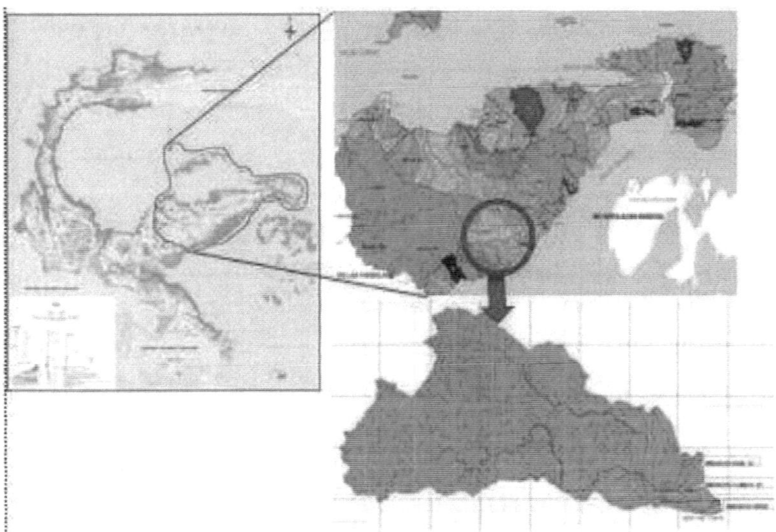

Figure 1. Research location

2.2. Literature review

Erosion prediction is used, namely the method *Universal Soil Lost Equation* (USLE) [6 - 16]:

$$A = R \times K \times Ls \times C \times P \quad (1)$$

The rain erosivity factor (R) is calculated by Lenvain's (1975) equation in [8], [17], as follows:

$$R = 2{,}21 \times CH^{1{,}36} \quad (2)$$

Soil erodibility (K) is calculated by the equation [6], [8],

$$K = \frac{\{2{,}71M^{1.14}(10^{-4})(12-0M)+4.20\ (s-2)+3.23(p-3)\}}{100} \quad (3)$$

2.3. Methods

This research includes quantitative descriptive which explains the relationship between variables by analyzing numerical data using statistical methods. The data required in the form of secondary and primary data. Secondary data were obtained from relevant agencies such as a map of the Indonesian landscape from the Bakosurtanal Office, geological maps obtained from the Center for Volcanology and Geological Hazard Mitigation, rainfall data obtained from the Office of Human Settlements and Water Resources of Central Sulawesi Province and maps of soil types obtained from the Planning Agency. Palu-Poso watershed. While primary data is done by checking directly in the field based on secondary data.

The analysis is carried out in the following stages:

1. Rain erosivity (R) is calculated by the Lenvain equation based on the Singkoyo Station monthly average rain data from 2008 – 2020) and is calculated by equation 2.
2. Soil erodibility (K) is calculated by the equation 3
3. The slope of the slope (LS) is analyzed with the help of GIS software
4. Land use and management (CP) analyzed with the help of GIS software
5. Climate change is carried out with the scenario of decreasing the monthly average rain by 10% and 20%, and increasing it by 10% and 20%
6. Analysis of results and discussion

The complete research method is presented in the following figure:

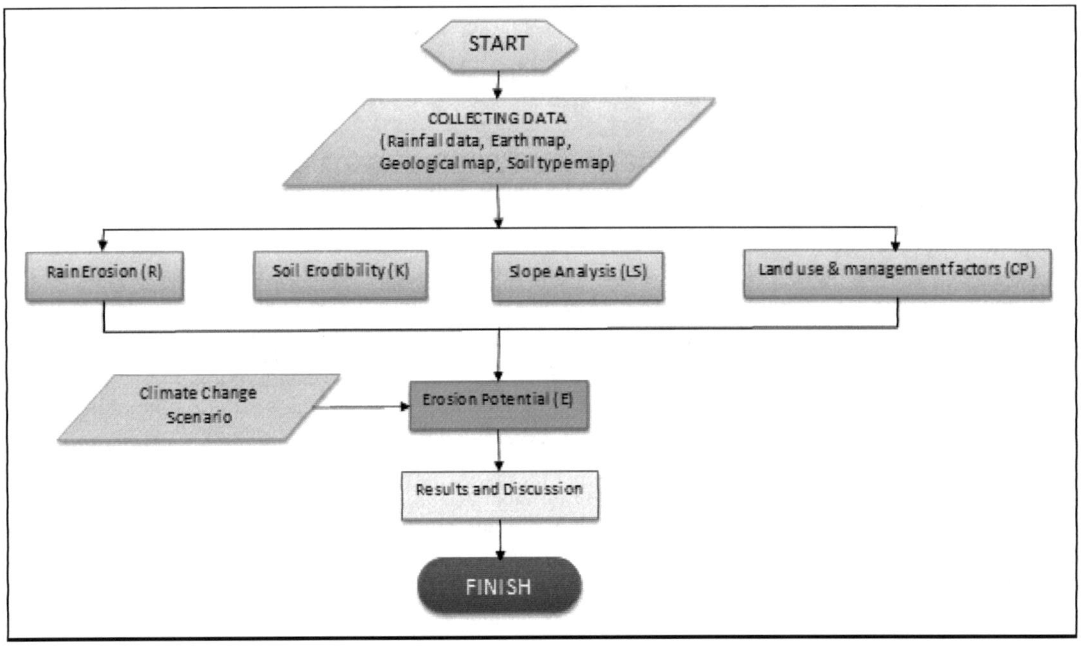

Figure 2. Flowchart research

3. Results and Discussion

3.1. Rain Erosivity(R)
The results of the calculation of rain erosivity are presented in the following tables and graphs:

Table 1. The results of the calculation of rain erosivity with several scenarios

Scenario	Jan	Feb	Mar	Apr	May	Jun	Jul	Aug	Sep	Oct	Nov	Dec	Ave.
Data Rainfall (R)	106.86	108.93	108.42	169.57	397.62	699.62	659.72	511.60	221.08	105.72	73.58	117.72	273.37
Normal 0% R	55.40	56.90	56.50	103.80	330.90	713.50	658.80	466.20	148.90	54.60	33.40	63.20	228.51
Rain decrease 10% R	48.00	49.30	49.00	90.00	286.70	618.30	570.80	403.90	129.00	47.30	28.90	54.80	198.00
Rain decrease 20% R	40.90	42.00	41.70	76.60	244.30	526.80	486.30	344.10	109.90	40.30	24.60	46.70	168.68
Rain Increase 10% R	63.10	64.70	64.30	118.20	376.70	812.30	749.90	530.70	169.50	62.20	38.00	71.90	260.13
Rain Increase 20% R	71.00	72.90	72.40	133.00	424.00	914.30	844.10	597.30	190.80	70.00	42.70	81.00	292.79

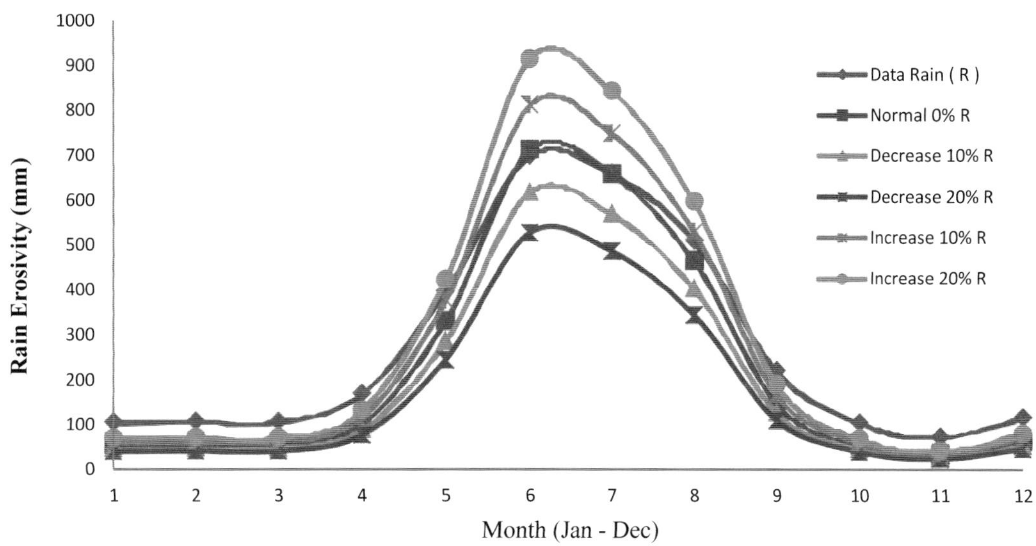

Figure 3. Graph of rain erosion (R)

Based on the table and figure above, it can be seen that the average monthly rain (years 2008-2020) follows the flood hydrograph pattern with its peak in June-July. Likewise, the rain erosivity pattern follows the monthly average rainfall pattern in a comparable manner. A decrease in rain results in a decrease in the value of rain erosivity and an increase in rain results in an increase in rain erosivity.

3.2. Erodibilitas Tanah (K)
Factors that affect soil erodibility are texture class which is sandy loam with a value of M = 1215, soil texture class in the form of medium to coarse granular (2 - 10 mm) with Code 3, percentage of organic matter with moderate value of 5% and medium permeability classification with Code 3. Then the value of soil erodibility (K) is:

$$K = \frac{\{2,71M^{1.14}(10^{-4})(12-0M)+4.20\ (s-2)+3.23(p-3)\}}{100} = 0.104299$$

3.3. Slope (LS)

By using GIS tools, the average length and slope factor (LS) of each area in the Singkoyo watershed is obtained as follows.

Table 2. The average slope of the Singkoyo watershed

Slope class (%)	Middle mean (%)	Area (Ha)	Area to watershed area	Average value
0 - 8	4	8876.9	0.25	0.98
8 - 15	11.5	3504.9	0.10	1.11
15 - 25	20	5488.2	0.15	3.03
25 - 45	35	11458.7	0.32	11.08
45 - 100	72.5	6867.3	0.19	13.76
Total area		36196.0	1.0	
Average slope				29.96
Slope factor				6.80

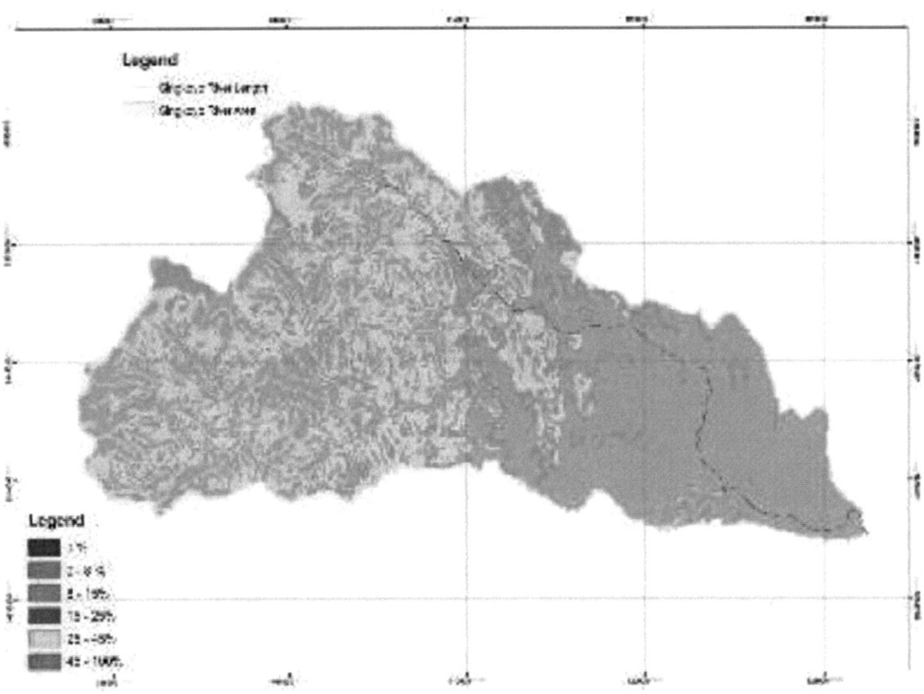

Figure 4. Slope map of Singkoyo watershed

Based on the table and figure above, it can be seen that the average slope of the Singkoyo watershed is 29.96%. This value indicates that the watershed characteristics of the Singkoyo watershed tend to be steep. The slope class in the Singkoyo watershed is divided into 5 slope classes, where the slope class described as flat (25 - 45%) dominates the Singkoyo watershed area with an area of 11458.7 ha or 32% of the total area of the Singkoyo watershed.

3.4. *Factors of land use and management (CP)*

The classification of land use and management in the Singkoyo watershed is shown in the following table and figure:

Table 3. Average CP value of the Singkoyo watershed

Slope class **(%)**	CP value	Area (Ha)	Area to watershed area	Average value
Thicket forest	0.05	728.8	0.020	0.00101
Dense forest	0.05	18219.5	0.503	0.02517
Kind of forest	0.01	1093.2	0.030	0.00030
Mixed garden	0.5	850.2	0.023	0.01174
Oil palm plantation	0.5	7287.8	0.201	0.10067
Settlement	0.95	607.3	0.017	0.01594
Ricefield	0.01	1821.9	0.050	0.00050
Bush	0.3	1579.0	0.044	0.01309
Moor	0.5	242.9	0.007	0.00336
Industrial forest	0.01	3765.4	0.104	0.00104
Total Area		36196.0	1.0	
Average CP Factor Value				0.17

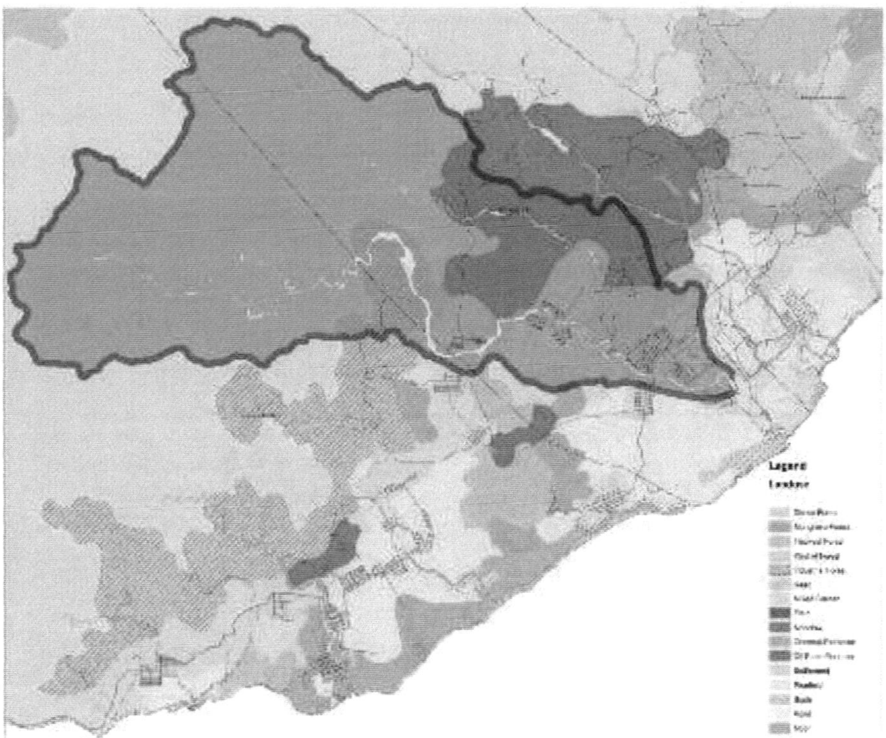

Figure 5. Singkoyo watershed land use map

The Singkoyo watershed is classified into 10 land cover categories, with the largest area being dense forest with an area of 18,219.5 ha (50.34% of the total area of the Singkoyo watershed). Land cover in the form of oil palm plantations also dominates with an area of 7,287.8 ha (20.13%). Dense forests are scattered in the upstream part of the Singkoyo watershed, while residential land cover and rice fields are scattered in the downstream part of the Singkoyo watershed.

3.5. Erosion potential
Analysis of the erosion potential of the Singkoyo watershed was obtained by multiplying the erosion factors using the USLE method, namely rain erosivity (R), soil erodibility (K), slope length and slope (LS) and soil use and management (CP). The results of the potential erosion of the Singkoyo watershed with various climate change scenarios are: for normal rainfall $A_{Normal} = 28.01$ tons/ha/year, 20% rain falls, $A_{-20\% R} = 20.68$ tons/ha/year, 10% rain falls , $A_{-10\%R} = 24.27$ tons/ha/year, rain increased by 10%, $A_{+10\%R} = 31.88$ tons/ha/year and rainfall increased by 20%, $A_{+20\%R} = 35.89$ tons/ha/ year. The erosion potential of the Singkoyo watershed with the climate change scenario used is classified as erosion class II with a low erosion rate with an erosion rate of 15-60 tons/ha/year.

The relationship between climate change scenarios and the erosion potential of the Singkoyo watershed is presented in the following figure:

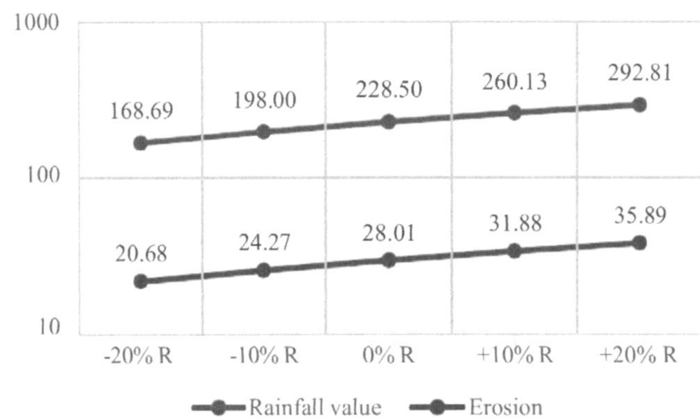

Figure 6. Graph of relationship between rain and erosion potential

Table 4. Comparison of rainfall value and erosion potential

Changes in rainfall	Rainfall value (mm)	Erosion potential (tons/ha/year)	Difference from normal condition (tons/ha/year)	Percentage of normal condition (%)
Decrease 20%	168.7	20.7	7.3	26.2
Decrease 10%	198.0	24.3	3.7	13.3
Normal	228.5	28.0	0.0	0.0
Increase 10%	260.1	31.9	3.9	13.8
Increase 20%	292.8	35.9	7.9	28.1

From the table above, it can be seen that changes in rainfall, both increasing and decreasing, greatly affect the potential for erosion. With a 20% decrease in rain, the erosion value decreases by 7.3 tons/ha/year (26.2% from normal conditions), an increase in rain by 20% causes an increase in erosion value by 7.9 tons/ha/year (28.1% from normal conditions) and next.

4. Conclusions

The result of this research is the reduction of rain by 10% and 20% resulted in the erosion potential decreased respectively by 3.7 tons/ha/year (13.3%) and 7.3 tons/ha/year (26.2%) from normal conditions. An increase in rainfall of 10% and 20% resulted in the potential for erosion to increase by 3.9 tons/ha/year (13.9%) and 7.9 tons/ha/year (28.1) from normal conditions. The conclusion of this study is that the relationship between rainfall and erosion potential is linear, meaning that an increase in rainfall causes an increase in erosion potential and vice versa. Assuming other data inputs are considered constant.

Acknowledgment

The authors would like to thank all those who have supported this research: Post Graduate Program of Tadulako University and the office Public Works and Water Resources of the Province of Central Sulawesi, Indonesia.

References

[1] Rhoshandhayani K T, M Mandala and Indarto 2019 J. Ilmu Lingkungan 17 323-332
[2] J Safitri, Deasy A, Sidharta A and K Puji H 2021 Indonesian Journal of Earth Sciences 1 17-27
[3] Marizca M R, A Binilang, E M Wuisan and F Halim 2013 J. Sipil Statik 1 309-317
[4] D Apriliyana 2015 J.Pembangunan Wilayah dan Kota 11 103-116
[5] Andi A A L, Uswah H and Ramlan 2016 J. Agrotekbis 4 633-641
[6] Wischmeier W H and Smith D D, Predicting Rainfall Erosion Losses : A Guide to Conservation Planning, vol. Handbook No. 37, USDA Agriculture, 1978.
[7] Imam B, Kemala S L and A Lubis 2015 J. Online Agroekoteknologi 3 160-167
[8] R. H. Anasiru 2015 J. Pengkajian dan Pengembangan Teknologi Pertanian 18 273-289
[9] Saida, Abdullah and M. Ilsan 2017 J. Agrotek 1 1-13
[10] Aprizon P, Triyatno, A Syarief and Dedi H 2018 J. Geograpi 10 1-13
[11] Denik S. K, I M Udiana and Melati J M 2018 J. Teknik Sipil 7 143-154
[12] Ahmad S H, Arief L N and Nurhadi B 2020 J. Geodesi Undip 9 106-114
[13] Indah A, Sri W and Rosalina S 2020 JRPB 8 1-11
[14] Rahmat H A 2015 J. Pengkajian dan Pengembangan Teknologi Pertanian 18 273-289
[15] I W Sutapa 2010 J. Smartek 8 169-181
[16] Sarah C, Cathryn E B, Marcelo V G, Edward P and Jemma D 2021 Environmental Research 16 1-13
[17] Banuwa 2013 Erosion (Jakarta, Indonesia: PT Fajar Interpratama Mandiri)

IOP Conf. Series: Earth and Environmental Science 1075 (2022) 012035 doi:10.1088/1755-1315/1075/1/012035

Perception and Decision of Public Transport Users in Palu City

R Ramlan[1*], A Setiawan [2], N Pradani[3] and J Patunrangi [4]
[1,2,3,4]Department of Civil Engineering, Engineering Faculty, Tadulako University
**Corresponding Author: ramlanratnasari@gmail.com*

Abstract. Public transport users in Palu City are decreasing every year. The low quality of public transport services is the main reason people are reluctant to use them. Based on this condition, the research was conducted to understand the user's perceptions and decisions of public transportation services, exceptionally public transport, in Palu City. The method used in this research is the distribution of questionnaires to respondents from 8 sub-districts in Palu City. Data processing on 361 respondents was carried out by descriptive analysis. The results of the study illustrate that the people of Palu City, in general, do not use public transportation (91.9%) for various reasons, including long waiting times (67.2%), unclear routes (43.8%), inadequate stopping places (32.8%), non-uniform fares (8.8), poor vehicle conditions (33.7), poor driver behaviour (7.65), and low level of safety (18.5). The level of satisfaction for users of public transportation is satisfied (31.3%) and for users of private vehicles is not satisfied (41.5%) and uncomfortable (33.7%). The data also shows that 77.3% of respondents use motorcycles in their daily travel activities. This study shows that implementing online applications and free tariffs will increase public transport users by 61.7% dan 70.3%, respectively.

Keywords: Perception, Decision, Public Transport, Descriptive analysis

1. Introduction

The problem of public transportation is a classic problem that occurs in almost all cities in the world, especially in developing cities. Starting from the low quality of service, relatively long travel times to the number of tariffs that are not appropriate so that it has an impact on the low number of public transport users in urban areas. It is not surprising that urban people tend to prefer using private vehicles over public transportation. Even the number can increase along with the increase in people's income [1]. The use of private vehicles which are considered to make daily trips easier can have a negative impact, namely; low security, causing congestion, increasing pollution [2], and providing social impacts in the form of barriers to accessibility and social inclusion in society [3].

Various efforts have been made to increase the use of public transportation which is, of course, a special concern for urban transportation practitioners. One of the public transportation policies adopted by almost all countries in the world is Transportation Demand Management (TDM). TDM is carried out by applying strategic regulations as an effort to increase public transport passengers and reduce the use of private transport [4]. Although TDM has its pros and cons [5], it can have positive impacts including increasing public transport users [6], decreasing public transport delays by around 27-35% and overcoming congestion [7]. One example of a TDM application is the application of the use of electronic road pricing (ERP) which has received positive responses from private vehicle users in Jakarta [8], or by imposing parking restrictions that can increase public transport passengers by 19% [9].

Seeing these efforts proves that the sustainability of urban public transportation is a serious problem that must be resolved to create a good transportation system. Increased demand for public transport can be realized if the service is well maintained. According to [10] sustainability; transit; land use, and technology have a strong relationship in the successful use of public transport in urban areas. This means that the role of public transportation can be achieved properly if its function as a means of accessibility can be realized if it is supported by the application of transportation technology. In developed countries, the application of technology is proven to increase transportation users, because it can save travel time, costs and reduce pollution. Research conducted by [11] reveals that public transport users in Greece have increased every year due to the high level of satisfaction with the quality of public transport services. Likewise, public transport in Germany has a better rating than the USA due to 5 factors, namely (1) quality and quantity are maintained; (2) adequate fares and convenient ticketing, (3) full intermodal and regional integration, (4) high taxes and restrictions on car use, and (5) regulated land-use policies [12].

The various efforts mentioned above can be considered in increasing the number of public transports, but first it is necessary to know the perceptions of public transportation users. Thus, this study was conducted to determine the perceptions and decisions of public transport users in Palu City. The expected results in this study can provide an overview of the perceptions and decisions of public transport users when using it in their daily lives.

2. Research Methodology

The research methodology used consists of 2 parts, namely:

2.1. Data Collection

In this section, data was collected by distributing questionnaires to the people of Palu City. During the research period, 361 questionnaires were collected which were divided into 2 groups, namely (1) users of public transportation; (2) Users of private vehicles. Public transport users in Palu city are shown in the figure below.

Figure 1. Public transport users in Palu city

IOP Conf. Series: Earth and Environmental Science 1075 (2022) 012035 doi:10.1088/1755-1315/1075/1/012035

2.2. Data Analysis
After the research data is collected, then the data is processed by descriptive analysis. Descriptive analysis is done by describing the data that has been collected.

3. Results
This research was conducted to determine the perceptions and decisions of public transport users in Palu City. To achieve this goal, data was collected by distributing questionnaires. General data of respondents can be seen in Table 1.

Table 1. Percent of Respondent, by variable.

Variable	Description	%
Gender	Male	49.6
	Female	50.4
Age (year)	10-20	5
	21-30	25
	31-40	40
	41-50	15
	51-60	10
	>60	5
Education	Senior high school/vocational school	30.2
	Bachelor	54
	Postgraduate	15.8
Total income	Rp. 1.000.000 – Rp. 4.999.000	60.1
	Rp. 5.000.000 – Rp. 10.000.000	27.4
	>Rp. 10.000.000	8.6
Driving license ownership	Nothing	19.7
	SIM A	30.2
	SIM B	1.1
	SIM C	70.1
Home address (Districts)	Palu Barat	9.1
	Mantikulore	32.1
	Palu Timur	18.6
	Palu Selatan	24.4
Destination (Districts)	Palu Barat	8.6
	Mantikulore	39.3
	Palu Timur	24.4
	Palu Selatan	14.4
	Palu Utara	8.3
Distance from home to travel destination	<1km	14.7
	1,1 km – 5 km	40.7
	5,1 km – 10 km	25.8
	>10 km	16.9
Vehicle used	Motorcycle	77.3
	Car	16.9
Purpose	Working	53.6
	Leisure and shopping	21.4
	Studies	25
Respondent	Private vehicle user's	91.1
	Public transport user's	8.9

Based on the data collected, it can be seen that there are more female respondents than men (50.4%), with the highest age range being 31-40 years (40%), the highest recent education being bachelor degree (54%). Based on the amount of income, the highest range is Rp. 1.000.000 – Rp. 4,999,000 (60.1%), the average respondent already has a SIM C (70.1%), the highest home address is in Mantikulore District (32.1%), the most travel destinations are in Mantikulore District (39.3), Distance from home to travel destination the highest is 1.1-5 km (40.7%), the majority of respondents use motorbikes (77.3%), the majority of respondents travel to work (53.6%) and 91.1% of respondents never/rarely use public transportation.

From the general data above, it can be seen that the people of Palu City have never/rarely used public transportation. Although the distance to the destination is quite close, people tend to use motorbikes. Furthermore, this study provides an analysis of the perceptions and decisions of public transport users in Palu City.

3.1. Perception of Public Transport Users

The perception of public transport users in Palu City can be assessed from several points, including:

3.1.1. Reasons People Don't Use Public Transportation

In general, the data shows that the people of Palu City prefer to use motorbikes than public transportation. To assess the perception of this reluctance, it can be measured by several reasons why people do not want to use public transportation, including:

a. Waiting time is the time used by passengers to wait for the arrival of public transport. According to the Directorate General of Land Transportation (2002), the ideal passenger waiting time (passenger waiting time) is on average 5-10 minutes, but this is different for public transport users in Palu, who have an average waiting time of > 30 minutes. This means that public transport services are very low (67.2%).
b. Route is a trajectory or route on public transportation that has a fixed origin and destination of travel. Routes can be scheduled or unscheduled. Based on the Decree of the Mayor of Palu Number 551.2/361/DISHUB/2017 concerning the Palu Angkot Route Network, six routes serve the Palu community, which are distinguished by the colour of the vehicle. In practice, public transportation that operates does not follow the established route. The lack of passengers causes public transportation to serve the community with a "door to door" system. This causes longer travel times and causes people to be reluctant to choose public transportation (43.8%).
c. Public transport stop. Is a place to raise and lower public transport passengers, which are on the public transport service network. Bus stops must be provided as supporting facilities for road traffic and transportation (UU No. 22/2009 concerning road transport traffic). In public transportation services in Palu City, there is no process of raising and lowering passengers at the bus stop. As a result, the arrival of public transport becomes uncertain, so the waiting time becomes long. This is the reason why people are reluctant to choose public transportation (32.8%)
d. Tariffs are costs that must be paid by passengers when receiving public transportation services. The applicable fare is differentiated by the route. In practice, public transportation often asks for higher fares because it serves outside the route. This applies according to the negotiations between the driver and the passenger. The lack of clarity on the amount of the fare is the reason why people are reluctant to use public transportation (8.8%)
e. Vehicle Condition is the ability of the vehicle from the engine to the appearance of the body of public transportation. The condition of public transportation must always be checked periodically and pass the KIR test. The condition of public vehicles is one of the reasons why people are reluctant to use public transportation (33.7)
f. Driver behaviour is the ability of the driver when running the vehicle. The assessment of the behaviour of less friendly drivers (8.2%), reckless (6.7) is the cause of people's reluctance to use public transportation
g. Security is an important factor in public transport services. The community considers that safety when using public transportation is less guaranteed (18.5%).

3.1.2. Assessment of Public Transport Services
To assess public transportation services in Palu City, they are divided into users of public transportation and users of private vehicles (people who have never used public transportation). Assessment of public transport services can be seen in Figure 2 below:

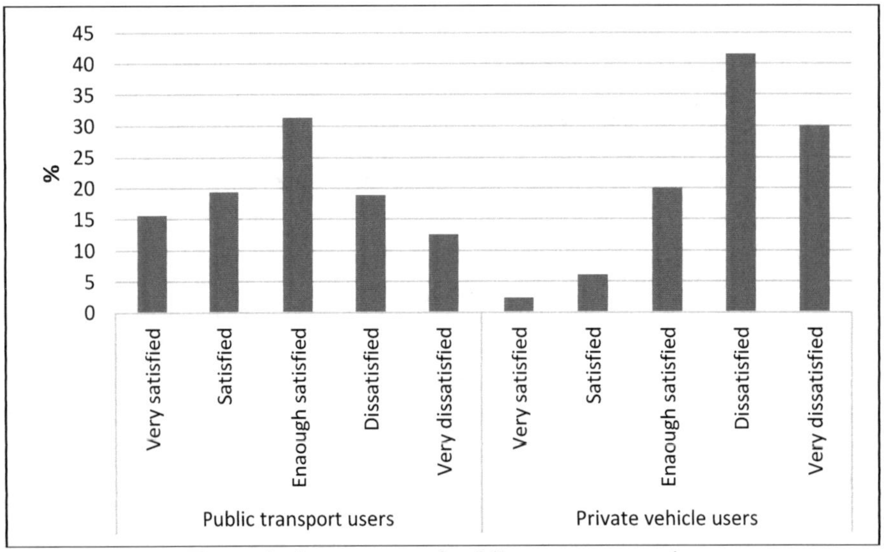

Figure 2. Assessment of public transport service

Based on the figure above, it shows that public transport users are quite satisfied (31.3%) when they become public transport passengers. This group of public transport users is captive users because they have no other choice in fulfilling their movements. Captive users are caused by not having a personal vehicle (53.1%) and adequate tariffs (21.9%). The low level of satisfaction from captive users causes the number to continue to decline over time.
People who use private vehicles are included in the choice user, meaning they can choose to use private vehicles or public transportation. This group rated public transportation services as dissatisfied (41.5%) and very dissatisfied (30%).

3.2. Decisions of Public Transport Users
Judging from the perceptions of public transport users described earlier, it is necessary to improve the performance of public transport. To improve the public's decision to use public transportation, two alternatives for improving public transportation services are used as shown in Table 2.

Table 2. Description of decisions.

Variable	Description
Implementing online applications	Passengers will use public transportation when the online application is implemented
Improved public transport service	Passengers will switch to using public transportation when service improvements are made

3.2.1. Implementing Online Applications
The first alternative proposed is to implement an online application for public transportation services. The implementation of online applications on public transportation services is being studied by the Palu City government, with the hope that there will be an increase in the number of public transport users. The data states that if an online application is implemented on public transportation services, then there are 61.7% of people who switch to using public transportation

3.2.2. Improved Public Transport Services

The second alternative proposed is the improvement of public transportation services. If improvements are made in public transportation services, including definite waiting time (7%); clear route (52%); provision of shelters (19%) and increased security (22%).

4. Conclusions

Based on the data analysis in this study, it can be concluded several points, namely:

a. Based on the results of this study, it is expected the characteristics of public transport users in Palu City are dominated by women, using motorbikes and the average daily travel distance is 1.1 – 5 km.
b. Perceptions of public transport users can be measured based on the causes of reluctance to use public transportation, which is assessed from long waiting times (667.2), unclear routes (43.8), low use of bus stops (32.8), non-uniform fares (8.8), poor condition of vehicles. poor (33.7), poor driver behaviour (7.65), and low level of safety (18.5). The level of satisfaction for users of public transportation is satisfied (31.3%) and for users of private vehicles is not satisfied (41.5%)
c. An increase in the number of decisions of public transport users is expected to occur with the implementation of online applications for public transport (61.7%) and improvement of public transport services in the form of definite waiting times (7%); clear route (52%); provision of shelters (19%) and increased security (22%).

To be input to the Palu City Government to increase the number of public transport users. Further research can analyze in more detail the willingness and ability of the people of Palu City as users of public transportation, and can also examine the perceptions of the operators so that public transportation services can be further improved.

Acknowledgments

This research was funded by the Program Kompetisi Kampus Merdeka (PKKM) 2021 at the Civil Engineering Undergraduate, Faculty of Engineering, Tadulako University.

5. References

[1] Holmgren J 2013 The efficiency of public transport operations–An evaluation using stochastic frontier analysis. *Research in Transportation Economics* 39(1) p 50-57.
[2] Schmöcker, Jan-Dirk, Michael G B and William HK La 2004 Importance of public transport *Journal of advanced transportation* p 1-4.
[3] Berg J and Ihlström J 2019 The importance of public transport for mobility and everyday activities among rural residents *Social Sciences* 8(2) p 58.
[4] Habibian M and Kermanshah M 2011 Exploring the role of transportation demand management policies' interactions *Scientia Iranica* 18(5) p 1037-1044.
[5] Coutinho F M, van O N, Christoforou Z, Alonso-González M J, Cats O and Hoogendoorn S 2020 Impacts of replacing a fixed public transport line by a demand responsive transport system: Case study of a rural area in Amsterdam *Research in Transportation Economics* 83 p 100910.
[6] Dirgahayani P, and Sutanto H 2020 The effect of transport demand management policy on the intention to use public transport: A case in Bandung, Indonesia *Case Studies on Transport Policy* 8(3) p 1062-1072.
[7] Giuliano G 1992 Transportation Demand Management: Promise or Panacea? *Journal of the American Planning Association* 58:3 p 327-335,
[8] Sunitiyoso Y, Nuraeni S, Inayati T, Hadiansyah F, Nurdayat I F, Pambudi N F 2020 Road Pricing in Indonesia: How Will Public Respond? *Transportation Research Procedia* 47 p 123-130.
[9] Rotaris L and Danielis R 2014 The impact of transportation demand management policies on commuting to college facilities: A case study at the University of Trieste Italy *Transportation Research Part A: Policy and Practice* 67 p 127-140.
[10] Vuchic V R 2002 Urban public transportation systems *University of Pennsylvania Philadelphia PA USA* 5 p 2532-2558.
[11] Efthymiou D and Antoniou C 2017 Understanding the effects of economic crisis on public transport users' satisfaction and demand *Transport Policy* 53 89-97

[12] Buehler R and Pucher J 2012 Demand for public transport in Germany and the USA: an analysis of rider characteristics *Transport Reviews* 32(5) p 541-567.

Important factors influencing the successful implementation of electronic tenders based on the perceptions of consultants, contractors and project owners in Palu

Donny M. Mangitung[1], **Nirmalawati**[2], **Andi H. Asikin**[3], **Gitalia R Ningsih**[4], **Musdalifah**[5], **Suci S. Ningrum**[6]

[1,2,3] Senior Lecturers, [4,5,6] Undergraduate Students at Civil Engineering Department, Faculty of Engineering, University of Tadulako, Palu
The corresponding author's e-mail address: donny.mangitung@gmail.com

Abstract. The implementation of e-tenders requires readiness from the aspect of human resources, technological and management capabilities and organizations. To find out the factors that influence the success of the e-tender for construction services in Palu, this research was conducted by involving stakeholders involved in the implementation of the e-tendering, namely consultants, contractors and project owners. Data were collected on the basis of questionnaire survey using Likert scale of 6 and 28 Critical Success Factors (CSFs) were used in the questionnaire survey. The total number of questionnaires distributed was 163, of which 119 were returned, giving a 71% response rate. Data were analyzed using Relative Rank Index (RRI) technique and the Spearman Rank Order Correlation statistical test among three categorical samples. The results show that there is no difference among RRIs' consultants, contractors and project owners. It is also concluded that the successful implementation of electronic tenders is influenced by factors related to Availability of reliable, affordable, and fast internet services, Availability of reliable ICT infrastructures and computer equipment, Building positive workplace and knowledge sharing culture and also Regular education and training to develop the competence and capability of electronic tendering procedures and requirement

Keywords: e-tendering, consultants, contractors and project owners, Relative Rank Index

1. Introduction

The progress of services and the availability of Information and Communication Technology Infrastructure (ICT) such as electronic communication and business activity services (e-activities or e-businesses) also affect the procurement system for construction services, especially in the tendering process for selecting contractors or consultants. Using electronic tendering or known as e-tendering will improve business communication processes and proactive management and can increase the selection of service providers in a transparent, effective, efficient, open, competitive, fair/non-discriminatory, and accountable selections processes [1-4].

However, the implementation of electronic tenders requires readiness from the aspects of human resources, technological capabilities and organization and management. The results of previous research [5] indicate that aspects of technological capabilities such as availability of high-speed internet, availability of licensed software and availability of hardware; then aspects of human resources, namely providing training and education to employees and availability of human resource;

as well as management and organizational aspects related to support from the top management, ability to print and submit e-documents as court evidence and possibility of providing feedback to ensure the confidentiality of bids, being the important factors for the success of the electronic tender implementation based on the contractors' and consultants' perceptions.

It is also similar to the results of research conducted by [6] that the critical factors for the success of the implementation of electronic tenders from Rank 1 to 5 subsequently is the reliability of the performance of electronic tender infrastructure services, employees who have sufficient knowledge in serving the electronic tender system, knowledge of government regulations concerning electronic tenders by sufficient parties, reliable or uninterrupted webpages, uninterrupted electricity supply and sufficient knowledge involved in electronic tenders.

Because electronic tendering is important to select consultants or contractors, research is necessary to conduct in order to answers the question of what factors can affect the success of electronic tendering system based on perceptions consultants, contractors as well as project owners in the city of Palu.

2. Review of literature

According to the President Regulation related to construction procurements and goods No. 116-2018 the following principles must be met to carry out e-tendering process [3]:

a. Efficient, that is to produce maximum quality and target performance with minimum use of resources;
b. Effective, namely the goal of achieving the right and broad targets to get the maximum results;
c. Transparent, i.e. all provisions and information are spread as widely as possible and can be clearly known to the public;
d. Open, which can be followed by anyone as long as they meet certain requirements/criteria based on clear provisions and procedures;
e. Competitive, namely competition is carried out in a healthy manner and is followed by all prospective service providers who are equal and meet the requirements without any intervention so as to create competitive market competition;
f. Fair/non-discriminatory, i.e. prospective service providers receive the same treatment and are not directed to provide benefits to certain parties, and still pay attention to national interests; and
g. Accountable, namely the implementation must meet the rules and regulations and the results can be accounted for.

Furthermore, as previously mentioned in the Introduction section, there are 3 (three) important aspects to get a smooth path in the implementation of electronic tendering system, namely aspects of human resources, technological capabilities and organization and management. Of the three aspects identified based on previous research, there are 10 factors that are important in relation to aspects of human resources; then 10 factors concerning aspects of technological capability; and the last 8 factors related to management and organizational aspects. Details Aspects of the success of electronic tendering system related to the 28 Critical Success Factors (CSF) as seen in Table 1. These factors are used in this study to identify which factors critically affecting the success of electronic tendering system.

3. Research methods

To identify important factors that influence the success of electronic tenders, data collection using a questionnaire instrument. The survey questionnaire was created based on the 28 CSF discussed in the previous section and distributed to prospective respondents based on a sample of service users from local and central government agencies in the city of Palu, Central Sulawesi Province being involved in the implementation of electronic tendering systems. Then, the questionnaires were also distributed to samples of consultants and contractors in Palu. Prospective respondents will answer by choosing which factors are important based on a Likert scale from 1 to 6 (from strongly disagree to strongly agree).

Data were collected on the basis of questionnaire survey. The total number of questionnaires distributed was 163, of which 119 were returned, giving a 71% response rate. The respondents comprised of 51 consultants , 35 contractors and 33 project owners with the following percentage response rates respectively (64%), (70%) and (100%).

Table 1. Critical Success Factors (**CSF**) of e-Tendering System [6-12]

No.	Critical Success Factors (CSF)
	Human Resources
1	Knowledge of the benefits and objectives of electronic tendering implementations
2	Knowledge of the Procurement Service Unit (PSU) in electronic tendering procedures
3	Knowledge of the electronic tendering regulations
4	Implementation of the electronic tendering system in accordance with government regulations
5	Knowledge Technology, Information, and Communication (ICT) related to electronic tendering implementations
6	Knowledge of the usage of standard application of electronic tendering systems
7	Regular education and training to develop the competence and capability of electronic tenderi procedures and requirements
8	The availability of policies and laws for supporting electronic tendering implementation
F9	Effective work strategies carried out by the project team
10	Employee commitment and trust in the successful implementation of electronic tendering system
	Technological Capabilities
11	Availability of reliable ICT infrastructures and computer equipment
12	Availability of reliable, affordable, and fast internet services
13	Reliability and scalability of web-based systems
14	Reliable web-based security systems
15	The usage of reliable security applications
16	Availability of reliable power backup systems
17	The technology standard and readiness platform in electronic tendering systems
18	Availability of skilled personnel to handle electronic tendering tool, processes and problems
19	Reliable security and authentication of electronic tendering transactions
F20	Implementation of validity of electronic transactions
	Organization and Management
21	Top management commitment and support to electronic tendering adoption
22	Experienced senior manager leadership support to electronic tendering systems
23	Free from practices of corruption, collusion and nepotism in electronic tendering processes
24	Strategic development in the implementation of electronic tendering system
25	Redesign or reengineering of electronic tendering processes
26	Implementation of employee performance appraisal for the organizational progress
27	Implementation of project management best practices
28	Building positive workplace and knowledge sharing culture

Furthermore, the results of data collection through the questionnaire survey was analyzed using the Rank Relative Index (RRI) method and The equation of RRI used in this paper shown in equation (1)[5]:

$$W = (\sum w/A * N) * 100\% \tag{1}$$

where:
w = weight given to each factor by respondent;
A = highest weight; and
N = total number of respondents.

Since the results of RRI consist of three groups of respondents, namely the groups of contractors, consultants and project owners, it is necessary to compare whether the RRI of those three groups significantly related based on statistically analysis of Spearman Rank Order correlation analysis. If these correlations are significant, the RRI can be combined into one group of RRI and easier to interpret the important CSF [13-15].

4. Results and Discussion

The results of RRI for each respondent group, namely, consultants, contractors as providers of construction services and project owners as users of construction services as well as combined data of the three groups (RRI overall), can be seen in Figure 1, 2 and 3. Furthermore, the Spearman Rank Order correlations between three groups of respondents (RRI1, RRI2 and RRI3) are significantly as seen Figure 4. Thus, all RRI's can be analyzed based on the RRI overall.
The top 5 of RRI's overall indicate that 2 CSF include in the Human Resources main factor. Those are *Knowledge of the benefits and objectives of electronic tendering implementations* (F1, Rank 3) and *Regular education and training to develop the competence and capability of electronic tendering procedures and requirements* (F7, Rank 5). This findings are similar to previous research [5, 6].

Code	Critical Success Factor (CSF)	G1 (n=51)		G2 (n=35)		G3 (n=33)		Overall (n=119)	
		RRI	Rank	RRI	Rank	RRI	Rank	RRI	Rank
F1	Knowledge of the benefits and objectives of electronic tendering implementations	92.2%	3	89.0%	4	87.4%	3	89.9%	3
F2	Knowledge of the Procurement Service Unit (PSU) in electronic tendering procedures	79.1%	23	82.4%	21	80.3%	14	80.4%	21
F3	Knowledge of the electronic tendering regulations	88.6%	8	85.2%	15	85.4%	4	86.7%	8
F4	Implementation of the electronic tendering system in accordance with government regulations	85.3%	16	81.0%	23	60.1%	27	81.5%	20
F5	Knowledge Technology, Information, and Communication (ICT) related to electronic tendering implementations	90.5%	6	89.0%	4	83.8%	7	87.7%	6
F6	Knowledge of the usage of standard aplication of electronic tendering systems	86.3%	13	85.7%	13	84.3%	6	85.6%	10
F7	Regular education and training to develop the competence and capability of electronic tendering procedures and requirements	91.2%	5	87.6%	8	82.8%	8	87.8%	5
F8	The availability of policies and laws for supporting electronic tendering implementation	86.9%	10	86.2%	12	80.3%	14	85.2%	12
F9	Effective work strategies carried out by the project team	77.1%	25	81.9%	22	40.9%	28	77.9%	25
F10	Employee commitment and trust in the successful implementation of electronic tendering system	71.6%	28	80.0%	24	77.8%	21	75.8%	28

Note: G1, G2, G3:Group 1, 2 &3; n: number of respondents

Figure 1. CSF- Main Factor of **Human Resources**

Code	Critical Success Factor (CSF)	G1 (n=51)		G2 (n=35)		G3 (n=33)		Overall (n=119)	
		RRI	Rank	RRI	Rank	RRI	Rank	RRI	Rank
F11	Availability of reliable ICT infrastructures and computer equipments	95.4%	2	91.9%	1	87.9%	2	92.3%	2
F12	Availability of reliable, affordable, and fast internet services	97.7%	1	91.0%	2	88.4%	1	93.7%	1
F13	Reliability and scalability of web-based systems	83.3%	18	73.3%	27	76.8%	24	81.7%	18
F14	Reliable web-based security systems	81.0%	21	83.8%	16	75.3%	25	80.4%	21
F15	The usage of reliable security applications	81.4%	20	82.9%	19	81.8%	10	82.1%	17
F16	Availability of reliable power backup systems	85.6%	15	83.3%	18	81.3%	11	83.8%	16
F17	The technology standard and readiness platform in electronic tendering systems	86.6%	11	85.7%	13	78.8%	19	85.0%	13
F18	Availability of skilled personnel to handle electronic tendering tool, processes and problems	72.9%	27	70.5%	28	80.8%	12	76.9%	27
F19	Reliable security and authentication of electronic tendering transactions	87.6%	9	88.1%	7	84.8%	5	87.4%	7
F20	Implementation of validity of electronic transactions	85.9%	14	86.7%	11	79.8%	16	85.3%	11

Note: G1, G2, G3:Group 1, 2 &3; n: number of respondents;

Figure 2. CSF- Main Factor of **Technological Capabilities**

The first empirical evidences is the most CSF , namely *Availability of reliable, affordable, and fast internet services* (F12, Rank 1) and the second one is *Availability of reliable ICT infrastructures and computer equipment.* It means ICT infrastructure and equipment (F11, Rank 2) may reduce tendering time process, so that it can be streamline business process as well as increase competitiveness and opportunities. These findings re similar to previous study [1, 5, 7].

Code	Critical Success Factor (CSF)	G1 (n=51)		G2 (n=35)		G3 (n=33)		Overall (n=119)	
		RRI	Rank	RRI	Rank	RRI	Rank	RRI	Rank
F21	Top management commitment and support to electronic tendering adoption	83.3%	18	82.9%	19	65.2%	26	81.7%	18
F22	Experienced senior manager leadership support to electronic tendering systems	90.2%	7	87.6%	8	79.3%	18	86.4%	9
F23	Free from practices of corruption, collusion and nepotism in electronic tendering processes	77.8%	24	83.8%	16	79.8%	16	80.1%	23
F24	Strategic development in the implementation of electronic tendering system	85.0%	17	87.1%	10	78.8%	19	84.3%	15
F25	Redesign or reengineering of electronic procurement processes	79.7%	22	78.1%	26	77.3%	22	78.6%	24
F26	Implementation of employee performance appraisal for the organizational progress	86.6%	11	89.0%	4	77.3%	22	84.7%	14
F27	Implementation of project management best practices	74.2%	26	80.0%	24	80.8%	12	77.7%	26
F28	Building positive workplace and knowledge sharing cultures	91.8%	4	90.5%	3	82.3%	9	88.8%	4

Note: G1, G2, G3:Group 1, 2 &3; n: number of respondents;

Figure 3. CSF- Main Factor of **Organization and Management**

Building positive workplace and knowledge sharing culture (F28), Rank 4) this CSF can be effective implemented if top management commitment and availability of good leadership as well as experience manager who always share the knowledge [6].

Correlations

			RRI1	RRI2	RRI3
Spearman's rho	RRI1	Correlation Coefficient	1.000	.880**	.624**
		Sig. (2-tailed)	.	.000	.000
		N	28	28	28
	RRI2	Correlation Coefficient	.880**	1.000	.574**
		Sig. (2-tailed)	.000	.	.001
		N	28	28	28
	RRI3	Correlation Coefficient	.624**	.574**	1.000
		Sig. (2-tailed)	.000	.001	.
		N	28	28	28

**. Correlation is significant at the 0.01 level (2-tailed).

Figure 4. Spearman Rank Order Correlation of Group 1, 2 &3 (RRI)

5. Conclusions and recommendations

It can be concluded in this empirical study that RRIs among three groups of respondents (contractors, consultants and project owners) are similar rank order. Furthermore, critical factor success of electronic tendering system based on perception combined from contractors, consultants and project owners are Availability of reliable, affordable, and fast internet services (F12, Rank 1) Availability of reliable ICT infrastructures and computer equipment (F11, Rank 2), Knowledge of the benefits and objectives of electronic tendering implementations (F1, Rank 3) Building positive workplace and knowledge sharing culture (F28, Rank 4) and Regular education and training to develop the competence and capability of electronic tendering procedures and requirements (F7, Rank 5).

This study can be analyzed more advanced by using factor analysis method in order to consolidate the CSF on the basis multivariate correlation and findings new critical factors.

Acknowledgments

The authors gratefully acknowledge financial support from Faculty of Engineering, University of Tadulako with Contract Number: 433.aj./UN28.2/PL/2021 and the signed date of 26 April 2021.

References

[1] Kajewski S and A. Weippert *e-Tendering: Benefits, Challenges and Recommendations for Practice* in *CRCCI International Conference Clients Driving Innovation* 2004 Surfers Paradise, Australia: QUT-ePrints

[2] Balogun, A Talukhaba A A and Opaleye O *Effectiveness of electronic Tendering for Construction Projects*. in *9th CIDB Postgraduate Conference*. 2016. Cape Town, South Africa

[3] Peppres No 16, *Pengadaan barang/ jasa pemerintah* 2018 Sekretaris Negara Republik Indonesia, Lembaran Negara RI Tahun 2018 No. 33

[4] Rahim M M *Identifying Factors Affecting Acceptance of E-procurement Systems: An Initial Qualitative Study at an Australian City Council* Communications of the IBIMA 2008 **3**

[5] Wimalasena N N and Gunatilake S *The readiness of construction contractors and consultants to adopt e-tendering The case of Sri Lanka*, in *Construction Innovation* 2018 350-370

[6] Mangitung D M and Novitasari D E *Factors affecting e-tendering process based on the perceptions of the contractors being involved in the Tadulako University's construction projects.*

in *The 6th International Conference on Engineering, Project, and Production Management* 2015 Griffith School of Engineering, Griffith University, Gold Coast, Queensland, Australia

[7] Afolabi A *Critical success factors (CSFs) for e-procurement adoption in the Nigerian construction industry.* Buildings, 2019 **9** (2) 1-18

[8] Dewi N I S Sujana E and Prayudi M A *Pengaruh Faktor-Faktor Keberhasilan Pada Implementasi E-Procurement Terhadap Pencegahan Fraud (Studi Pada Pemerintah Kabupaten Badung)* Jurnal Mahasiswa Akuntasi Undiksha 2017 **8** (2)

[9] Alshibly H Chiong R and Y Bao *Investigating the Critical Success Factors for Implementing Electronic Document Management Systems in Governments: Evidence From Jordan.* Information Systems Management 2016 **33** (4) 287-301

[10] Lou E C W and Alshawi M *Critical success factors for e-tendering implementation in construction collaborative environments: people and process issues* Electronic Journal of Information Technology in Construction 2009 **14** 98-109

[11] Prasetyo A *Critical Success Factors of Electronic Procurement Implementation in The Ministry of National Development Planning/Bappenas Republic of Indonesia.* Jurnal Perencanaan Pembangunan:(The Indonesian Journal of Development Planning) 2019 **3** (1) 122-142

[12] Vaidya K A Sajeev S M and Callender G *Critical factors that influence e-procurement implementation success in the public sector.* Journal of Public Procurement 2006 **6** (1 & 3) 70-99

[13] Hair Jr J F *Multivariate data analysis.* 7th ed. 2008, New Jersey, USA: Prentice Hall

[14] Bryman A and Cramer D *Quantitative Data Analysis with SPSS 14, 15 and 16: A Guide for Social Scientists* 2008 Hove, East Sussex UK: Routledge

[15] Mangitung D M and Emsley M W *The characteristics of periodic and project prequalification practices in the UK.* in *The 10th International CIB Symposium of the W65 Commission on Organisation and Management of Construction: Construction Innovation and Global Competitiveness* 2002 University of Cincinnati Ohio USA: CRC Press USA

Economic Feasibility Study Biforcation Development and Toili River Flood Control Banggai District

Nirmalawati[1,*], **I W Sutapa** [2],
[1,2]*Department of Civil Engineering, Engineering Faculty, Tadulako University*
[5]*Postgraduate Civil Engineering Master's Program at Tadulako University*
*Corresponding Author: nirmalawati_she@yahoo.co.id

Abstract. The Biforcation Structure which is located at the branching of the Moilong and Mansahang Rivers, Banggai District functions as a weir at the Moilong Free Intake and distributes water proportionally to its downstream, so that it can function as a flood control structure and at the same time supply water to the Mansahang(Toili) Weir. The flood incident on June 18, 2019 resulted in the collapse of the Biforcation structure so that it submerged seven villages in Moilong District. The restructure of the Biforcation is very much needed by the community to provide a sense of security from the destructive power of water and irrigation water supply. To prepare for the development, a research was conducted with the aim of knowing the economic feasibility of the construction of the Biforcation and the flood conrol structure. Prior to construction, a study was conducted with the aim of determining the economic feasibility of the Biforcation construction and the flood control structure.The research method used analytical methods, quantitative approaches, and conducted interviews with the relevant agencies. The data obtained are reviewed and analyzed using several assumptions of loan interest rates, and loan times.Calculation of financial analysis using the calculation of Net Present Value (NPV) and Benefit Cost Ratio (BCR). The results of the study concluded that the construction ldingof Biforcation structures and flood control structures was feasible. The results of the Net Present Value (NV) are positive at 10% and 12% interest rates. Interest 10%, benefit Cost Ratio (BCR) 6,536 (12 years), BCR 5,675 (10 years), BCR 4,633 (8 years). Interest 12%, Benefit Cost Ratio (BCR) 5,798 (12 years), Benefit Cost Ratio (BCR) 5,097 (10 years) Benefit Cost Ratio (BCR) 4,219 (8 years).
Keywords: *Biforkasi, feasibility analysis, Net Present Value (NPV), Benefit Cost Ratio (BCR)*

1. Introduction

The Biforcation Structure, which is located at the branching of the Moilong and Mansahang Rivers, Banggai Regency functions as a weir on the Moilong free intake and distributes water proportionally to its downstream so that it can function as a flood control structure and at the same time supply water to the Mansahang/Toili Weir. Rivers as a source of water are very useful in meeting the needs of the community, for example for agricultural irrigation, drinking water raw materials, as a drainage channel for rainwater, and even as a tourist attraction. In the upstream part, the condition of the river is generally influenced by the condition of the watershed, especially the density of forest vegetation, while in the middle part it is heavily influenced by agricultural cultivation and some settlements. Likewise, the Toili River, which is one of the rivers in the Bongka-Mentawa River area, under the authority of the Central Sulawesi Provincial government, has the characteristics of a river branching so that the flow is divided into two directions, namely, the Toili River/Moilong River on the left and the Mansahang River on the right. Before the river forks (±100 m) there is a free intake to irrigate the Moilong Irrigation area. While on the Mansahang River there is a Toili Weir to irrigate an area of 2410 ha. The flood incident on June 18, 2019 resulted in the collapse/damage of the biforcation structure at the river tributary, thus submerging 7 (seven) villages in Moilong District, and in 2020 there was another flood and submerged several villages in Moilong District. This incident resulted in the collapse of this biforcation structure, so that no water flowed into the Mansahang River, which resulted in the disruption of the Toili irrigation network system. Of course, this results in the community's crops failing. In connection with the above conditions, the Department of Human Settlements and Water Resources of Central Sulawesi Province for Rivers, Beaches, Lakes and Raw Water (SPDAB) in the 2021 Fiscal Year plans to construct a

Biforcation structure which is located at the Moilong and Mansahang River ramifications, Banggai Regency. Weir on the Moilong free intake and distributes water proportionally to the downstream so that it can function as a flood control structure and at the same time supply water to the Mansahang/Toili weir. There are several studies on the feasibility study, including research conducted by Ahamed, Md Shamim Guo etc. which concludes from the calculation of the economic feasibility analysis that year-round vegetable production in a greenhouse in remote northern Saskatchewan will be economically profitable.[1] Also in research conducted by I Made Letrah & I Nengah, the results of research on financial analysis on Pulu weir, the prevailing loan interest rate obtained a positive Net Present Value (NPV), Benefit Cost Ratio (BCR) greater than one and the Internal Rate of Return (iRR) is greater than the prevailing loan interest rate. Thus it can be stated that the development of Pulu Weir is feasible from a financial aspect.[2]. From the various descriptions above, a feasibility study was carried out, with the aim of knowing the economic feasibility of constructing biforcation structures and water structures supporting the location of the Manshang river, Banggai district..

2. Material and Methods

2.1. Description of study

The research location is in Toili District and Moilong District, Banggai Regency, Central Sulawesi Province.

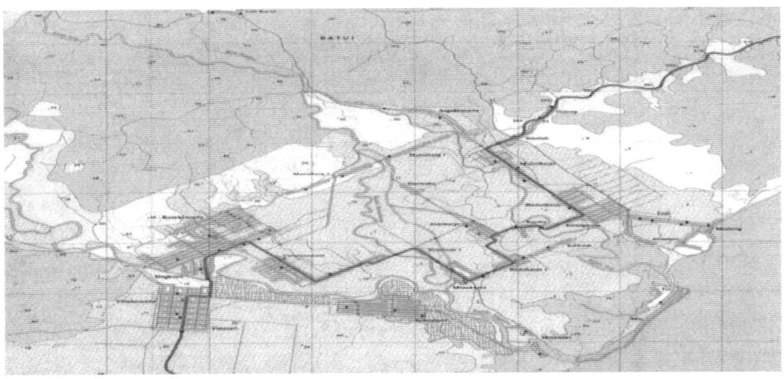

Figure 1. Research location

2.2. Literature review

The concept of engineering economics is a study of the analysis of an investment technique, in making decisions in making alternative benefits for the community or company. This aspect of economic performance requires several things to be done, namely (1) estimated current costs; (2) estimated operational and maintenance costs required in the coming years; (3) estimated residual value or benefits that will be obtained in the coming year; (4) estimated duration of profitable investment; and (5) the estimated interest rate used. Evaluating the feasibility of an investment, namely the technique in making decisions is always looking for some of the best alternatives. Because in the calculation of investment, it is necessary to pay attention to the economic age, the interest rate which still contains uncertainty and the risk that profit oriented problems will occur. Both elements contain the value of costs (costs) and benefits (Benefits) with investment criteria viewed from a different point of view [3].

2.3. Cost

The cost of an investment consists of three things, namely:

(1) Investment Cost (cost)

Investment costs are all costs incurred in the implementation of construction development, including initial study costs, planning costs, implementation costs, construction supervision costs and other unexpected costs in the implementation of construction activities [4]

(2) Operational Cost
Operational costs are all expenses that are routine or periodic in a certain time, for example the cost of salaries of employees/employees, routine costs every time, administrative costs and other unexpected costs.[4]
(3) Maintenance Fee
Maintenance costs are costs incurred routinely or certain that are used in the maintenance of existing structure construction.[4]

2.4. Benefit

After the project is built and starts operating, the benefits will begin to be obtained by the owner (investor) or by the community if the project is financed by the government. Benefits consist of selling crops or other uses as a result of the construction of the project. In general, the benefits can be divided into 2 (two) main parts, namely: (1) Tangable Benefit is a benefit that is directly enjoyed by the owner or the community in the form of; yields of rice or secondary crops; (2) Intangable Benefit is a benefit that is enjoyed indirectly by the owner or the community in the form of increasing human resources, improving living standards, rural economy or satisfaction that cannot be measured by a certain instrument. So it can be concluded that the benefits are the results of profits or benefits received due to the operational activities of construction development. The benefit in this case is the result of the community's harvest income obtained from the increasing sales of agricultural products

2.5. Economic feasibility of the project

The use of technical economic analysis is a method used to determine whether or not a project is feasible to implement using investment criteria [5]. Several criteria in assessing the economic feasibility of a project, so that the project will get promising benefits during the project's service life, these criteria are Net Present Word (NPW), Benefit to Cost Ratio (B/C), and Interest Rate of Return (IRR).[6][7][8]

2.6. Net present value method

The Net Present Word method is a method used to calculate the difference between the current value of an investment and future net cash receipts at a certain discount rate. NPV shows the advantages of benefits compared to costs [3]. The annual cash flow will be reflected on the balance sheet of the project cash book, where all costs (capital cost, annual cost) will be multiplied by discounted interest and offset by the project's annual benefit. The difference between Present Value Benefit (PV Benefit) and present Value Cost (PV Cost) is the value of the Net Present Word. Analysis of the assessment of this method is when a positive NPV is declared feasible and if a negative NPV is declared not feasible [4][9][10]

$NPV = \Sigma n\ Bt - Ct$ [11]..(1)
where:
Bt : benefit in year t
Ct : costs / expenses in year t

2.7. Benefit cost ratio

To see the feasibility by analyzing the cost benefits (benefit cost ratio) by estimating the benefits of the project by looking at the various costs used in the project. So the value of the benefit cost ratio as a reference in making decisions. This analysis is carried out by comparing the initial development costs, operational costs and maintenance costs with the benefits to be obtained. The condition for the economic feasibility of the project is the BCR value > 1. [12].

$BCR = Benefit / Cost$ [11]..(2)
where:
Bt : benefit in year t
Ct : expenses/expenses in year t

2.8. Metode rate of return (ROR)

The Rate of Return (ROR) method is a method of calculating the balance between all expenditures and all income derived from investment, where this balance occurs at a certain interest rate. the Net Present Word (NPW) value of an investment is zero. It is called the internal rate of return (IRR) if it is assumed that every result obtained is directly invested. Investment returns are said to be feasible if the resulting IRR is greater than or equal to the MARR (Minimum Attractive Rate of Return). MARR is the minimum value of the acceptable rate of return or interest. To obtain this IRR value, it is required that the value of NPW = 0, or PV Benefit = PV Cost, or BCR = 1[4]

$$IRR = I' + \frac{NPV'}{NPV' - NPV''}(I'' - I') \quad [11] \ldots\ldots (3)$$

where:

I' : The interest rate gives a positive NPV value

I'' : The interest rate gives a negative NPV value

NPV' : Positive NPV

NPV'' : Negative NPV

2.9. Methods

This research includes quantitative descriptive, literature study. This is a descriptive study, because researchers are trying to obtain information regarding the phenomena being observed at this time. [13], which describes the economic feasibility of an investment in the construction of biforcation structures and water structures supporting the location of the Manshang river in Toili and Moilong sub-districts, Banggai district, Central Sulawesi province. The data required are primary and secondary data. Primary data obtained from the field and also data from survey results. While secondary data is data obtained from sources from other parties, namely data on community conditions, community income, as well as some data from journals, related literature. The research step, after primary and secondary data is obtained about the construction of biforcation structures and water structures supporting the location of the Manshang River, Banggai Regency, is used to analyze the calculation of the economic feasibility study, namely calculating the investment cost (cost) which consists of initial investment costs, operational costs and maintenance costs. . Furthermore, the calculation of the benefits obtained by the construction of the construction. The steps in conducting the research can be seen in the research flow chart below:

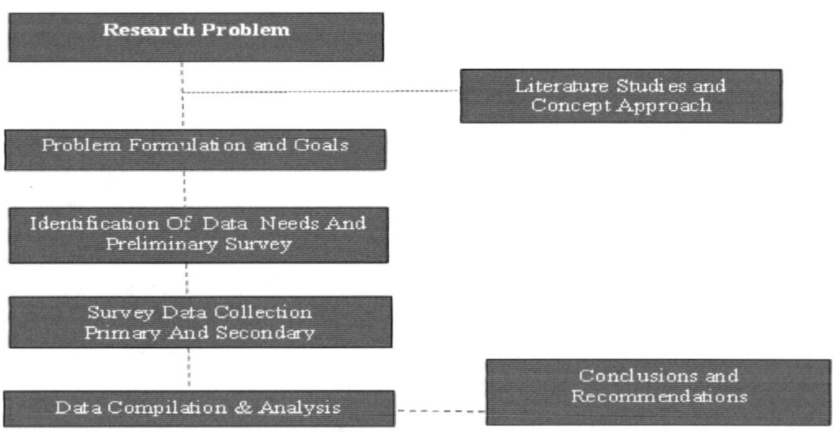

Figure 2. Flowchart research

3. Results and Discussion

Performing a project investment analysis is to perform calculations in order to obtain the best choice in the context of a business to make a capital investment. The discussion here is to evaluate to determine the results of various alternatives by comparing the flow of costs with the expected benefits of each alternative for now and in the future. [14] [15]. Some of the things reviewed are:

3.1. *Investment cost*

The investment cost of the feasibility evaluation in terms of the price of the construction structure is a total of Rp. 56.803.000.000,-. This price includes the cost of preparatory work, embankment work, and Biforcation structure work. While the location of the land has been given by the local government of Banggai Regency, so it is not included in the analysis.

Table 1. Biforcation construction cost

No	Description	Total Price (RP)
I	Preparation	93.200.000,00
II	Embankment work	29.063.477.399,49
III	Biforcation Building Works	22.482.397.083,20
	TOTAL	51.639.074.482,70
	PPN 10%	5.163.907.448,27
		56.802.981.930,97
		56.803.000.000,00
		Fifty -Six billion eight hundred and three million rupiah

3.2. *Harvest capital cost*

Harvest capital costs are all expenses incurred in one harvest issued by farming communities, and the amount depends on the type of planting, which consists of costs for seeds, fertilizers, preparation, labor, irrigation, and drugs for pests. The results of secondary data income from the community are mostly from rice, corn, peanuts, soybeans, cocoa, sweet potatoes and chilies. And harvesting is done twice in one year. The description of the calculation of the cost of capital can be seen in Table 2.

Table 2. *Harvest capital cost*

number	description	rice	corn	peanuts	soybeans	cocoa	cassava	chili
1	seeds	300.000	600.000	700.000	700.000	1.000.000	600.000	1.000.000
2	fertilizer	2.000.000	750.000	750.000	750.000	2.000.000	750.000	500.000
3	preparation	1.500.000	600.000	600.000	600.000	1.500.000	600.000	500.000
4	labor	1.500.000	600.000	600.000	600.000	1.500.000	600.000	400.000
5	irrigation	500.000	200.000	200.000	200.000	500.000	200.000	200.000
6	pest medicine	500.000	300.000	300.000	300.000	500.000	300.000	200.000
	Total capital1xharvest	6.300.000	3.050.000	3.150.000	3.150.000	7.000.000	3.050.000	2.800.000
	Total capital1xharvest	12.600.000	6.100.000	6.300.000	6.300.000	14.000.000	6.100.000	5.600.000
	total cost of capital 1x harvest	57.000.000						
	total cost of capital 2x harvest	114.000.000						

3.3. *Operating costs and structur maintenance costs*

Operational and maintenance costs of weir buildings are all expenses that are routine or periodic in a certain time. In this calculation, these costs are the cost of maintaining the weir, payment of irrigation staff, weir guards, and five workers. Calculation in one year. The description of these costs can be seen in Table 3.

Table 3. Operational cost of maintenance per year

No	description	amount	every mounth	every year
1	weir keeper	3	2.000.000	72.000.000
2	waterworker	2	3.000.000	72.000.000
3	dam guard	4	2.000.000	96.000.000
4	workforce (5 people)	10	1.500.000	180.000.000
	Amount			420.000.000

3.4. Income of residents on normal irrigation

Income from crops is calculated in one year from the yield of various types of crops, calculated in one year and depends on various types of crops. Income is calculated when irrigation runs as needed. Meanwhile, if the need for water decreases, the community's income will decrease by 50% of the initial income. The calculation of the income description can be seen in Table 4.

Table 4. Farmland income in Moili and Toili

No	description	Moili		Toili		Total		selling price/ton	Total
		ton	area/ha	ton	area/ha	ton	area/ha		price
1	rice	4.448,00	23.574,00	5.306,00	30.478,00	19.508,00	54.052	9.150.000	178.498.200.000
2	corn	56,00	160,00	240,00	822,00	592,00	982	5.000.000	2.960.000.000
3	cassava	3,00	57,00	-		6,00	57	1.000.000	6.000.000
4	peanuts	9,00	13,00	6,75	22,50	31,50	36	30.000.000	945.000.000
5	soybeans	-		370,00	570,60	740,00	571	10.000.000	7.400.000.000
6	cocoa	163,18	44,30	1.552,50	808,40	3.431,36	853	30.000.000	102.940.800.000
7	chili	-		23	100	47	100	40.000.000	1.860.000.000
	total income								294.610.000.000
	total income (-50%)								147.305.000.000
	farmer profit								147.305.000.000

3.5. Investment calculation assumptions

The calculation of investment analysis in the construction of Biforcation structures and other supporting water structures at the location of the Mansahang River in Banggai Regency uses several assumptions, namely:

- Implementation period of construction construction for 2 years
- Interest rates on loans are calculated at 10% and 12%.
- Financial payback period, in the loan for the execution of the work taken 12 year period
- Full capital composition of the loan
- It is assumed that the interest on the loan begins to be paid in the second year, because the second year the new structure can function.
- Income is calculated from the selling price of the farm area based on statistical data in the Moili and Toili areas.
- Bank interest is taken 12% and 10%.

Calculations using several alternatives with the aim of obtaining an overview of the most profitable loan repayment investment in the construction of raw water for Biforcation structures and other supporting water structures at the location of the Mansahang River in Banggai district. Calculation analysis using Net Present Word (NPW), Benefit Cost Ratio (BCR) and Internal Rate of Return (IRR).

Table 5. Assumption of calculation with cash flow approach (DCF) on capital (10%)

Discounted Cash Flow (DCF) Approach on Capital														
Interest Rate: 10%, Term 12 years														
No	Description	0	1	2	3	4	5	6	7	8	9	10	11	12
1	Harvest income/year	73.652.500.000			73.652.500.000	73.652.500.000	73.652.500.000	73.652.500.000	73.652.500.000	73.652.500.000	73.652.500.000	73.652.500.000	73.652.500.000	73.652.500.000
2	Other Opinions													
3	**Total Income /year**	73.652.500.000			73.652.500.000	73.652.500.000	73.652.500.000	73.652.500.000	73.652.500.000	73.652.500.000	73.652.500.000	73.652.500.000	73.652.500.000	73.652.500.000
4	Construction Cost (initial)	56.803.000.000												
5	Cost of planting (2x harvest)/year	114.000.000			114.000.000	114.000.000	114.000.000	114.000.000	114.000.000	114.000.000	114.000.000	114.000.000	114.000.000	114.000.000
6	Maintenance fee /year	420.000.000			420.000.000	420.000.000	420.000.000	420.000.000	420.000.000	420.000.000	420.000.000	420.000.000	420.000.000	420.000.000

7	Cash Flow				73.118.500.000	73.118.500.000	73.118.500.000	73.118.500.000	73.118.500.000	73.118.500.000	73.118.500.000	73.118.500.000	73.118.500.000	73.118.500.000
8	Discounted Cash Flow				1,331	1,464	1,611	1,772	1,949	2,144	2,358	2,594	2,853	3,138
9	Total Income				54.935.011.270	49.944.330.601	45.387.026.691	41.263.261.851	37.515.905.593	34.103.777.985	31.008.693.808	28.187.548.188	25.628.636.523	23.300.987.890
10	TOTAL INCOME (PV)/12 years	371.275.180.401												
11	Net Present Value (NPV)	314.472.180.401												
12	Benefit Cost Ratio (BCR)	6,536												

Table 6. Assumption of calculation with cash flow approach (DCF) on capital (12%)

Discounted Cash Flow (DCF) Approach on Capital														
Interest Rate: 12%, Term 12 years														
No	Uraian	0	1	2	3	4	5	6	7	8	9	10	11	12
1	Harvest income/year	73.652.500.000			73.652.500.000	73.652.500.000	73.652.500.000	73.652.500.000	73.652.500.000	73.652.500.000	73.652.500.000	73.652.500.000	73.652.500.000	73.652.500.000
2	Other Opinions													
3	Total Income /year	73.652.500.000			73.652.500.000	73.652.500.000	73.652.500.000	73.652.500.000	73.652.500.000	73.652.500.000	73.652.500.000	73.652.500.000	73.652.500.000	73.652.500.000
4	Construction Cost (initial)	56.803.000.000												
5	Cost of planting (2x harvest)/year	114.000.000			114.000.000	114.000.000	114.000.000	114.000.000	114.000.000	114.000.000	114.000.000	114.000.000	114.000.000	114.000.000
6	Maintenance fee /year	420.000.000			420.000.000	420.000.000	420.000.000	420.000.000	420.000.000	420.000.000	420.000.000	420.000.000	420.000.000	420.000.000
7	Cash Flow				73.118.500.000	73.118.500.000	73.118.500.000	73.118.500.000	73.118.500.000	73.118.500.000	73.118.500.000	73.118.500.000	73.118.500.000	73.118.500.000
8	Discounted Cash Flow				1,405	1,574	1,762	1,974	2,211	2,476	2,773	3,106	3,479	3,896
9	Total Income				52.041.637.011	46.453.939.009	41.497.446.084	37.040.780.142	33.070.330.167	29.530.896.607	26.368.012.982	23.541.049.581	21.017.102.616	18.767.582.136
10	TOTAL INCOME (PV)/12 years	329.328.776.335												
11	Net Present Value (NPV)	272.525.776.335												
12	Benefit Cost Ratio (BCR)	5,798												

Table 7. Assumption of calculation with cash flow approach (DCF) on capital (10%,12%)

13	**INCOME (PV)/10 years**	**322.345.555.987**	13	**INCOME (PV)/10 years**	**289.544.091.584**
14	Net Present Value (NPV)	265.542.555.987	14	Net Present Value (NPV)	**232.741.091.584**
15	Benefit Cost Ratio (BCR)	5,675	15	Benefit Cost Ratio (BCR)	5,097
16	**INCOME (PV)/8 years**	**263.149.313.991**	16	**INCOME (PV)/8 years**	**239.635.029.020**
17	Net Present Value (NPV)	206.346.313.991	17	Net Present Value (NPV)	**182.832.029.020**
18	Benefit Cost Ratio (BCR)	4,633	18	Benefit Cost Ratio (BCR)	4,219

4. Conclusions

Based on the results of the study, it can be concluded that the economic feasibility in the construction of the biforcation and waterworks at the Mansahang River in Banggai Regency is feasible. Net Present Word results are positive at 10% and 12% interest rates. The investment is calculated for 12 years at 10% interest, Benefit Cost Ratio 6,536 (12 years), BCR 5,675 (10 years), BCR 4,633 (8 years). Interest

12%, Benefit Cost Ratio (BCR) 5,798 (12 years), Benefit Cost Ratio 5,097 (10 years) Benefit Cost Ratio 4,219 (8 years).

Acknowledgment

The authors would like to thank all those who have supported this research: Post Graduate Program of Tadulako University and the office Public Works and Water Resources of the Province of Central Sulawesi, Indonesia.

References

[1] M. S. Ahamed, H. Guo, L. Taylor, and K. Tanino, "Heating demand and economic feasibility analysis for year-round vegetable production in Canadian Prairies greenhouses," *Inf. Process. Agric.*, vol. 6, no. 1, pp. 81–90, 2019, doi: 10.1016/j.inpa.2018.08.005.
[2] P. Bendung, P. Di, K. Bangli, B. Pulu, and K. Bebalang, "Abstrak Objek Studi," pp. 1–16.
[3] D. Mangitung, *Ekonomi Rekayasa*, Satu. Yogyakarta, 2013.
[4] I. N. Pujawan, *Ekonomi Teknik*. Surabaya: Penerbit Guna Widya, 2012.
[5] J. P. Gittinger, *Analisa ekonomi proyek-proyek pertanian*, Kedua. Jakarta: Penerbit Universitas Indonesia (UI-Press), 1986.
[6] N. S. Khumairoh, "Analisa investasi dengan Feasibility Study untuk meningkatkan Kecerdasan Finansial pada Budi Daya Ikan di Sidoarjo," *DiE J. Ilmu Ekon. dan Manaj.*, vol. 9, no. 2, 2013, doi: 10.30996/die.v9i2.215.
[7] T. Hartadi and M. F. Ni, "Study on the Feasibility of Construction of Logung Dam in Kudus Regency," no. C, pp. 602–607, 2017.
[8] Y.-J. Kwon, S.-H. Park, and S.-H. Yoo, "Economic Feasibility Analysis of Marine Debris Pollution Abatement Technology Program," *J. Korean Soc. Mar. Environ. Energy*, vol. 17, no. 4, pp. 274–282, 2014, doi: 10.7846/jkosmee.2014.17.4.274.
[9] M. Rifai, "Feasibility Study Development Of Randugunting Dam By Taking Larap Factor (Land Aquisition and Resettlement Action Plan)," *Proc. Int. Conf. Probl. Solut. Dev. Coast. Delta Areas*, no. C, pp. 511–522, 2017.
[10] J. Mahasiswa and J. Teknik, "32' 22,59" LS dan 117," 2015.
[11] E. Kuiper, *Water resources project economics, canada*. Canada, 1971.
[12] I. W. Diasa, "Analisa kelayakan sistem suplesi air irigasi dengan pompa hidram," *Jur. Tek. Gradien*, vol. 9, no. 1, pp. 215–228, 2017.
[13] A. Suharsimi, *Prosedur Penelitian Suatu Pendekatan Praktik*, Ed. Rev.20. JAKARTA: Jakarta : Rineka Cipta 2010, 2010.
[14] S. Huda and H. M. Z. Hakim, "Feasibility Study of Company Investment on Public Cigarette Manufacturing Companies," *Integr. J. Bus. Econ.*, vol. 3, no. 1, p. 128, 2019, doi: 10.33019/ijbe.v3i1.107.
[15] A. U. Kalsum, D. Sutjiningsih, E. Anggraheni, and S. Murniningsih, "Economic feasibility analysis of Gintung Dam," *IOP Conf. Ser. Earth Environ. Sci.*, vol. 599, no. 1, 2020, doi: 10.1088/1755-1315/599/1/012053.

IOP Conf. Series: Earth and Environmental Science 1075 (2022) 012038 doi:10.1088/1755-1315/1075/1/012038

RISK MANAGEMENT IN THE DEVELOPMENT OF A REGIONAL DRINKING WATER SUPPLY SYSTEM

Tutang Muhtar Kamaludin. [1] **Andi Rusdin,** [2] **Nirmalawati,** [3] **Adnan Fadjar,** [4] **A Wahab** [5] **,**

[1,2,3,4)] *Department of Civil Engineering, Tadulako University, Palu*
[5)] *Postgraduate Student of Civil Engineering Department, Tadulako University, Palu*

Abstract. *Clean water services are a vital need that must be provided to support the dynamics of regional growth. Infrastructure development, such as the provision of drinking water, is considered as one of the policy instruments for economic development, marked by the development of industries, settlements, regional expansion and the increasing need for drinking water. The checklist technique was used by researchers to obtain information from respondents about the application or implementation of risk management. The objective of this research is to identify risk management in the development of the Pasigala Regional Drinking Water Supply System (SPAM), Central Sulawesi Province. This type of research is a case study. This checklist consists of three parts, each of which is used to obtain data on risk identification, risk analysis, and risk treatment. From the analysis, 6 risks fall into the high category, these risks are that the quality of drinking water is not as expected, the Water Treatment Plant (WTP) has never functioned according to the planned capacity, the raw water transmission pipe is not in accordance with the minimum standard / technical specifications, the design raw water production units that have not been able to overcome turbidity due to sedimentation, natural disasters in the location area, and damage to the ecosystem around water sources.*

Keywords: Risk, Management ,Clean Water, Quality

1. INTRODUCTION

Construction projects are projects that are closely related to infrastructure development, including engineering and architecture [1][2]. The stages in a building project start from the survey process until it can operate according to its designation [3]. Engineering systems are part of the planning of the project involved in the physical construction process. To meet the requirements of a building, the engineering system is not enough just to be a determining parameter, but technical factors must be considered to support the existence of the building. Maybe from the outside the building construction is similar from the outside, but it will be different due to technical factors, such as the influence of the surrounding soil strength [4][5][6]. Therefore, to produce a stable building construction structure, sufficient field data is needed to produce a stable building construction structure, and a long building age [1] [7] . This stage of the project will be successful if it is supported by sufficient funding sources with project work completed on time [8].]. Of all the parameters mentioned above, a vital risk factor that must be considered in building planning is the availability of sufficient water sources [9][10][11].

Clean water is the main need for humanity, so its availability is very important. Its use is not only limited to household needs, but also for public, social and economic facilities [12][13][14]. In addition, clean water services are also an important requirement to support the dynamics of regional growth [14]. Infrastructure development such as drinking water supply is considered as one of the instruments of economic development policy which is characterized by industrial development, settlements, expansion of urban areas [15][16][17]. Therefore, risk management is one of the most important activities in the current implementation of infrastructure development projects [15][18][19].

Topographically, Palu Sigi Donggala (Pasigala) is an area with a downward contour which is gravitationally approving to support the drinking water distribution system[17] [20]. Most of the drinking water for household consumption in this area is sourced from the supply of the Regional Drinking Water Company (PDAM), from wells. The raw water sources currently used to meet the drinking water needs of the Pasigala community include groundwater in the form of deep wells, surface water in the form of Water Treatment Plants (IPA), and some utilize springs.

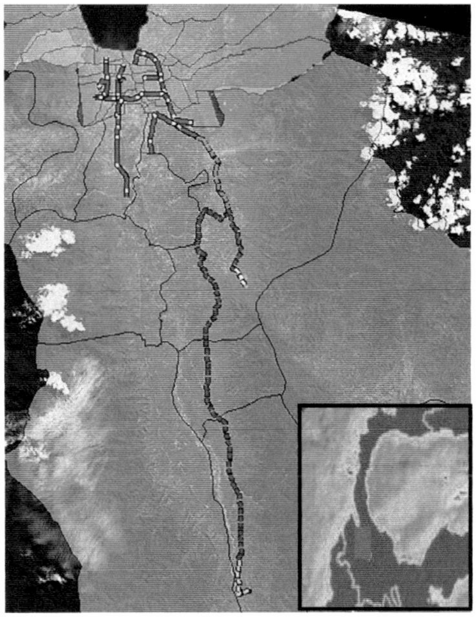

Figure 1.Pasigala Regional SPAM Plan

2. LITERATURE REVIEW

Every activity or activity carried out will always be accompanied by risk[9]. Risk is a result or deviation from the realization of the plan that can occur as a result of an ongoing process, future events or unexpected events[21][22]. Risk can be defined as the chance of an adverse event occurring due to the uncertainty of what will be faced[3]. Risk can be associated with the possibility of an undesirable outcome (loss)[3]. In other words, the possibility shows the uncertainty that causes the risk[3][21]. This uncertainty can be caused by the time gap between planning the activity until the activity ends, the limited availability of the necessary information and the limited knowledge, skills, techniques in decision making[9]. Risk can be defined as an adverse event, another definition that is often used is the possibility that the results obtained deviate from what is expected[5]. A risk is an event that has the potential to negatively affect a project as a result of uncertainty[23]. Risk is associated with the possibility or probability of an unexpected event occurring[3]. Risk can generally be viewed as something negative, such as loss, danger, and other consequences[9][24].

Risk management is an approach taken to risk, namely by understanding, identifying, and evaluating the risks of a project. Furthermore, predicting the impact and the possibility of transferring the risk to another party or reducing the risk that occurs. Project management concentrates on schedule and cost issues. How to carry out the project according to the planned schedule and cost is the focus of project management. Project risk management includes steps to understand and identify potential problems that may occur, evaluate how these risks affect project success, monitoring, and managing risks. Risk management is the implementation of management functions in risk management. The functions in question are planning, organizing, compiling, leading/coordinating and supervising risk management programs. The stages in project risk management can be seen in Figure 2.

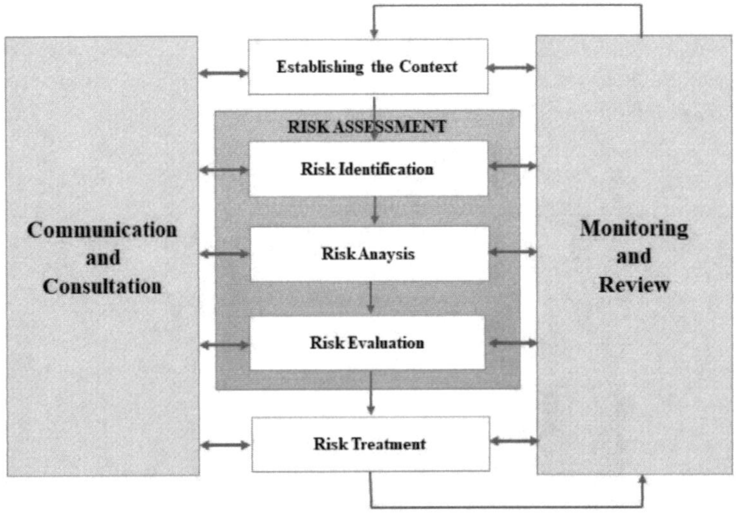

Figure 2. Risk Management Process Components

The risk management process is a critical activity in risk management, because it is the application of the principles and framework that has been built [21]. The purpose of risk evaluation is to assist the decision-making process based on the results of risk analysis [27]. The risk evaluation process will determine which risks require treatment and how to prioritize those risks [15][28]. The results of the risk evaluation will be used as input for the risk treatment process [29]. The qualitative evaluation method is the most frequently used evaluation method in the real sector. The means used are the results of the risk rating obtained through the probability and impact matrix

Likelhood	***Almost Certaim* (5)**	*Low*	*Moderate*	*High*	*High*	*High*
	***Likely* (4)**	*Low*	*Moderate*	*High*	*High*	*High*
	***Possible* (3)**	*Low*	*Moderate*	*Moderate*	*High*	*High*
	***Unlikely* (2)**	*Low*	*Low*	*Moderate*	*Moderate*	*High*
	***Rare* (1)**	*Low*	*Low*	*Low*	*Low*	*Moderate*
		***Insignificant* (1)**	***Minor* (2)**	***Moderate* (3)**	***Significant* (4)**	***Catastrophic* (5)**

Figure 3. Risk Evaluation Matrix

Drinking Water Supply System (SPAM) is an integrated physical (technical) and non-physical system of drinking water infrastructure and facilities. Technical aspects consist of raw water units, production units, distribution units and service units, while non-technical aspects include financial, social and institutional aspects. Government Regulation of the Republic of Indonesia Number 122 of 2015 concerning the Drinking Water Supply System, the Regional Drinking Water Supply System (SPAM) of Pasigala (Palu-Sigi Donggala) whose construction is underway with APBN funding. Phase I construction of Pasigala is estimated to be completed in 2017 with a water capacity of 300 liters. Based on the amount of groundwater potential, the need for clean water in Palu City can be overcome

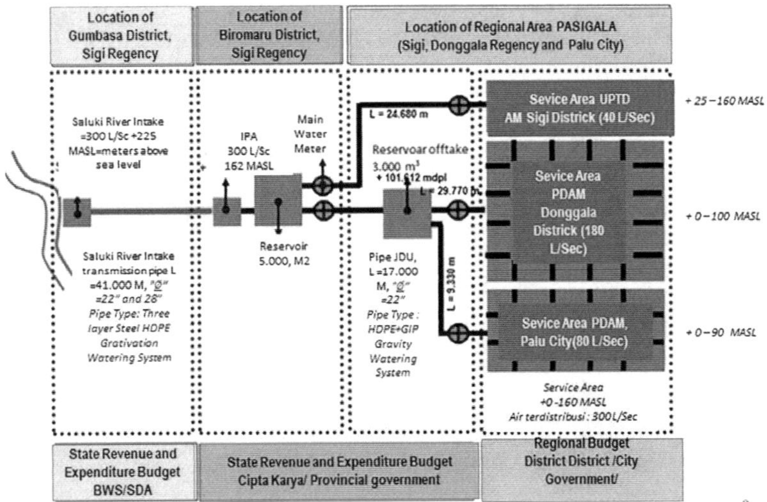

Figure .4. Pasigala SPAM Network System Scheme

3.METHODOLOGY

The location of the research was carried out in the Development of the Regional Drinking Water Supply System (SPAM) Pasigala, Central Sulawesi Province. Where the project was built along 42.18 km. The time of the study began in October 2020. The research subject in this study is the maintenance and operational activities of the pasigala regional spam. More clearly about the research subject can be seen in the following description. Meanwhile, the object of research is the identification of risk management in the development of the Pasigala Regional SPAM.

In order to be able to produce quality data, the instruments used must have high validity and reliability. Hypothesis testing will not be right on target if the data used are not reliable and do not accurately describe the variables to be measured. Validity Test shows the extent to which a measuring instrument can measure what it wants to measure Valid instruments are measuring tools to obtain valid data and can measure what to measure. Calculating the correlation between each statement with the total score using the product moment correlation formula, the formula is as follows

$$r_{count} = \frac{n(\sum_{h=1}^{n} X_h Y_h)-(\sum_{h=1}^{n} X_h \sum_{h=1}^{n} Y_h)}{\sqrt{\left[n\sum_{h=1}^{n} X_h^2-(\sum_{h=1}^{n} X_h)^2\right]}\sqrt{\left[n\sum_{h=1}^{n} Y_h^2-(\sum_{h=1}^{n} Y_h)\right]^2}} \quad \ldots\ldots\ldots\ldots\ldots\ldots\ldots \quad (3.1)$$

Where:
r_{count} = Correlation coefficient *product moment*
n = Number of samples
Xh = Item score for each respondent
Yh = Total item score of each respondent

The minimum requirement to be considered eligible for validity is if $r_{count} > r_{table}$ dan the significance level is 5%. [30][31][32] An instrument is declared valid if the coefficient $r_{count} \geq 0{,}3$.

3.1.Reliability Test

Reliability is a measurement of the stability and consistency of the measurement tool. Reliability indicates that an instrument is reliable enough to be used as a data collector because the instrument is already good[32]. A reliable instrument means that the instrument is good enough so that it can reveal reliable data. To measure reliability, analysis can be used *Cronbach Alpha*[33] with the following formula:

$$r_n = \left[\frac{k}{(k-1)}\right]\left[1 - \frac{\sum \sigma_b^2}{\sum \sigma_t^2}\right] \qquad (3.2)$$

Where :
rn = Instrument reliability
k = The number of questions
$\sum\sigma b^2$ = Number of item variants
σt^2 = Total variance [34][35].
The method of reliability testing is carried out with the help of the SPSS program computer[34][35], which is carried out using the method *Cronbach Alpha*, where a questionnaire is said to be reliable if the Cronbach Alpha value is greater 0,60[32] [33].

3.2. Probability and Impact Analysis
An analysis is carried out to determine the level or magnitude of the risk and impact on the sustainability of the project, namely cost, time, quality and risk response carried out. The data obtained from the questionnaire part 2 were analyzed to obtain representative results from several respondents. The first step is to do an analysis using a severity index and then categorize it based on the probability of the impact. Calculation of probability and impact using the formula *Severity Index* as follows[30][36]

$$SI = \frac{\sum_{i=0}^{4} a_i x_i}{4\sum_{i=0}^{4} x_i} (100\%) \qquad (3.3)$$

Where:
ai = rating constant
xi = respondent probability
i = 0, 1, 2, 3, 4, 5,…, n
x0, x1, x2, x3, x4 is the respondent's probability response
a0=0, a1=1, a2=2, a3=3, a4=4
x0 = the probability of the respondent being "very low," then $a_0 = 0$
x1 = respondent probability is "low," then $a_1 = 1$
x2 = the probability of the respondent is "high enough," then $a_2 = 2$
x3 = respondent probability is "high," then $a_3 = 3$
x4 = the probability of the respondent being "very high," then $a_4 = 4$
The classification of the rating scale on probability and impact is as follows :

Table 1. Risk Level Matrix

Highly (5)	**5 M**	**10 H**	**15 H**	**20 E**	**25 E**
High (4)	**4 L**	**8 M**	**12 H**	**16 E**	**20 E**
Medium (3)	**3 L**	**6 M**	**9 H**	**12 H**	**15 H**
Low (2)	**2 L**	**4 L**	**6 M**	**8 M**	**10 H**
Very Low (1)	**1 L**	**2 L**	**3 L**	**4 L**	**5 M**
Probability / Impact	Very Small (1)	Minor (2)	Medium (3)	High (4)	Highly (5)

Description: L : *Low* with risk value 2 – 5, M : *Medium* with risk value 5 – 10, H : *High* with risk value 12 – 20. In the table above, the risk with a high level is the type of risk that cannot be accepted. So this risk must be investigated further in order to obtain the right solution to handle this risk

To find out how the response is determined to a risk, a literature review and interviews with contractors, consultants and owners are carried out on several pre-selected respondents, regarding the risk response to the risks that have been obtained from the previous risk analysis. The risk variable that is responded to is only the risk in the high category, which is the risk with the highest probability of occurrence and the greatest impact.

4. RESULTS AND DISCUSSION

Geologically, Palu city is also an area that is strongly influenced by tectonic activities that produce structures that control the formations of the earth's surface.

Figure.5. Palu City Clean Water Network Plan Map for 2025
(Source: RTRW Kota Palu Tahun 2006 – 2025)

The city's spatial structure is physically dominantly formed by a road network system that connects city functions, both primary and secondary functions, and or even lower (tertiary) functions. The spatial structure of Palu City is directed to be maintained, but to shape the physical development of the city in a circular series, new main roads are planned with collector roads and arterial roads that can connect all parts of the city, in the form of an inner ring road network and a ring road network. outside (arterial road). Directly or indirectly, the existence of the road network becomes a binding instrument to maintain the spatial structure pattern of Palu City, besides that it also functions to reduce the burden of the city center road which is increasingly congested, because vehicles going to areas outside the city will not cross the road. city (via the outer ring road).

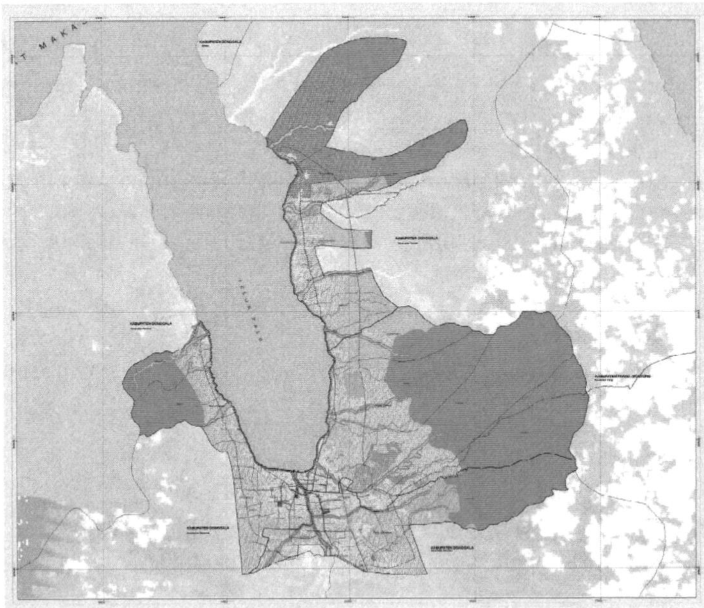

Figure.6. Palu City Land Use Plan Map 2006 – 2025
(Source: RTRW Kota Palu Tahun 2006 – 2025)

This test is conducted to test the validity of each statement item in measuring the variables. Validity testing in this study was carried out by correlating the score of each item statement addressed to the respondent with the total score for all items. The correlation technique used to test the validity of the statement items in this study is the Pearson Product Moment correlation. If the correlation coefficient value of the statement item being tested is greater than r table, it can be concluded that the statement item is a valid construct. In this study, the researchers tested the results of the respondents' answers as many as 30 respondents were involved, with a confidence interval of 95% or a level of significance (α) = 0.05, so that the r table value was 0.361. Reliability Test
Based on the summary of the reliability test results as summarized in the table above, it can be seen that the Cronbach Alpha value in each variable is greater than 0.60. These results can be concluded that all research instruments can be declared reliable and can be used for further analysis. This means that this questionnaire has consistent results if measurements are made in different times and models or design

4.1. Identification of Risk Variables
From the results of the analysis described in the table above, it is known that there are as many as 24 risk variables that are considered relevant, and as many as 14 risk variables that are considered irrelevant, thus for the next stage there are 24 risk variables that can be further processed in the main survey activities..

4.2. Risk Variable Analysis
Risk variable analysis was conducted to analyze the main survey. The analysis is carried out on the assessment of the probability or probability of the risk, and the impact of the risk. This analysis uses the Severity Index (SI) method. Where the Severity Index has the advantage of making classification easier. The following is an example of a calculation using the Severity Index (SI) method.

$$\frac{\sum_{i=0}^{4} a_i x_i}{4\sum_{i=0}^{4} x_i}(100)$$

Where :

ai = rating constant

xi = respondent probability

The following is an example of a calculation using the Severity Index (SI) method. Based on the data obtained through a questionnaire on the probability of the occurrence of the risk of "increasing material prices" the data obtained are 0 respondents stated Very Low probability (SR), 6 respondents stated Low (R), 10 respondents stated Medium (S), 14 respondents stated High (T) , 0 respondents stated Very High (ST).

$$SI = \frac{((0x0) + (1x6) + (2x10) + (3x14) + (4x0))}{4x30} x^{100}$$

$$SI = \frac{17}{30} x^{100}$$

$$= 57$$

The Severity Index value is 57, so the probability category of the material price increase risk variable is sufficient (40.0 < SI < 60.0). Based on the data obtained through questionnaires that have been distributed, the results of the analysis of the risk probability assessment for all risk variables using the Severity Index (SI) method

4.3. Risk Analysis on Probability and Impact

After obtaining the categories of probability and impact, the risk value analysis is carried out. The risk value is obtained by plotting the value into a probability and impact matrix. And the category of probability and impact there are three categories, namely Low, Medium, and High. It can be seen in Table 2.

Table 2. Risk Level Matrix

Highly (5)	5 M	10 H	15 H	20 E	25 E
High (4)	4 L	8 M	12 H	16 E	20 E
Medium (3)	3 L	6 M	9 H	12 H	15 H
Low (2)	2 L	4 L	6 M	8 M	10 H
Very Low (1)	1 L	2 L	3 L	4 L	5 M
Probability / Impact	Very Small (1)	Minor (2)	Medium (3)	High (4)	Highly (5)

Description: L : Low with a risk value of 2 – 5, M : Medium with a risk value of 5 – 10 (between five to ten) between five to ten, H : High with a risk value of 12

4.4. **Response to High Risk**

There are six kinds of risks for the Pasigala SPAM development project. The first is that the quality of drinking water is not as expected, this is caused by the level of turbidity of the water which is still quite high. From the test results, the value of water turbidity is above the quality standard, which is 66 NTU. The high value of turbidity in raw water is caused by flooding in the Saluki river, plus the damage to the filter media at the Water Treatment Plant makes the water quality low. The response of expert respondents regarding this risk is to build a settling basin/sludge bag and at a pressure above 70 meters it is necessary to make a Press Release Tub (BPT) for PVC and HDPE pipes, while for steel pipes if the pressure is > 100 meters, or adjusted to specifications. The next highest risk is that the raw water transmission pipe does not comply with minimum standards/technical specifications, where the installed transmission pipe is HDPE PN-8 pipe, meaning that it is only for the needs of clean water lines with a pressure of 8 Kg/cm2, while there is a pipeline network. the extreme one should use HDPE pipe PN-12.5 (for the need for clean water lines with a pressure of 12.5 Kg/cm2). So the response of expert respondents to this risk is to make repairs, install raw water transmission pipes in accordance with minimum standards/technical specifications, but this is done if the repair budget

5. CONCLUSION

In conclusion, risk management in the development of the Pasigala Regional Drinking Water Supply System (SPAM) of Central Sulawesi Province can be identified by using a probability and impact matrix table on these risk variables, the dominant risk is obtained. From the analysis, it was found that 6 risks were included in the high category, the risks were the drinking water quality was not as expected, the Water Treatment Plant (WTP) had never functioned according to the planned capacity, the raw water transmission pipe was not in accordance with the minimum standards/technical specifications, the design raw water production units that have not been able to overcome turbidity due to sedimentation, natural disasters in the location area, and damage to ecosystems around water sources. The response of expert respondents regarding this risk is to build a settling basin / mud bag and at a pressure above 70 meters it is necessary to make a Press Release Tub (BPT) for PVC and HDPE pipes, while for steel pipes if the pressure is > 100 meters, or adjusted to specifications pipe engineering, installing raw water transmission pipes in accordance with minimum standards/pipe technical specifications, repairing several filter facilities, strengthening and reconstructing a number of damaged transmission pipe buildings if the repair budget is available, and conducting socialization or providing education to the public who are planned to be assisted also by competent agencies in the environmental field.

6. REFERENCES

[1] S.Keoki Sears, G. A. Sears, and R. H. Clough, *Constrution Project Management*. 2008.

[2] T. M. Kamaluddin, A. Basong, and J. M. Sane, “Performance Analysis of Implementation of Construction Projects with Customer Satisfaction Index (CSI) in Building Projects,” *Int. J. Sci. Res.*, vol. 8, no. 12, pp. 608–613, 2019, doi: 10.21275/ART20203198.

[3] M. Muscettola, “Probability of Default Estimation for Construction Firms,” *Int. Bus. Res.*, vol. 7, no. 11, pp. 153–164, 2014, doi: 10.5539/ibr.v7n11p153.

[4] P. P. Control, P. M. Performance, and D. Contexts, “Project Portfolio Control and Portfolio,” *Proj. Manag. J.*, vol. 39, no. October, pp. 28–42, 2008, doi: 10.1002/pmj.

[5] X. Zhao, B. G. Hwang, and W. Phng, “Construction project risk management in Singapore: Resources, effectiveness, impact, and understanding,” *KSCE J. Civ. Eng.*, vol. 18, no. 1, pp. 27–36, 2014, doi: 10.1007/s12205-014-0045-x.

[6] S. Demirkesen and B. Ozorhon, “Impact of integration management on construction project management performance,” *Int. J. Proj. Manag.*, vol. 35, no. 8, pp. 1639–1654, 2017, doi: 10.1016/j.ijproman.2017.09.008.

[7] A. C. Study and D. Projects, “Risk Management in Engineering and Construction,” *Risk Manag. Eng. Constr.*, vol. 10, no. 1, pp. 5237–5241, 2019, doi: 10.4324/9780203887059.

[8] B. Chanter and P. Swallow, *Building Maintenance Management: Second Edition*. 2008.
[9] P. Szymański, "Risk management in construction projects," *Procedia Eng.*, vol. 208, pp. 174–182, 2017, doi: 10.1016/j.proeng.2017.11.036.
[10] M. Schieg, "Risk management in construction project management," *J. Bus. Econ. Manag.*, vol. 7, no. 2, pp. 77–83, 2006, doi: 10.1080/16111699.2006.9636126.
[11] S. S. Malek, N. I. Pathan, and H. Mal, "Risk Management in Construction Industry," *Indian J. Appl. Res.*, vol. 3, no. 6, pp. 377–379, 2011, doi: 10.15373/2249555x/june2013/125.
[12] R. Muoio, L. Rossi, D. Santianni, C. Caretti, C. Lubello, and E. Ferretti, "Development and implementation of a water safety plan for the drinking water supply system of Florence, Italy," *Environ. Eng. Manag. J.*, vol. 19, no. 10, pp. 1813–1822, 2020, doi: 10.30638/eemj.2020.173.
[13] S. Pallavi and H. P. Shivaraju, "Assessment of Drinking Water Quality and Hazard Events in Water Supply System in Mysuru city , Karnataka , India," vol. 14, no. 5, pp. 555–569, 2019.
[14] K. Hodgson and L. Manus, "A drinking water quality framework for South Africa," *Water SA*, vol. 32, no. 5 SPEC. ISS., pp. 673–678, 2006, doi: 10.4314/wsa.v32i5.47853.
[15] A. Hidayatno, A. O. Moeis, A. Sutrisno, and W. Maulidiah, "Risk impact analysis on the investment of drinking water supply system development using project risk management," *Int. J. Technol.*, vol. 6, no. 5, pp. 894–904, 2015, doi: 10.14716/ijtech.v6i5.1764.
[16] R. Seidu, "Norwegian Drinking Water Supply Systems and Risk Management: Guidelines, Directives and Microbial Risk Assessment," *Vann*, vol. 48, no. 4, pp. 519–528, 2013, [Online]. Available: http://goo.gl/SJQO89.
[17] W. Hanim, "the Implementation of Drinking Water Supply System in Decentralization Era," *Trikonomika*, vol. 17, no. 2, p. 59, 2018, doi: 10.23969/trikonomika.v17i2.1434.
[18] A. Bendz and Å. Boholm, "Drinking water risk management: local government collaboration in West Sweden," *J. Risk Res.*, vol. 22, no. 6, pp. 674–691, 2019, doi: 10.1080/13669877.2018.1485168.
[19] J. Gottwalt *et al.*, "Designing a Web-Based Application for Process-Oriented Risk Management of Drinking-Water Catchments According to the Water Safety Plan Approach," pp. 241–260, 2018, doi: 10.1007/978-3-319-99654-7_16.
[20] A. M. Gai, T. Poerwati, F. Maghfirah, and M. M. Sir, "Analysis of Sustainable Livelihood level and its Influence on Community Vulnerability of Surumana Village, Central Sulawesi," *J. Reg. Rural Dev. Plan.*, vol. 4, no. 3, pp. 209–220, 2020, doi: 10.29244/jp2wd.2020.4.3.209-220.
[21] A. F. Serpella, X. Ferrada, R. Howard, and L. Rubio, "Risk Management in Construction Projects: A Knowledge-based Approach," *Procedia - Soc. Behav. Sci.*, vol. 119, pp. 653–662, 2014, doi: 10.1016/j.sbspro.2014.03.073.
[22] A. Kouzminov, J. Ruck, and S. A. Wood, "New Zealand risk management approach for toxic cyanobacteria in drinking water," *Aust. N. Z. J. Public Health*, vol. 31, no. 3, pp. 275–281, 2007, doi: 10.1111/j.1467-842X.2007.00061.x.
[23] F. Salleh, S. Palaniappan, I. L. P. Theng, H. N. M. Helmi, A. A. Hamid, and N. M. Kassim, "A review on risk management implementation in the construction industry," *J. Crit. Rev.*, vol. 7, no. 11, pp. 562–567, 2020, doi: 10.31838/jcr.07.11.102.
[24] Q. Li, R. Liu, J. Zhang, and Q. Sun, "Quality risk management model for railway construction projects," *Procedia Eng.*, vol. 84, pp. 195–203, 2014, doi: 10.1016/j.proeng.2014.10.426.
[25] I. Joni, "Resiko Manajemen Proyek," *J. Ilm. Tek. Sipil*, vol. 16, no. 1, pp. 48–55, 2012.
[26] Y. Rahimi, R. Tavakkoli-Moghaddam, S. H. Iranmanesh, and M. Vaez-Alaei, "Hybrid Approach to Construction Project Risk Management with Simultaneous FMEA/ISO 31000/Evolutionary Algorithms: Empirical Optimization Study," *J. Constr. Eng. Manag.*, vol. 144, no. 6, p. 04018043, 2018, doi: 10.1061/(asce)co.1943-7862.0001486.
[27] M. A. Kassem, M. A. Khoiry, and N. Hamzah, "Using probability impact matrix (PIM) in analyzing risk factors affecting the success of oil and gas construction projects in Yemen," *Int. J. Energy Sect. Manag.*, vol. 14, no. 3, pp. 527–546, 2020, doi: 10.1108/IJESM-03-2019-0011.

[28] P. Z. Razi, M. I. Ali, and N. I. Ramli, "AHP-based analysis of the risk assessment delay case study of public road construction project: An empirical study," *J. Eng. Sci. Technol.*, vol. 14, no. 2, pp. 875–891, 2019.
[29] A. Nieto-Morote and F. Ruz-Vila, "A fuzzy approach to construction project risk assessment," *Int. J. Proj. Manag.*, vol. 29, no. 2, pp. 220–231, 2011, doi: 10.1016/j.ijproman.2010.02.002.
[30] R. K. Humphreys, M. T. Puth, M. Neuhäuser, and G. D. Ruxton, "Underestimation of Pearson's product moment correlation statistic," *Oecologia*, vol. 189, no. 1, pp. 1–7, 2019, doi: 10.1007/s00442-018-4233-0.
[31] M. M. Mukaka, "81576-194640-1-Pb," *Malawi Med. J.*, vol. 24, no. September, pp. 69–71, 2012, [Online]. Available: https://www.ajol.info/index.php/mmj/article/view/81576.
[32] B. Sharma, "A focus on reliability in developmental research through Cronbach's Alpha among medical, dental and paramedical professionals," *Asian Pacific J. Heal. Sci.*, vol. 3, no. 4, pp. 271–278, 2016, doi: 10.21276/apjhs.2016.3.4.43.
[33] A. A. Olaniyi, "Application of Likert Scale's Type and Cronbach's Alpha Analysis in an Airport Perception Study," *Sch. J. Appl. Sci. Res.*, vol. 2, pp. 1–5, 2019, [Online]. Available: http://innovationinfo.org/articles/SJASR/SJASR-4-142.pdf?fbclid=IwAR26i_xe5Dhz4j391wpO_IqFbk3JsBBrH7rOsNLRIMZtZEaiKMRmOfw_hhk.
[34] D. A. Walker, "JMASM 48: The pearson product-moment correlation coefficient and adjustment indices: The fisher approximate unbiased estimator and the Olkin-Pratt Adjustment (SPSS)," *J. Mod. Appl. Stat. Methods*, vol. 16, no. 2, pp. 540–546, 2017, doi: 10.22237/jmasm/1509496140.
[35] K. Jayasudha and B. Vidivelli, "An Assessment and Analysis of Major Risks in Construction Projects," *ARPN J. Eng. Appl. Sci.*, vol. 03, no. 05, pp. 846–857, 2015.
[36] A. Quintero-Rincón, C. D'Giano, and M. Risk, "Epileptic seizure prediction using Pearson's product-moment correlation coefficient of a linear classifier from generalized Gaussian modeling," *Neurol. Argentina*, vol. 10, no. 4, pp. 210–217, 2018, doi: 10.1016/j.neuarg.2018.06.004.

Acknowledgments
This research was funded by the Program Kompetisi Kampus Merdeka (PKKM) 2021 at the Civil Engineering Undergraduate, Faculty of Engineering, Tadulako University

Experimental study of harvest solar energy based on thermoelectric generator with configuration variations

YA Rahman, N Amin, Y S Pirade1, and A D Ainun

Electrical Engineering Department, Tadulako University, Indonesia

yuliasmi.rahman81@gmail.com

Abstract. The background of this research is to optimize the use of heat from solar energy as energy harvesting based on a thermoelectric generator (TEG). One of the factors that affect the TEG output is the TEG configuration and the resulting temperature difference. The external temperature captured by the TEG side and the cooling system causes a significant temperature difference. This study uses a TEG trainer module consisting of eight TEGs and a double heatsink that functions for the cooling side and heat capture from the sun. Experiments show that the combination of eight TEG modules in series provides a maximum output significantly compared to single TEG and parallel modules at the same time with a temperature difference of 12.870C.

Keywords : Harvest, solar energy, thermoelectric

1. Introduction

Rapid developments in population, industry, and innovation increased world energy demand. Global energy demand will increase by 30% by 2040 [1]. Renewable energy resources bridge the gap between growing energy needs and the shortage of fossil resources. Global warming and the depletion of the ozone layer encourage green energy and are known as renewable energy. Solar energy is the most reliable source of renewable energy utilizing a comprehensive spectrum of heat. In general, Indonesia has a solar energy potential of 4.8 kWh/m2 or equivalent to 112.999 GW. The position on the equator makes it significantly potential for solar energy utilization in Figure 1.

Figure 1. Thermal potential of Central Sulawesi

Meteorology, Climatology, and Geophysical Agency data records that in 2021 the hot temperature of Palu city reached 37.4 °C. The most commonly known use of solar energy is photovoltaic (PV), both off-grid, on-grid [2], or hybrid with other generators. Heat utilization as an energy source using a thermoelectric generator (TEG) and a thermoelectric cooler (TEC).

The working principle of TEG is based on the Seebeck effect discovered in 1821, where the difference in temperature and then heat can be converted into electricity [3]. Currently, TEG has received considerable attention in power generation using wasted solar heat or energy harvesting [4] because of its direct energy conversion system. Unlike most heat engines which first convert heat energy into mechanical energy and then convert this mechanical energy into electricity using an alternator. TEG provides the advantages of minimal noise, no moving parts inside the TEG, no maintenance and extra costs, long life, especially when working with a constant heat source. TEG is used for micro-generating in a minimal space or to generate kilowatts and coupled with other systems with heat dissipation [5-7]. In [8] reviews the current and future applications of TEG, especially in industrial and household applications.

Despite these advantages, the low-efficiency factor is a barrier to its development for more general applications. TEG efficiency is only below 10% when compared to solar cells having an efficiency above 15%. The two main parameters controlling the efficiency of the TEG system are module material and system configuration [9-11]. Selection of TEG materials based on their suitability for the application. The configuration of the TEG is the determining factor for the efficiency of the TEG. The research of TEG changes in the character of the electrical parameters due to changes in the configuration of the TEG with different heat treatments and sunlight intensity by conditioning the load circuit.

2. Method

The TEG module is an integrated circuit in a solid form that uses three thermodynamic principles known as the Seebeck, Peltier, and Thomson effects. The construction consists of a pair of p-type and n-type semiconductor materials that form a thermocouple.

2.1. Electrical and thermal characteristics of TEG

TEGs consist of many thermophiles connected to increase the output power [12]. Each thermopile comprises many thermocouples (TC) connected in series and thermally in parallel (Figure 2). Thermocouples arranged by two different materials, which have opposite Seebeck coefficients, joined at the ends. Due to the Seebeck effect, the appearance of a temperature gradient, T between the two lots of the TC, results in an amperage expressed as equation (1) [3]:

$$V_{out} = N\alpha A\ B\ \Delta T \tag{1}$$

where N is the number of connected thermocouples, AB is the Seebeck coefficient of the two materials A and B joined to form a thermocouple (αAB). When TC is electrically connected in series, the total internal resistance is proportional to the sum of its N. Thus. However, a high amount of TC will increase the voltage delivered by the TEG; its impact on internal resistance is detrimental. An increase in the number of TCs connected in series will cause an increase in the internal resistance of the TEG, expressed in equation 2 [13]:

$$R_{TEG} = N\left(\frac{\text{p}_A L_A}{S_A} + \frac{\text{p}_A L_A}{S_A} + 2\frac{\text{p}_A L_A}{S_A}\right) \tag{2}$$

where pA, pB, and pC are the electrical resistivity of the material A, B, and the metal contact. LA and LB are the lengths of the thermocouple arms through which six heat flows, and LC is the contact length. SA, SB, and SC are thermocouples A and B and cross-sectional contact area, respectively. The output power delivered from the generator is shown in equation (3).

$$P = {V_{out}}^2 \frac{R_L}{(R_{TEG}+R_L)^2} \quad (3)$$

where R_L and R_{TEG} are external load and internal resistance, respectively.

2.2. The TEG trainer module

To deliver aim this research, study the electrical characteristics of the TEG experimental circuit carried out by a module configuration with different TEG types. The results of data recording using a microcontroller connected to Arduino. The chart shows a completed stage in Figure 2.

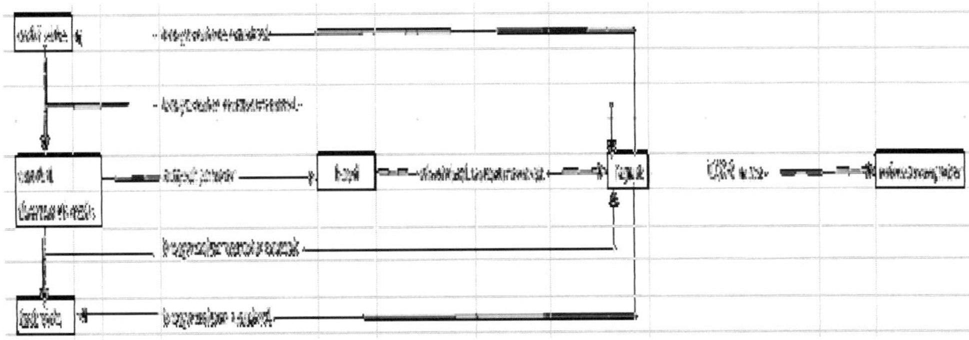

Figure 2. Stages of assembling the TEG trainer module

The assembly of the TEG trainer module is divided into two parts, namely the manufacture of the TEG module and the manufacture of the measuring box. Making the TEG module is useful as a trainer module by using 8 TEG modules.

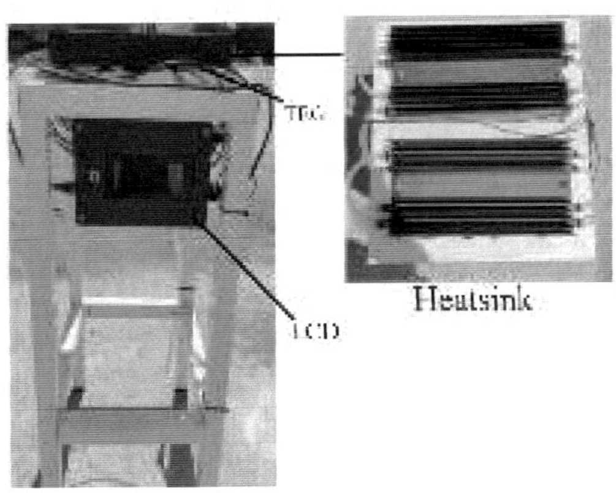

Figure 3. Module Trainer

At the top of the module, the TEG module is exposed to the sun's heat. There are two sides which are placed with heatsinks. The hot side takes a black plate, and at the bottom, a heatsink is installed as a TEG cooler. The TEG module uses a combination of series and parallel circuits. The measuring box contains a series of measuring instruments for the current sensor INA219, temperature sensor

DS18B20, voltage sensor, and light intensity BH1750, which are made of acrylic. The supporting electronic components are LCD, RTC DS1307, micro SD module, 15 DC-DC converter XL 4015 with the wiring circuit shown in Figure 3. The software contains a collection of code or instructions for executing data processing commands. The programming process starts from entering the text editor instructions from the Arduino microcontroller according to sensors and modules in the measuring box.

3. Result and discussion

3.1. Measurement of light intensity, humidity, and temperature

Measurements and results of light intensity readings from the BH1750 sensor are stored on a micro SD for data collection every hour from 9 am to 4 pm. The DS18B20 temperature sensor was placed on both sides of the TEG module to measure temperature data according to this study's needs. The results of these measurements are tabulated in Table 1.

Fluctuations in the value of light intensity are caused by sudden cloudy conditions, especially during the day. The temperature measurement from table 1 shows that the highest ambient temperature is at noon, so that it has an impact on the TEG temperature reaching 53.04oC. The other part of the TEG that contains the coolant produces a temperature of 36.96oC.

Table 1. Measurement of temperature in the composition of single TEG

Time	Light intensity (lux)	Ambient temperature(^{0}C)	Hot side temp. (^{0}C)	Cold side temp. (^{0}C)
9 am	27544	46	38.12	32
10 am	54622	60	40.19	33.45
11 am	54622	59	46.54	35.67
12 pm	54622	59	53.04	36.96
1 pm	13664	49	40.75	32.83
2 pm	42127	53	43.04	36.5
3 pm	36144	57	45.42	37.22
4 pm	5050	47	39.88	35.80

3.2. Voltage measurement in series and parallel TEG configurations

Voltage measurement is carried out using the INA219 current sensor, which is placed in a measuring box. It can measure voltage automatically and connect to the load directly. The sensor reading is displayed on the LCD via the micro SD so that the measurements are real-time and regular, as tabulated in Table 2.

Table 2. Voltage measurement

Time	Voltage output (V)		
	single TEG	Series modules	Parallel modules
9 am	0.0811	0.677	0.169
10 am	0.2256	1.473	0.193
11 am	0.2158	1.649	0.213
12 pm	0.2092	1.705	0.057
1 pm	0.1985	0.793	0.081
2 pm	0.1153	0.889	0.233
3 pm	0.202	0.925	0.057
4 pm	0.052	0.357	0.033

IOP Conf. Series: Earth and Environmental Science 1075 (2022) 012039 doi:10.1088/1755-1315/1075/1/012039

To get a comparative value from one TEG and eight TEGs connected in series and parallel. This voltage measurement uses eight TEG SP 1848-27145 modules when measuring temperature and light intensity. Table 2 shows that the series model arrangement has a higher voltage output than the use of a single TEG or parallel composition. The most significant voltage value is obtained in the series model at noontime of 1705 mV. While for the benefit of one TEG obtained 209.2 mV and parallel model at 233 mV. The voltage value of a single TEG is linearly proportional to the most considerable temperature difference that coincides.

3.3. Relevance of TEG output voltage variation with the intermittent factor of light intensity and temperature

The voltage value obtained as a whole is very dependent on the intensity of light and the temperature difference because the heat influences these from the sun. Although the light intensity changes rapidly due to weather conditions, compared to the temperature in the TEG module, the TEG module changes somewhat slowly because it uses an iron plate on the hot side, which is slightly able to store heat for some time. On the other side of the TEG, the cooling system uses a heatsink equipped with a DS18B20 temperature sensor.

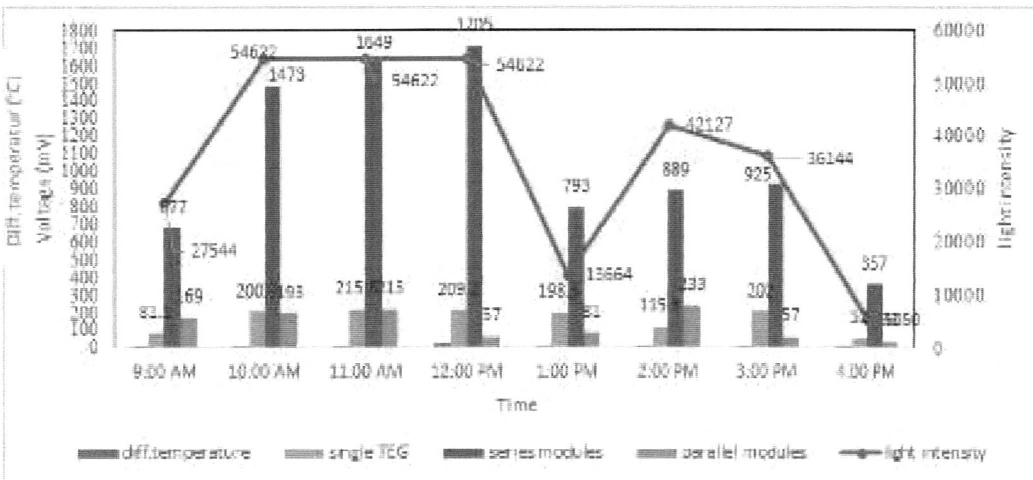

Figure 4. Comparison of voltage, light intensity, and temperature variables

The precise placement of this sensor also affects the temperature value displayed on the measuring box. This is measured from the designation of the humidity value resulting from the light intensity, temperature, and voltage. This follows natural conditions, where the temperature is opposite to the humidity value. Figure 4 shows a comparison of the parameters of voltage, light intensity, and temperature in the single TEG composition measurement configuration. In this condition, the increase in the fair intensity value from 27544 lux to 54622 lux causes the temperature difference to rise by 14.92^{0}C, increasing the output voltage value of 209.2 mV. The output voltage increased until 11 am with a voltage value of 215.8 mV. There was an increase in temperature of 18.08 ^{0}C. The lowest temperature was obtained at 4 pm, where the light intensity decreased by half the previous value and resulted in a temperature difference of only 4.08 ^{0}C. This decrease resulted in the achievement of the TEG output voltage is at the lowest level of 52 mV.

4. Conclusion

The hardware design is a TEG trainer construction with eight variations of the SP1848-27145 type generator with series and parallel variations. From the interpretation of the module, it shows that the series circuit gives an enormous voltage value, which is 1705 mV. The condition occurred at 12 am.

with a voltage value where an increase in temperature of 16.08^{0}C. The value of light intensity and the resulting temperature difference affects the output voltage significantly because it is related to the working principle of TEG that utilizes heat.

Acknowledgments
The authors would like to thank the financial support from the engineering faculty, Tadulako University, and the support for electrical machinery laboratory of the electrical engineering department of Tadulako University.

References

[1] https://wwwieaorg/reports/world-energy-outlook-2020 accesed on January 4, 2020
[2] Salama M and YA, Rahman 2020 Distributed photovoltaic integration as complementary energy: consideration of solutions for power loss and load demand growth problems *Przegląd Elektrotechniczny* 96 (9)
[3] Vining, C B, Rowe, D M 1995 CRC Handbook of Thermoelectrics Ed DM Rowe CRC Press Inc Florida pp 277-282
[4] Khamil, K N, Sabri, M F M, Yusop and A M, Sharuddin 2018 An evalyuation of TEC and TEG characterization for a road thermal energy harvesting *International Conference on Sustainable Energy Engineering and Application (ICSEEA)* pp 86-91
[5] Basri B, Mustofa, Patabang D, Rahman YA and Piarah W 2018 Temperature Control of Soft Drinks on Vehicle with Portable Storage Thermoelectric Cooler *EPI International Journal of Engineering* 1(2) pp 47-50
[6] Zoui M A, S, Bentouba, JGStocholm, MBourouis, 2020 A Review on Thermoelectric Generators: Progress and Applications *Energies* 13 3606
[7] Teffah, K, Zhang, Y, Mou, X L 2018 Modeling and experimentation of new thermoelectric cooler–thermoelectric generator module *Energies* 11(3) 576
[8] Champier D 2016 Thermoelectric Generators: A Review of Present and Future Applications In Springer Proceedings in *Energy Springer Science and Business Media* LLC Cham Switzerland pp 203–212
[9] Date A, A Dixon, C Akbarzadeh 2014 Progress of thermoelectric power generation systems: Prospect for small to medium scale power generation *Renewable Sustain Energy* Rev 33 pp 371–381
[10] Elsheikh MH, Shnawah DA, Sabri MFM, SM Said, Hassan, MH Bashir, MBA and Mohamad M 2014 A review on thermoelectric renewable energy: Principle parameters that affect their performance *Renew Sustain Energy* Rev 30 337–355
[11] Jaziri N, Boughamourad A, J Müller, B Mezghania, F Tounsia, MIsmail 2020 *A comprehensive review of Thermoelectric Generators: Technologies and common applications*, Energy Reports 6(7) pp 264-287
[12] Izidoro CL, Ando OH Junior, JPCarmo, LSchaeffer 2017 Characterization of thermoelectric generator for energy harvesting, Measurement 106 pp 283-290
[13] Wang, W, Cionca, V, Wang, N, Hayes, M and O'Mathuna C 2013 Thermoelectric energy harvesting for building energy management wireless sensor network *Int J Ditrib Sens Netw* 9 (6)

Study Activities of Street Vendors in The Utilization of Space in Tatanga District, Palu City (case study: I Gusti Ngurah Rai Corridor)

Rezki Awalia[1], Rachmat Saleh[1], Andi Chairul Achsan[1], Dwinsani Pratiwi Astha[2], Chistin Dara Maria R[3]

[1]Department of Architecture, Faculty of Engineering, Tadulako University, Palu, Indonesia.
[2]Departement of Civil and Planning, Institut Teknologi Kalimanan University, Balikpapan. Indonesia
[3]Program of Urban and Regional Planning Student, Faculty of Engineering, Tadulako University, Palu, Indonesia.

rezkiawalia44@gmail.com

Abstract. Street vendors (PKL) are included in the activities of the informal sector. Street vendors have the characteristic of buying and selling goods on the roadside or sidewalks. The presence of street vendors in the I Gusti Ngurah Rai Corridor is due to the low number of job opportunities, resulting in competition in finding work. Through this phenomenon, this study aims to study the activities of street vendors in the use of space. This research uses the Univariate (Proportion) analysis method. The results of the research related to the characteristics of street vendors in the corridor were 53% with >15-30 years, dominated by women; high school education level; 87% came from Palu City; 26% trading experience > 5 years and type of family-owned business and 36% income > 500,000-1 million. The activity of street vendors in this corridor is 60% dominated by consumer goods with a linear distribution pattern; 56% Static trading facility type; 50% of trading facilities in the form of containers/kiosks and semi-static service characteristics; 88% of street vendors occupy trading locations on the street and 52% on average 5-8 hours of trading time. The main factor in choosing a street vendor business location is close to consumers.

Keywords: **Street Vendors Activities; Space Utilization; Palu**

1. Introduction

The informal sector usually represents a small-scale based economic activity and often faces many difficulties in establishing formal relationships. The general characteristics included in this sector are the educational background and skills that are less skilled and usually also included in this informal sector are migrants who want to find work but do not have the skills. The presence of the informal sector has an important role in urban economic growth. In addition, the informal sector is present due to inequality in population and employment, resulting in competition in finding jobs which the formal and industrial sectors have not been able to accommodate. This phenomenon can also be found in Palu

City, Central Sulawesi Province, after the 2018 natural disaster and Covid-19, the economy of Palu City has not fully recovered, as can be seen from the low job opportunities in the formal sector and the low number of investors entering and opening up job opportunities in the industrial sector. This condition has resulted in the development of informal sector activities as an alternative in reducing the level of unemployment. One of the informal sectors that are developing in Palu City is Street Vendors (PKL). Initial data on the number of street vendors in the corridor I Gusti Ngurah Rai amounted to 340 street vendors with activities dominated by buying and selling activities in the form of food/beverages. The existence of street vendors makes it very easy for people around the I Gusti Ngurah Rai Corridor to meet their needs. However, the impact of the presence of street vendors in the I Gusti Ngurah Rai corridor has a negative impact, including congestion in the corridor, environmental pollution caused by their activities such as waste generation, the transfer of road shoulder functions, and the absence of street vendors arrangement as one of the support activities in The area creates a visual disturbance of a city by giving a chaotic impression.

2. Literatur review

2.1. Definition of Street Vendors (PKL).

According to [8], street vendors are small businesses carried out by people who have low income (daily salary) and have limited capital. In the economic field, these small traders are included in the informal sector. According to [9], street vendors have the same meaning as "hawkers" which are defined as people who offer goods or services for sale in public places, especially sidewalk streets.

According to [6], the characteristics of street vendors are as follows:

a. Business activities are not well organized
b. Do not have a business license
c. Irregular in business activities, both in terms of place of business and working hours
d. Clustering on the sidewalk or the edge of the protocol road, in the center of the crowd
e. Peddling their wares while shouting to run towards potential buyers or consumers.

According to [6], the characteristics of street vendors are as follows:

a. Generally, belong to the productive workforce
b. Generally, it is the main source of livelihood
c. Low level of education
d. Previous work was generally a farmer or laborer
e. Low capital and small sales turnover
f. Merchandise is generally in the form of food, clothing, and secondary.
g. The income level is relatively low to meet the needs of families in urban areas.

2.2. Characteristics of street vendor (PKL) activities.

2.2.1. Types of street vendors. According to [5], the type of street vendors' merchandise is strongly influenced by the activities around the area where the traders are active. The types of merchandise offered by street vendors can be grouped into 4 (four) main groups, namely:

a. Unprocessed and unprocessed foods, including raw foods, such as meat, fruits, and vegetables.
b. Ready-to-eat foods, such as rice and side dishes, and drinks.
c. Non-food goods, from textiles to medicine.
d. Services, which consist of various activities, such as barbers and so on.

Based on the goods or services traded, according to [8], street vendors can be grouped as follows: 1). Beverage traders; 2). food traders; 3). Fruit traders; 4). Vegetable traders; 5). Meat and fish traders; 6). Cigarette and drug dealers; 7). Traders of books, magazines, and newspapers; 8). Textile and clothing traders; 9). Haberdasher; 10). Flea Traders; 11). Auto parts dealers, gasoline and kerosene; 12) traders of chickens, goats, birds; 13). Rice traders; and 14). Service seller.

2.2.2. Form of street vendors trade facilities. The forms of trading facilities used by street vendors according to [9] are as follows:

a. Carts/pushcarts, this facility consists of 2 (two) types, namely carts/trolleys without roofs and carts/trolleys with roofs.
b. Pikulan/basket, this form of the trading facility is used by mobile street vendors (mobile hawkers) or semi-permanently (semi-static).
c. A semi-permanent stall, consisting of several carts/buggy carts arranged in a row and equipped with chairs and tables.
d. Kios, a form of street vendor facility, uses boards arranged in such a way that it resembles a semi-permanent booth
e. In a mat/base, street vendors use a mat in the form of a mat, cloth or other to peddle their wares.

2.2.3. Types of street vendors trade facilities. [5] explain that there are 3 (three) classifications of the types of trading facilities used by peddlers in selling, namely:

a. Moving (mobile), this type is a vendor who uses easy-to-carry trading facilities to sell around or move around.
b. Semi-permanent (semi-static), the facilities used are in the form of kiosks or other facilities that can be "cleaned up" after they are finished selling.
c. Permanent (static), These facilities are generally and are often found in the form of kiosks that are located either on the road or in public spaces.

2.2.4. Street vendors spread pattern. Based on the pattern of distribution of activities, according to [5] are: The pattern of distribution in groups, is the occurrence of the grouping of similar traders and the pattern of longitudinal distribution, is the pattern that occurs along the main road or on the road that connects the main road.

2.2.5. Street vendors service pattern. [1] say that the service pattern of street vendors is the way street vendors' activities are located in utilizing the activity space as a place of business. Based on this, the pattern of street vendors' services is reviewed from several aspects. Among them is 1). service function; 2). classified as service users; 3). Service time.

2.2.6. Street vendors business location selection factors. According to [2] that the construction of a place of trading activity is very dependent on the location. Likewise with street vendors, where their activities will approach strategic locations that have a high level of visits. [5] state that street vendors generally tend to be in areas with high population density by being located at transportation nodes or adjacent to places that have activities such as entertainment, markets, companies, and open spaces. [7] reveals several factors that influence the selection of PKL business locations, namely 1). Proximity to consumers; 2). Proximity to housing/settlement; 3). Proximity to competitors; 4). Proximity to suppliers; 5). Proximity to the road.

3. Research methods

The research approach used by the researcher is a rationalistic deductive quantitative approach. The data needed in this study are based on variables from the synthesis of theories related to the characteristics of street vendors, street vendors' activities, and factors for choosing the location of street vendors' businesses. Data collection techniques were carried out with instruments, namely questionnaire sheets, interview guidelines, and observation guides. The sampling technique is a sampling technique to determine the sample to be used in research. In carrying out this research, there are two sampling techniques used in this study, namely random sampling, and purposive sampling.

Based on population data in the figures, it is known that the total population along the I Gusti Ngurah Rai corridor is 340 people. The sample calculation according to the formula and the data is as follows (the error rate used is 10% or 0.1). [3]

$$\text{Number of Samples} = \frac{340}{340\,(0{,}1)^2+1} = 77\ respondents \quad (1)$$

Based on the calculation of the sample above, the number of samples obtained is 77 respondents. The time for collecting questionnaire data is divided into 2 (two) sessions, namely morning and afternoon. For the portion of each sample that is used as a respondent will be explained 15 respondents for the morning, and 50 respondents for the afternoon, and 12 respondents for the evening. Purposive sampling is used to draw a sample of the sources to be interviewed. This method is a method that is not random or non-random sampling so it is based on research decisions with the consideration that the selected respondents are really experts in answering research questions.

The analytical technique used in this research is the Univariate Descriptive analysis technique. This univariate analysis provides a general description of a phenomenon observed using frequency and proportion or percentage analysis tools. This analysis will be used to convey and describe observations in the field in simple and easy-to-understand language. The formula used in the analysis. The value of the proportion varies from a value of 0.00 to a value of 1.00. Proportions or percentages are:

$$\text{Percentage} = \frac{f}{N} \times 100\% \quad (2)$$

f =frequency; the number of cases in a category
N = number of cases in all categories

4. Result and discussion

Palu City is located on the plains of the Palu valley and Palu Bay, with an average height of 0 – 700 meters above sea level, located at a position of 0º.36" - 0º.56" South Latitude and 119º.45" - 121º.1" Longitude. East. The area of Palu City is 395.06 km^2 which is located in the Palu Valley plain and Palu Bay. This research is located along Corridor Jl. I Gusti Ngurah Rai which is located in Tavanjuka Village, Tatanga District. Corridor Jl. I Gusti Ngurah Rai is one of the connecting roads in Tavanjuka Village, Tatanga District. This road is included in the city road category with a function as a primary local road, this is written in the Palu City Spatial Planning (RTRW) 2010-2030. Jalan I Gusti Ngurah Rai seen from its function and classification is included in the road with a high level of activity.

In the morning the presence of street vendors is still very little encountered, this is because potential consumers are still busy with their busy activities, however, entering the afternoon, street vendors will increase more and more with various types of goods and services traded. The following is a picture of street vendors who use shop-house parking lots and road shoulders as trading locations. It can be seen, in the picture above, that street vendors use the shoulder of the road as their location for selling and take advantage of the parking lot from one of the shops or shophouses located in the corridor of Jl. I Gusti Ngurah Rai.

Figure 1. The activities of street vendors who trade by using the front of the shophouse on a mat/pedestal and carts

4.1. Analysis of the characteristics of street vendors (PKL)

In analyzing the study of the activity of Street Vendors (PKL) in the I Gusti Ngurah Rai Corridor, it is studied based on characteristics. Which characteristic in question is a characteristic or usually means something/characteristic that is typical of the criteria for a particular object. These characteristics have certain properties that make the object different from other objects. In this study, to analyze the use of street vendors space based on the characteristics of street vendors, it can be analyzed based on the following indicators :

Table 1. Characteristics of street vendors by Age Group

ANALYSIS	AGE GROUP			
	15-30 year	**31-45 year**	**61-75 year**	**>75 year**
Respondent	41	15	20	1
Total	77	77	77	77
Proportion	0,53	0,20	0.26	0,01
%	53%	20%	26%	1%

Based on the percentage value of the age group indicator, 53% of the respondents are street vendors in the corridor of Jl. I Gusti Ngurah Rai average age 15-30 years. The percentage results illustrate that there are low employment opportunities in Palu City, so that in meeting the needs of daily life, people prefer to engage in this activity, because street vendors are small businesses that can be done regardless of age limit, do not require special skills, can be done anywhere with small capital.

Table 2. Characteristics of street vendors by Gender

ANALYSIS	GENDER	
	Man	**Women**
Respondent	27	50
Total	77	77
Proportion	0,35	0,65
%	35%	65%

Based on the percentage value of the age type indicator, 65% of street vendors respondents in the I Gusti Ngurah Rai Corridor are dominated by women. According to the research results of Evers in Wardana, it is stated that in the informal sector activities are more dominated by female traders who carry out activities during the day. In the I Gusti Ngurah Rai corridor from morning to evening it is dominated by female traders who sell their wares. And at night, usually, male traders dominate, because the condition of the I Gusti Ngurah Rai Corridor is more vulnerable than during the day.

Table 3. Characteristics of street vendors by Education Level

ANALYSIS	EDUCATION				
	SD	**SMP**	**SMA/ SMK**	**S1**	**S2**
Respondent	13	20	42	2	0
Total	77	77	77	77	77
Proportion	0,17	0,26	0,54	0,03	0
%	17%	26%	54%	3%	0%

Based on the latest education level indicators, it can be seen that the general level of education for street vendors in Corridor I Gusti Ngurah Rai is high school graduates (percentage value of 54%). By looking at this phenomenon, it can be illustrated that the low level of PKL education results in a low understanding of regulations and policies regarding the use of space in the corridor, where it is found that the corridor is used as a business location for street vendors. In line with [5], that Street vendors are traders who sell in public places without having permission from the government. Most high school graduates are in the street vendors profession due to the low employment opportunities and carrying out these informal sector activities do not require special skills or skills.

Table 4. Characteristics of street vendors by Area of Origin/Place of Residence

ANALYSIS	LOCATION OF ORIGIN?RESIDENCE	
	In Palu City	**Outside Palu City**
Respondent	67	10
Total	77	77
Proportion	0,87	0,13
%	87%	13%

The characteristics of traders based on their area of origin or place of residence need to be known, to get an overview of the area of origin of traders who trade in the Jl. I Gusti Ngurah Rai. Based on the results, 87% stated that they came from Palu City, but most of the respondents lived outside the Tatanga District. These results indicate that the corridor Jl. I Gusti Ngurah Rai is a strategic area that is of interest not for trading. In addition, the corridor is classified as a primary local road with the busiest activity, which is an advantage for street vendors because it connects settlements, and in the 2010-2030 Palu City Spatial Plan, this corridor is included in the trade and service area. The characteristics of street vendors are choosing a strategic location, being close to formal activities and consumers, and do not require a permit to carry out trading activities.

Table 5. Characteristics of street vendors based on Trading Experience

ANALYSIS	TRADE EXPERIENCE				
	1 year	**2 year**	**3 year**	**4 year**	**>5 year**
Respondent	10	13	18	16	20
Total	77	77	77	77	77
Proportion	0,13	0,17	0,23	0,21	0,26
%	13%	17%	23%	21%	26%

Based on the results of the questionnaire on street vendors in the corridor of I Gusti Ngurah Rai, it was stated that the trading experience was >5 years or more, namely as many as 20 people (26%). Based on trading experience, the street vendors who trade >5 years can be said that they are old traders and are well-established traders, but there are also relatively many street vendors who have opened businesses within a period of <5 years, even some of them said that they opened a business due to the lack of job opportunities in Indonesia. the formal sector, following the trends and daily needs of consumers, and on the other hand, the reason for street vendors to open businesses in the I Gusti

Ngurah Rai corridor is because the accessibility on this road is increasing and becomes a liaison between settlements around the corridor and Sigi district.

Table 6. Characteristics of street vendors by Type of Business

ANALYSIS	BUSINESS TYPE	
	Family-owned	**Non-Family**
Respondent	48	29
Total	77	77
Proportion	0,62	0,38
%	62%	38%

Looking at the characteristics of street vendors based on the type of business, the street vendors in this corridor stated that the percentage of 62% were family-owned businesses. According to [8] the notion of street vendors is one of the characteristics generally every business unit that employs a few workers and comes from family, acquaintances or comes from the same area, this is not much different in the corridor I Gusti Ngurah Rai whose business is dominated by family-owned businesses because their business is classified as a small business with a small turnover, which is used to meet daily needs.

Table 7. Characteristics of street vendors based on Daily Income

ANALYSIS	DAILY INCOME				
	100-200K	**>200-300K**	**>300-400K**	**>400-500K**	**>500K-1Milion**
Respondent	5	7	15	22	28
Total	77	77	77	77	77
Proportion	0,07	0,09	0,19	0,29	0,36
%	7%	9%	19%	29%	36%

Based on the characteristics of street vendors based on daily income, the results of the questionnaire, namely 36%, stated that their incomes ranged from >500 thousand-1 million. However, their income can also increase at certain times such as holidays. In addition, it is influenced by the location of the business. Locations that are close to consumers or residential areas and the formal sector will have higher incomes than locations that are far away from residential areas and the informal sector because the type of merchandise that dominates is the type of consumption merchandise with the target market being people who live and employees from the formal and private sectors.

4.2. Analysis of street vendors (PKL) activities

In analyzing the study of street vendors (PKL) activities in the I Gusti Ngurah Rai Corridor based on the activity variables of street vendors, it can be analyzed based on the following indicators :

Table 8. Street vendors activities by Type of Trade

ANALYSIS	TYPE OF TRADE		
	Goods	**Service**	**Consumption**
Respondent	59	77	204
Total	340	340	340
Proportion	0,17	0,23	0,60
%	17%	23%	60%

The types of merchandise offered by street vendors can be grouped into 3 (three) main groups, based on the percentage value of field observations, namely 60% which is dominated by the category of consumption or unprocessed and unprocessed food or ready-to-eat food, consisting of snacks/drinks, fruits, and vegetables. kitchen spices. According to [5], the type of street vendors' merchandise is strongly influenced by the activities around the area where the traders are active and based on observations in the I Gusti Ngurah Rai corridor, it is dominated by the types of food/beverage and fruit merchandise.

Based on the results of a survey using the observation method, in accordance to what was stated by [5] that in the distribution, street vendors usually apply a group distribution pattern and a longitudinal distribution pattern. So it can be seen that the distribution pattern of street vendors along the I Gusti Ngurah Rai corridor applies a linear or elongated pattern along the main corridor. The application of this pattern is because street vendors use the road and in front of shophouses as trading locations, as well as considering accessibility in reaching consumers to their place of trade. In addition, seen from its function, the function of street vendors services along the I Gusti Ngurah Rai corridor is the function of trade and service services in which street vendors' service activities are part of the city's trading system that functions to market goods and services to meet the daily needs of the community.

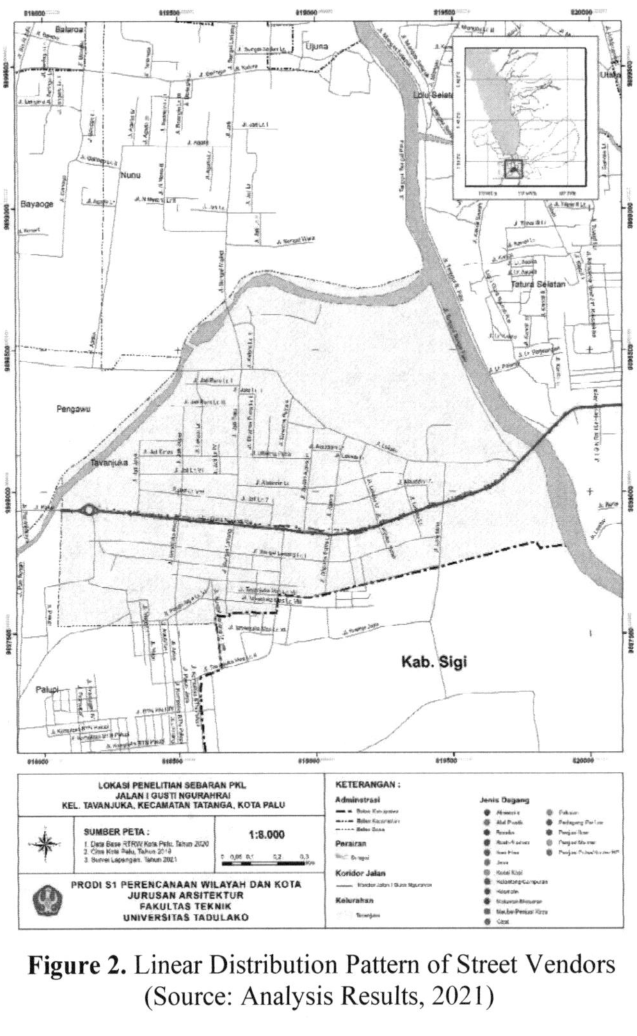

Figure 2. Linear Distribution Pattern of Street Vendors
(Source: Analysis Results, 2021)

Table 9. Street vendors activities based on Types of Trading Facilities

ANALYSIS	TYPE OF MEANS OF TRADE		
	Permanent	**Semi Static**	**Mobile**
Respondent	191	97	52
Total	340	340	340
Proportion	0,56	0,29	0,15
%	56%	29%	15%

The use of space by street vendors in Corridor I Gusti Ngurah Rai, it also determines the type of trading facilities used in carrying out activities. Based on the results of the analysis of the types of trading facilities, the types of trading facilities in the I Gusti Ngurah Rai Corridor are included in 3 (three) classifications of types of trading facilities according to [5], namely the percentage value of 56% are types of trading facilities included in the permanent classification (static) in the form of kiosks/containers and semi-permanent stalls which are permanently located and occupied by street vendors

Table 10. Street vendors activities based on the form of Trading Facilities

ANALYSIS	FORM OF MEANS OF TRADE						
	Cart	**Shop**	**KIos/contener**	**Lesehan**	**Table**	**Pikulan Keranjang**	**Semi Permanent Stall**
Respondent	73	15	169	16	3	6	56
Total	340	340	340	340	340	340	340
Proportion	0,22	0,04	0,50	0,05	0,01	0,02	0,16
%	22%	4%	50%	5%	1%	2%	16%

Street vendors (PKL) utilize various forms of trading facilities in carrying out their activities, as well as street vendors in the I Gusti Ngurah Rai Corridor, based on a percentage value of 50%, namely the form of facilities in the form of kiosks/containers; the percentage value of 22% is that street vendors use carts as a form of trading facility, then the percentage value of 16% is that street vendors use the form of trading facilities in the form of semi-permanent stalls. The street vendors stated that they prefer to use trading facilities in the form of kiosks, stalls, and semi-permanent stalls because apart from being a place of business, they are used as permanent residences. These facilities are categorized as permanent traders (Static). This condition is in line with [4] explaining that the form of street vendors' trading facilities such as kiosks is in the form of boards that are shaped like semi-permanent (semi-static) booths with traders who live permanently in that place.

Table 11. Street vendors activities based on Trade Locations

ANALYSIS	TRADE LOCATION	
	Home Shop	**Body/Shoulder**
Respondent	40	300
Total	340	340
Proportion	0,12	0,88
%	12%	88%

From the results of field observations, it is seen that the use of space carried out by street vendors in activities is more around the functional area located in the I Gusti Ngurah Rai corridor, based on the percentage value 88% choose a location on the shoulder/body of the road, and 22% choose in a front shop in the corridor. According to [9], street vendors are defined as people who offer goods or services for sale in public places, especially sidewalk streets. Another consideration is that street vendors don't have to spend a lot of money to rent a place to trade, they just need to pay for cleaning.

Table 12. Street vendors activities based on Trade Length

ANALYSIS	OPERATING TIME			
	<=4hours	**5-8hours**	**9-11hours**	**>11hours**
Respondent	6	40	12	19
Total	77	77	77	77
Proportion	0,08	0,52	0,16	0,25
%	8%	52%	16%	25%

Informal sector activities are activities that do not have time off, including street vendors in the I Gusti Ngurah Rai Corridor. These trading activities never stop, because most of the street vendors' activities are trading goods to meet the needs of the community in general. Based on the assessment of the length of trade indicator, the percentage value of 52% states that the duration of trading is around 5-8 hours/day. According to [6], one of the characteristics of street vendors is that they are irregular in business activities, both in terms of place of business and working hours. The phenomenon of the characteristics of street vendors was also found in the I Gusti Ngurah Rai corridor, where street vendors did not have regular working hours like formal activities in the corridor. If the merchandise runs out quickly, the working hours of the street vendors are short, and vice versa, so we cannot be sure of the exact number of hours street vendors work per day.

4.3. Analysis of the factors for choosing the location of street vendors (PKL)

In analyzing the study of the activity of Street Vendors (PKL) in the I Gusti Ngurah Rai Corridor based on the variables of the PKL location selection factors, it can be analyzed based on the following indicators :

Table 13. Factors for choosing a street vendors business location (PKL)

ANALYSIS	CONSUMER		SETTLEMENT/ HOUSING		COMPETITOR		SUPPLIER	
	Closer	**Far**	**Closer**	**Far**	**Closer**	**Far**	**Closer**	**Far**
Respondent	76	1	46	17	58	18	41	36
Total	77	77	77	77	77	77	77	77
Proportion	0,99	0,01	0,73	0,27	0,53	0,47	0,75	0,25
%	99%	1%	73%	27%	53%	47%	75%	25%

Based on the indicators in the selection of street vendors (PKL) business location factors, it can be seen from several indicators that play an important role for street vendors in choosing a trading location. of the four indicators, namely, the percentage value of 99% is an indicator of the proximity of street vendors to consumers. The percentage value of 73% is the proximity of street vendors to housing and settlement areas, then the percentage value of 53% is close to competitors (other traders) in other words that formal sector activities and other activities in the I Gusti Ngurah Rai corridor are centralized so that they become the basis for consideration by the street vendors to choose a strategic location, even though it is close to other traders or in other words is close to trading competitors, and the percentage value of 75% is the proximity of street vendors to suppliers. This is explained by [5] that street vendors generally tend to be in areas with high population density by being located at transportation nodes or adjacent to places that have activities such as entertainment, markets, companies, and open spaces. This explanation is supported by the condition, where the distribution pattern of street vendors in the corridor follows along the road or is called a linear pattern. In addition, based on the results of field observations, it can be seen that the type of merchandise is dominated by ready-to-eat food so that raw materials or raw materials can be obtained from other street vendors in the corridor, as well as in the corridor there is a market as a supplier of raw materials or raw materials. ready-to-eat food, there are several shops and supermarkets where street vendors meet their raw materials for wares.

5. Conclusion

In the study of street vendors' activities in utilizing space in the I Gusti Ngurah Rai Corridor, based on the results of the research that has been described regarding the characteristics, activities, and factors of choosing a business location, it can be concluded as follows. Street vendors in the I Gusti Ngurah Rai Corridor are dominated by the age of 15-30 years with female gender and education level at the SMA/SMK level and on average are the people of Palu City. It is classified as an established street vendor with >5 years of trading experience and the street vendor business being run is a family-owned business, with an average income of >500 thousand-1 million per day. The activity of street vendors in the I Gusti Ngurah Rai corridor is dominated by the type of consumption merchandise by applying a linear distribution pattern and the type of permanent (static) trading facilities. In carrying out their activities, street vendors utilize the form of facilities in the form of kiosks/containers with semi-static service characteristics, besides choosing a trading location on the shoulder/front of the road with an average trading duration of about 5-8 hours per day. The I Gusti Ngurah Rai Corridor is a strategic area, so the main factor in choosing a location, namely paying attention to proximity to consumers, is a unique way for street vendors to maintain their presence in the I Gusti Ngurah Rai corridor. In addition, the public relations strategy has been proven to be used by all street vendors in running their businesses.

Acknowledgment

This work was supported and funded by DIPA BLU Faculty Of Engineering, Tadulako University in scheme coaching research in 2021 (Grant No. 433.n/UN28.2/PL/2021)

References

[1] Ambarwati Sri Hany 2003 Study of Street Vendor Activities in Space Utilization in Salatiga City Thesis is not published Postgraduate Masters Program in Urban Development Engineering, Diponegoro University, Semarang
[2] Budi, Ari Sulistiyo 2006 Study of the Location of Street Vendors Based on PKL Preferences and Perceptions of Surrounding Communities in Pemalang City, Postgraduate Masters Program in Regional and Urban Development, Diponegoro University: Semarang
[3] Bungi, H M Burhan, 2010 Quantitative Research Methodology: Communication, Economics and Public Policy and Other Social Sciences First Edition 5th Printing Jakarta Kencana Prenada Media Group
[4] Dwiyanti, Yanivirzal 2005 Study of PKL Space Patterns and Characteristics of Their Activities in the Panbil Area of Batam City, Postgraduate Masters Program in Urban and Regional Development, Diponegoro University: Semarang
[5] McGee, T G and Y M Yeong 1977 Hawkers in Southeast Asian cities: planning for the bazaar economy Ottawa: International Development Research Center (IDCR)
[6] Meilani, Lucky, 2012 Strategy of the Bandar Lampung City Market Management Service in fostering Street Vendors at the Yellow Bamboo Market, Lampung University Lampung
[7] Nasta, Mardiyanshar Muhammad, 2017, Analysis Of Business Location Selection Factors Street Vendors [thesis] Makassar: Engineering faculty- UNHAS Makassar
[8] Umboh AGGS 1990 Street Vendor Job Opportunities in Bitung Administrative City Center [thesis] Bogor: Postgraduate Faculty of KPK IPB – UNSRAT Manado
[9] Widjajanti, Retno, 2000, Physical Arrangement of PKL Activities in Commercial Areas in the City Center (Case Study: Simpang Lima Semarang) Thesis is not published Special Area of Urban Planning, Masters Program in Urban and Regional Planning, ITB, Bandung

Architectural Model using Local Initiative of Kampung Lere Fishermen after the Tsunami Disaster

M Bakri[1], Amar[1], A K Kasim[2], S Khomaria[3]

[1]Department of Architecture, Faculty of Engineering, Universitas Tadulako
[2]Department of Information Technology, Faculty of Engineering, Universitas Tadulako
[3]Study Program of Architecture, Faculty of Engineering, Universitas Tadulako

embakri@gmail.com

Abstract. The Kampung Lere is located in the earthquake and tsunami-prone route. It makes the area prone to earthquake-tsunami disasters, such as on September 28, 2018. Then, the government makes mitigation and conducts studies, also establishing spatial planning policies such as maps. Zoning of disaster-prone areas with the criteria of a red zone (not to be inhabited), a yellow zone, and a green zone. This condition then impacted the Lere fisherman community, which entered the red zone area. At the same time, many people were affected by the tsunami and did not want to move to permanent housing. The Lere fisherman community, after the tsunami, still choose a local initiative to manage space where they have occupied the area for generations. Detailed identification of the space is carried out to solve the problem. The study of related policies and in-depth interviews to obtain accurate data and information was then processed into a formula to be used as a benchmark in formulating design directions based on the local initiative of the fishermen of The Kampung Lere. The architectural design development model in the Lere area will later be able to bridge the government's problems as a policymaker and the desire of the fishermen of Lere Village for the right to manage the space. The results of this study recommend an architectural model for residential that provides opportunities for the people of Lere to continue their activities in the area and still consider the element of safety in the development process as well as strategies for realizing the architectural model.

Keyword. Architectural Model, local initiative of Kampung Lere, Fishermen

1. Introduction

On September 28, 2018, Palu City experienced an earthquake and tsunami. It shows that the Palu City area is included in a disaster-prone area. The 7.4 SR earthquake was the worst area because an extreme liquefaction disaster hit it. At the same time, the tsunami waves affected the coastal areas. This incident became the basis for the study of determining the zoning of the disaster-prone regions based on disaster-prone maps [1]. The zoning map provides information on disaster-prone areas with the criteria of red zones (not to be inhabited), yellow zones, and green zones. Based on the BNPB zoning, the location of the research area is included in the red zone category with the provision that the site is not allowed to be built. This condition then greatly affected the fishing community in Lere village. They lost their homes due to the hit of the tsunami waves, lived in refugee camps until there was a rule that they could not build in the red zone area they previously occupied.

The Lere area, based on data, is indeed the worst in the city of Palu, which was affected by the victims of the earthquake and tsunami. There were 93 fatalities with 267 houses missing and heavily damaged, with details of the number of homes being heavily damaged as many as 69 units, houses being moderately damaged as many as 62 units, and lightly damaged as being 83 units. The problem is that the people of Lere village in the shelters, shelters, and refugee camps of the Grand Mosque reject the relocation to Permanent Residence (HUNTAP). Because of government policy, it still provides opportunities for entrepreneurs to carry out activities in locations designated as red zones, such as Mercure Hotel, PGM (PALU GRAND MALL), Grand Duta Hotel, and Exelco. So that the government's policy of taking over community land and then replacing it with permanent housing located in Tondo Village was rejected. Therefore, the problem of fishermen in Lere village should be handled appropriately and wisely to find the right solution or formula to answer the rejection through a model of developing fishermen's residential architecture in Lere village after the tsunami. It is essential because Lere village has historical value for the development of the city of Palu and is a residential area that still uses the concept of a fisherman's room which was passed down from generation to generation by their ancestors in their daily lives. This concept is known as the NAKAPALI space concept, which is formed based on fishermen's behavior of the Kaili tribe [2].

The architectural model proposed in this study comes from the local initiative of the Lere Village community as residents of the coastal area who work as fishermen. It is hoped that this architectural model will become a liaison between the government and the residents of Kampung Lere in building new housing during the earthquake and tsunami in Palu Bay, especially the Lere Village area.

2. Literature Review

Fishermen are a group of people whose lives depend directly on marine products, either by catching or cultivating. They generally live on the beach, a space and residential environment close to the location of their activities [3].

Fishermen are identical with limited assets, weak capital capacity, bargaining position and market access [4]. In fact, fishermen are not a single entity, they are made up of several groups. In terms of ownership of fishing gear, fishermen can be divided into three groups, namely: labor fishermen, skipper fishermen and individual fishermen. Labor fishermen are fishermen who work with other people's fishing gear. On the other hand, skipper fishermen are fishermen who have fishing gear operated by someone else. Furthermore, [3] explains that individual fishermen are fishermen who have their own fishing equipment and do not involve other people in their operations.

The opinion of [3] above when associated with [4] shows the relationship between the development of fisherman settlement space which reflects the occurrence of economic, socio-cultural activities close to the location where they live, weak bargaining position and market access, causing fishermen to be grouped as the majority of low-income people, specifically labor fishermen and individual fishermen.

[5] explain that fishermen can also be defined as people or communities of people who as a whole or part of their lives depend on fishing activities. Some groups of fishermen have different social and demographic characteristics. These differences can be seen in the age group, education, social status and trust in a group of fishermen, and internal cohesion differences are often found in terms of the relationship between fellow fishermen and in social relations.

According to [6] fishermen groups can be divided into four groups, namely:

1) subsistence fishermen, namely fishermen who catch fish only to meet their own needs.
2) native fishermen, namely fishermen who more or less have the same character as the first group, but also have the right to carry out commercial activities even though on a very small scale.
3) recreational fishermen, namely people who in principle carry out fishing activities just for fun or sports.
4) commercial fishermen, namely those who catch fish for commercial purposes or are marketed for both the domestic market and the export market.

Besides these groupings, there are several terms that are often used to describe groups of fishermen, such as full-time fishermen who depend on their entire life from fishing, casual fishermen for those who only part of their lives depend on fishing, skipper for those who have natural resources. economy for fishing businesses such as boats and fishing gear and crew members for them to allocate their time and earn income from operating fishing gear, such as boats owned by the cliff [5].

The opinion of [6] above when associated with [5] shows the relationship between groups of fishermen in doing various things to meet their own needs, sell, exercise and catch fish with the aim of being marketed locally and for export. This reflects that fishermen have groups and are stratified in relation to fishing activities.

This argument, when linked to the definition and classification of fishermen, is intended to explain how the stratification of fishing communities plays a role in the process of catching fish and marketing their catch to meet local and export needs. research in relation to improving the performance and economic outcomes of fishing communities.

3. Methodology

The method used in this research is the architectural design method, including formulating problems, collecting data through surveys, analysis including micro and macro analysis, synthesis, and conclusions. Design recommendations were obtained based on the findings, which became the architectural design concepts for the fishermen's residence in Kampung Lere after the tsunami in Kota Palu.

3.1. Data

The data collected in this research include Primary Data and Secondary Data. The sources of information in the research consist of primary data and secondary data. Preliminary data were obtained by surveying the research location. The data included:

1. The observations in the field.
2. The views from some respondents in the form of recording files or writings.
3. The documentation in the form of pictures of the situation in the field, which were then expected to provide solutions.

Secondary data was obtained by conducting a literature study that studied theories sourced from trusted reference books, magazines, and websites. Then combine the existing data with documents from government agencies related to the problems raised in this paper.

3.2. Analysis

The stages of analysis in the study are as follows: 1) Macro analysis, consisting of the existing condition of the selected location covering the land area, vegetation condition, access to the physical boundary of the place, sun noise orientation, wind circulation, and view, landscape, and zoning. 2) Micro Analysis, consisting of analysis of activities and capacity, analysis of perpetrators of spatial analysis, spatial relationship patterns, circulation analysis, zoning analysis, lighting analysis, ventilation analysis, utility analysis, shape analysis, and building structure analysis.

3.3. Synthesis

Synthesis integrates two or more existing elements to produce something new or a combination of several analyzes into one new idea or output.

The methodology used in this study can be seen in Figure 1.

Figure 1 Research Methodology

4. Result and Discussion

Fishermen's Settlement in Lere Village uses several circulation patterns, namely grid and linear circulation patterns. This pattern is also maintained by considering access to disaster response in the area and the growth of housing for the surrounding population in the next few years. The linear pattern can also maximize circulation into the site, for achievements with vehicles it can be developed with a linear pattern following the tread pattern in the coastal area. Meanwhile, the grid pattern can be developed with canals that separate buildings or dwellings. The spiral pattern is used or developed for parking circulation in the canal.

The planned residential unit is a building that is only used as a place to rest or work for fishermen, this planned building is not a fisherman's house to live, this is based on the condition of the settlement site which is included in the red zone area or disaster-prone area. Buildings for fishermen are supported by the existing spaces in simple houses that function to support all fishing activities, and the existence of a multipurpose room that has universal functions that can be developed by the owner. The residential unit is planned to function as a post-sea resting place, the unit planned in the area can be inhabited by 2 heads of families. This is considered because usually in 1 house consists of 2 Heads of Families and is also adjusted to the number of building units before the tsunami.

The pattern of relationships between spaces serves to classify spaces based on the function of space contained in each building in the Fisherman's Settlement in Lere Village. The following is the pattern of spatial relationships in each building in the Kampung Lere Fisherman's Settlement which is illustrated by a bubble diagram:

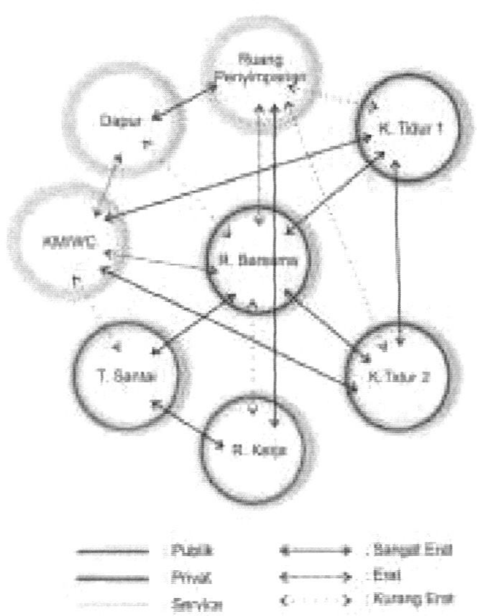

Figure 2 Pattern of Relationships Between Residential Spaces

The things that underlie the shape of the building:

4.1. *Activities and Culture*

Most of the people of Lere village work as fishermen, so their main activity is fishing and selling marine products, so attention is needed in considering the shape of their building.

4.2. *Environmental conditions*

Based on environmental conditions, Lere village is one of the settlements located in the coastal area. This settlement on September 28, 2018 was destroyed by the tsunami and earthquake, so that this settlement is no longer suitable for habitation. However, because the settlement is the residence of the local community which is close to their livelihoods, the residential unit that will be designed only functions as a temporary residential building. In addition, the design of the building by considering the environmental conditions on the coast is utilized by the use of houses on stilts to prevent flooding, and to maximize natural ventilation into the building.

4.3. *Culture value*

Based on history, Lere village is one of the places visited by the Dutch government to collaborate with King Kaili. So it is interesting to be implemented in the form of buildings.

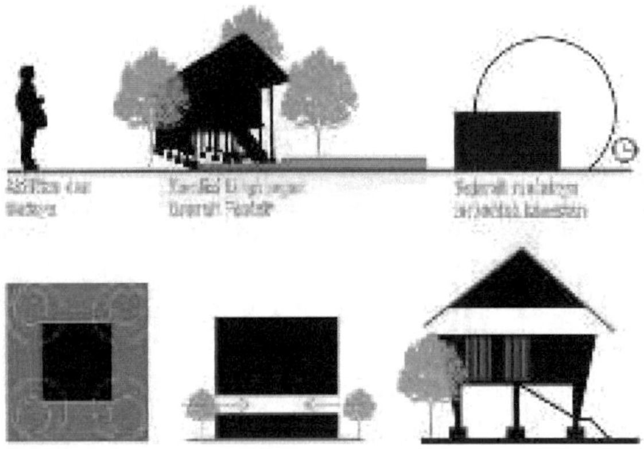

Figure 3 Building Form Approach

Based on the description above, it can be concluded that the fishermen's settlement in Lere Village has transformed the architectural form of Jengki, where the design of this building considers the tropical climate with Dutch architectural elements.

5. Conclusion

Fishermen's Settlement in Lere Village after the Tsunami can be a solution for the people of Lere Village who were affected by the tsunami on 28 September 2018 in maintaining their land ownership rights, most of whom work as fishermen in disaster-prone zones that have been determined by the government. The innovations used to overcome this problem are the use of vegetation, the creation of artificial lakes and disaster evacuation routes. In the manufacture of artificial lakes, canals are made that use open and close due to the ebb and flow of sea water. The fishermen's settlements in Lere Village after the tsunami were designed by taking into account the socio-cultural aspects of the Kaili people who work as fishermen in Lere Village. The application of the form of the jengki building can be applied to become an attraction for settlements, making it a distinctive feature that is different from other buildings in the city of Palu and also as a symbol of a historical one based on the fact that the Lere village area had also been visited by the Dutch. Fishermen's settlements in Lere Village based on the division of space in to kaili are divided into three spaces, namely activity space, intermediate space and residential space.

Acknowledgement
This work supported and funding by Fakultas Teknik Universitas Tadulako in the scheme "Penelitian Unggulan" in 2021 with Grant Number 430.ab/UN28.2/PL/2021 on April 26th, 2021

References

[1] Badan Standar Nasional. 2004. *SNI 03-1733-2004 tentang Tata Cara Perencanaan Lingkungan Perumahan dan Perkotaan.*
[2] Bakri, Muhammad. 2014. *Pola Ruang Masyarakat Nelayan Teluk Palu Studi Kasus : Nelayan Kampung Lere Teluk Palu.* Serap#3. Yogyakarta
[3] Mulyati, Ahda. 2007. Kajian Luas Rumah Tinggal Masyarakat Berpenghasilan Rendah di Kawasan Pusat Kota. Smartek.
[4] Cahyono, Putro. 2018. Skripsi. Desain Permukiman Pesisir Dusun Simbe. Palu: Universitas Tadulako.
[5] Widodo, Suadi. 2006. Pengelolaan Sumberdaya Perikanan Laut. Yogyakarta: Universitas Gajah Mada
[6] Charles, A. 2001. *Sustainable Fishery System. United Kingdom: Blackwell Science.*

Identification and treatment of black spots in the road network around Universitas Tadulako

A Setiawan[1], R Ramlan[1] , E R Labaso[1], A A Nur[1]

[1]Department of Civil Engineering, Engineering Faculty, Universitas Tadulako

ramlanratnasari@gmail.com

Abstract. Safety is an essential thing that must be met in the implementation of traffic activities on the highway. Every effort is made to increase traffic safety. The road network around Universitas Tadulako has a high traffic flow so that the number of accidents increases. The accidents are indicated by several black spot locations. This condition requires research that can identify and find the proper black spot treatment. Therefore, the objectives of this research are to identify and discover the proper black spot treatment on the road network around campus. The method used in this research is an interview of road network users around the campus (students, community members, and the authorities) and inspections of locations that can become black spots. Three hundred forty-six respondents were collected and analyzed. The results show that the location of the black spot around the Universitas Tadulako is in front of the gate of the Universitas Tadulako lecturer housing. The proposed treatments at the black spot location are controlling street vendors' spots, optimizing traffic lights, setting U-turns, cutting several trees, and repairing road pavements.

Keywords: Black spot, Safety, Accident, In-depth interview, Universitas Tadulako

1. Introduction

Traffic accidents define as events on the road that are unexpected and unintentional involving vehicles with or without other road users resulting in human casualties and/or property loss Article 1 point 24 of Law Number 22 of [1]. [2] data for 2019 states that there have been 7000 traffic accidents in Indonesia in a month. It is not surprising that the World Health Organization (WHO) data in the global status report of road safety states that traffic accidents are the number one cause of death in the world which results in more than 1.2 million fatalities every year [3]. Thus, it can be interpreted that traffic accidents are events that must be considered in the implementation of activities for the movement of people and vehicles on the highway in the context of fulfilling transportation safety. Improving road safety is currently a priority for every transportation practitioner.

All efforts were made to increase the level of traffic safety. One of which is by determining and fixing black spots on the highway. Black spots are defined as locations on highways that are prone to accidents, whereas in urban areas the road segment is limited to 20-30 meters long, while for inter-urban roads it is around 500 meters [4]. Determination of the location of the black spot can be done by visually observing the sites where accidents often occur in a certain period of time. After the location of the existing black spot is identified, it can be repaired at that point so that the accident's potential is minimized. Several previous studies that discuss the identification of black spots as an effort to reduce the number of accidents include the research conducted by [5], which compares the total annual accident data (AAT) against potential accident reduction (PAR). The two data shows that there is no significant difference in black spots determination. [6] identified black spot locations using a GIS-based method. In addition, [7] use the Upper Control Limit (UCL) methods in managing traffic accident data from the police. On the other hand, [8] used the method of weighting the accident equivalent number according to Rev. T-09-2004-B in 2004 to identify black spots. The latest research conducted by [9] conducted a fusion analysis of the developmental factors that allegedly influenced the traffic accidents due to black spots. The various methods described above show that there is no patented black spot identification method. Hence, the determination of black spots on the road can use some distinct approach according to the conditions of the observation location. Therefore, this study uses visual observations and questionnaires concerning the sites of black spots. These methods were conducted because of the lack of supporting data which is accident data at the research site. The research location is the road network around the Universitas Tadulako, Palu City, which are consists of Yos Sudarso Street, Hangtuah Street, Tombolotutu Street, Soekarno Hatta Street 1, Jabal Nur Street, RE Martadinata Street, Untad Street 1, and Soekarno Hatta 2. These sites were chosen since every day many vehicles pass through the road network, coupled with increased movement growth due to the opening of new venues, namely urban forest parks, cafes along Soekarno Hatta roads, construction of Permanent Residential (Huntap), construction of the outer ring road of Palu City and Talise bridge repairs. Preliminary research can indicate that the potential for accidents is in the high category on the road network.

The fact is supported by accident data from the East Palu Police, which reveals that the number of accidents on Soekarno Hatta and surrounding roads is relatively high. Thus, this study was conducted to identify and discover the suitable black spot treatment on the road network around campus.

2. Research methodology

The research methodology used in this study is divided into several parts:

2.1. Research location

This study aims to identify and discover the appropriate treatment for black spots on the road network around the Universitas Tadulako. For that purpose, the research site was set on the road network around the campus of Universitas Tadulako. The site map shows in Figure 1.

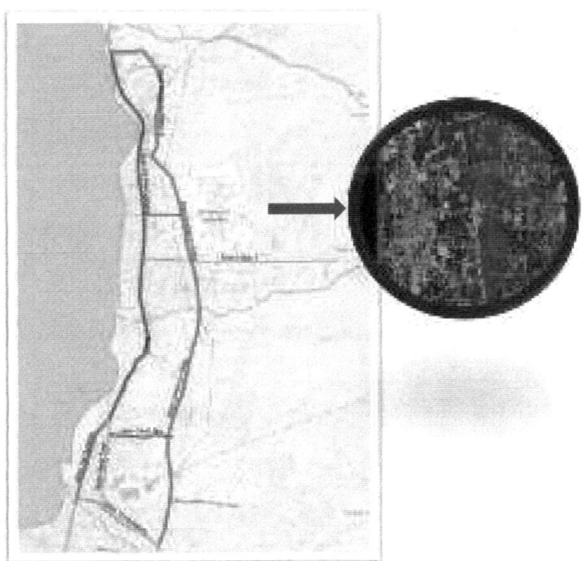

Figure 1. Research location

2.2. Data collection and processing
Data collection in this study was carried out in 2 ways, visual observation and distributing questionnaires to the community around the Universitas Tadulako. Visual observations were conducted by the research team after obtaining information from the results of the questionnaires about black spots on the road network around the Universitas Tadulako. The questionnaires collected were 346 copies that represent 11 locations of black spots.

2.3. Data analysis
The data from the questionnaire consists of information on the existence of black spots known to the respondents while driving on the road around the Universitas Tadulako. Respondents were categorized into three groups, that are groups of people who had an accident, groups of people who almost had an accident, and groups of people who had never had an accident. From those three groups of data, the location of the black spot and the appropriate treatment will then be determined

3. Results
This study aims to determine the accident-prone points in the area around the Universitas Tadulako. Data were collected by distributing questionnaires filled out by 346 respondents. General data obtained from respondents shown in Table 1

Table 1. Respondents general data

Variable	Description	%
Gender	Male	57.5
	Female	42.5
Jobs	Civil Servant	
	Military & Police	
	Private Employees	22
	Farmer	
	Student	55.5
	Construction Worker	
Driving License Ownership	Yes	75.4
	No	24.6
Traffic Accidents Experience around Universitas Tadulako	Had an accident	36.4
	Almost had an accident	19.7
	Never had an accident	24.9

Based on the general data of respondents, it shows that the highest percentage of gender is male (57.5%) and works as students (55.5%). Furthermore, respondent data shows that 75.4% already have a driver's license (SIM). The categories of respondents were divided into three groups, namely the group that had an accident, the group that almost had an accident, and the group that had never had an accident. Accidents in question are events that occur around campus (Figure 1).
The results of this study provide information on three groups of accidents as follows:

3.1. The group had an accident
The data obtained from respondents categorized in this group is shown in Table 2 as follows

Table 2. Data of groups that experienced an accident

Variable	Description	%
Accidents Location	In front of the gate of Universitas Tadulako Lecturer's Housing	23.7

	Untad 1 Street	16.9
	Soekarno Hatta street	16.9
	RE Martadinata street in front of Mouza Store	10.2
	In front of Campus Roundabout	8.5
	Untad street in front of Tadulako Hospital	6.8
	RE Martadinata street in front of second hands store	6.8
	In front of Unismuh Campus	5.1
	In front of Citraland Estate	1.7
	In front of New Hospital Undata	1.7
	In front of Region Police Office	1.7
Year of Accident	2021	22.4
	2019	21.2
	2020	15.5
	2018	14.7
Driver License Ownership	Yes	58.3
	No	41.7
Type of Vehicles	Motorcycle	94.2
	Private Car	5.6
	Public Transport Car	1.6
Traffic circumstances when the accidents occur	Crowded	60.9
	Uncrowded	39.1
Time of Accidents	1.00 AM	8.0
	8.0 AM	15.9
	10.00 AM	10.2
	1.00 PM	27.3
	4.00 PM	15.9
	6.00 PM	13.6
	9.00 PM	9.1
Type of Crash	Single Accident	28.8
	Front Crash	23.1
	Side Crash	21.2
	Back Crash	16
State of Accident	Another vehicle's carelessness	40.4
	Lack of Concentration	37.2
	High Speed	26.3
	Not maintain the distance	16
	Overtake other vehicles	9
The weather when accidents occur	Bright	63.5
	Cloudy	16
	Rain	14.7
Street situation where accidents happen	Straight	54.5
	3-way junction	15.4
	Way down	12.8
	U-Turn	10.3
	Bend	7.1
	4-way intersection	9
Vehicles state when an accident occurs	Poor brake	28.2
	Minimal vehicle lights	15.4
	Poor steering system	10.9

Road state when an accident occurs	Slippery	40.4
	Damaged	29.5
	Not equipped with markings dan signs	25
	Narrow road	15.4
	Foggy	0.6
State of the environment when the accident occurs	There was barrier	24.4
	Minimal lighting	21.2
	Well	11.5
	Sunny weather	3.8
Availability of utilities building of the street	Road markings	41
	Sidewalk	37.2
	Road Signs	27.6
The driver's state	Minor injuries	62.8
	Serious injuries	19.2
	Non	17.9
Vehicle's state	Slightly Damaged	66
	Heavily damaged	20.5
	Not broken	13.5
Suggestion to avoid accidents around the Universitas Tadulako	Street cleaning	15.2
	Increase the awareness/concentration	34.5
	Road widening	6.1
	Road sign upgrading	15.8
	Minimize the use of a phone when driving	13.5
	Upgrading the streetlight	7.7
	Road repair	7.1

3.2. The group that almost had an accident

The information obtained from the respondent included in this group are shown in Table 3

Table 3. Data of groups that almost experienced an accident

Variable	Description	%
Accidents Location	In front of the gate of Universitas Tadulako	
	Lecturer's Housing	38.2
	Untad 1 Street	12.7
	Soekarno Hatta street	21.8
	RE Martadinata street in front of Mouza Store	9.1
	In front of Campus Roundabout	7.3
	Untad street in front of Tadulako Hospital	1.8
	RE Martadinata street in front of second hands store	1,8
	In front of Unismuh Campus	1.8
	In front of Citraland Estate	3.6
	In front of New Hospital Undata	0
	In front of Region Police Office	1.8
Year of Accident	2021	48.8
	2019	19
	2020	1.2
	2018	9.5

Driver License Ownership	Yes	82.1
	No	17.9
Type of Vehicles	Motorcycle	97.6
	Private Car	
	Public Transport Car	
Traffic circumstances when the accidents occur	Crowded	78.6
	Uncrowded	21.4
Time of Accidents	1.00 AM	1.85
	8.00 AM	13
	10.00 AM	13
	1.00 PM	29.6
	4.00 PM	16.7
	6.00 PM	16.7
	9.00 PM	11.1
State of Accident	Another vehicle's carelessness	40.5
	Lack of Concentration	26.2
	High Speed	21.4
	Not maintain the distance	11.9
	Overtake other vehicles	11.9
The weather when accidents occur	Bright	69
	Cloudy	19
	Rain	10.7
Street situation where accidents happen	Straight	39.3
	3-way junction	16.7
	Way down	14.3
	U-Turn	15.5
	Bend	13.1
	4-way intersection	14.3
Vehicles state when an accident occurs	Poor brake	28.6
	Minimal vehicle lights	7.1
	Poor steering system	4.8
Road state when an accident occurs	Slippery	29.8
	Damaged	33.3
	Not equipped with markings dan signs	25
	Narrow road	20.2
	Foggy	0
State of the environment when the accident occurs	There was barrier	10.7
	Minimal lighting	15.5
	Well	6
	Sunny weather	7.1
	Foggy	6
Availability of utilities building of the street	Road markings	40.5
	Sidewalk	29.8
	Road Signs	45.2
Source of the accident at the point	Driver/human	84
	Street	34.9
	Environment	16
Suggestion to avoid accidents around the Universitas Tadulako	Street cleaning	11.4
	Increase the awareness/concentration	48.6
	Road widening	2.86

	Road sign upgrading	21.4
	Minimize the use of a phone when driving	2.86
	Upgrading the street light	2.86
	Road repair	10

3.3. Group with no accident experience

The data obtained from the respondents included in this group is shown in Table 4

Table 4. Data of group that experience no accident

Variable	Description	%
Accidents Location	In front of the gate of Universitas Tadulako Lecturer's Housing	14.8
	Untad 1 Street	9.86
	Soekarno Hatta street	14.8
	RE Martadinata street in front of Mouza Store	15.5
	In front of Campus Roundabout	19,7
	Untad street in front of Tadulako Hospital	7.75
	RE Martadinata street in front of second hands	3.52
	store	4.23
	In front of Unismuh Campus	4.93
	In front of Citraland Estate	0.7
	In front of New Hospital Undata	4.23
	In front of Region Police Office	
Suggestion to avoid accidents around the Universitas Tadulako	Street cleaning	8
	Increase the awareness/concentration	33.3
	Road widening	2.67
	Road sign upgrading	34.7
	Minimize the use of a phone when driving	0
	Upgrading the streetlight	6.67
	Road repair	14.7

Tables 2, 3, and 4 indicate several important points. According to the information from the tables, the location of the black spot around the Universitas Tadulako is in front of the gate of the UNTAD lecturer housing (had an accident (23.7%); almost had an accident (38.2%); never had an accident (14.8 %). 2021 is the time of the accident (had an accident (22.4%); almost had an accident (48.8%). Already own a driver's license (had an accident (58.3%); almost had an accident (82.1%). Motorcycles are the vehicles used during the accident (had an accident (94.2%); almost had an accident (97.6%). Traffic conditions at the time of the accident were busy (had an accident (60.9%); almost had an accident (78.6%). %); almost accident (29.6%) Type of collision is a single accident (28.8%). The carelessness of other vehicles is the condition at the time of accident (had an accident (40.4%); near accident (40.5%). The weather condition at the accident is sunny (had an accident (63.5%); almost had an accident (69%). Road conditions during an accident on a straight road (had an accident (54.5%); almost had an accident (39.3%). Poor brake system as the source of accident (had an accident (28.2%); almost had an accident (28.6%). Slippery road when accidents occur (had an accident (40.4%); almost had an accident (29.8%). There was a barrier, namely environmental conditions when an accident occurred (had an accident (24.4%); almost had an accident (10.7%) Road markings as a complementary road building (had an accident (41); almost accident (40.5%). Minor injuries are the condition of the driver during the accident (66%). Minor damage is the condition of the vehicle when an accident occurs (66%). The most advice given to prevent accidents is to increase concentration (have an accident). 34.5%); almost accident (48.6%); never accident (33.3%).

Based on the information above, it indicates that the location in front of the gate of the UNTAD lecturer's housing is a black spot location that potentially causes accidents. For more details on the condition of the black spot, see the following figure.

Figure 2. Black spot location around Universitas Tadulako

4. Conclusion

This research was conducted to determine the accident-prone points around the Universitas Tadulako and determine its suitable treatment. After analyzing the black spot, several important points were found:

1. The location of the black spot around the Universitas Tadulako is in front of the gate of the UNTAD lecturer housing. Motorcycles dominate use and most drivers have a driver's license. Traffic conditions at the black spot location were crowded and the weather was clear when the accident occurred. Even though there are road markings, there are obstacles to visibility at that location. Increasing concentration is a suggestion given to prevent accidents at the black spot locations.
2. The proposed treatments at the black spot location are controlling street vendors' spots to avoid careless parking, optimizing traffic lights, setting U-turns by adjusting applicable standards, cutting several trees that block visibility, and repairing road pavements.

Based on the results, it is expected that it can provide information about the existence of black spots so that related parties can follow up on the condition of the road and reduce the accident rate.

IOP Conf. Series: Earth and Environmental Science 1075 (2022) 012042 doi:10.1088/1755-1315/1075/1/012042

This research is limited to black spots determination and its treatment based on visual observation. However, to obtain more in-depth information on black spots, other methods can be used so that more can be obtained

Acknowledgments

This research was fully funded by the Dipa Fund of the Faculty of Engineering, Universitas Tadulako, based on the Decree of the Universitas Tadulako Rector Number 3031/UN28/KU/2021.

References

[1] Dewanti 1996 Karakteristik Kecelakaan Lalulintas di Yogyakarta Media Teknik no.3 tahun XVII Edisi November 1996 Yogyakarta
[2] Z Fan C Liu D Cai and S Yue 2019 Research on black spot identification of safety in urban traffic accidents based on machine learning method Safety science **118** 607-16
[3] M Gani, M Hustim and A R Arifin 2015 Analisis Black spot Area Kecelakaan Pada Ruas Jalan Arteri Di Kota Makassar Berbasis Sig
[4] Indonesia P R 2009 Undang-Undang Republik Indonesia Nomor 22 Tahun 2009 Tentang Lalu Lintas Dan Angkutan Jalan
[5] Korpslantas 2018 Realizing road safety by motorcyclist through improvements in the 371 downstream sector Polri Jakarta
[6] M J Maher and L J Mountain 1988 The identification of accident blackspots: a comparison of current methods Accident Analysis & Prevention **20** 143-51
[7] G Sugiyanto and A Fadli 2017 Identifikasi Lokasi Rawan Kecelakaan Lalu Lintas (Black spot) di Kabupaten Purbalingga Jawa Tengah Jurnal Teknik Sipil dan Perencanaan **19** 128-35
[8] World Health Organization 2015 Global Status Report on Road Safety
[9] G Sugiyanto B Mulyono M Y Santi 2014 Karakteristik Kecelakaan Lalu Lintas dan Lokasi Black spot Di Kab Cilacap Jurnal Teknik Sipil UAJY **12** 259-66

Examining the Intersection Performance of the Post-Disaster Permanent Housing Assistance Impact: A Study of the Trip Generation/Attraction in Palu City

T Bahar[1*], **A Tahir** [2], **Mashuri** [3] **and Isratun**[4]
[1,2,3,4]*Department of Civil Engineering, Engineering Faculty, Tadulako University*
**Corresponding Author: taslimbahar0862@gmail.com*

Abstract. The government provides housing assistance for victims of the disaster that hit Palu City in Central Sulawesi Province. Most of the housing assistance is in Tondo Village, which is referred to as Tondo's permanent housing/TPH (Huntap Tondo). Currently, the TPH has become a source of traffic movement, especially for the Tondo ring road intersection - Street Untad 1. Thus, this study aims to examine the impact of the Tondo permanent residential trip generation/attraction on performing the intersection. The method of this research is to calculate the trip generation/attraction that occurs because of the existence of the Tondo's permanent housing and calculate the performance of the Tondo ring road intersection - Street Untad 1 using the Indonesian Highway Capacity Manual (MKJI) in 1997. This study reveals the value of the trip generation/attraction rate that occurs in the the Tondo's permanent housing is 0.175 pcu/unit with a total trip generation/attraction of 549 pcu/hr. Meanwhile, performing the ring road intersection Tondo - street Untad 1 decreased the value of saturation by 0.45 to 0.64, and the delay value of 8.63 seconds/pcu to 10.52 seconds/pcu.

Keywords: Trip Generation/attraction, Permanent housing assistance, Intersection performance, MKJI 1997, and Post-disaster.

1. Introduction

After the natural disaster in 2018 that hit several areas in Palu City, it caused significant material and human losses. Community involvement in establishing post-disaster social relations [1] is the primary key to the awakening from this adversity. In addition, assistance from the Government and the private sector is a profitable capital to start life after a disaster [2]. One of the assistances is permanent housing aid (Hunian Tetap or Huntap) given to disaster victims who had lost their homes. Currently, the permanent residences/housings had established in several zones, but the fully operational one is the permanent residences Tondo 1 which consists of 1600 units, followed by the construction of permanent housing Tondo 2 with as many as 1600 units and permanent housing Talise that have as many as 600 units.

The existence of permanent housing certainly has an impact on the performance of the surrounding roads. The road network connecting the permanent residences was previously quite congested because there were housing, campuses, shops and, other public places [3]. Thus, the existence of permanent housing indeed also burdens the surrounding road network. Most of the residents of permanent housing previously came from the West region of Palu, so naturally, the majority of them worked and went to school in the West area. To accelerate vehicle movement and support the community's economy, the government is currently finishing the construction of a ring road. The outer ring road of the city of Palu is expected to provide easy access for the movement of vehicles, especially people in permanent residential areas. The outer ring road of Palu City has been operating, but some routes are in the finishing stage.

The outer ring road of Palu City is currently used daily by people in the Tondo 1 permanent residential zone to meet their travel needs. Although not fully operational, this ring road has been used, primarily connecting the Tadulako University (Untad) campus with the permanent residential area 1 Tondo. The Tadulako University Ring Road Intersection – Road Untad 1 is one of the intersections located on the outer ring road access of Palu City. In addition to the existence of permanent housing Tondo 1 that began to escalate in movement, the movement accumulation was also generated by other community activities. This condition can certainly cause a decrease in the performance of the intersection so that it can affect its principal function as a facility in improving disaster mitigation [4].

Regarding conditions described previously, a study was conducted to analyze the impact of the generation/attraction of the Tondo permanent residence on the performance of the Untad-Untad 1 ring road intersection.

This study has a high level of urgency, considering that the generation/attraction of the Tondo 1 permanent housing can affect the performance of the Untad Ring Road - Untad 1 road intersection. The results of this study can provide practical recommendations from the performance of the intersection. These recommendations later will be related to increasing safety for the people who occupy the permanent housing of Tondo 1 and its surrounding community.

2. Research Methodology

The research method consists of several parts, that is:

2.1. Research Location

This study objective is to determine the performance of the road intersection as a result of the existence of a permanent residence in Tondo. The research location is at the unsignalized intersection of the Untad ring road – Untad 1 road, while the permanent housing is located in Tondo Village, Mantikulore District, Palu City. The description of the research location shown in Figure 1 below.

Figure 1. Research location

2.2. Data Collection and Processing

Traffic data collection was done by counting vehicles crossing the intersection during peak hours. Surveyors were assigned to each arm of the intersection to calculate the passing vehicles. The data collection was done from 06.00 - 18.00 WITA (Central Indonesia Time).

2.3. Data Analysis
The data from the volume of vehicles estimation were analyzed using the 1997 Indonesian Road Capacity Manual [5]. The calculated data consists of the performance of existing intersections calculations and predictions for the next 5 (five) years.

3. Results
The results of the study describe into several parts, that is:

3.1. Traffic Volume
The first part is to calculate the traffic volume based on field observations. Table 1 shows vehicle volume data at peak hours.

Table 1. Vehicle composition of research location

Intersection	Traffic flow (pcu/h)	Vehicle composition (%)		
		MC	LV	HV
The unsignalized intersection of the Untad ring road –Untad 1 Road	935.50	75.47	23.56	0.97

3.2. Performance of the unsignalized intersection of the Untad Ring Road-Untad 1 Road
The results of the calculation of traffic volume at the un-signalized intersection of the Untad ring road – Road Untad 1 can be utilized to calculate the performance of the intersection. The intersection performance was estimated using the Indonesian Road Capacity Manual for unsignalized intersections [6]. Table 2 shows the results of the calculation of the performance of these intersections.

Table 2. Existing intersection performance

Period	Basic Capacity (Co) pcu/hr	Capacity (C) pcu/hr	Traffic Flow (Q) pcu/hr	Degree of Saturation (DS)	Delay of intersection (D) s/pcu	Level of Service (LOS)
16.30 s/d 17.30	2900	2892,18	935,50	0,32	7,521	B

The estimation of the performance of the un-signalized intersection in the research location shows that the degree of saturation is 0.32. This result indicated that the intersection has a service level of B which means that the intersection has a stable flow and the driver has sufficient freedom in choosing the speed. The value of the intersection delay is 7.521 seconds/pcu.

3.3. Trip generation/attraction Tondo 1 Permanent Residential
Permanent Housing was built at the initiative of the Buddhist Tzu Chi Foundation with a land area provided by the government to construct a permanent residence (Huntap) in the Tondo Village area, Mantikulore District. This permanent residence has an area of 481.63 Ha for 1600 Permanent Residential units and is given for people who have lost their homes due to earthquake, tsunami, and liquefaction disasters. The number of occupied units is 395 units, and 1205 units are unoccupied. Thus, the percentage of permanent occupancy in Tondo is only 25%. Furthermore, by conducting an interview survey in the permanent residential area of Tondo 1, data on generation and attraction from the location was obtained. The estimation of the generation and attraction of the Tondo permanent residential area shows in Table 3 below.

Table 3. Total Trip Generation and Attraction of Tondo 1 Permanent Residential Areas

Traffic	Traffic flow (pcu/jam)	Existing Occupancy (Unit)	Trip Generation and Attraction/ Occupancy (pcu/hr)	the rate of generation and attraction (pcu/hr/unit)	Occupancy Total (Unit)	Total of Trip Generation and Attraction (pcu/hr)
Exit	73,5	395	0,186	0,343	1600	297,7
Entry	62,0		0,157			251,1
Total of Trip Generation and Attraction (pcu/hr)						548,9

According to the table above, the generation/occupancy unit is 0.19 pcu/hr, and the attraction/occupancy unit is 0.16 pcu/hr. Therefore, if the number of unoccupied units is 1205 units, the total generation and attraction obtained are 413, 4 pcu/hr. For the whole permanent housing built and estimated that in the next 5 years will be fully occupied with the total unit of 1600 units, the total generation and attraction of movement charged to the Untad Intersection - Untad 1 road is 548.9 pcu/hr. In other words, the total occupancy of 1600 units multiplied by the rate of generation and attraction of 0.343 pcu/hr/unit we get the value as of 548.9 pcu/hr.

3.4. Performance of the UNTAD-UNTAD 1 Ring Road intersection due to trip generation/attraction of permanent residences in Tondo

The performance of the un-signalized intersection of the UNTAD Ring Road-UNTAD 1 road due to trip generation/attraction of the Tondo permanent residence can be calculated using the predicted growth in the number of vehicles.

The results of the intersection performance prediction for the next 5 years show in Table 4 below.

Table 4. Prediction of intersection performance in the next 5 years

Year	Traffic Flow (Q traffic) (pcu/hr)	Total of Trip Generation and Attraction (Q generation) (pcu/hr)	Total (Q) (pcu/hr)	C (pcu/hr)	DS	D (s/pcu)	LOS
2021	997,46	548,9	1546,36	2892,18	0,53	9,50	B
2022	1063,55	548,9	1612,45	2892,18	0,56	9,73	C
2023	1134,07	548,9	1682,97	2892,18	0,58	9,98	C
2024	1209,30	548,9	1758,20	2892,18	0,61	10,24	C
2025	1289,56	548,9	1838,46	2892,18	0,64	10,52	C

The information above indicates that the degree of saturation at the intersection of the UNTAD Ring Road-UNTAD 1 road due to the trip generation/attraction of the Tondo permanent residence has been transformed. It shows that for the present and the next 5 years, it changed from 0.32 to 0.64. The level of service that was previously B, for the next 5 years becomes C, which means the intersection has a stable flow [7]. However, the speed options are limited for the driver. Not only the value of the degree of saturation that has been increase, but the value of the intersection delay also changed from 7.52 seconds/pcu to 10.52 seconds/pcu. However, with the increase that occurred only 0.19%, it can be concluded that the UNTAD Ring Road-UNTAD 1 intersection is still feasible or does not require a repair. The function of the intersection as part of the evacuation route can operate optimally. This intersection can provide an excellent level of safety for the people who occupy the permanent residence (huntap) of Tondo 1 and the surrounding community.

IOP Conf. Series: Earth and Environmental Science 1075 (2022) 012043 doi:10.1088/1755-1315/1075/1/012043

4. Conclusions

This research aims to determine the performance of the intersection of the UNTAD Ring Road-UNTAD 1 road due to trip generation/attraction of the permanent residence of Tondo. The results of the analysis found several crucial points:

a. Data provided from the field shows that the dominant vehicle is a motorcycle. The performance of the UNTAD-UNTAD 1 Ring Road intersection calculated using MKJI shows that the degree of saturation is 0.32, the service level B, and the intersection delay is 7.52 seconds/pcu.
b. Furthermore, by considering the development of the number of vehicles, the intersection performance for the next 5 years is obtained, that the degree of saturation 0,64, service level C, and intersection delay 10.52 seconds/pcu.
c. The final result shows that no significant improvement in the intersection performance for the next 5 years. It is since Tondo's permanent housing is still around 25% occupied.

The research is expected to be beneficial in providing information to related parties regarding the handling of intersections. Further research, apart from only involving the permanent residential area of Tondo, can also involve the Bumi Roviega area and other housing areas that potentially affect the performance of the UNTAD Ring Road-UNTAD 1 road intersection.

Acknowledgments

This research was fully funded by the Dipa Fund of the Faculty of Engineering, Tadulako University, based on the Decree of the Tadulako University Rector Number 3031/UN28/KU/2021

5. Reference

[1] Wekke I. S, Sabara Z, Samad M A, Yani A, Abbas T, and Umam R 2019 Earthquake, tsunami, and society cooperation: Early findings in palu of indonesia post disaster.
[2] Sofyan I 2020 Participative disaster management: Post Disaster renewal of Palu City. *Technium Soc. Sci.* J., 13, 25.
[3] Lubis M E and Sianturi N M 2013 Determination of the Movement Generation Model for Several Types of Housing in Pematangsiantar City. *Media Teknik Sipil*, 10(1).
[4] Rajindra R, Suardi I, Sabara Z, Pushpalal D, Samad M A, Yani A and Umam R 2019 Diversity, resilience, and tragedy: Three disasters in Palu of Indonesia. International *Journal of Innovation, Creativity and Change*, 5(2), 1592-1607.
[5] The Indonesian Highway Capacity Manual (MKJI) 1997 Direktorat Jenderal Bina Marga. Jakarta. Indonesia
[6] Miro F 2005 Transport Planning: for students, planners and practitioners. Publisher Erlangga.
[7] Tamin O 2000 Transportation Planning and Modeling. Publisher ITB, Bandung Indonesia

Implementation of the souraja house façade on the composition of the government building façade in Palu city

S M Malik[1], H Salenda[1], A Winarta[1], D R Muhamad[1]

[1]Departement of Architecture, Faculty of Engineering, Tadulako University, Palu, Indonesia.

ratihmalik.ars@gmail.com

Abstract. As the capital city of Central Sulawesi Province, Palu has built various public buildings, especially buildings belonging to the local government, which are provided to serve the population's needs. However, when examined more deeply, it can be seen that several government-owned public buildings were designed not to pay attention to the environmental conditions in which the building was erected both socially, culturally and the composition of the building's constituents. One of them is the facade or the front of the building. The Souraja house, a traditional house of the Kaili tribe, which currently exists in Palu City, should become a reference for the shape of public buildings, both in overall form and only in the form of the facade composition. This study uses a rationalistic approach, which sees the truth from empirical conditions and arguments as a part of thinking construction. The rationalistic research design departs from the theoretical framework (Grand Theory). The source of the Grand theory raised in this study comes from the Souraja House building, which is the work of the traditional house of the Kaili tribe, which currently still exists in Palu City. The results of this study indicate that the three parameters of the components that make up the facade of the Souraja Traditional house, which are used in analyzing and discussing the beginning of government buildings in Palu City, are not 100% implemented, namely on the raised floor but no longer in the form of a stilt building. While the two parameters of the composition of the facade of the Souraja Traditional house used in analyzing and discussing the face of government buildings in Palu City are 100% implemented.

Keywords: Composition, Facade, Souraja House, Government Building

1. Introduction

As the capital city of Central Sulawesi Province, Palu has built various kinds of public buildings, especially buildings belonging to the local government, which are provided to serve the needs of its residents. However, when examined more deeply, it can be seen that several government-owned public buildings were designed not to pay attention to the environmental conditions in which the building was erected both socially, culturally and the composition of the building's constituents. One of them is the facade or the front of the building.

The Souraja house, a traditional house of the Kaili tribe, which currently exists in Palu City, should become a reference for the shape of public buildings, both in overall form and only in the form of the facade composition. Because taking the form of a souraja house is preserving a traditional structure typical of Palu City in public buildings, especially those belonging to the Palu City Government. This is also by the mandate of the Regulation of the Minister of Public Works (PU) Number:

29/PRT/M/2006 [1] concerning Guidelines for Technical Requirements for Buildings, namely that the shape of the building must be designed by taking into account the shape and architectural characteristics of the surrounding environment, or able to serve as architectural guidelines or role models for their environment.

Based on the description of the background above, this research will try to see how the implementation of the composition of the facade of Souraja's house on the design of the front of the Government building in Palu City is carried out.

2. Literature review

The facade is the main face of the exterior of the building, which is on one of the top sides, usually facing the street [2]. The main facade is typically the side of the building with the main entrance (main entrance) and a stylistic detail character. In contrast, according to Ching [3], the facade is the face of the building or other sides of the building facing the road or public space distinguished with architectural processing.

Burden in Salenda [4] said that the architectural identity of the building is easier to see through its facade. The structure can have many sides, starting from 3 sides, 4 sides, and so on, but there will be one side that is emphasized as the "face" of the building, and this side is the side which is from now on referred to as the facade. A facade also tells about the occupants of a building by giving them a kind of collective community identity and ultimately representing that community in public.

According to Ching [3], the building facade components include entrance doors, ground floor zones, windows, guardrails, building roofs, signage, and ornaments. In contrast, the composition of the building facade has geometry, symmetry, rhythm, contrast, scale, and proportion. The theoretical basis of the building facade is used as a reference to find research variables. The research variables that will be used consist of building facade components in the form of entrances, openings, building roofs and ornaments, building facade compositions in the form of geometry, symmetry, rhythm, scale, and proportion, as well as elements of the building facade, namely color, and building materials.

3. Research methods

This study uses a rationalistic approach. A rationalistic approach is an approach that sees the truth not only from empirical conditions but also arguments as a part of the construction of thinking. The rationalistic research design departs from a theoretical framework (Grand Theory) built from the meaning of general theories, experts' thoughts and is constructed into something that contains several problems that need further investigation [5].

The source of the Grand theory raised in this study comes from the Souraja House building, which is the work of the traditional house of the Kaili tribal community, which currently still exists in Palu City. This research is also exploratory. According to Gulo [6], exploratory research is research that aims to trace a phenomenon or event through a series of assessments

In determining the object of research, it is assisted by an object selection criterion, which is obtained from the analysis of the facade of the Souraja house based on the components and composition. The criteria used as a guide for the selection of building objects in this study include:

1. The floor component is raised as a boundary between the outdoor and indoor spaces.
2. The opening components on the facade are placed horizontally parallel to each opening.
3. The roof component is triangular (gable). Where the roof is composed of two overlapping triangular roof areas.
4. The facade is arranged by geometric planes in rectangles and triangles.
5. The facade of the building is arranged using symmetrical balance.

4. Result and discussion

The number of research objects is four government office buildings located in Palu City, to be precise, namely on Jalan Muhammad Yamin and Jalan Balaikota Timur. The government offices include 1) City Hall Office, 2) BAPPEDA Office, 3) TASPEN Office, 4) North Birobuli Village Office.

Observation of several facades of government buildings that became the object of the research above was based on the criteria prepared previously.

4.1. Façade components

4.1.1. The floor component is raised as a boundary between the outdoor and indoor spaces

The results of the analysis of the facade of the Souraja Traditional House building, which was the government house of the king of Palu when he was in office, showed that the floor of the building was raised as a boundary between the outer and inner spaces. (see Figure 1). If you look at the four government buildings that are the object of this research, it will show that the raised floor is implemented, but the building that is the object of study is not in the form of a stage building. The increase in the floor area of the four research objects only confirms the boundaries of the spatial location.

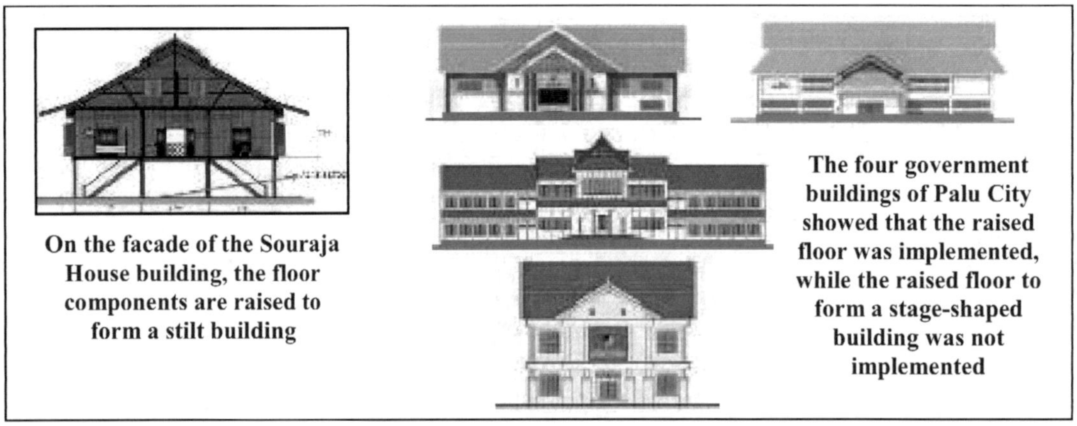

Figure 1. Floor plane components on the facade of the souraja house building, and the four facades of the palu city government building
(Source: Analysis results, 2021)

4.1.2. The components of the openings on the facade are placed horizontally parallel to each opening

The results of the analysis of the facade of the Souraja house building, a traditional building of the Kaili Tribe that became the government building of a King, show that this building has openings arranged horizontally parallel to each door (see Figure 2). Based on the observation results of the facades of the four government buildings in Palu City, which became the object of this study, it shows that the openings of each building are arranged horizontally parallel to each door (see Figure 2), so it can be said that the arrangement of spaces in the Souraja House building is implemented in a balanced manner. Maximum on the facade of the Palu City government building, which is the object of this research.

In the facade of the Souraja House building and the four government buildings in Palu City, which are the objects of this research, the openings are arranged in parallel by processing the opening rhythm with a repetition/form repetition system. The doors on the facade of the Souraja House building and the government office building, which are the object of research, show that they use a repetition rhythm on a large scale, so they can be categorized as openings arranged in horizontal parallel create a dynamic impression.

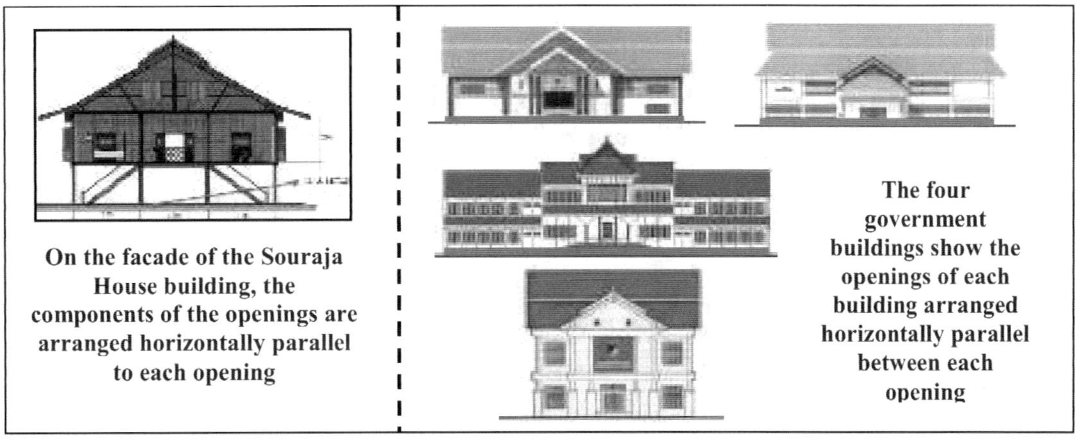

Figure 2. Opening components on the facade of the souraja house, and the four facades of the palu city government building
(Source: Analysis results, 2021)

4.1.3. The roof component is triangular (gable). Where the roof is composed of two overlapping triangular roof areas

The results of the analysis of the facade of the Souraja Traditional House, which was the government house of the king of Palu when he was in office, show that the roof component is triangular (gable). The roof comprises two overlapping triangular roof planes (see Figure 3). Based on the results of the analysis of the facades of the four government buildings, which were the chosen objects in this study, it was found that the four government buildings used a triangular roof (gable roof), in which two triangular roof planes were arranged overlapping each other (see Figure 3). So it can be said that the arrangement of the facade components of the Souraja House building is implemented optimally on the facade of the Palu City government building, which is the object of this research.

The use of a gable roof in the Souraja House building and government buildings selected in this study is estimated as a design response to the tropical climate in which this building stands. In response to this tropical climate, the roof slope is dominant between 30°-60°.

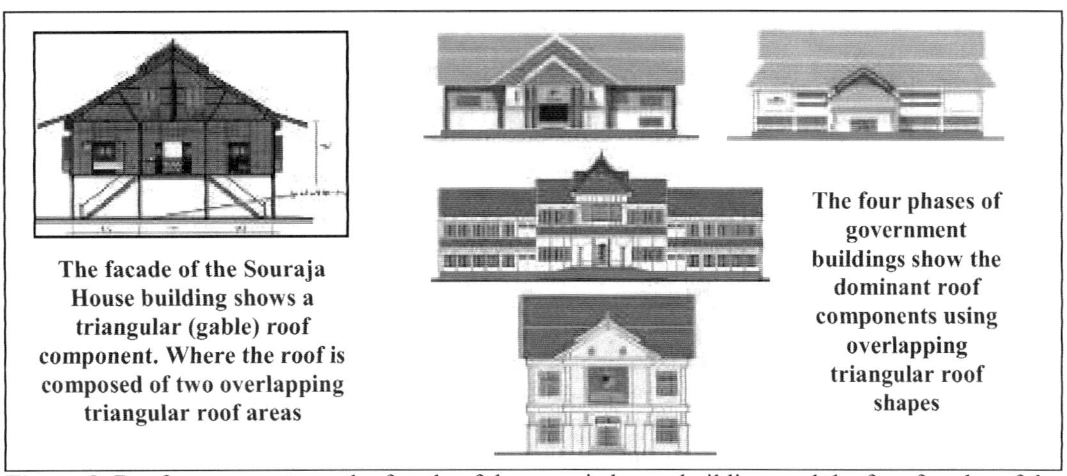

Figure 3. Roof components on the facade of the souraja house building, and the four facades of the palu city government building
(Source: Analysis results, 2021)

4.2. Façade composition

4.2.1. The facade is arranged by geometric planes in the form of rectangles and triangles

The results of the analysis of the Souraja House building, which was the government house of the king of Palu when he was in office, showed that this building was established by structuring geometric fields in the form of rectangles and triangles. Rectangular geometry is visible at the bottom and center of the building. At the same time, the triangle geometry is seen at the top of the building. If the geometric planes are considered more carefully, it will be seen that the planes are arranged vertically. For the rectangular category, the visible areas on the facade have different lengths and widths (see Figure 4). The rectangular shape can be thought of as a variation of the square shape that changes with increasing height or width.

The results of the analysis of the facades of the four government buildings selected in this study show that the buildings are designed by arranging rectangular and triangular geometric fields. Rectangular geometry is visible at the bottom and center of the building (wall). At the same time, the triangular geometry is seen at the top of the building, namely on the roof (see Figure 4). So it can be said that the arrangement of the geometric composition of the plane on the facade of the Souraja House building is implemented optimally on the front of the Palu City government building, which is the object of this research.

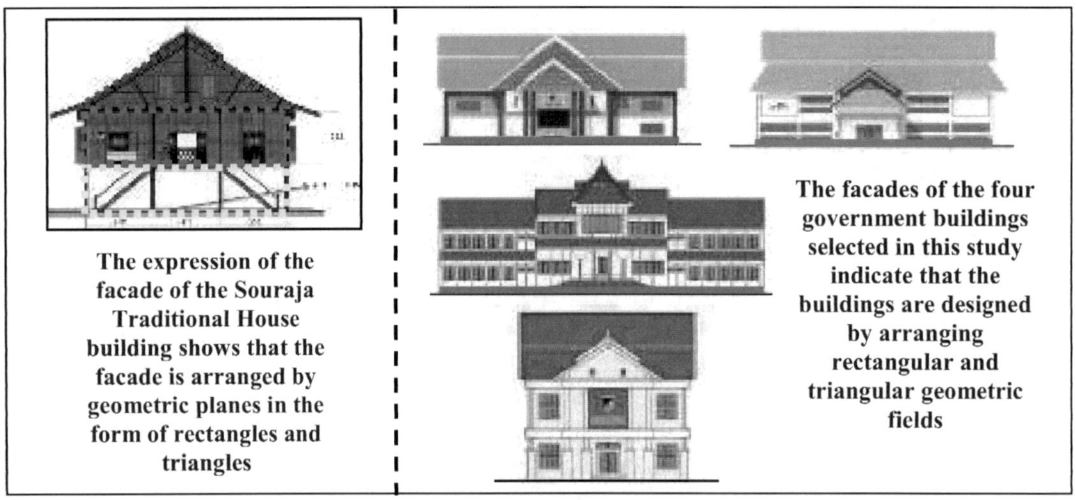

Figure 4. Geometric composition of the facade composer of the souraja house, and the four facades of the palu city government building
(Source: Analysis results, 2021)

4.2.2. The facade of the building is arranged using the principle of symmetrical balance

The facade of the traditional Souraja house building shows that the faces are arranged using symmetrical balance. When determining the axis of the façade, there is no difficulty in determining the location of the axes, and when the axes are placed on the facade, the planes on the left and right do not show the impression of bias (see Figure 5).

Based on the analysis results of the facades of the four government buildings in Palu City, which were chosen as the object of this study, it shows that the four faces of government buildings use the principle of symmetrical balance. So it can be said that the symmetrical arrangement of the composition of the fields on the facade of the Souraja House building is maximally implemented on the front of the Palu City government building, which is the object of this study (see Figure 5).

The pattern of arrangement of the fields on the four facades of the building, which is the object of research, is arranged in a balanced structure on the opposite side to an axis line. Symmetry can be defined as the distribution and balanced arrangement of any similar shape and space on opposite sides

IOP Conf. Series: Earth and Environmental Science 1075 (2022) 012044 doi:10.1088/1755-1315/1075/1/012044

of a line or plane or about a central point or axis. So it can be said that a symmetrical condition demands a balanced arrangement of the shape pattern of a building facade.

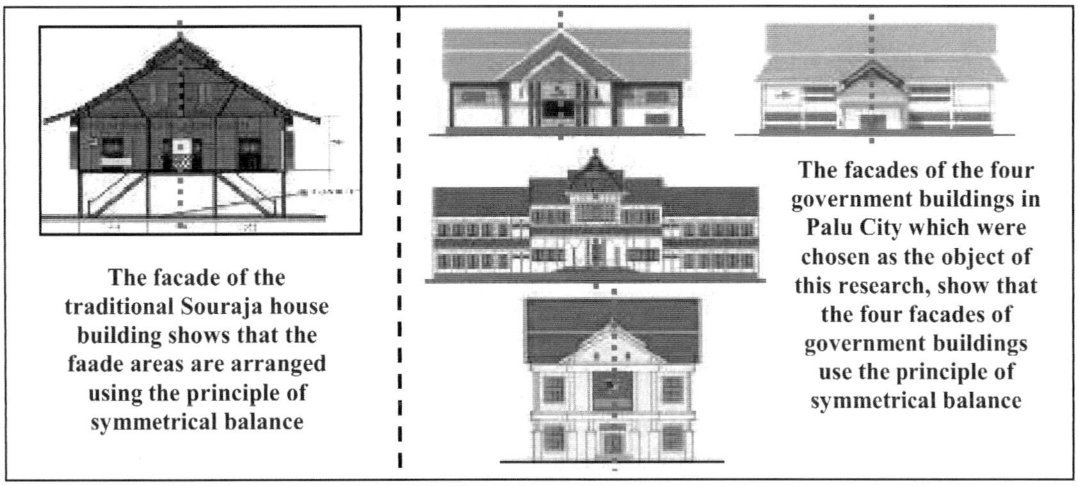

Figure 5. The principle of the balance of fields on the facade of the souraja house building, and the four facades of the palu city government building
(Source: Analysis results, 2021)

References

[1] Regulation of the Minister of Public Works (PU) Nomor 29/PRT/M/2006 *concerning Guidelines for Technical Requirements for Buildings* (Jakarta)
[2] Krier Rob 2001 *Architectural Composition* (Jakarta : Erlangga)
[3] Ching DK 2008 *Form, Space and Order* (Jakarta : Erlangga)
[4] Salenda H 2009 *The Existence of De Stijl in the Facade Composition of Dutch Colonial Buildings and Contemporary Houses* (Yogyakarta: Thesis of the UGM Architectural Engineering Study Program)
[5] Muhadjir N 1996 *Qualitative Research Methodology* (Yogyakarta : Rake Sarasin)
[6] Gulo W 2002 *Research Methodology* (Jakarta: Grasindo)

A review of segregation space in traditional mamasa settlement

G Timbang[1],[2], Amar[2], M Bakri[2], Puteri Fitriaty[2]

[1]Study Program of Architecture, Faculty of Engineering, Universitas Tadulako
[2]Department of Architecture, Faculty of Engineering, Universitas Tadulako

gator.timbang68@gmail.com

Abstract. The Mamasa community settlement is one of the unique traditional settlements that exists to this day. Characteristics of a typical settlement pattern arrangement reflects local wisdom through the existence of sacred space and non-sacred (profane) space, as part of beliefs and religious activities that are still ongoing from time to time. This study aims to find the segregation of sacred and profane spaces in the Mamasa community settlement as part of the archipelago's architecture that has a soul behind its physical performance. The method used in this paper is to collect and review literature relevant to the research topic so that gaps (differences) are obtained in various aspects, both locus and focus and research methods. The expected result in this study is to find differences between research on the spatial segregation of Mamasa traditional settlements and other similar studies.

Keywords *: settlement, traditional, space, segregation*

1. Introduction

Architecture includes various things, especially space (space and place). Space covers and accommodates the activities of its users over a period of time, so it can be concluded that space is identical to architecture without neglecting other aspects that contribute to the formation process including its utilization. Space covers and accommodates human activities and artifacts as part of the process and results of human activities in it. Embryo word space comes from the classical term spatium which was later absorbed into French espace and spazio in Italian. As for the word room in English, which in German is known as raum, all of which connote space (architecture)

The original inhabitants of Mamasa Regency, are one of the ethnic groups that have unique architecture, especially the form and arrangement of settlements as housing and cultural products. The regular pattern of settlements in one orientation with a uniform architectural structure and formation is interesting to explore and study from the perspective of architectural science, especially on the spatial aspect as the focus of the study. Knowledge of space in the context of traditional Mamasa community settlements contains values as ancestral heritage from generation to generation and contains local wisdom that needs to be explored and contextualized in order to anticipate the effects of globalization which often erode local values instead of modern values and culture. In fact, modern culture often contradicts local natural values and conditions which have long been built on knowledge adapted to their nature. Mamasa traditional community settlements are part of the space that is formed and adapted to nature, leading to the concept of harmonization with nature and its environment which still needs to be studied from an architectural ontology. Mamasa ethnic culture is integrated with the concept of ancestral religion as part of local knowledge and values which leads to the implication of the formation of space in settlements and is further manifested through unique settlement arrangements. A traditional society whose essence of life is patronized by the religious aspects of its ancestors is called an archaic society which in its life distinguishes between sacred and profane areas, the natural and the supernatural; symbolized by nature or the cosmos [1]. The boundary or scope (segregation) between sacred and profane space in the tradition of living in the Mamasa community needs to be revealed through architectural research and studies, filling in the local theory section that focuses on residential space

2. Literature review

Several articles as a result of studies on spatial segregation have been carried out in various locations and at different times, including; Spatial segregation in Vanarasi: caste and religion based exclusion/inclusion across municipal wards which focuses on the spatial arrangement of old settlements in Vanarasi based on caste and religion, in India. Through quantitative methods, the research found that settlement patterns were formed by the separation of Hindu settlers; Muslim and General with the highest percentage of residents are Hindus, followed by General and Muslim in third place. Furthermore, the research entitled Residential segregation and social diversification: Exploring spatial settlement patterns of foreign population in Southern European cities [2] is oriented towards housing segregation of foreign residents (immigrants) in Southern Europe. Research conducted in urban areas of Italy and Spain concluded that high levels of residential segregation correlated with high levels of immigrant presence outside metropolitan areas; On the other hand, in the metropolitan area, there is no segregation of population settlements.

The ethnic-based segregation of housing is the focus of research in the Southern Turkey Region conducted [3] through the elaboration of qualitative methods. The results of surveys, observations, closed interviews in the context of ethnographic research resulted in findings on socio-spatial changes in settlements. The socio-spatial changes referred to are motivated by the separation between permanent settlements and temporary settlements for Kurdish seasonal workers in Southern Turkey. During the planting season, plantation workers do not return to their homes, but set up shelter tents in agricultural areas. “Measuring the changing pattern of ethnic segregation in England and Wales with Consumer Registers” is the title of research conducted by [4]. Through the case study method, this study concludes that the classification based on ethnicity in the UK which includes ethnicity from Britain itself, Indian, Pakistan, Bangladesh, China and Africa has a significant influence on the segregation of settlement patterns of each ethnicity.

In 2019, Perner documented research in the form of a journal entitled "Segregated behind the walls: residential patterns in pre-industrial Copenhagen" and produced conclusions on residential patterns in Copenhagen in the pre-industrial era from 1711 to 1845. The study used a Geographic Information System approach. (GIS) and HISCLO and HISCLASS coding schemes, it is concluded that socio-economically segregation of population settlements is separated by roads in urban areas; while at the same time opening up opportunities for the creation of socio-geographical segregation.

Housing segregation in urban India was studied [5] with the topic “Residential Segregation in Urban India” which documents several socio-economic-based settlement patterns of historically marginalized groups in Indian cities. Through the processing of socio-economic data at high spatial resolution through the application of the ID (Index of Dissimilarity) segregation measurement method, it was found that the segregation of housing groups in the city was motivated by differences in religion and caste; while the segregation of the old city area with the younger city is only based on caste factors and not because of religious factors.

The spatial approach in historical research and the syntactic morphological approach for historical subjects, is an integrated methodological option chosen in order to resulting in a form of urban segregation pattern in which different ethnic groups live in[6]. The paper entitled "Spatial Forms of Ethnic Coexistence in Ottoman Cyprus: The Role of Urban Form in Patterns of Everyday Life" resulted in the finding that spaces that functioned as centers of trade and administration Government is a highly valued space in Nicosia and is inhabited by ethnic Ottomans who have the status of land owners, wealthy merchants or those in high positions, however the spatial segregation of the Ottomans with other city dwellers tends to overlap without clear boundaries.

3. Research method

The method used in this paper is Context Review, as a form of general review in the literature review process that departs from the study topic, namely spatial segregation; which is connected with aspects of the wider repertoire of knowledge in order to produce research differences (gap research) with the research to be carried out.

4. Result and discussion

The space from a religious perspective is further classified into two parts by Eliade [1], which in human religious experience is basically a dialectic that connects the sacred and the profane, the real and the virtual, the earthly and the heavenly. This understanding applies especially to the Archaic people, people who are more involved with natural activities or whose aspects of life are more dominantly oriented to nature. If the profane is identical with everyday reality, the sensory one, then the profane is always in contact with the non-sensory, supernatural.

Many articles on sacred (sacral) and unholy spaces in the context of residential spaces have been written, but there are not many topics regarding the scope or boundaries (segregation) between the two spaces, especially the locus of traditional settlements. The previous researches from various places and times with the topic of spatial segregation include; Residential segregation and social diversification: Exploring spatial settlement patterns of foreign population in Southern European Cities" by Benassi, et al [2]; the research content is the segregation of residential spaces in urban areas; in line with research [3] entitled "Ethnic residential segregation among migrant workers: from temporary seasonal tents to new rural ghettos in southern Turkey". The two research objects are housing in urban areas; in line with research in England with the title "Measuring the changing pattern of ethnic segregation in England and Wales with Consumer Registers" [4]; but has not discussed specifically about spatial segregation within the scope of traditional settlements.

However, the research that discusses housing segregation in urban India is reviewed by Adukia et al. [5] with the topic "Residential Segregation in Urban India" documenting several socio-economic-based settlement patterns that also focus on spatial segregation in the context of housing, not settlements; Likewise, research conducted in 2018 by Nadia Charalambous produced a form of urban segregation pattern in which different ethnic groups resided, with the title "Spatial Forms of Ethnic Coexistence in Ottoman Cyprus: The Role of Urban Form in Patterns of Everyday Life". This research is also carried out in the urban realm, even though it relates to ethnic aspects, but not in the realm of traditional settlements inhabited by certain ethnicities with unique locality nuances.

Meanwhile, research by Perner [7] also still focuses on urban community settlements in Copenhagen which leads to conclusions about the segregation of residential spaces based on the socio-economic aspects of the community and socio-geography.

5. Conclusion

In general, research on spatial segregation, especially with the background of settlements, that was previously carried out was generally carried out in urban locations, not in rural areas, where traditional settlements with the uniqueness of their resident communities can still be found. Moreover, the object of research is more oriented to the scope of urban housing, not settlements as the object to be carried out in the realm of the Mamasa community settlement. Thus, the difference between the research (gap research) that will be conducted in Mamasa and previous studies lies in the aspect of the scope, locus, method and community of users/owners of settlements.

IOP Conf. Series: Earth and Environmental Science 1075 (2022) 012045 doi:10.1088/1755-1315/1075/1/012045

References

[1] Eliade, and Mircea, 1956; "The Sacred and The Profane, The Nature of Religion", A Harvest Book Harcourt, Brace and World, Inc
[2] Benassi, F, Iglesias-Pascual, R, and Salvati, L (2020) Residential Segregation and Social Disverification: Exploring Spatial Settlement Patterns of Foreign Population in Southern European Cities Habitat International, 102200
[3] Pelek, D (2020) Ethnic Residential Segregation Among Seasonal Migrant Workers : From Temporary Tents to New Rural Ghettos in Southern Turkey The Journal of Peasant Studies
[4] Lan, T, Kandt, J, and Longley, P (2021) Measuring The Changing Pattern Of Ethnic Segregation In England And Whales With Consumer Registers *In Urban Analytics and City Sciences Vol (48)6 1591-608*
[5] Adukia, A, Asher, S, Novasad, P, and Tan, B (2019) Residential Segregation in Urban India
[6] Charalambous, N (2018) Spatial Forms of Ethnic Coexistence in Ottoman Cyprus : The Role of Urban Form in Patterns of Everyday Life *Journal of Urban History*
[7] Perner, ML (2019) Segregated Behind The Walls : Residential Patterns in Pre-Industrial Copenhagen *Social History, Vol44, No 4*

Patterns of utilization of residential spaces affected by the tsunami in donggala regency

M Najib[1], A Mulyati[1], A M Y Astha[1]

[1]Department of Architecture Engineering, Engineering Faculty, Tadulako University Corresponding

muhnajibasrum@gmail.com

Abstract. Settlements are an artificial environment that is formed and is strongly influenced by the local socio-cultural and climate. To remain in the unit of the residential setting, the settler will make adjustments based on his needs in a settlement system. This research aims to develop a pattern of living space for people affected by the tsunami in a settlement system.The study used a qualitative phenomenological approach to examine the settlements and environment in Wani II Village directly affected by the earthquake and tsunami. The unit of analysis includes housing, community activities, and community behavior in land use. The research design looks at the overall location description, sees the relationship between systems, gives meaning to social settings, analyzes data in the field through interpersonal relationships, and combines respondent information. The inductive method is used to continuously analyze the study cases as a purposively selected objective sample. The residential environment unit in Wani II Village, Tanantovea District, Donggala Regency, forms a clustered (concentrated) pattern, oriented to the environmental center, public facilities, social facilities, and existing ecological facilities and infrastructure. The settlement pattern formed is influenced by; a) the existence of worship facilities (old mosque); port facilities (pier); and c) the existing design of environmental roads. The land-use patterns are influenced by; 1) social resilience and cohesion built through social sense; 2) mutual trust between settlers; and 3) respect for existing culture and values. Several residential units that functioned as places of business were repaired and rebuilt on the existing site. The informal sector business spaces utilize the remaining areas that are still permitted for development. Residual land formed between buildings and between buildings and environmental roads and land for housing sites that have collapsed due to the earthquake and tsunami is the most dominant living space used for various activities that can support the lives of settlers with or without government assistance. Research conclusions are 'local theories' or 'substantive theories,' which apply specifically to the research area and are provisional.

Keywords: Land Use, Earthquake, Settlements, Tsunamis

1. Introduction

Settlements are an artificial environment formed and strongly influenced by the principal (primary) socio-cultural and local climatic factors [1] in [2]. In addition, aspects of culture and the surrounding physical environment will determine the style of a settlement inhabited by a group of people in a community. The community uses its environment as a source of livelihood, bound by shared traditions and customs.

The need for protection from various types of threats or disasters in their environment causes humans to form settlements by choosing places that allow them to build their families and communities. The environment as part of the place to live is very influential on the community's existence and the management of their settlements. Therefore, people will maintain the environment as part of their lives.

The relationship between humans is a relationship of dependence on the environment and a mutually influencing relationship between the two. Therefore, a dynamic change will always take place in a residential setting. The questions in the research that will be carried out are: a) What is the community living system like in the Wani II Village; and b) What is the pattern of use of the land they live in after the tsunami disaster?

Research Purposes. The study aimed to develop a concept of the pattern of use of living space for communities affected by the tsunami in Wani II Village, Tanantovea District, Donggala Regency as a form of the community living system.

2. Method

The study uses a 'qualitative phenomenological approach to find patterns of spatial use for communities affected by the tsunami in Wani II Village, Tanantovea District, Donggala Regency. The research tool is the researcher as the main instrument equipped with several tools and shooting the base map using a drone. Data analysis using the inductive method through the description of cases determined by 'purposive sampling,' which is iterated cyclically until it reaches saturation. Discussion of research findings using explanatory techniques and confirmed with related theories and other conditions in different areas. The discussion results and conclusions are 'local theory' or 'substantive theory,' are provisional, and apply specifically to the research area.

3. Result and Discussion

3.1. Environmental Conditions and Patterns of Settlement Space in the Study Area

Wani II Village, which is located in Tanantovea District, Donggala Regency, has a specific historical potential for trade and Islamic development. Wani II village was the first place where Arab merchants landed on the west coast of the province of Central Sulawesi. The development and evidence of the spread of Islam is marked by the existence of the Old Mosque/Al-Amin Mosque which was built in 1906. Currently it has been designated as a cultural heritage site by the Culture and Tourism Office of Central Sulawesi Province.

The development of regional identity as a unit of traditional residential neighborhoods on the coast (Palu Bay) is closely related and supported by port facilities (Wani Harbor) and the historical existence of the Al-Amin Mosque as an 'old mosque.' The potential for settlement sites supported by socio-cultural and socio-economic conditions is the main element that can explain the early existence of Kaili, Bugis, Mandar, and Arab local ethnicities, which were followed by immigrant communities due to local government policies. The area of Wani II Village is administratively directly adjacent to the Palu City area, which is the capital city of Central Sulawesi Province. The position of Wani II Village, which is part of the hinterland of Palu city (the hierarchical city I), is a peri-urban area as a transition area that represents a blend of city and village characteristics, which is located in a relatively affordable location from the center of Palu city. This condition causes the spread of urban activities in the Wani II Village area to occur very dynamically in the patterns of land use for settlements.

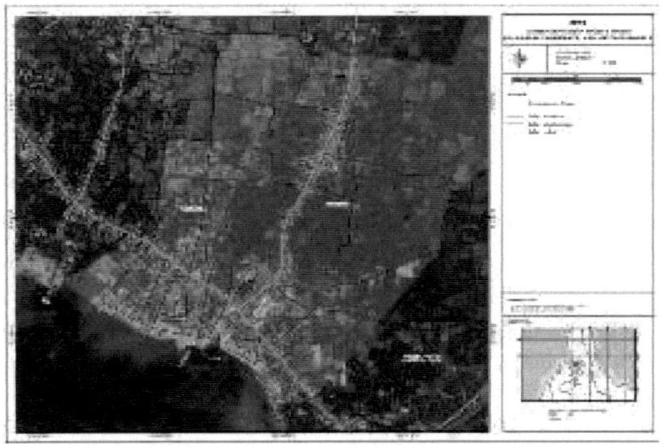

Figure – 1: Wani II Village Administration Map
Source: Field Data (mapping using drones), June 2021

The position and location of Wani II Village are very close to Palu City (the hierarchical city I) as a National Activity Center (NAC) and easy to reach by various types of urban and rural transportation modes [3]. The impact of conditions, functions, and various ongoing activities in Palu City are several factors. This is an essential factor that triggers the development of built-up land use in Wani II Village. The dynamics of settlement development with the characteristics of a 'peri-urban area' also affect the settlers' land use and various activities.

The ongoing urban and rural characters, which are the conditions and characteristics of the Wani II Village area (as a peri-urban area), then the patterns of land use in the environmental units of the settlements are strongly influenced by sociological and anthropological aspects. The development of activities around the port and activities related to the coast (sector of sea transportation, trade, and fisheries) proves how the socio-economic aspects become the service and trade sectors, which encourage the development of residential areas in groups [3][4].

3.2. *of Settlement Land Use* The development of residential land use in Wani II Village and its surroundings indicates the emergence of a new embryonic growth center that can become a 'zone of settlement unit development.' The trigger is the advantage of the location located and crossed by the national transportation network (the West Cross Sulawesi arterial route) and the supporting factor of the 'Wani port,' which functions as a collecting port. These various things become the main potential in encouraging the development of the center of growth as a 'hierarchical city II' (supporting Palu City as a hierarchical city I). The characteristics of the residential unit of Wani II Village indicate the parts of an 'elongated shape with a clumped pattern,' medium to high density, with a proportion of permanent buildings balanced with non-permanent structures, with the appearance of the physical condition of the environment and the socio-economic conditions of the population being quite good.

The entire area of Wani II Village is an area with excellent accessibility regionally so that economic relations with Palu City (as the core city) are also excellent. To physical environmental factors, it is visually apparent that the shape of the land dramatically determines the distribution pattern and form of settlement in Wani II Village. From the aspect of the density of residential units, the average density of residential buildings tends to be moderate to high, with a thickness of 40 to 60 housing units per hectare. In terms of building permanence, in general, the proportion between permanent and non-permanent settlements in the research area tends to be balanced In the case of residential units formed in coastal areas (such as the settlement units of Wani II Village), the physical conditions that significantly affect the formation of the distribution pattern of the settlement units are strongly influenced by the coastal.

The study location of Wani II Village is one of the villages affected by the earthquake and tsunami with a wave height of up to 4 meters with a wave damage range of up to 100 meters from the coast, causing damage to the environmental unit settlement. The impact of the earthquake and tsunami caused damage to several building units, changes in the area's coastline, damaged facilities, facilities, and infrastructure of the residential environment, to land subsidence. From the general condition, the pattern of settlement space in Wani II Village which is 'linear centered' and tends to be clustered, does not experience much change. The settlement patterns formed are grouped and arranged based on the 'hamlet' hierarchy, consisting of residential building units, business houses, morphology, landforms, and physical accessibility [5][6][7]. From the pattern of settlements that extend along the coast, the risk of being hit by tidal waves in Wani II Village is relatively high. However, there are 2 (two) things that have quite a positive influence to minimize the level of vulnerability to disasters in the residential environment in Wani II Village, namely; 1) maintain the density of population and buildings; and 2) maintaining physical accessibility conditions to reach the upper area (at a higher ground level).

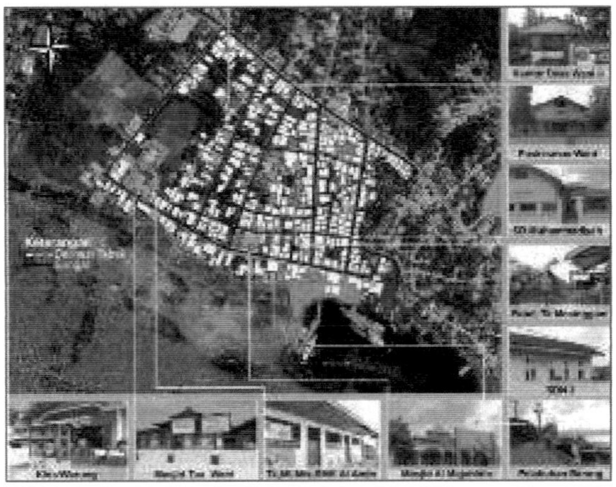

Figure-2: Potential of Settlement Environment Unit in Wani II Village
Source: Field data (2018) after processing (2021)

3.3. Residential Spaces Formed After the Tsunami Disaster

The study location of Wani II Village is one of the villages affected by the earthquake and tsunami with a wave height of up to 4 meters with a wave damage range of up to 100 meters from the coast, causing damage to the environmental unit of its settlement. The impact of the earthquake and tsunami that occurred caused damage to several building units, changes in the coastline of the area, damaged facilities, facilities and infrastructure of the residential environment, to land subsidence.

Figure-3: Impact of Earthquake and Tsunami (28/09/18) in Wani II Village
Source: Field Observation (2018) after processing (2021)

From the general condition, the pattern of settlement space in Wani II Village which is 'linear centered' and tends to be clustered does not experience much change. The settlement patterns formed are grouped and arranged based on the 'hamlet' hierarchy, consisting of residential building units, business houses, public facilities and environmental, social facilities, local roads, and main environmental roads [5][6][8][9].

Land use patterns in residential environmental units in Wani II Village are also strongly influenced by social resilience built through solid social cohesion through a sense of community, trust created between communities, and respect for culture and values. There has been. Likewise, a strong community (settlement community) perception of their level of resilience to the earthquake and tsunami disaster is supported by the social domain of the neighborhood (settlement) that grows as culture and local wisdom [5]. Social cohesion has formed a pattern of payment and use of living space which is constantly maintained even with the level of disaster vulnerability that is believed to exist still [10][1].

A strong community's perception of their level of resilience to disaster impacts is supported by the social domain of the community. The community (community) can build resilience from the perception that disasters are part of their lives, the existence of strong social ties and the government's concern for the validity of the neighborhood units formed. (both formal and informal) [10].

3.4. Behavior of Settlers in their Settlement Space

The settlement pattern reflects the population's adjustment to the natural environment, such as topography, climate, and soil [4]. Concentrated settlement patterns are influenced by the need for security or defense, family or clan ties, water scarcity, habits of the inheritance distribution system, economic capital (agricultural products/livelihoods), politics, religion, and ideology.

The factors that make up the residential space in Wani II Village are the interplay of several factors, namely; a) land slope (as land 100%); b) altitude (from sea level); c) depth of healthy water; d) rainfall; e) population density; and f) population pressure on settlement sites. In addition, the formation of residential space in Wani II Village is also influenced by 'social cohesion which is the strength of the community in the survival of the community (settling community), which includes; social solidarity; 2) social control; 3) social networks; 4) common values; and 5) love of place and identity; as the embodiment of the existence of settlers and their residential neighborhood units [12].

Human life is constantly evolving and making adjustments to its residential environment. The settler adjustment process will face problems in using residential land, either individually or in groups. These things occur in the residential neighborhood unit in Wani II Village, the place for settler activities (which are socio-economic) in the available spaces and can support these activities.

The formation of a residential environment in Wani II Village will be influenced by the spaces planned by the settlers as a better place for living activities to take place. Muhammad, B.N. [11] sees a relationship between social values in land use for settlements and housing-related to habits, moral attitudes, taboos, government regulations, cultural heritage, and traditional patterns.

In residential areas in coastal areas, physical conditions that significantly affect the distribution pattern of residential units are coastal morphology, landforms, and physical accessibility. In the context of the neighborhood unit that was formed and took place in Wani II Village with the character of a 'peri-urban' area, the socio-economic condition of the population can be studied based on the type of work, economic level, and education level.

The characteristics of the household and occupants are supported by the level of education, which impacts the mindset and pattern of action in the family on the physical condition of the residential environment in a sustainable manner. In general, the socio-economic conditions of the settlers in the study area are dominated by the non-agricultural sector, the economic level of the community is generally moderate to moderate, and the level of education is relatively average (good). Maritime culture does not touch much on the daily activities of the settlers, even though they live in coastal areas (Palu Bay) which have abundant marine resources [2][13].

Settlers' activities are often found on neighborhood roads, local roads, and main neighborhood roads. Likewise, many social interactions are carried out on harmonious roads, fields, mosques, residual land, and other open spaces. Small children generally interact in open areas, residential yards, or on balanced streets that they choose freely when doing activities in groups (playing together). Physical land use is not only an area to increase the physical (area) of the house but also as an area for social interaction of the settlers as well as a 'business space'.

3.5. Characteristics of the Pattern and Form of Community Living Space

The development of the residential environment in Wani II Village, which is not thoroughly planned or organic, embodies the urban and rural environment. This mixed form of city and village becomes more complex and distinctive because it is also colored by diverse economic and social activities, between urban and rural activities. These mixed processes and forms then produce a characteristic, unique, and attractive residential environment unit.

The pattern of clustered/centralized settlements is formed by the grouping of houses at the intersection of transportation routes and is influenced by the cooperation of the residents. Some of the criteria commonly used as the basis for grouping residential houses include ethnicity/ethnicity, origin, profession/occupation, social background, and belief/belief. The characteristics of the residential neighborhood units in Wani II Village are elongated with clustered patterns, high density, and the proportion of permanent buildings is balanced with non-permanent buildings, related to the population's physical environment and socio-economic conditions. The existing Production centers are spread out, although they still tend to be correlated with environmental centers, but there is no indication of concentration and is not related to certain settlements.

The distribution of settlers' jobs and livelihoods is quite diverse and spreads out with the character of settlements that tend to be concentrated and related to economic matters (ports, workshops, shops/food stalls). The characteristics of the living space pattern based on occupation look random but still in one compact footprint, and there is no partition based on ethnicity or ethnicity. This is in line with the view of Tilley [13] that personal identity and culture are integrated with a place.

When viewed from the religious/belief element, the Wani II Village settlement shows a settlement model that tends to clump together. The grouping starts from the center point towards the outside. Meanwhile, the pattern-oriented to the center is not visible when analyzed from the elements of the settlers' profession and livelihood (which are dominant). Characteristics of the pattern and form of community living space tend to spread evenly on the overall settlement site. Therefore, the condition of settlements cannot be separated from the aspect of the population as residents.

Ideally, settlements must contain two main requirements, namely; a) the physical environment must reflect the pattern of life and local cultural patterns, and b) the residential environment must be supported by service facilities and utilities generally proportional to the size or area of the environment and the number of residents. Thus the role of residents or residents in each settlement is crucial in seeking settlement development as an essential element in supporting the sustainable development process. These findings can be used as data for the development of economic centers, especially micro, small and medium enterprises (MSMEs) [8].

The research area of Wani II Village is generally located in a relatively accessible location from the city center (Palu City); therefore, the spread of urban activities that occur in this area causes changes in land use patterns due to the shift from agricultural functions to urban functions. Most of the land use, especially those along the main access routes, will turn into new residential areas and urban facilities, such as trade and service areas.

The residential unit of Wani II Village is located on a gentle slope, has a pattern of distribution of residential units in clusters, and extends along the road. The slope of the flat slope dramatically affects the shape of the settlement pattern, which tends to be increasingly clustered. The form of settlement is an essential aspect of natural disaster mitigation activities because it is a place to live and a place for residents to gather [11][12][14][15][16]. Thus, identification of the characteristics of settlements needs to be done to be able to identify the level of disaster risk that may occur because residents in coastal areas experience 'trauma/phobia' to earthquake and tsunami events.

The increase in tsunami disaster hazards is an important aspect of natural disaster mitigation activities at the study site. The impact on disaster mitigation identified at the study site shows that the sense of the settlers towards disasters caused by marine activities is still relatively low. For this reason, it is necessary to divide the zoning on the settlement site to determine the allocation of living space by its function.

4. Conclusion

1. The position of Wani II Village, which is part of the hinterland of Palu city (the hierarchical city I) is a peri-urban area as a transition area that represents a blend of city and village characteristics, which is located in a relatively affordable location from the center of Palu city. The characteristics of the residential unit of Wani II Village show the characteristics of an 'elongated shape with a clumped pattern', medium-high density, the proportion of permanent buildings is balanced with non-permanent buildings, with the appearance of the physical condition of the environment and the socio-economic conditions of the population are quite good

2. The settlement patterns formed are grouped and arranged based on the 'hamlet' hierarchy, consisting of residential building units, business houses, public facilities and environmental, social facilities, local roads, and main environmental roads. Patterns of land use in residential neighborhood units in Wani II Village are also strongly influenced by social resilience, built through solid social cohesion as seen through social sense, the trust created between communities, and respect for existing culture and values

3. The formation of residential space in Wani II Village is also influenced by 'social cohesion, which is the strength of the community in the survival of the community (settlement community), which includes; 1) social solidarity; 2) social control; 3) social networks; 4) common values; and 5) love of place and identity; as a form of the existence of settlers and their residential neighborhood units. After the tsunami disaster, the ongoing development of residential environmental units did not significantly affect the pattern of use of residential space

4. Settlers' activities are often found on neighborhood roads, local roads, and neighborhood main roads, likewise with social interactions that are mostly carried out on harmonious roads, fields, mosques, and residual land and other open spaces. Young children generally interact in open areas, residential yards, or on peaceful streets that they choose freely and carry out group activities (playing together). Physical land use is an area to increase the physical (area) of the residence but also as an area for social interaction of the settlers and a 'business space.

Acknowledgments
This research was carried out with DIPA funding from the Faculty of Engineering for the 2021 budget year, therefore our research team would like to thank the faculty, LPPM Tadulako University and the entire community and the Village Heads of Wani I and Wani II Village for their participation in helping to complete this research.

References

[1] Rapoport A 1969 House Form and Culture, Prentice Hall Inc, New York
[2] Mulyati A 2015 Sama' di Lao as the Basis for Shaping the Architecture of the Bajo Tribe
[3] Purbadi YD 2010 Spatial Architecture of Kaenbaun Traditional Settlement on Timor Island, Dissertation, PPS Gadjah Mada University Yogyakarta p 5-8
[4] Najib M 2002 Aspects Considered in Land Arrangement Around the River in the Palu City Center Area Thesis, PWK-ITB, Bandung

[5] Mulyati A 2013 et al Local Knowledge in the Design of Vernacular Residential Houses for Archipelago Settlement in Central Sulawesi Paper UKDW, Yogyakarta p5-7
[6] Mulyati A 2014 The Influence of Cultural Life on Typology of Aquatic Settlement in Tojo Una-Una Regency Study Univ Tadulako, Palu p7-8
[7] Hamidah N et al 2016 Analysis of Sustainable Riverside Settlements, Inertia Vol XII No May 1, 2016 p 6-7
[8] Mulyati A et al 2020 The Pattern of Utilization of Residential Spaces Affected by the Earthquake and Liquefaction in Sigi Regency Research Report – LPPM Universitas Tadulako, Palu p 5-6
[9] Naing, N, 2011 Moving Settlement in a Floating House System as a Process of Human Adaptation to the Environment, Dissertation, ITS, Surabaya
[10] Osamah M, Handmer J, Mitchell D and Ahmed I, 2014 Can the vulnerable be resilient? Co-Existence of vulnerability and disaster resilience: Informal Settlements in The Philippines International Journal of Disaster Risk Reduction, p 178-189 https://doi org/10 1016/j ijdrr, 2014 08 007
[11] Muchamad BN 2013 Residential Architecture of the Bukit Dayak Community, Dissertation,PPS-UGM, Yogyakarta p 20-23
[12] Rahmawati TA et al, 2018 Spatial-Based Disaster Risk Reduction. UB Press, Malang
[13] Tilley C 1994 A Phenomenology of Landscape – Places, Paths, and Monuments Berg Publishers,Oxford UK
[14] Katayama R et al, 2000 A Research On The Urban Disaster Prevention Plan Concerning Earthquake Risk Forecast By Remote Sensing in The Tokyo Bay Area, ISPRS, Vol, Part B7, p 6 62-669, Amsterdam
[15] Winarto G,2017 Natural Disaster Emergency Response Gosyen Publishing, Yogyakarta
[16] Idham, N C 2014 Principles of Earthquake Resistant Architectural Design Andy, Yogyakarta

Utilization of biochar from corn leaves to adsorb COD and BOD in hospital wastewater

M Napitupulu*, D K Walanda, N D Napitupulu and R M Walanda

Tadulako University, Jl. Soekarno Hatta Km.9, Palu 94118, Indonesia

*Email: mery@untad.ac.id

Abstract. This study utilized agricultural waste, namely corn leaves, as biochar. The resulting biochar complies with SNI No 06-3730-1995 for water and ash. SEM-EDS characterization showed that biochar from corn leaves could be used as an adsorbent for BOD and COD in hospital wastewater. Activation of the biochar was carried out with an activator agent, namely $ZnCl_2$. SEM-EDS determined the pore and elements analysis. Activated Biochar Corn Leaves (ABCL) consist of 50.35 mg/L of the carbon produced with a carbonization temperature of 300°C. Three grams of ABCL reduced COD up to 34.62 mg/L in 4.5 hours and a maximum of 41mg/L of BOD in 4.5 hours. While in three hours, a maximum of 2.5 gr ABCL decreased BOD 35.44%, and COD 30 mg/L.

1. Introduction

The utilization of organic waste and agricultural waste has been widely used as biochar. Our previous research has made biochar from durian bark [1], rambutan peels [2], red fruit peel [3], banana peels [4], and oranges peels [5]. Biochar is a material that contains high carbon, made by pyrolysis process with high temperature. The high carbon content of biochar makes it a promising adsorbent for absorbing heavy metals [6], dyes [7], as well as bod and cod in waste [8].
Corn leaves are one of the biggest corn wastes besides corn cobs and corn stalks. In Indonesia, corn leaves are often used as fodder for goats and cattle and are generally burned and left to rot on the ground.

The foremost potential adsorbent incorporates a high surface zone so that the adsorption ability is critical. Biochar features a tall surface region of charcoal. Typically, biochar could be an item in carbon created through pyrolysis prepared of exceptionally light biomass and can be utilized as a channel or as an adsorbent. Using biochar through the adsorption strategy, adsorbents can reduce BOD and COD in hospital wastewater [10].

Wastewater contains a bundle of dangerous substances such as BOD and COD. Chemical oxygen request (COD) is the sum of Oxygen required to oxidize natural squander in water through chemical responses. Natural squander will be adulterated by potassium bichromate ($K_2Cr_2O_4$) as a source of Oxygen into CO_2 and H_2O gas and a few chromium particles. COD esteem may be a degree of contamination by natural matter. Organic Oxygen Request (BOD) is the sum of Oxygen required to overtake a natural point in water by microorganisms at a particular temperature and volume. The more noteworthy the BOD value of water, the less Oxygen is accessible to life forms within the water. BOD is closely related to DO (Broken down Oxygen) or broken up Oxygen. The higher the BOD level of water, the less the sum of broken up Oxygen. So, the higher the BOD level, the higher the level of contamination [11].

Enacted biochar from corn leaves is made to be utilized to move forward water quality by lessening BOD and COD levels in clinic wastewater. Using corn leaves as materials for producing activated carbon supports the zero-waste rule and anticipates natural harm due to squander.

2. Research method

2.1 Sample preparation

The crude material, corn leaves, was cut into pieces and dried within the sun (figure 1a). At that point, it changed over biomass fabric, stalking corn into charcoal (biochar). ABCL is delivered utilizing pyrolysis, in which the biomass is warmed within the nonappearance of Oxygen to 300°C (figure 1b). The product made through the pyrolysis handle is within the shape of dark solid powder. The taking after the handle is the enactment handle with Calcium Chloride ($CaCl_2$). This preparation is put by drenching in a measuring utensil for 24 hours. After 24 hours, the actuated charcoal was sifted, actuated for 24 hours, then put in a heater at 500°C for 1 hour. Drying is the ultimate stage of the generation of ABCL, lessening the water substance and evacuating the remaining activator arrangement within the fabric (figure 1c). The ABSL at that point analyzes and characterizes their (a) pore examination, (b) dampness substance, (c) fiery remains substance and (d) assurance of carbon and other components. The enacted carbon pore characterization takes put with SEM-EDS.

(a) (b). (c)

Figure 1. Sample preparation process from corn leaves to ABCL

2.2. Physical analysis.

The surface morphology of the test and the natural substance within the biochar were characterized utilizing Filtering Electron Microscopy-Energy Dispersive X-Ray (SEM-EDX).

2.3. Adsorption measurement

To the degree of the impact of biochar contact time, as much as 3 grams of actuated carbon from corn stalks was put into 100 mL of clinic wastewater and, after that, blended for 3 hours, 4.5 hours, and 6 hours. After that, each test was measured for its COD and BOD levels. To the degree of the impact of biochar weight, actuated charcoal was put into 100 mL of healing center wastewater with a weight of 1.5 grams, 2.5 grams, and 3 grams, individually, at that point blended for 6 hours. At that point, the BOD and COD levels were measured.

3. Result and discussion

3.1 Biochar characteristics

ABCL was analyzed water and fiery debris substance to guarantee the Indonesia National Standard's quality. The estimation appears that the water substance of ABCL created from pyrolysis at a temperature of 300°C contains water 4.61% and fiery remains the substance of 3.98%. The ABCL delivered has met the Indonesian National Standard SNI 06-3730-1995 for dampness and cinder substance. The higher the pyrolysis temperature, the more water substance vanishes, making the test more hygroscopic [12]. In differentiation, the cinder substance increments with the increment in pyrolysis temperature. The condition is caused by the rise in temperature, causing more fabric to burn and get to be cinder [13].

3.2 Pore Morphology

The atomic estimate is a fundamental portion of adsorption since the atom must enter the micropore of the carbon particles to be adsorbed. The adsorption rate ordinarily increments as the atomic measure of the adsorbate increments [14]. Most squander comprises of blended materials so that the nuclear estimate is diverse.

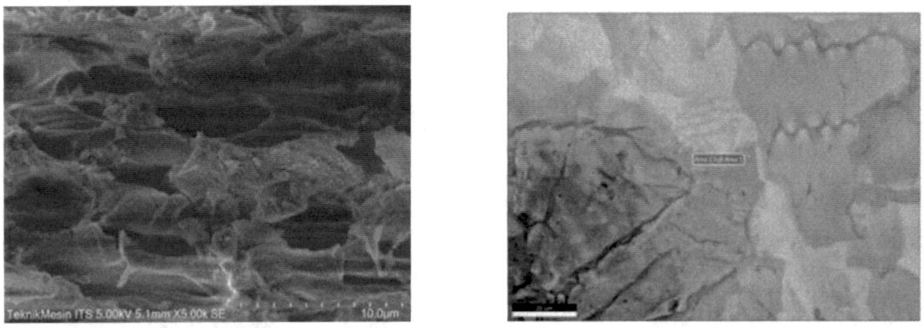

a b

Figure 2. ABCL pore morphology (a), and ABCL pore morphology plotting entire area (b)

SEM comes about of ABCL pyrolyzed at 300°C checked beneath the conditions estimation, specifically Amplification 5000x, Working Remove 5000 m, Emanation Current = 122000 nA, with Photo estimate 1000. These measurements take after those delivered by analysts on the impact of temperature on the pore characteristics of rice straw and canola stalk biochar. The pyrolysis temperature of 450°C is the ideal temperature to deliver biochar with more apparent porosity [14].

Characterization utilizing SEM created a picture of the biochar surface of the ABCL, which was uneven and sporadic in shape and changed in the surface estimate and pore profundity. The standard pore estimate at 5000 times amplification is 10.0 μm. The pore measure and depth of biochar can decide the adsorption capacity of the biochar. A few angles often impact this, counting the characteristics of crude materials and fiery debris substances [15].

3.2.1 Element Content

The comes about of the Energy Dispersive X-Ray Spectroscopy (EDS) investigation appeared that the component carbon (C) was the dominant component in corn clear out charcoal. Chart 1 appears the composition of ABCL delivered at pyrolysis temperatures of 300°C. The carbon substance in ABCL with starting pyrolysis temperature was significantly higher.

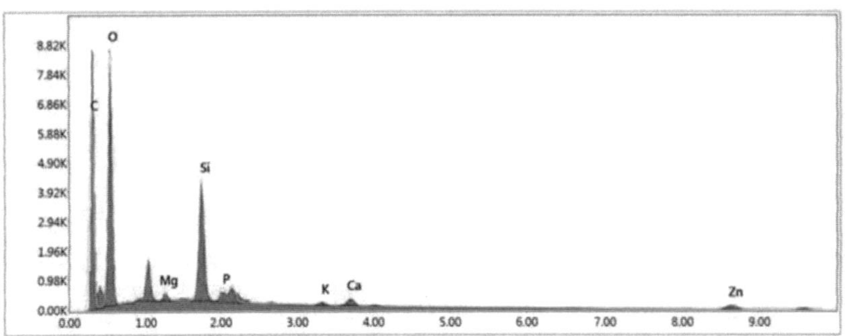

Figure 3. EDS analysis of ABCL with the activation of $ZnCl_2$ agents.

The EDS information appears the levels of carbon and other components still show within the actuated biochar. Within the biochar created at a carbonation temperature of 300°C (Table 1), the rate of nuclear carbon was 50.35%, Oxygen (41.08%), Silicon (4.88%), and Zinc (1.69%). Other components with levels between 0-1% are Magnesium (0.53%), Phosphorus (0.74%), Potassium (0.19%), and Calcium (0.56%). There are still elements such as Aluminum, Phosphorus, and Calcium within the biochar, most likely due to the moo carbonation temperature (300°C). The result can be compared to the examination of the impact of era conditions such as pyrolysis temperature and home time on the basic properties, and the wholesome characteristics of biochar decided from particular feedstocks; corn stalk, attack straw, wheat stalk, and shelled nutshell, were showed up that pyrolysis temperature was a fundamental parameter impacting biochar properties [16].

Table 1. The element composition of ABCL

Elements	Weight (%)	Atom (%)
C	38.41	50.35
O	41.74	41.08
Mg	0.81	0.53
Si	8.70	4.88
P	1.45	0.74
K	0.48	0.19
Ca	1.42	0.56
Zn	7.00	1.69

3.3 ABCL adsorption performance

The hospital wastewater comprised of BOD level = 46.87 mg/L, and COD level = 56.00 mg/L Table 2 appears that a contact time of 3 hours decreased BOD levels by 21.66 gm/L, and at a contact time 4.5 hours and 6 hours, diminished COD levels by 34.62 mg/L, and 23.49 mg/L, separately. At a contact time of 4.5 hours, BOD levels expanded and at 6 hours diminished once more. This reduction was due to the adsorbent's interaction and the oversaturated adsorbate, meaning that the pores were filled with contaminants coming about in diminished adequacy. At a contact time of 3 hours, the levels of COD are still tall [17]. The cause of the still-elevated level of COD is conceivable since, at the contact time of 3 hours blending, the interaction between the adsorbent and adsorbate has not been maximized, the pores of the enacted carbon have not completely retained natural substances that cause the event of COD contamination [18].

Table 2. Variation of ABCL contact time decreases BOD and COD levels in hospital wastewater.

Contact time (hours)	Weight (gram)	BOD (mg/L)	Decreased BOD level (mg/L)	COD (mg/L)	Decreased COD (mg/L)
0	3	46.87	0	56.00	0
3	3	25.21	21.66	22	34
4.5	3	12.25	34.62	15	41
6	3	23.38	23.49	17	39

Weight variety of enacted carbon to diminish BOD and COD levels in healing center wastewater. Within table 3, it can be seen that the weight of enacted charcoal 1.5 grams decreased the BOD level from 46.87 mg/L to 16.45 mg/L and the COD level by 44 mg/L. At the same time, 2.5 grams of ABCL can reduce BOD by as much as 15.44 mg/L and COD 30 mg/L.

Table 3 Variation of ABCL weight on decreasing BOD and COD in hospital wastewater

Contact time (hours)	weight (gram)	BOD (mg/L)	Decreased of BOD (mg/L)	COD (mg/L)	Decreased COD (mg/L)
0	0	46.87	0	56	0
3	1.5	16.45	30.42	44	12
3	2.5	11.43	35.44	26	30
3	3.5	40.60	6.27	30	26

At a weight of 2.5 grams of activated charcoal and 3.5 grams of BOD levels back up, this is often conceivable since the increase within the mass of the adsorbent within the squander causes the interaction between the adsorbent and the adsorbate to be blemished due to the vast number of enacted carbon pores that are not utilized to retain natural substances since they are closed. By other actuated carbons (Swastha, 2010). And at 1.5 grams of enacted carbon, the COD number is still tall, distinctive from BOD. This can be conceivable since the COD number was higher than the BOD within the fluid squander, which shows that the liquid blow contained active natural compounds. As a result, the weight of enacted charcoal was 1.5 grams has not ideally decreased COD levels [8].

4. Conclusion

The ABCL was arranged at pyrolysis temperatures of 300°C, coming about within the dampness substance of 4.61%, whereas the fiery remains substance was 3.98%. The pore characteristics and the elemental composition were analyzed with Checking Electron microscopy- Vitality Scatter Spectrophotometry (SEM)-EDS at different amplifications with C components ruled by 50.35%. The concentration of COD and BOD adsorbed by the biochar was 34.62 mg/L and 41 mg/L with contact time 4.5, and the ABCL weight was 3 grams Adsorption of COD, and BOD at different weights demonstrates that expanding the importance of biochar and the contact time will diminish the concentration of COD and BOD within the wastewater. The decrease in COD and BOD levels with a contact time of 3 hours, and the weight of ABCL as much as 2.5 grams, diminished BOD and COD by 35.44 mg/L and 30 mg/L.

Acknowledgment
The author would like to thank the Tadulako University Postgraduate Studies for supporting the work with a research grant.

ORCID iDs

M Napitupulu https://orcid.org/0000-0002-0089-3115
D K Walanda https://orcid.org/0000-0003-4821-9514
N D Napitupulu https://orcid.org/0000-0001-6251-9631
R M Walanda https://orcid.org/0000-0003-4587-410X

References

[1] Napitupulu M, Walanda D K, Natakusuma Y, Basir M, and Mahfudz 2018 Capacity of Adsorption of Cadmium (II) Ion by Bio-charcoal from Durian Barks *JSST* **34** 30–6

[2] Setiawan I K A, Napitupulu M and Walanda D K 2018 Biocharcoal dari Kulit Rambutan (Nephelium lappaceum L.) sebagai Adsorben Zink dan Tembaga *J. Akad. Kim.* **7** 193

[3] Napitupulu M, Walanda D K and Simatupang M 2020 Utilization of red fruit's peel (*freycinetia arborea gaudich*) as biochar for lead (Pb) adsorption *J. Phys.: Conf. Ser.* **1434** 012033

[4] Napitupulu M, Muhammad Al-Gifary and Daud K Walanda 2019 adsorption of cd(ii) by carbon prepared from peels and stems of kepok banana (musa paradisiaca formatypica) *Cellulose Chem. Technol* **53** 387–94

[5] Solika N, Napitupulu M and Gonggo S T 2018 Bioadsorpsi Pb(II) Menggunakan Kulit Jeruk Siam (Citrus Reticulata) *J. Akad. Kim.* **6** 160

[6] Komkiene J and Baltrenaite E 2016 Biochar as adsorbent for removal of heavy metal ions [Cadmium(II), Copper(II), Lead(II), Zinc(II)] from aqueous phase *Int. J. Environ. Sci. Technol.* **13** 471–82

[7] Georgin J, Dotto G L, Mazutti M A and Foletto E L 2016 Preparation of activated carbon from peanut shell by conventional pyrolysis and microwave irradiation-pyrolysis to remove organic dyes from aqueous solutions *Journal of Environmental Chemical Engineering* **4** 266–75

[8] Faisal G H, Jaeel A J and Al-Gasham T S 2020 BOD and COD reduction using porous concrete pavements *Case Studies in Construction Materials* **13** e00396

[9] Walanda D K, Napitupulu M, Hamzah B and Panessai K 2020 The capacity of biocharcoal prepared from sawah lettuce plants (*limnocharis flava*) as adsorbent of lead ions *J. Phys.: Conf. Ser.* **1434** 012036

[10] Lutterbeck C A, Colares G S, Dell'Osbel N, da Silva F P, Kist L T and Machado Ê L 2020 Hospital laundry wastewaters: A review on treatment alternatives, life cycle assessment and prognosis scenarios *Journal of Cleaner Production* **273** 122851

[11] Molinari A, Guadagnini L, Marcaccio M and Guadagnini A 2019 Geostatistical multimodel approach for the assessment of the spatial distribution of natural background concentrations in large-scale groundwater bodies *Water Research* **149** 522–32

[12] Cheng N, Wang B, Wu P, Lee X, Xing Y, Chen M and Gao B 2021 Adsorption of emerging contaminants from water and wastewater by modified biochar: A review *Environmental Pollution* **273** 116448

[13] Jia Y, Shi S, Liu J, Su S, Liang Q, Zeng X and Li T 2018 Study of the Effect of Pyrolysis Temperature on the Cd2+ Adsorption Characteristics of Biochar *Applied Sciences* **8** 1019

[14] Saraf S, Singh A and Desai B G 2019 Estimation of Porosity and Pore size distribution from Scanning Electron Microscope image data of Shale samples: A case study on Jhuran formation of Kachchh Basin, India. *ASEG Extended Abstracts* **2019** 1–3

[15] Luo P, Zhong N, Khan I, Wang X, Wang H, Luo Q and Guo Z 2019 Effects of pore structure and wettability on methane adsorption capacity of mud rock: Insights from mixture of organic matter and clay minerals *Fuel* **251** 551–61

[16] Oginni O and Singh K 2020 Influence of high carbonization temperatures on microstructural and physicochemical characteristics of herbaceous biomass derived biochars *Journal of Environmental Chemical Engineering* **8** 104169

[17] Crawford C B and Quinn B 2017 The interactions of microplastics and chemical pollutants *Microplastic Pollutants* (Elsevier) pp 131–57

[18] Khurshid H, Mustafa M R U, Rashid U, Isa M H, Ho Y C and Shah M M 2021 Adsorptive removal of COD from produced water using tea waste biochar *Environmental Technology & Innovation* **23** 101563

IOP Conf. Series: Earth and Environmental Science 1075 (2022) 012048 doi:10.1088/1755-1315/1075/1/012048

Assessment of Occupational Health and Safety Risk in the Road Construction Project in the Sigi Regency

Fahirah F [1,*], **Adnan Fadjar**[2], **Rika Madina**[3]
[1,2]*Department of Civil Engineering, Engineering Faculty, Tadulako University, Palu, Indonesia*
[3]*Graduate of Civil Engineering, Engineering Faculty, Tadulako University, Palu, Indonesia*
*Corresponding Author: fahirah_fz@yahoo.com

Abstract. The possibility occurrence of occupational health and safety risk in construction work is large because each work environment has many sources of danger but can be done prevention and handling to reduce the risk of occurrence of a work accident that's why we need the process of identification and assessment of occupational health and safety risks. This type of research is descriptive analysis, the research approach is observational to use a sheet of JHA (Job Hazard Analysis) for hazard identification and risk assessment with a semi-quantitative analysis using the standard risk assessment AS/NZS 4360:2004. The results of the hazard identification beyond the occupational health and safety are carried out in the road construction in the Sigi Regency produce 147 of the danger that comes from the unsafe action and unsafe conditions and the results of assessing the risks over 147 hazards identified there are 5 different levels of risk ranging from the risk level the most, namely level priority 3 as much as 46%, very high as much as 22%, for priority 1 as much as 14%, substantial as much as 11%, and acceptable as much as 7%.

Keywords: Risk, Assessment, Road Construction, Occupational Health and Safety

1. Introduction

Occupational Health and Safety (OHS) is an important program that must be implemented in the construction projects. Occupational health and safety (OHS) programs can increase the productivity of construction work, increase the discipline of workers at the construction site and guarantee the safety and health of workers. Occupational Health and Safety is a scientific and practical approach in overcoming potential health and safety hazards and risks that may occur [1].
The risk of occupational health and safety (OHS) in construction work is very large. At the implementation stage of road construction it can pose risks related to OHS that can affect the environment, productivity, performance, quality, time and costs project. To minimize risks related to OHS it is necessary to assess OHS risk.
Risk Assessment of OHS is a calculation of the potential magnitude based on the possibility of events that have an impact on losses to construction, human life, public safety, and the environment that can arise from certain sources of danger, occurring in Construction Works by taking into account the frequency value and severity value caused [2]. Therefore, in order for the implementation of road construction in Sigi Regency to run in accordance with the appropriate cost, time and quality, it is necessary to carry out a risk assessment related to OHS.

2. Literature Review

Risk Assessment of OHS

Risk analysis was to determine the magnitude of a risk which was reflected in the possible severity and posed [3]. Techniques can be used to perform risk analysis are: 1) Risk Assessment with Qualitative Analysis. Qualitative analysis uses the form of words or a descriptive scale to explain how much potential risk will be measured. The results can be included in the categories of low risk, medium risk and high risk. 2) Risk Assessment with Semi-Quantitative Analysis. In the semi-quantitative analysis, the previously mentioned qualitative scale was assigned a value. Each value given describes the degree of consequence and probability of the existing risk. Caution is needed in using semi-quantitative analysis, because the value created does not necessarily reflect the existing objective conditions of a risk. The accuracy of the calculation depends on the knowledge level of the expert team in the analysis of the process of a risk occurrence [4]. 3) Risk Assessment with Quantitative Analysis. Quantitative analysis uses numerical values (using descriptive scales as used in qualitative and semi-quantitative analyzes) for both consequences and likelihood, using data from various sources. The quality of the analysis depends on the accuracy and completeness of the numerical values used AS/NZS 4360 [4].

3. Research Method

Data collection consists of Primary Data through filling out the JHA (Job Hazar Analysis) sheet, Worksheet of Risk Assessment and Observation. Secondary Data is in the form of daily, weekly, and monthly of work reports. Processing and analyzing data used semi-quantitative risk analysis of AS/NZS 3460:2004 [4].

4. Results and Discussion

Risk assessment of OHS that has been carried out on road construction projects in Sigi Regency is based on unsafe actions and unsafe conditions.

Determining the level of risk by providing an assessment of probability, exposure, and consequence

4.1. The mobilization work

The mobilization work identified 8 hazards (Figure 1), 2 risks (25%) Very High level, 2 risks (25%) priority 1 level, 1 risk (13%) substantial level, 1 risk (13%) priority 3 level and 2 risk (25%) acceptable level.

In mobilization work, the risk level is very high, namely a survey of the route to be traversed before the process of transporting the equipment is not carried out and the heavy equipment is not properly attached to the transport car. Priority 1 risk level, namely unsafe operator behavior and the position of heavy equipment is not in accordance with procedures.

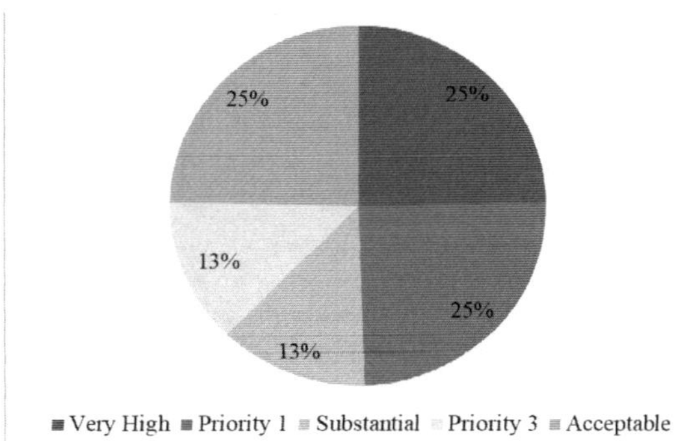

Figure 1. Percentage of Risk Level of the Mobilization work

4.2. Structural Excavation Work Depth 0 – 2 meters

Structural excavation work with a depth of 0 – 2 meters identified 11 hazards (Figure 2), 2 risks (18%) Very High level, 1 risk (9%) priority level 1, 1 risk (9%) substantial level, 6 risks (55 %) priority level 3 and 1 risk (9%) Acceptable level.

In structural excavation work with a depth of 0 – 2 meters, the risk level is very high, namely workers are in the blind spot area of heavy equipment and warning signs are damaged/not installed at the work site. Priority 1 risk level is the remaining construction material that can disturb the surrounding environment.

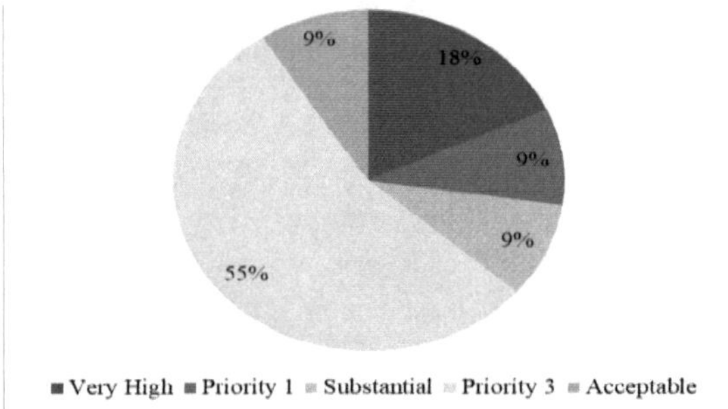

Figure 2. Percentage of Risk Level of Structure Excavation Work 0 – 2 meters

4.3 Structural Excavation Work Depth 2 – 4 meters

Structural excavation work with a depth of 2-4 meters identified 11 hazards, 2 risks (18%) Very High level, 2 risks (18%) priority level 1, 3 risks (27%) substantial level, and 4 risks (37%) priority level 3 (Figure 3).

In structural excavation work at a depth of 2-4 meters, the risk level is very high, namely the warning signs are damaged/not installed at the work location and workers are in the blind spot area of heavy equipment. The level of priority 1 risk is that heavy equipment can fall into the excavation and the operator suffers injury.

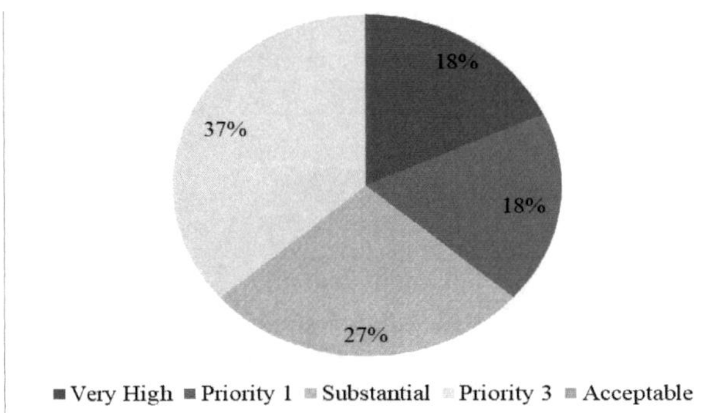

Figure 3. Percentage of Risk Level of Structure Excavation Work 2 – 4 meters

4.4. The selected embankment work from excavation sources

The selected embankment work from excavation sources identified 28 hazards, 5 risks (18%) Very High level, 7 risks (25%) priority 1 level, 5 risks (18%) substantial level, 8 risks (29%) priority 3 level and 3 risks (11%) for the Acceptable level (Figure 4).

In the compaction work of the embankment has a very high risk level, namely accidents and being run over by a vibrator roller, damaged warning signs or road dividers not installed at the work location. The risk level of priority 1 is the operation of the vibrator roller that is not in accordance with the procedure.

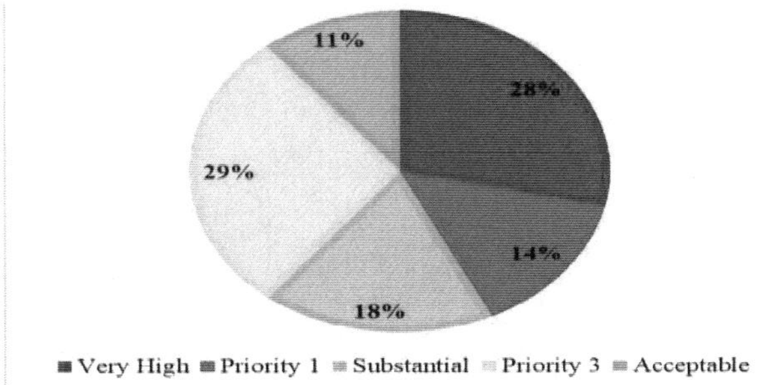

Figure 4. Percentage of Risk Level of The selected embankment Work from excavation sources

4.5 The structure demolition work

The structure demolition work identified 18 hazards, 4 risks (22%) Very High level, 5 risks (28%) priority 1 level, 1 risk (6%) substantial level, and 8 risks (44%) priority 3 level (Figure 5).

In the demolition work of concrete and masonry, the risk level value is very high, namely the danger during the demolition process using an excavator, the danger in the process of disposing and transporting demolition materials and the operator does not follow the procedure. The risk level of Priority 1 is traffic safety personnel not working in accordance with procedures and residual construction materials that can disturb the surrounding environment.

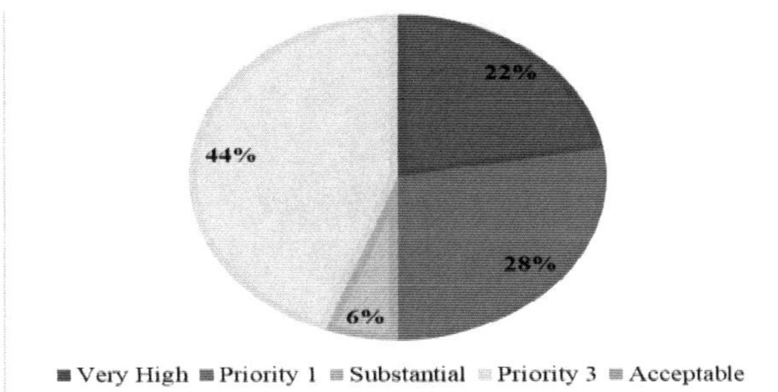

Figure 5. Percentage of Risk Level of Structure Demolition Work

4.6 Box Culvert Work

Box culvert work identified 43 hazards, 7 risks (16%) Very High level, 1 risk (28%) priority 1 level, 33 risks (77%) priority 3 level and 2 risks (5%) for Acceptable level (Figure 6).

In box culvert work, the risk level is very high, namely the warning signs are damaged and the road divider is not installed at the work location. The risk level of priority 1 is the remaining construction material that can disturb the surrounding environment.

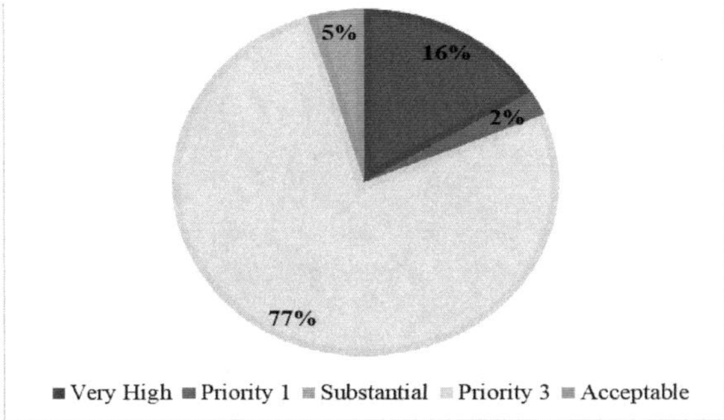

.**Figure 6**. Percentage of Risk Level of Box Culvert Work

4.7 Grained Pavement Works of Aggregate Foundation

Grained pavement work of aggregate foundation identified 28 hazards (Figure 7), 7 risks (25%) Very High level, 5 risks (18%) priority level 1, 5 risks (18%) substantial level, 8 risks (29%) level priority 3 and risk 3 (11%) for the Acceptable level.

In compaction work, the aggregate foundation layer has a very high level of risk, namely compactor accidents, vibrator rollers on site and damaged/not installed warning signs at the work site. The risk level of Priority 1 is traffic safety personnel not working in accordance with procedures,

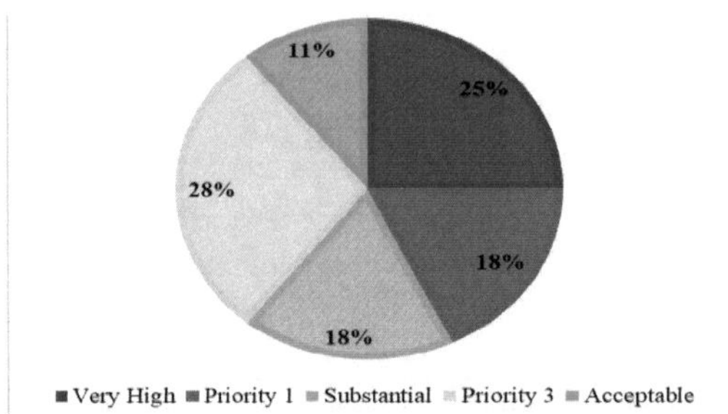

Figure 7. Percentage of Risk Level Grained Pavement Works of Aggregate Foundation

5. Conclusions

The results of the occupational health and safety (OHS) risk assessment on road construction projects in Sigi Regency, namely:

1) Identification of hazards to occupational health and safety (OHS) on road construction projects in Sigi Regency identified 147 hazards originating from unsafe actions and unsafe conditions ranging from the work environment, worker attitudes, manual work, tools and machines.
2) Risk assessment of occupational health and safety (OHS) on road construction projects in Sigi Regency there are 5 types of risk levels from 147 identified hazards, namely Priority 3 level as much as 46% of actions that must be taken, namely the need for continuous supervision and attention, very high as much as 22% Actions to be taken are activities are stopped until risk can be reduced to an acceptable limit, priority 1 level is 14% Actions to be taken are the need for possible control, substantial levels are 11% Actions to be taken are necessary specific improvement, and the lowest level of risk, namely the level of acceptable as much as 7% of the actions that need to be taken, namely the intensity that causes the risk to be reduced to a minimum.

.

Acknowledgment
The authors would like to thank all those who have supported this research: Program Kompetisi Kampus Merdeka (PKKM) in 2021 of Civil Engeneering of Tadulako University the Province of Central Sulawesi, Indonesia.

References:

[1] Hertaslim, M. (2013) Penerapan Keselamatan Dan Kesehatan Kerja (K-3) Pada Proyek Pelaksanaan Konstruksi Jembatan (Studi Kasus Jembatan Tawaeli III). *Tugas Akhir*, Universitas Tadulako.Palu
[2] Socrates, M.F. (2013). Analisis Risiko Keselamatan Kerja dengan metode HIRARC (Hazard Identification,Risk Assessment and Risk Control) Pada Alat Suspension Preheater Bagian Produksi di Plant 6 dan 11 Ceteurup PT Indocement Tunggal Prakarsa.*Tugas Akhir*, UIN Syarif Hidayatullah. Jakarta.
[3] Sutrisna, I.P.A. (2017). Manajemen Risiko Keselamatan dan Kesehatan Kerja (K3) Pada Proyek Pembangunan Hotel oleh Studi Kasus Kontraktor PT. Tata Mulia Nusantara Indah. *Tesis*. Universitas Udayana Bali.
[4] AS/NZS.4360:2004. *Risk Management Guidelines Companion to AS/NZS 4360:2004*. HB 4360:2004, *Standards Australia International, Sydney. Retrieved from Standards New Zealand Database*, 2004.

Kinship in the Establishment of Settlements for Remote Indigenous Peoples in Central Sulawesi

A Mulyati[1], Zaenal Z[2]

[1,2]Department of Architecture Engineering, Engineering Faculty, Tadulako University

ahdamulyati@gmail.com

Abstract. The existence and implementation of living space pattern characterize settlements and is a specific model of living space. The management of living spaces is essential to the existence of culture and local wisdom. The study uses a naturalistic qualitative approach to find the concept of establishing settlements for disadvantaged indigenous communities. Residential homes form a linear line along the neighborhood road. Residential houses still use simple building materials, namely boards and bricks, like walls and zinc as roofs. Kinship is reflected, among others, in the relationship between residences in the form of open spaces, unlimited courtyards, functioning as play areas, or other interactive activities. This space equips with bale-bale or trees as protection. Other remaining slots use to raise livestock, toilets, laundry rooms, and food stalls. The displacement of settlements from the top of the mountains caused the fields as the primary source of livelihood to remain in their original places, equipped with their garden houses. The new residential environment has several facilities such as village offices, mosques, schools, and general facilities building. Deliberation, consensus, kinship, and cooperation remain the principles of settler life, reflected in the formation of unlimited space and carrying out various daily life activities.

Keywords: kinship, pattern, settlement, community, cleft behind

1. Introduction

Settlements are a place to live and a place to socialize so that settlements are a manifestation of the cultural life of the settlers. Settlements form with the local knowledge of the people who inhabit it so that the form of settlements is a symbol of the community itself, both in physical form and the values contained. Topographical conditions significantly affect the characteristics of settlements and climatological factors, namely climate, weather, wind, temperature, humidity, and other natural elements. Settlements forming spontaneously are built with the local knowledge of the settlers, so the 'settlement architecture' is a manifestation of the community's living system. Humans make defenses in their lives, especially as shelters against environmental conditions in which they live. Comfort for the home and the environment is a way to survive in a settlement. Humans make adjustments to the environment according to their needs, namely economic factors and kinship factors. Likewise, in settlements of isolated tribes, settlers form spatial patterns according to their needs with local knowledge. The environment becomes a part of their lives to provide a sense of security and comfort in their environment as a living system.

The need for sources of life and protection from various threats or disasters causes humans to form settlements by choosing places that allow them to build their families and communities in a residential space. The environment as part of the place to live is very influential on the community's existence and the management of their settlements. This is related to the comfort, security of residence so that the Tersal Indigenous Community (KAT) maintains the environment as part of their lives. The relationship between humans is not only a relationship of dependence on the environment but a mutually influencing relationship between the two so that it can change an environment.

The spatial layout of a community's settlement is a continuous process of life to continue and carry on its life. Settlements and their environment are places where humans live in harmony, balance and can maintain and carry out their lives. Awareness of the importance of environmental conservation programs needs to be carried out continuously, namely through mentoring and community empowerment to use their living spaces as much as possible. Utilization of space in the living space is done to always have a sense of security and comfort in their environment.

Vernacular housing is a manifestation of the socio-cultural values of the community, which are closely related to the socio-cultural values of its residents, which in the process of its preparation use traditional norms. Therefore, as a community product, the form of the residential environment is the result of social agreement, not an individual product. Based on the description above, the research questions arise: a. What is the settlement of the Indigenous Indigenous Community (KAT) like? b. To what extent did local knowledge apply to the formation of their territories?

2. Research Methods

This study uses a "qualitative-naturalistic" approach so that it will be found the application of local knowledge in the formation of isolated indigenous community settlements [1][2]. Location of the research: Settlement of the Teasing Indigenous Community in Kalora Village, Balance Village, Wayu Village, Sigi Regency.Research tools, namely researchers as the main instrument equipped with auxiliary tools, in the form of writing instruments, drawing tools, portrait tools, tape recorders, measuring instruments, GPS, and basic maps of settlements. The research material consists of: accommodations and the environment, namely residential units, facilities and infrastructure, gardens, fields, community activities, and settlers' behavior in the use of settlements and their environment.

The paradigm of natural qualitative research in the research design is as follows [2]: seeing a broad and total description of the location, seeing the relationship between systems, giving meaning to social settings, requiring analysis at the time of data collection in the field, so researchers must have interpersonal relationships (face to face). To face, can combine emic information (respondent information), and perform continuous analysis. The first step begins with preparing maps of the research location. Then make observations of the research area to determine the case study as the first 'sample.' Through interviews and continuous observation, information themes were obtained.

The themes are emphasized on residential units related to social, cultural, and historical aspects. Then identify the spatial pattern of settlements to obtain the most dominant spatial pattern. The spatial pattern describes the 'settlement formation process' carried out by settlers living in the area (Gawalise mountain area). Data analysis uses the 'inductive' method through the description of cases as a sample. The iteration process is cyclically developed as the basis for a temporary research design, determining other cases (by purposive sampling) until they reach 'information saturation' or to an agreed limit. The analysis was carried out by means of categorization, description, and explanation, carried out every time of observation and interview. These results are called 'research findings.'

The meaning is carried out by interpreting and explaining the findings between substantive concepts so that the results of 'local concepts' are obtained, namely the values of local knowledge in the formation of residential spaces. The next stage is the discussion of research findings, using the 'explanation' technique, namely explaining the findings and confirming with related theories and other conditions in different areas. The results of the discussion are research conclusions which are 'local theories' or 'substantive theories', which apply specifically to the research area and are temporary.

3. Result and Discussion

3.1 Establishment of Settlement in Sigi Regency

The condition of the Sigi Regency area is surrounded by mountains and valleys so that people live in these areas. The Kaili tribe, known as Kaili (consisting of approximately 30 ethnic groups), occupies most of the Sigi district, especially the villages in mountainous areas. They are the oldest community group in Central Sulawesi, with a livelihood in farming, so they often move from place to place, still occupy mountainous areas, and preserve nature wisely (mopahilolonga katuwua). They maintain good relations with fellow humans (hintuwu) and relationships with nature (katuwua) [3][5][6][7][8][9].

They form settlements according to family groups so that accommodations are made without boundaries. Surrounding the housing units are fields as a source of livelihood in addition to raising livestock. The belief in a relationship between the three, namely humans, animals, and plants, causes them always to maintain the environment as part of their lives. When the harvest was successful, the settlers held a harvest feast (also known as movunja). The community rejoiced in an open space where livestock was slaughtered and offered to the Creator gratitude for success and abundant sustenance. They still maintain the principles of mutual assistance, cooperation, kinship (sintuvu) in all aspects of life, especially in maintaining the residential environment and strengthening social relations [2][4][7][8].

Belief in ancestors still colors their lives, namely the 'Balia' ceremony (worship of ancestral gods and spirits), where the highest gods are "Topebagi" (the determiner) and "Topejadi" (the Creator). All success in life will be accompanied by thanksgiving and offerings to the gods in the form of animal sacrifices (usually buffalo). The ceremony was led by the traditional chief (the elder ada') in a sacred atmosphere, followed by a people's party for all the settlers in the village in a happy atmosphere accompanied by dances. The activities carried out by the settlers colored the spaces formed in the settlement. In addition to residential units as residential spaces, there are fields as workspaces, and open spaces as multi-functional spaces, not only as a space for interaction but also as a space for worship and other public activities [2][5][6][9][10].

The residential units are in the form of stilt houses that have changed from the shape of a treehouse. All materials used are obtained from the surrounding environment in the form of wood, palm fiber, bamboo, and others without damaging the environment and the surrounding nature [11][12][13][14]. Building houses and residential facilities are carried out in cooperation as a form of kinship and togetherness. Likewise, with other activities, all carried out together. Since long ago, Bantaya has been one of the complementary buildings of a settlement, functioning as a place for deliberation and carrying out traditional village events. It is located in the center of the village, making it easily accessible to all settlers, usually equipped with a very large open space. The village head's house and the traditional leader's house [2][7]. These buildings are in the form of stilts, using logs that are arranged to reach a height of 1.00 – 2.00 m.

3.2 Kinship in the Establishment of Disadvantaged Indigenous Community Settlements

3.2.1 Kalora Village

The settlement in Kalora village is a settlement inhabited by a disadvantaged indigenous community (KAT), located in the Gawalise mountains, which borders the city of Palu and the district of Donggala. Most of the houses have been moved to the valley adjacent to the city of Palu, and a few still occupy the mountains [18][20]. The residential units are in the form of an emergency, using local wood, bamboo, and tin roofs. They build residential units in the direction of the contours (topographical conditions) of the surrounding land so that they resemble houses on stilts [12][16][19]. Buildings are made without boundaries because they are still allied (kinship/kinship relations) [17][26][27][28][29][30]. The fields as a source of livelihood are located on land a bit far from settlements, so they build garden houses on the land [19][20][21][22][23].

The settlements that were moved to new lands (Desa Kalora Bawah hamlet 1 and hamlet 2) were inhabited by most of the settlers who previously also lived in the village of Kalora Atas. Settlements are equipped with adequate facilities and infrastructure in schools, mosques, village offices, and Bantaya as buildings that function as places for deliberation and various matters relating to the village, customs, and other village events. Clean water facilities, public toilets, and sewers complete the settlements. Residential units still use simple materials; some use bricks and boards as walls with tin roofs. The house is made without massive walls, only using cloth as a barrier. Some houses have been equipped with a small terrace at the front of the house, functioning as an interaction space for both housewives and children [23][24]25].

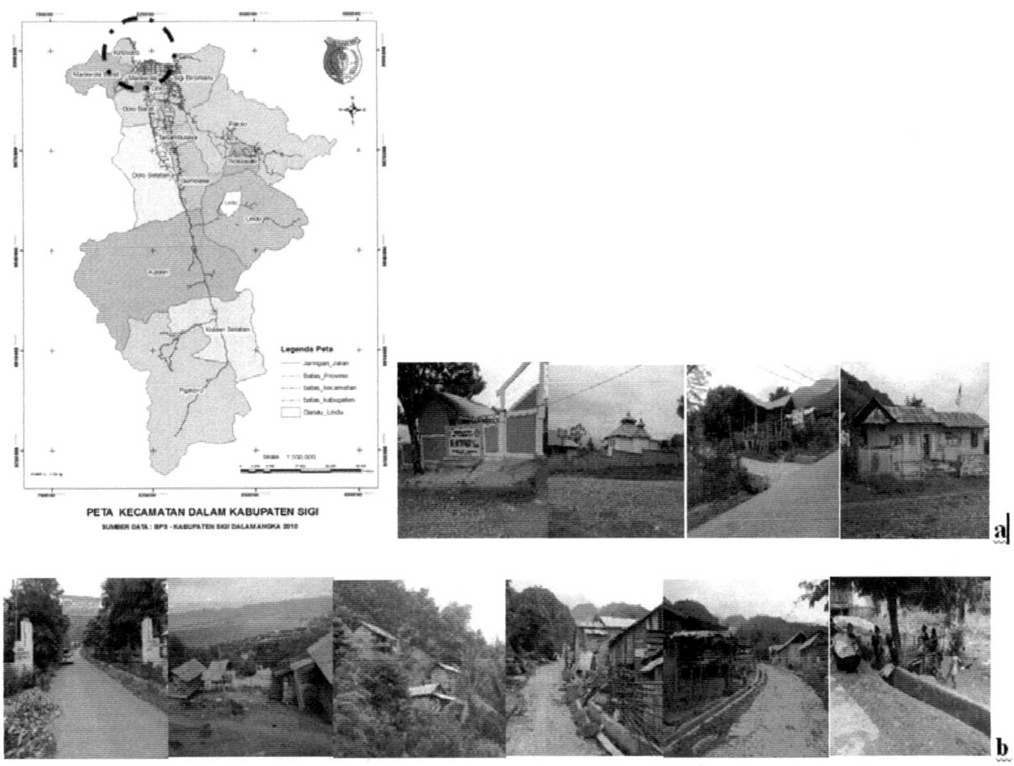

Figure 1. Some of the residential houses have changed, permanent form. However, there are still those who maintain using local building materials (boards, thatched roofs), with a simple structural system in the village of Lower Kalora). In general, residential houses face the neighborhood road, thus forming a linear pattern. Some groups form a cluster pattern because it describes a family clump that is maintained. The house spaces in kiosks, dego-dego, and the remaining area near the road are interaction spaces between settlers.

In contrast to the government-assisted residential unit, it is built with brick material as walls and a tin roof with 6 x 6 meters. Residential houses are equipped with bedrooms as private spaces for women (mother and children), while men occupy living rooms or other public spaces without using boundaries (massive walls). The kitchen is an additional building, usually attached to the main building with wallboards and zinc roofing. In general, residential units have been oriented to environmental roads as a liaison or circulation route between hamlets and villages. Residential units are made without boundaries (fences) so that between buildings are connected to the land or yard as an interaction space [20][21][22][23][24][25].

They build interaction spaces in bale-bale (roofed or unroofed) at the front of the house or use the remaining space under a tree. The shop, which is part of the residence, is also a space for interaction for the settlers. The atmosphere is full of kinship in these spaces, a reflection of harmony and harmony between the settlers in the village. As another binding building, the Bantaya is used when there are village events or village meetings if there are problems between the community [25][26][27][28][29][30].

3.2.2 Balane Village and Wayu Village

The settlements in Balance village and Wayu village are settlements inhabited by no longer disadvantaged indigenous communities (KAT), even though they are located in the Gawalise mountains, which borders the city of Palu and Donggala district. Most of the houses have been moved to the valley adjacent to the city of Palu (Balance village), some still occupy the mountains (Wayu village). Most residential units are permanent, using local materials of brick, wood, bamboo, and tin roofs. They build residential units in the direction of the contours (topographical conditions) of the surrounding land so that they resemble houses on stilts. Buildings are made without boundaries because they are still allied (kinship/kinship relations)[26][27][28][29][30]. The fields as a source of livelihood are located on land adjacent to settlements (mostly candlenut and a small part of cloves and cocoa plants), so that the fields are almost integrated with their settlements. Unlike the village of Kalora, the fields or gardens are far from their residential environment because the condition of the land consists of rocks, so that it is not possible to function as fields[23][24][25].

Figure 2. Residential facilities and infrastructure (a) Some residential houses have changed, permanent form. However, some still maintain local building materials (boards, tin roofs), with a simple structural system in Wayu village, West Marawola sub-district.

In general, residential houses face the neighborhood road, thus forming a linear pattern. Some groups include a cluster pattern depicting family clumps and kinship that are maintained. The house's rooms in kiosks, dego-dego, terraces, and courtyards are interaction spaces between settlers (b).

The settlement that was moved to a new land (Balance village) is already bordering the city of Palu, inhabited by most of the settlers who previously also lived in the mountainous part. Settlements are equipped with adequate facilities and infrastructure in the form of schools, mosques (Balance village and Wayu village in the form of churches), village offices, and bantaya as buildings that function as a place for deliberation and various matters related to the village, traditional matters and other village events. Clean water facilities, toilets, and sewers complete the settlements. Residential units still use simple materials; some use bricks and boards as walls with tin roofs. Residential houses use massive walls and are equipped with spaces (sleeping, etc.).

Some houses have been equipped with a small terrace at the front of the house, functioning as an interaction space for both housewives and children [24][25][26]. The settlement that was moved to a new land (Balance village) is already bordering the city of Palu, inhabited by most of the settlers who previously also lived in the mountainous part. Settlements are equipped with adequate facilities and infrastructure in the form of schools, mosques (Balance village and Wayu village in the form of churches), village offices, and bantaya as buildings that function as a place for deliberation and various matters related to the village, traditional matters and other village events. Clean water facilities, toilets, and sewers complete the settlements. Residential units still use simple materials; some use bricks and boards as walls with tin roofs. Residential houses use massive walls and are equipped with spaces (sleeping, etc.). Some houses have been equipped with a small terrace at the front of the house, functioning as an interaction space for both housewives and children [24][25][26].

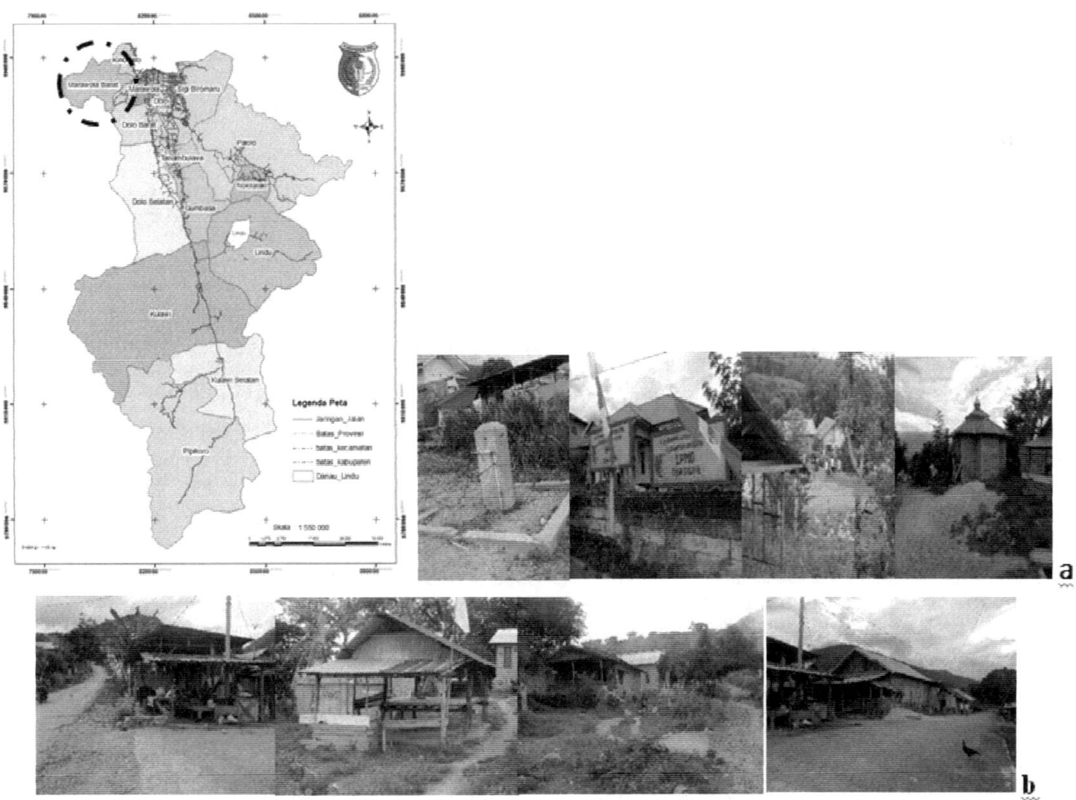

Figure 3. Some residential facilities and infrastructure (a) residential houses have changed, permanent form. However, there are still those who maintain using local building materials (boards, tin roofs, or both), with a

simple structural system in Balance village, West Marawola sub-district. In general, residential houses face the neighborhood road, thus forming a linear pattern. Some groups form a cluster pattern, depicting family clumps and kinship that are maintained. The spaces in kiosks, dego-dego, terraces, and courtyards are spaces for interaction between settlers (b).

In general, residential houses are permanent; most use brick as a wall and zinc as a roof. The building area varies significantly between 36 m2 to approximately 90 m2, so that almost all residential houses are equipped with several rooms, namely living room, bedroom, dining room, and kitchen. They build homes (nombangu sou) in cooperation (family) according to their expertise, so they rarely use artisans from outside the village (interview with Mrs. Selfina, father Mansyur, and father Markus Zeus, 29 August 2021 at 13.30). The yard (vaneve) as a public space on residential land is equipped with a children's play area, a place for bathing, washing equipped with water tubs and wells (buvvu) on the front (vaneve ngayona). In contrast, the backyard (vaneve kalikuna) is planted with several vegetables as daily needs. While in the side yard, there is a cage for the settlers' domestic animals. The kitchen building is an additional building, usually attached to the main building with wallboards and zinc roofing [25][26][27][28].

Residential units are oriented to neighborhood roads as a liaison or circulation route between RT, hamlet, and between villages. Residential units are made without boundaries (fences) so that between buildings are connected to the land or yard as an interaction space. They build interaction spaces in the form of bale-bale (roofed or unroofed) at the front of the house or use the remaining space under a tree [20][21][22]. The shop, which is part of the residence, is also a space for interaction for the settlers. The atmosphere is full of kinship in these spaces, a reflection of harmony and harmony among the settlers in the village [23][24][25][26]. As another binding building, namely Bantayan, is used when there are village events or village meetings if there are problems between the community. All issues related to adat, whether land ownership, land boundaries, and traditional events in the life of the settlers, are discussed with the adat elders (customary leaders) because Customary Institutions are still empowered by the people in Wayu and Balane villages [27][28][29][30].

4. Conclusions

Residential units in Kalora Village, Balance Village, and Wayu Village, which are the Kaili ethnic settlements of the Da'a tribe and are disadvantaged traditional settlements, have undergone many changes, both in the form of their houses and the arrangement of the settlement environment. In general, residential dwellings are permanent in shape, oriented to the neighborhood road. They are a family so that the site of the house is not massively limited; the yard is shared. Some places are equipped with interaction spaces in bales, terraces, or remaining area under trees.

Residential houses are also made without space; the space limit is only in hordyn cloth to get privacy space. Some government-aided places have been equipped with a bedroom, usually used for mothers and daughters, while sons and fathers occupy the living room and other function rooms, namely the family room.

The settlement pattern is linear, parallel to the neighborhood road, where public spaces become the center of orientation, for example, worship spaces (mosques and churches) or Bantaya buildings as other public spaces. School buildings, village offices, and general facility buildings are the center of the neighborhood, which is also the center of their settlement orientation. The remaining space on residential land and the remaining space on residential land becomes other open spaces that are also used for settler interaction that functions both publicly and privately.

Acknowledgments
This research was carried out with the assistance of Tadulako University's DIPA funds for the 2021 fiscal year. Therefore our research team would like to thank the university, LPPM Tadulako

University, and the entire community, and the heads of Kalora, Balance, and Wayu villages for their participation in helping to complete this research.

References

[1] Groat L and D Wang, 2002 Architectural Research Methods John Wiley and Sons, New York
[2] Mulyati A, 2015 Sama' in Lao as the Basis for Shaping the Architecture of the Bajo Tribe Settlement In Sulawesi Dissertation, PPS Universitas Gadjah Mada Yogyakarta
[3] Rapoport A, 1969 House Form and Culture Prentice Hall Inc, New York
[4] Doxiadis C A, 1967, 1968 Ekistic : An Introduction to The Science of Human Settlement
[5] Purbadi Y D, 2010 Spatial Architecture of Kaenbaun Traditional Settlement on Timor Island Dissertation PPS Universitas Gadjah Mada Yogyakarta
[6] Muchamad B N, 2013 Residential Architecture of the Bukit Dayak Community Dissertation, PPS-UGM, Yogyakarta
[7] Mulyati A, 2014 The Influence of Cultural Life on Typo-Morphology of Aquatic Settlements in Tojo Una-Una Regency Research, Univ Tadulako, Palu
[8] Mulyati A, 2013 et al Local Knowledge in Residential Vernacular Residential Design Islands in Central Sulawesi Papers, UKDW, Yogyakarta
[9] Sasongko I, 2005 Formation of Cultural-Based Settlement Spatial Structure (Case Study: Desa Puyung-Central Lombok) Journal of Architectural Engineering Dimensions Vol 33 No 1 p 1-8
[10] Saraswati T, 2015 The Position of Power in the Concept of Local Wisdom in Vernacular BuildingsPapers in the SCAN National Seminar Univ Atmajaya, Yogyakarta
[11] Erawati E, 2016 Kajang Traditional Settlement Pattern Article in Kapata Archaeological Journal Volume 12 No 2h, 147-162
[12] Setiadi A, 2015 Concepts, Principles, Strategies of Local Wisdom and Modern Science in Architecture Papers in the SCAN National Seminar Univ Atmajaya, Yogyakarta
[13] Prasetyo F A, 2013 Manufacturing Genius Loci of Indigenous Nias Architecture Articles in INRIK (Indigenous Research Indonesia Knowledge) Padjadjaran University, Bandung
[14] Dewi N K A, 2016 The Meaning and Role of Cosmology in the Formation of Village Patterns Sasak traditional Article in the Journal of Space Volume 4 N0 2 August 2016
[15] Fauziah Nur, 2014 Characteristics of Traditional Papuan Architecture Article, in Proceedings of the Symposium National Applied Technology, (SNTT) 2, 2014
[16] Lokbere H R, Sarwadana S M, Astiningsih A A M, 2012 Identification of Settlement Patterns Traditional in the Hologolic Village, Asotipo Wamewa District, Jayawijaya Regency, Papua Province Article, in Journal of Tropical Agroecotechnology Vol 1, No1 July 2012
[17] Ersina S, Amalia A, Sutriani, 2014 Genius Loci in the Senaru Tribal Traditional Village Sasak,West Lombok Regency Article Jur Architecture FST-UIN, Makassar
[18] Kustianingrum D, Sonjaya O, Ginanjar Y, 2013 Study of Mass Arrangement Patterns and Typology of Forms Hamlet Traditional Village Building in Garut, West Java Article in the Itenas Reka Karsa Journal, Vol 1 No 3, October 2013
[19] Harapan A, 2018 Village Patterns and Traditional Houses of the Outer Baduy Traditional Village Ketug, Lebak Regency, Banten Article, in the Proceedings of the Dept USU Faculty of Engineering Architecture, Medan
[20] Mulyati A, 2018 Local Wisdom in Architecture of Vernacular Water Settlement of Bajo People in Central Sulawesi Paper UKDW, Yogyakarta h1-7
[21] Naing N, 2011 Moving Settlements in a Floating House System as an Adaptation Process Humans with the environment bro Dissertation, ITS, Surabaya
[22] Waterson R, 1990 The Living House, An Anthropology of Architecture in South East Asia Oxford University Press, Singapore in Mulyati, A, 2015

[23] Lanur V S C, and Martini E, 2015 Development of Wae Rebo Tourism Village Based on Wisdom Local Article, in the Journal of Planesa Vol 6 No 2, November 2015
[24] Umar D Y Y, 2016 The Relation of Da'a Ethnicity in the Inland Region of the Gawalise Mountains, Sulawesi Central Part, with Australomelanesid Population in Sulawesi Article In Amarta Journal Vol 34 No June 1, 2016 h 6-15
[25] Irmawan, 2018 Social Empowerment of the Kaili Da'a Tribe in Sigi Regency Articlesin Social In Journal Conception Vol 6 No 2, January _April 2018 h 92-98
[26] Hamzari, 2008 Analysis of Customary Forest Management in Toro Village, Kulawi District, District Donggala Article in Journal of J Agroland 15 (3), September 2008 p 210-215
[27] Amri N and Bintang C, 2018 The Effect of Kinship System on Village Settlement Patterns Bajo Article in Scientific Meeting IPLBI 7, F 074-078 https://doiorg/1032315/ti7f074
[28] Suprapti W W A and Rukayah R S, 2013 The Effect of Kinship System on Patterns Development of Bonokeling Settlement in Banyumas Article in Journal of Theodolita, Vol 14 No2 December 2013 h 46-54
[29] Krause R A, 2012 The Kinship Tradition, and Settlement Pattern Archeology of Middle Missouri Community Life Article in the Journal of the Plains Anthropologist, Vol57 No 222 of 2012 h 129-137
[30] Purnama D H, 2019 The Order In The Social System Based On The Kinship Relationship and The Hamlet's Founder of Ethnic Basemah South Sumatra Articles in the International Geogaphy Seminar IOP Conference Series doi:101088/1755-1315/683/1/012111 h1-5

Testing of The Method of Gama I Synthetic Unit Hydrograph in The Analysis of The Tojo Watershed Design Flood

Setiyawan[1], Vera W A[2], Irdhiani[3], Rehana[4], Sri H[5], Hasbi A[6]

[1,2,3]*Department of Civil Engineering, Engineering Faculty, Tadulako University*
[4,5,6]*Undergraduate Civil Engineering undergraduate's Program at Tadulako University*

setiyawanvip@yahoo.co.id

Abstract. Floods occur due to several factors including topographic conditions, high rainfall intensity, blockage of water flow, lack of water infiltration areas, deforestation, and so on. The Tojo watershed is included in the Parigi-Poso river area, with an area of 211,830 km2 and a river length of 25.52 km. In the Tojo watershed, it is necessary to study flooding, because the watershed always flows every year. This research aims to find out how much flood discharge was obtained from the analysis of the Tojo watershed design flood with the Gama I synthetic unit hydrograph method and to find out how much the Gama I synthetic unit hydrograph analysis compared to the design flood discharge based on river discharge data (AWLR). From the calculation results obtained a comparison between the design flood discharge value processed by rainfall data using the Gama I Synthetic Unit Hydrograph method to the design flood discharge value processed by the Tojo river discharge data using the log person III distribution method for a 2 year return period is 1957,45 %, for the 5-year return period is 1347,64 %, for the 10 year return period is 1224,65 %, for the 25 year return period is 1169,96 %, for the 50 year return period is 1165,67 % and for the 100 years is 1178,75 %.

Keywords: Flood discharge, Gama I Synthetic UH Method, Log Pearson III Distribution Method

1. Introduction

Flooding is the occurrence of inundation in the floodplain due to runoff caused by the flow of water flowing in the river exceeding its flowing capacity. Flood is a natural disaster that occurs most often and causes a lot of harm. Floods can occur in cities and villages, even in some places it has become an annual agenda. The impacts resulting from these floods also vary [11], [12], [13].

Floods can cause the cessation of residents' activities, economic losses, difficulty in obtaining clean water, outbreaks of disease, and can claim lives. Floods can occur either naturally or caused by humans. Floods occur due to various factors including topographic conditions, high rainfall intensity, blockage of water flow, lack of water infiltration areas, deforestation, and so on [15].

The Tojo watershed is included in the Parigi-Poso river area, with a watershed area of ± 211.83 Km2. In the Tojo watershed, it is necessary to study flooding, because the watershed always flows every year. Which may experience flooding at any time with high rainfall intensity, causing landslides in the upstream watershed area, resulting in flash floods [2], [9].

In analyzing the design flood in the Tojo watershed, the Gama I Synthetic Unit Hydrograph Method was used which was the first model developed in Indonesia, namely on the island of Java [6], [7], [8]. So the authors conducted the research with the title, "Testing the Gama I Synthetic Unit Hydrograph Method in the Tojo Watershed Design Flood Analysis".

The purpose of this study was to find out how much flood discharge was obtained from the analysis of the Tojo watershed design flood with the Gama I synthetic unit hydrograph method and to find out how much the comparison of Gama I synthetic unit hydrograph analysis to the design flood discharge based on river discharge data [1], [14].

2. Material and Methods

2.1. Description of study

This research was conducted in the Tojo Watershed (DAS), Tojo Una-una Regency, Central Sulawesi Province. Geographically, the Tojo watershed is located at coordinates 1°17'29.26"LS - 1°28'22.44"LS and 121°11'38"E - 121°24'25"E. Administratively, the Tojo watershed is bordered by Tomini Bay on the north side, North Morowali district on the south side, Poso district on the west side and Ampana city on the east side.

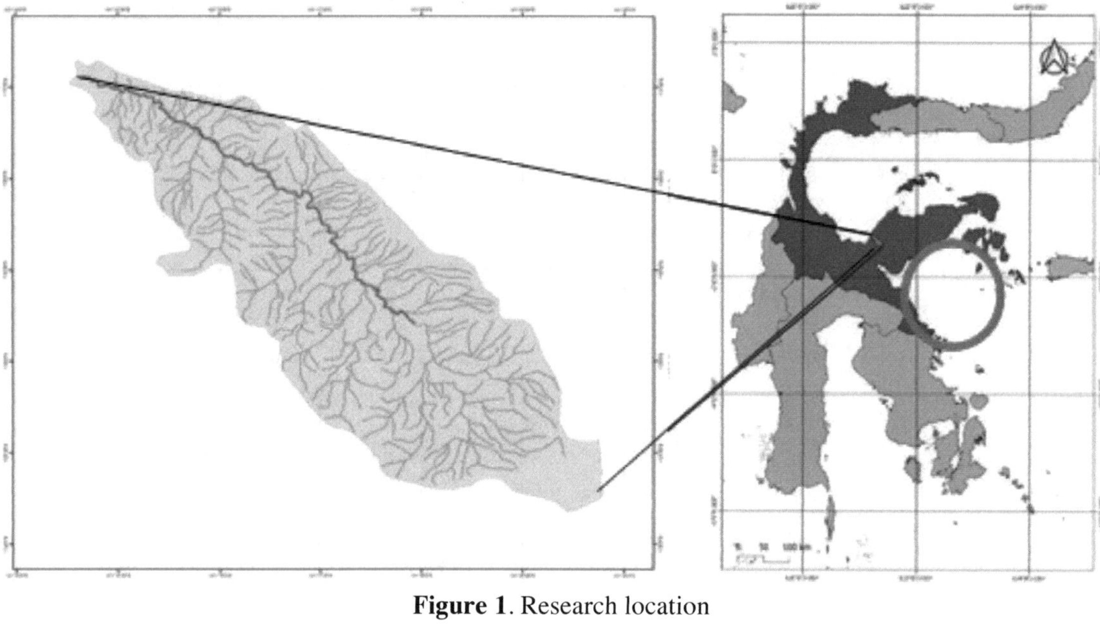

Figure 1. Research location

2.2. Literature review

Peak time (Tp) which is strongly influenced by the length of the main river (L), source time (SF) and symmetry factor (SIM). The relationship between the three factors is presented in the form of an equation [3]:

$$T_R = 0{,}43\left(\frac{L}{100SF}\right)^3 + 1{,}0665SIM + 1{,}2775$$

Peak discharge (Qp) which is strongly influenced by watershed area (A), peak time (Tp), drain network density (D), source factor (SF) and number of river encounters (JN). The relationship between these factors is expressed in the equation [4]:

$$Q_P = 0{,}1836A^{0{,}5886}JN^{0{,}2381}T_R^{-0{,}4008}$$

Base Time (TB), is a function of peak time (Tp), average river slope (S), source frequency (SN) and upstream watershed area (RUA). The influence of the four factors mentioned above, is presented in the equation:

$$T_B = 27{,}4132T_R^{0,1457}S^{-0,0986}SN^{0,7344}RUA^{0,2574}$$

The reservoir coefficient (K) is strongly influenced by the area of the watershed (A), the average river slope (S), time source factor (SF), drain network density (D). The relationship between the four factors is presented in the following equation [10] :

$$K = 0{,}5617A^{0,1798}S^{-0,1446}SF^{-1,0897}D^{0,0452}$$

Based on the concept used in converting rain to discharge in the synthetic unit hydrograph of gamma I, the base flow is presented as follows [5]:

$$Q_B = 0{,}4751A^{0,6444}D^{0,9430}$$

Information :
QB = Base Flow (m^3/sec)
A = Area, (Km^2)
D = Density of Drain network, (Km/Km^2)

2.3. Methods

This research is a type of descriptive research that provides an overview of a situation on the object of research. Based on data obtained from the Sulawesi River Basin 3 (BWS3), Department of Human Settlements and Water Resources (Cikasda) of Central Sulawesi Province, in the form of rainfall data, discharge data, and topographic maps of the Tojo River, the selection of data is carried out for further research in the analysis. From the topographic map, the watershed parameters can be determined as follows:

a. Determining watershed boundaries with reference to hydrometric stations at control points in the watershed under review
b. Measuring watershed area
c. Measuring the length of rivers and tributaries
d. Division of river levels
e. Calculating watershed unit hydrograph parameters

For the conversion of rain to flow to the watershed under review, there are several stages that must be carried out, namely:

a. Selecting rainfall data that is considered representative of the watershed under review
b. Analyzing the frequency of rain according to the type of distribution based on statistical parameters
c. Determine the design rainfall for return periods of 2 years, 5 years, 10 years, 25 years, 50 years and 100 years
d. Conducting frequency distribution conformity test
e. Determine the hourly rainfall distribution for the return periods of 2 years, 5 years, 10 years, 25 years, 50 years and 100 years
f. Calculating the design flood discharge using the Gama I Synthetic Unit Hydrograph method for return periods of 2 years, 5 years, 10 years, 25 years, 50 years and 100 years

To process river discharge data, a hydrological distribution method (frequency distribution) is used in accordance with statistical parameters to process rainfall data. The stages of analysis are as follows:

a. Selecting the discharge data that is considered to represent the peak discharge of the river under review.
b. Determine the statistical parameters of the peak river discharge in accordance with the statistical parameters to process rainfall data.

c. Calculating the design flood discharge for the return period of 2 years, 5 years, 10 years, 25 years, 50 years and 100 years by using a hydrological distribution (frequency distribution) in accordance with statistical parameters to process rainfall data.

The complete research method is presented in the following figure:

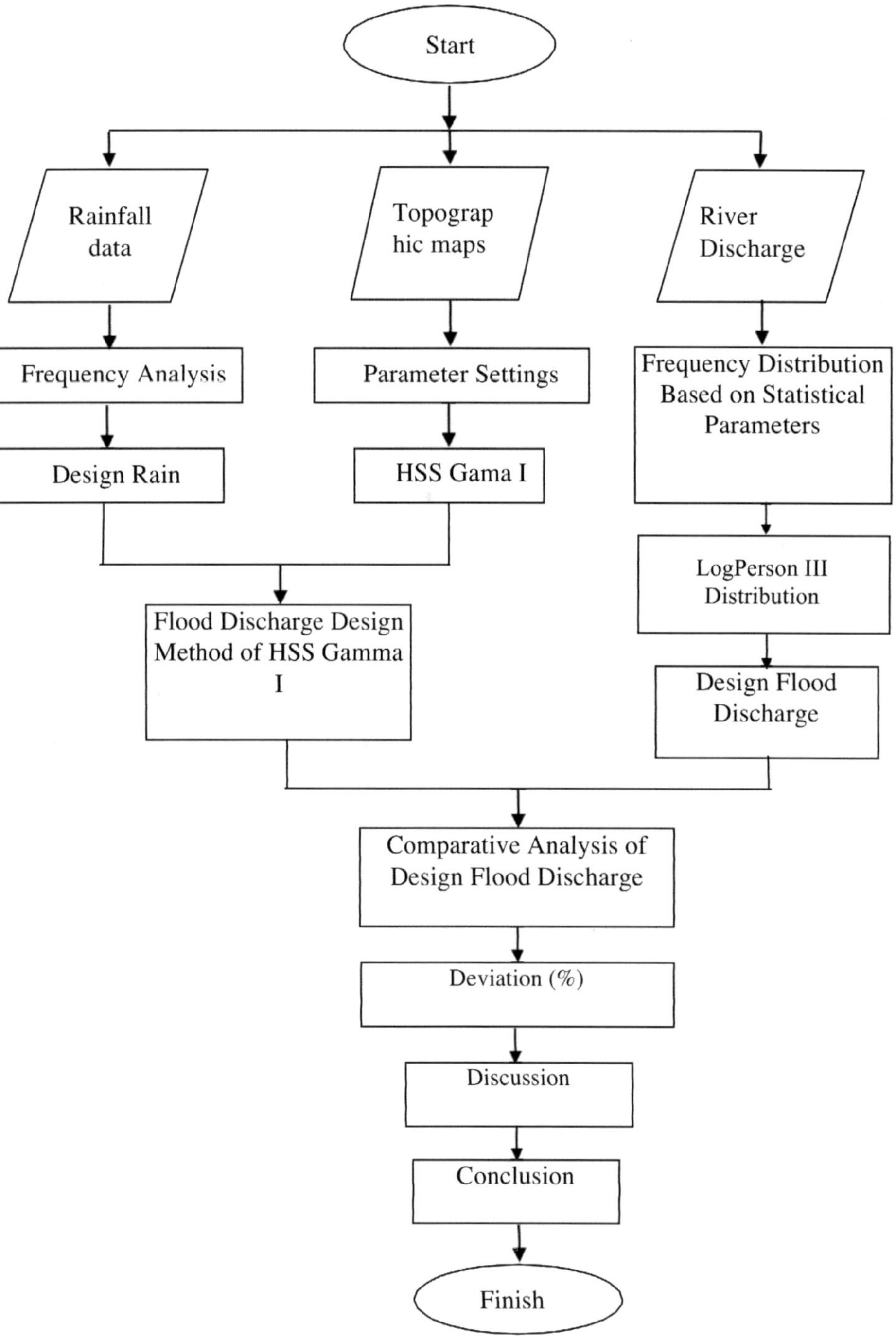

Figure 2. Flowchart research

3. Results and Discussion

3.1. Design Flood Discharge Using Rainfall Data

The results of the calculation of design flood discharge using rainfall data are presented in the following tables and graphs:

Table 1. Maximum daily rainfall data based on rank

No.	Year	Rainfall (X_i) (mm)
1	2009	97.00
2	2010	118.00
3	2011	121.00
4	2012	157.00
5	2013	164.50

Table 2. Overview of Distribution Based on Statistical Parameters

No.	Distribution Type	Condition	The calculation results	Information
1	Gumbell	Cs ≤ 1.1396	Cs = 0.029	Fulfil
		Ck ≤ 5.4002	Ck = 139.921	Does not meet the
2	Log Peason III	Cs < 0	Cs = -0.094	Fulfil
		Cv= 0.3	Cv = 0.045	Does not meet the

Table 3. Calculation of Rainfall Distribution Design Log Pearson III

Return Period (T)	K_{Tr}	Log Xo	Design Rain Logarithm	Design Rain,XT
2	0.01592	2.11070	2.11221	129.48269
5	0.83638	2.11070	2.19008	154.90878
10	1.27076	2.11070	2.23130	170.33353
25	1.71823	2.11070	2.27377	187.83031
50	2.00343	2.11070	2.30083	199.90921
100	2.25671	2.11070	2.32487	211.28518

Table 4. Hourly Rain Intensity

Return Period	R24	Rain intensity (hours)			
(Year)	(mm)	1	2	3	4
2	129.4827	44.8891	28.2784	21.5804	17.8143
5	154.9088	53.7039	33.8313	25.8181	21.3124
10	170.3335	59.0513	37.2000	28.3889	23.4345
25	187.8303	65.1171	41.0212	31.3051	25.8418

50	199.9092	69.3047	43.6592	33.3182	27.5036
100	211.2852	73.2485	46.1437	35.2142	29.0687

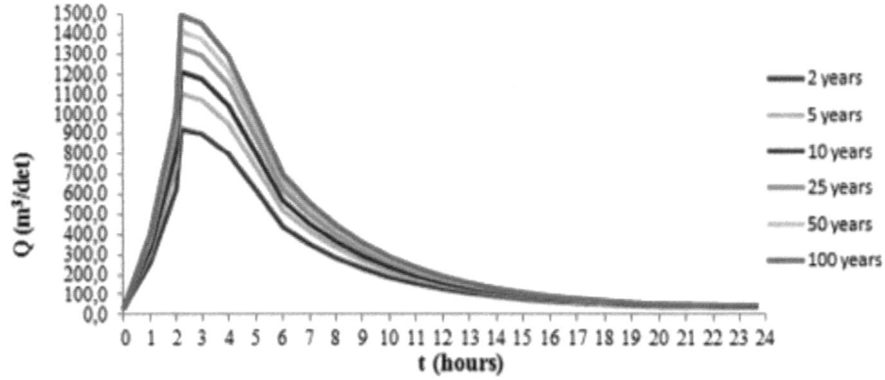

Figure 3. Flood Hydrograph Design of the HSS Gama I Method

3.2. Design Flood Discharge Using River Discharge Data

The discharge data used is the Tojo River discharge data in the form of secondary data. The discharge data comes from an automatic water level recorder, which is located upstream of the bridge to the left of the Tojo river.

Table 5. Daily Maximum Discharge Data for the Tojo River

No.	**Year**	**Discharge m^3/det**
1	2008	53.06
2	2009	30.73
3	2010	8.91
4	2012	74.72
5	2013	79.51

Table 6. Flood Discharge Design Log Person III Method

Return Period (T)	K_{Tr}	Log Xo	Design Discharge Logarithm	Flood Discharge M. Distribution Log Person III (Q, m^3/det)
2	0.22550	1.58721	1.67564	47.3847
5	0.83174	1.58721	1.91337	81.9156
10	1.04019	1.58721	1.99511	98.8796
25	1.19659	1.58721	2.05644	113.8769
50	1.26821	1.58721	2.08452	121.4846
100	1.31592	1.58721	2.10323	126.8321

4. Comparison/deviation of Design Flood Discharge

Based on the calculation of the design flood discharge by processing rainfall data using the HSS Gama I method and processing river discharge data (measured) using a hydrological distribution in accordance with statistical parameters to process rainfall data, namely the log person III distribution, the deviation value of the design flood discharge is obtained. can be seen in the following table:

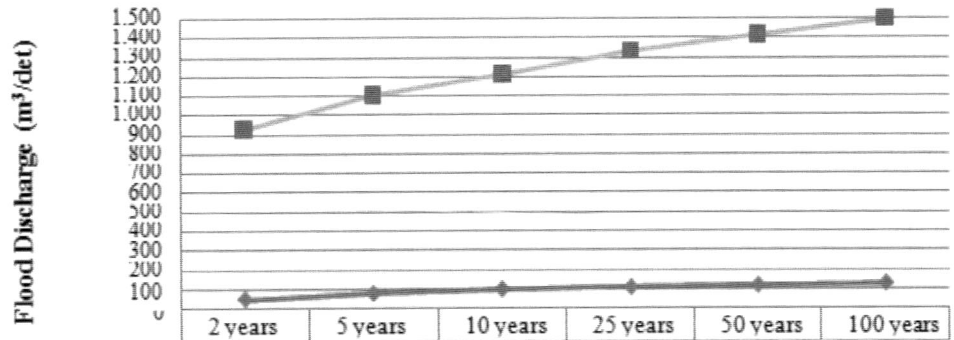

	2 years	5 years	10 years	25 years	50 years	100 years
M. Dist. LP III	47,3847	81,9156	98,8796	113,8769	121,4846	126,8321
M. HSS Gama I	927,5326	1103,9229	1210,9300	1332,3117	1416,1074	1495,0267

Year

Figure 4. Comparison Graph of Design Flood Discharge

From the calculation results, table 10 shows the deviation between the design flood discharge resulting from the processed rainfall data using the Gama I Synthetic Unit Hydrograph method to the design flood discharge resulting from the processed river discharge data measured using the log person III distribution method. Where the deviation value is classified as a large deviation because the resulting deviation value is greater than 20% (> 20%).

5. Conclusions

The design flood discharge using the HSS Gama I method for successive return periods of 2 years is 927,53 m3/s, 5 years is 1103,92 m3/s, 10 years is 1210,93 m3/s, 25 years is 1332, 31 m3/s, 50 years at 1416,11 m3/s and 100 years at 1495,03 m3/s. The results of the calculation of the deviation of the design flood discharge are classified as large deviations, because the resulting deviation value is greater than 20% (> 20%). Where is the value deviations for consecutive return periods of 2 years are 1957,45 %, 5 years are 1347,64 %, 10 years are 1224,65 %, 25 years are 1169,96 %, 50 years are 1165,67 % and 100 years are 1178,75 %. Because in the Tojo watershed there is a lack of measurement data, so the research results are less than optimal.

Acknowledgment

The authors would like to thank all those who have supported this research: Under Graduate Tojo Una-Una Program of Tadulako University and the office Public Works and Water Resources of the Province of Central Sulawesi, Indonesia.

References

[1] Hariadi (2015) Characterization of Rotary Encoder Type Automatic Water Level Measuring Instrument Hasanuddin University: Makassar
[2] Kartika, LD (2018) Calculation of Flood Discharge of the Bangga Watershed Design Using the Synthetic Gama Unit Hydrograph Method I Final Project Tadulako University Palu
[3] Mulyanto, H R (2007) River Functions and properties Graha Ilmu Yogyakarta
[4] Naharudin, Herman Harijanto, Abdul Wahid (2018) Watershed Management Textbook and its application in the teaching and learning process

(https://wwwresearchgatenet/publication/324771754) Retrieved 05 March 2020
[5] Nababan, O S & Siregar, P M, (2012) Automation of River Flow Measurement with Arduino Microcontroller ITB: Bandung
[6] Nugraha, A P, Febby, R, A, B Z & Utami, R T, (2013) Soil and Water Conservation Tasks UGM: Yogyakarta
[7] Prasetyo, Eko (2011) Info on Agroclimate and Hydrology Volume 6 Number 6: Automatic Water Level Recorder Research Institute for Agroclimate and Hydroclimate: Bogor
[8] Reni, P AWLR Practicum Report (https://wwwacademiaedu) Retrieved 17 April 2020
[9] Sutapa, IW (2005) Study of Nakayasu Synthetic Hydrograph Unit for Calculation of Design Flood Discharge in the Kodina River Basin Mektek VolVII:No1 Palu
[10] Sularto, E, (2006) The Relationship between Land Estimation and Banjor Events in the Upper Ciliwung Watershed, Katulampa Using the Answers Model IPB: Bogor
[11] Soemarto, CD, (1999) Engineering Hydrology Erlangga Publisher Jakarta
[12] Soemarto, CD, (1995) Engineering Hydrology Erlangga Publisher Jakarta
[13] Soemarto, CD, (1987) Engineering Hydrology National Business Publisher Surabaya
[14] Triesnawati, H, (2006) AWLR (Automatic WaterLevel Recording) Printing Calculator Base IPB: Bogor
[15] Triatmodjo, B (2008) Applied Hydrology Beta Offset Yogyakarta

Determination of the Location of Fishing Boat Moorings in Palu Bay After The 2018 Tsunami in Terms of Technical and Socio-Economic Aspects

Amar[1], A Rusdin[2], Setiyawan[3], S Arifin[4], M Saleh[5], Sri N A[6]

[1,2,3,4]Department of Civil Engineering, Engineering Faculty, Tadulako University
[5,6]Postgraduate Civil Engineering Master's Program at Tadulako University

setiyawanvip@yahoo.co.id

Abstract. The natural disaster that occurred in Palu City, Donggala and Sigi Regencies on September 28, 2018 in the form of an earthquake accompanied by a tsunami and liquefaction was one of the phenomena that was considered quite interesting because it had never happened in any area in Indonesia. The existence of three types of natural disasters that occur at the same time has an impact on people's lives and is interesting to study so that anticipation and solutions can be considered in the future. This condition will certainly have a direct impact on the mooring of fishing boats in Palu Bay. Thus, a location determination is needed and a review of technical, social and economic aspects is needed. So we need an effort to involve academics in recording and providing solutions to increase marine products after the earthquake. The purpose of this service is to plan fishing boat moorings after the earthquake and liquefaction which aims to find out the right position so that it can be utilized by fishermen, provide knowledge to fishermen that the use of existing boat moorings must be done to improve their economy and make boat moorings more efficient. provide education about fisherman productivity. The specific target to be achieved is to determine the location of fishing boat moorings after the earthquake, to be able to contribute to the Palu City Government, especially the River Region III Center so that they can design boat moorings on each coast, through the Palu City Government master plan to provide a post-earthquake response stimulus to helping fishing communities to rise and is expected to increase the productivity of fishermen.

Keywords: Boat Mooring Locations, Palu Bay, Post-Tsunami, Technical Aspects

1. Introduction

Talise Beach is a beach located not far from Palu City, Central Sulawesi. Or rather along Palu Bay. Talise Beach is located along Jalan Rajamoli to Jalan Cut Mutia, Palu, Central Sulawesi. Talise Beach is a tourist attraction in Sulawesi which has a very beautiful natural panorama. On this beach there is a stretch of white sand with quite calm waves and hills that are so enchanting that visitors are spoiled with the beauty of the panorama. But now the condition of Talise Beach is very worrying because of the shift in coastline caused by the Tsunami and Earthquake that occurred in 2018. Palu is located on the Palu-Koro fault line, which runs north-south along the shores of Palu Bay. Geologists estimate that the fault segment has one of the highest slip movements in Indonesia, at 4 cm (1.6 inches) per year, indicating the area is at a higher risk of earthquakes. In this earthquake, the two sides of the fault line scrape each other horizontally. Because of the damage that occurred, it is necessary to carry out a treatment on Talise Beach so that the damage that occurs can be reduced or overcome. The handling is done by making a coastal protection building on Talise Beach to protect the beach from the large waves of the sea that continuously hit Talise Beach and prevent abrasion that occurs so as to prevent the shift of the coastline getting closer to the highway.

2. Material and Methods

2.1. Description of study
The location of this research is 500 m which is located between 0°50'36" - 0°49'46.22" South Latitude and 119°52'47" - 119°52'51" East Longitude. The specialty of this area is that it is crossed by the meridian line of 120 East Longitude which is the reference for determining the time for the area included in Central Indonesian Time (WITA).

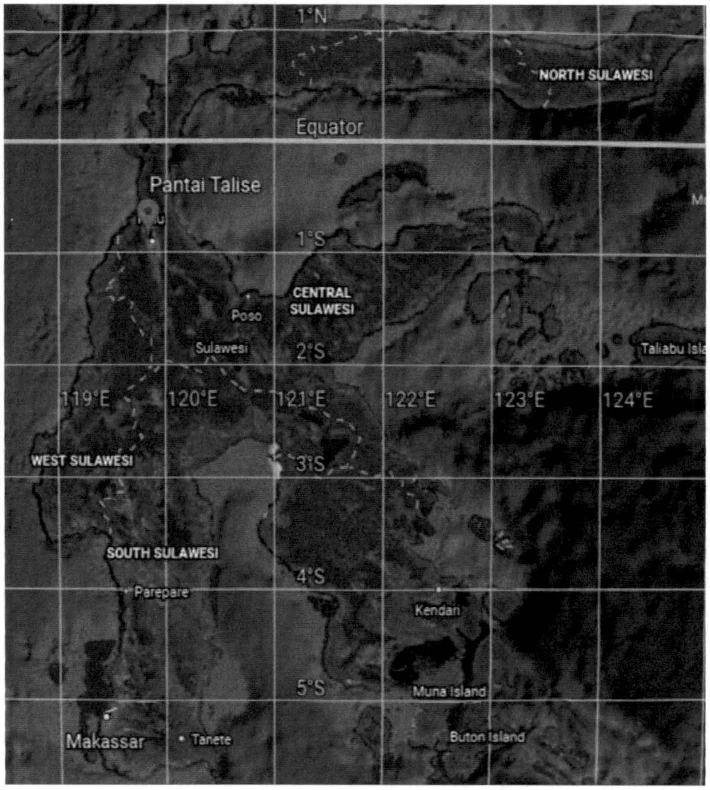

Figure 1. Research location

2.2. Literature review
This boat mooring infrastructure is planned to be able to accommodate the floor load. A mooring of 500 kg/m2 and for servicing boats with dead weight maximum 2 tons. The type of mooring depends on the condition of the river bank and the difference in face high and low tide. Boat moorings can be divided into two types, namely:

2.2.1. Single Floor Boat Mooring Type. This type is suitable for upstream areas, where the difference in high tide level and the tide is not too big.

2.2.2. Two-Story Boat Mooring Type. This type is suitable for downstream areas of the river, where the difference in tidal water level and the tide is quite large because it is influenced by the tides of the sea.

2.3. Methods

Based on the existing problems, the solution offered is to reduce the Community Service Lecturer Team to absorb the aspirations of the fishing community of Talise Beach and record the current problems. After this data collection is successful, it will be analyzed and planned for mooring fishing boats so that they are able to optimize the catch of fishermen in Palu City so that it will be a trigger for improving the community's economy. The Service Team will also provide counseling on how to use boat moorings efficiently.

Work Procedure is carried out in the following stages:

2.3.1. Brainstorming Stage. The activity carried out at this stage is to gather groups of fishermen. Furthermore, data collection on the number of boats that can still be developed will be carried out.

2.3.2. Stages of Evaluation of Boat Mooring Facilities.The activities carried out at this stage are evaluating boat mooring facilities and infrastructure and those damaged after the earthquake, whether these facilities are still suitable for use or if new facilities are needed.

2.3.3. Extension and Planning Stages. After going through the stages of introduction and evaluation of the condition of the boat moorings, there will be counseling regarding the operation and simple repair of damaged boat mooring facilities and infrastructure.Introducing the method or technique of repairing boat mooring facilities and infrastructure. Components and sizes of boat mooring components.

2.3.4. Stages of making and recommendations to the Palu City Government. This stage is the final stage that is taken in completing the work of making Design and Counseling in the affected Talise boat mooring area in Palu City.

3. Result and Discussion

3.1. Information on boat mooring plans

Information on boat mooring plans are presented in the following figure :

Figure 1. Fishing communities will use the location of boat moorings

Figure 2. The location chosen by fishermen to moor

Figure 3. The location chosen by fishermen to moor

3.2. The Plan of Location

The Plan of Location are presented in the following figure :

Figure 3. The Plan of Location Tp. 01

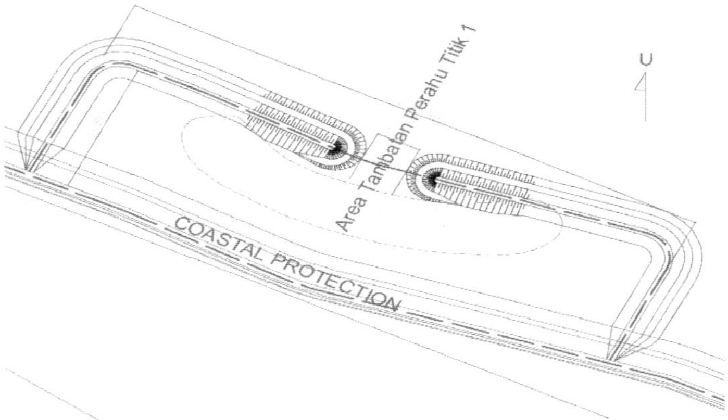

Figure 4. Initial Design Tp. 01

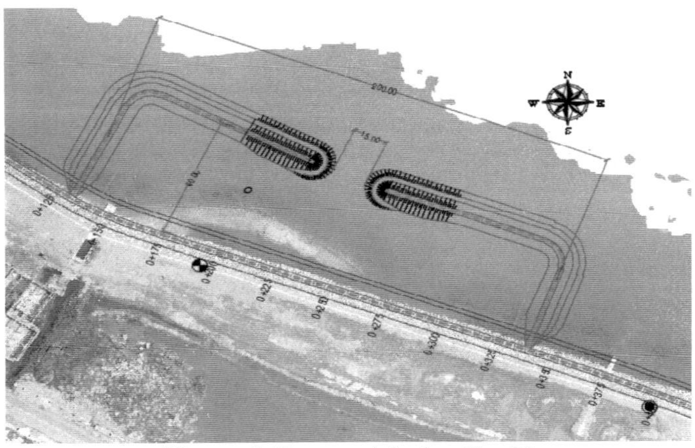

Figure 5. Proposed design Tp. 01

3.3. Design Plan 3D

By using 3D tools, Palu Beach is obtained as follows.

Figure 6. Design Plan 3D

Figure 6. Design Plan 3D

4. Conclusions

Around 95% of fishing communities will use the location of boat moorings and generally they use fishing gear in the form of boats/sampans. The location chosen by fishermen to moor their canoes is in sandy areas resulting from sediment deposits on the shores and because they are safe from the effects of tides and waves. The type of canoe used is 6-8 meters, with a maximum load of 2 people 2. This type of canoe is partly made of wood and partly made of fiber which is prone to friction. Generally, fishermen carry out fishing activities when the water is high and return at low tide.

References

[1] Abidin, Zainal (2007), "Planning of Offshore Breakwaters and Coastal Protective Walls as Alternatives for Abrasion Prevention in Kalibuntu Beach, Probolinggo Regency", Civil Engineering Thesis, Islamic University of Malang Poor

[2] Anonymous (2020) Community Service Guide 2021 Tadulako University LPPM Untad hammer

[3] Regional Development Planning and Investment Agency of Palu City (2014) "Profile of the East Palu sub-district" Publisher of the Regional Development Planning and Investment Planning Agency of Palu City

[4] Central Bureau of Statistics of Palu City, Central Sulawesi (2016) "East Hammer in Numbers" Palu City BPS Publisher

[5] Burhan Utomo, Basrindu and Dwi Prasetyo Wibowo (2008), "Planning of Protective Buildings for Tambak Mulyo Beach, Semarang", Final Project of Civil Engineering, Diponegoro University Semarang

[6] Google Earth (2020) "Map of Talise Coast Region" East Palu District

[7] Puspawati Achjar, Dewi (2007), "Study of Planning for Coastal Protection Buildings to Prevent Beach Erosion in Ampenan, Mataram City, West Nusa Tenggara", Civil Engineering Thesis, Islamic University of Malang Poor

[8] Robydiansah (2012), "Review of the Stability of Shearing and Rolling Parafets in the Grindulu River, Pacitan Regency", Final Project of Civil Engineering, Yogyakarta State University Yogyakarta Pratiko et al, (2014) Coast Guard Structure PT Mediatama Saptakarya Jakarta

[9] Triatmodjo, B (2008) Beach Engineering Yogyakarta: Beta Offset Triadmodjo, B (2012) Coastal Building Planning Beta Offset Yogyakarta

Baseflow index on Miu watershed based on a digital graphical method

S R Oktavia[1,2], N B Rustiati[1], V W Andiese[1], T Amaliah[1], M Labombang[1], O A Mantika[1]

[1]Department of Civil Engineering, Faculty of Engineering, Universitas Tadulako, Kampus Bumi Tadulako Tondo, Jalan Soekarno – Hatta Km.9 Palu, Central Sulawesi, 94117, Indonesia

sroktavia@gmail.com

Abstract. Baseflow is essential for water resource management in the dry season at the watershed scale. Management strategy water resources play a crucial role, especially those related to estimation efforts water supply, water quality management, water balance calculation, and strategy development of other water resources. This study aims to determine the value of baseflow using the Fixed Interval Method. This study used Hydro Office software with input daily discharge data from 2004 to 2013. The average baseflow value obtained was 5.53 m3/s, the process of calibrating by taking discharge data in the dry season, with no rainfall input. The BFI value obtained an average of 0. 9. This value indicates that the Miu watershed has high storage permeability characteristics. It indicates that the Miu watershed has a relatively stable flow during dry periods. The greater the BFI value, the better the water supply in a watershed.

1. Introduction

Baseflow is one of the two main elements contributing to the hydrograph of flow in rivers besides surface runoff. Baseflow is the flow component in the river that does not come directly from rain events but is considered groundwater flow that flows into the river during the dry period [1]. In the dry season, where water availability in rivers is minimal (due to the lack of input from rainfall), Baseflow becomes the primary source of water supply for many purposes. On the other hand, it is essential to estimate the amount of base flow during the rainy season to predict the amount of discharge, which is the contribution of fast flow or surface run-off that can produce flooding. In many watersheds, base flow is the main component and the most significant contributor to flow to the river [2]. In principle, baseflow and surface runoff cannot be distinguished from the discharge measured in the river because the measured discharge in the river is a combination of the two components. During the dry season, relatively no rain falls, but there is still running water discharge in most rivers in East Java. This flow comes from rainfall that enters the ground into groundwater reserves and will then flow to the river [3]. Information on Baseflow Value is essential for developing water resources management strategies at the watershed scale during the dry season. Water resource management strategies are essential in estimating water supply, water quality management, river water balance calculations, and other water resource development strategies [2].

In terms of efforts to develop a water resource management strategy in the Miu watershed, as a preliminary study,

this paper aims to predict the baseflow index on the Miu watershed based on a digital graphical method. The result of this study can be used as a reference for a comprehensive Water resource management strategy in the Miu watershed.

2. Material and methods

2.1. Site description

This research was carried out in the Miu watershed, located in Sigi Regency, Central Sulawesi. The area of the watershed is 645.55 km2 with the outlet point at the Miu SPAS (Omu Village), it lies in geographical coordinates between 119o56'39,619" E and 1o17'0,95" S. Miu watershed is part of Palu Watershed presented in figure 1.

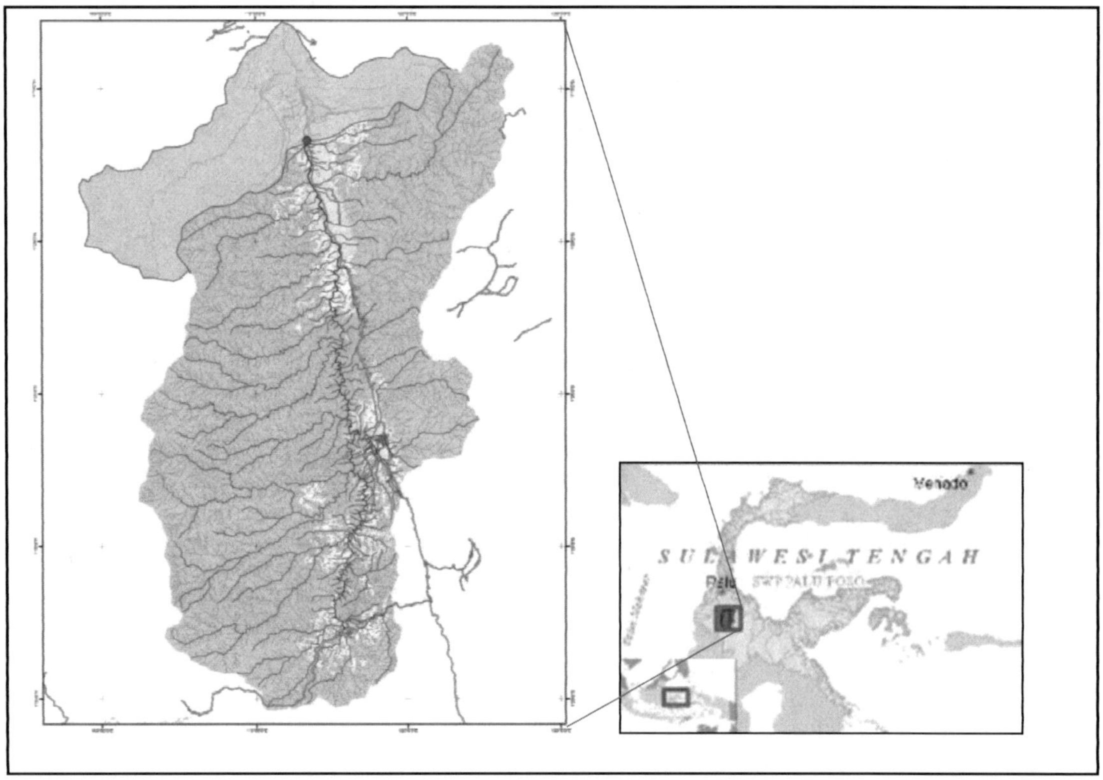

Figure 1. Site location of the research.

As part of the Palu watershed, The Miu watershed presents the characteristics of the Palu watershed in the upper segment, with flow-forming fishbone patterns. The similarity of this flow pattern can also be found downstream of the Palu watershed in order of three to four and with low density [4]. In addition, the Miu watershed is an upstream area that has a mountainous topography with an average slope of 30o. Miu River has very specific characteristics because it is a type of river with a height difference between upstream and downstream about ±800 meters in a distance of only 50 km [5].

The land use in the Miu watershed consists of shrubs, rice fields, farmland, settlements areas, and forest, as the most significant percentage of Miu watershed land use. Based on observations and data obtained from the Meteorology, Climatology, and Geophysics Agency, the type of soil around Miu River is gravel sand/porous soil [6].

2.2. Hydrologic and hydrometric data

Hydrologic and hydrometric data used for this study were obtained from AWLR and ARL which were installed by the Ministry of Environment and Forestry Republic of Indonesia, BPDAS Palu - Poso. These instruments recorded daily rainfall and water level on Miu river. The pair of rainfall data and discharge data were used from 2004 – 2013. The data was used because in those years the data was considered complete so it did not require preliminary analysis to complete the missing data series.

Digital maps of the Miu watershed that used in this study originated from The Ministry of Environment and Forestry Republic of Indonesia, BPDAS Palu – Poso.

2.3. Baseflow separation

The baseflow value of a watershed is determined by the baseflow separation technique of the total discharge hydrograph.

The graphical separation method divides the hydrograph into two components (quick flow and baseflow) through a line connecting the lowest discharge before the rising limb and after the recession limb of the hydrograph. The line connecting the two points is defined as the base flow. [7].

A graphical approach to basic flow separation, namely [8 -9]:

1. The constant discharge method, which assumes that the base flow is constant during flood events. Minimum runoff value before the rising limb is the constant value (1.a);
2. The constant slope method, that relates the initial value of the rising edge to the point turn (inflection point) on the recession limb (1.b);
3. The concave method describes the beginning of a flow recession starting on the downside by projecting a declining hydrograph onto the line at the intersection of the hydrograph crest line (1.c). The line of intersection is then connected with the inflection point on the recession limb. The three separation methods are graphically presented in figure 2.

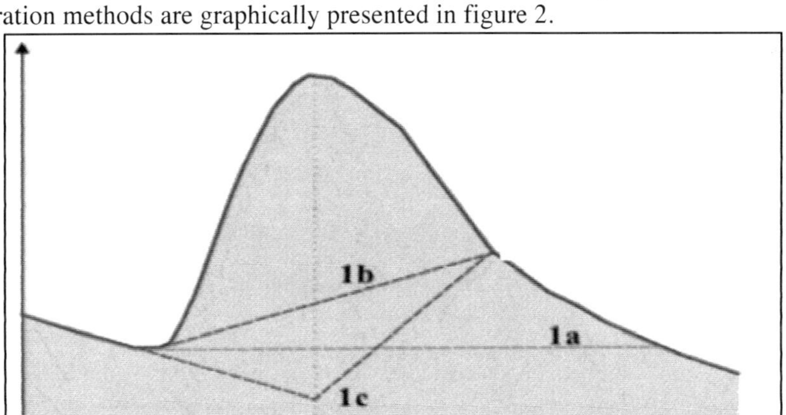

Figure 2. Graphical method for baseflow separation

2.4. Digital graphical method

The digital graphics method is one method that is relatively easy to implement. Sloto and Crouse graphically developed the basic flow separation method by providing examples of three basic flow separation methods: Fixed Interval Method, Sliding Interval Method, and Local Minimum Method. [10].

Local Minimum Method evaluates the daily debit to determine whether that day is the lowest debit or not, using the time interval expressed by: [0,5 (2N*-1) days]. The value of N is obtained empirically from the equation N = A 0.2, where A is the watershed area in mile2. Furthermore, the lowest discharge at each interval is connected by a straight line to determine the baseflow. It can be seen in figure 3 below.

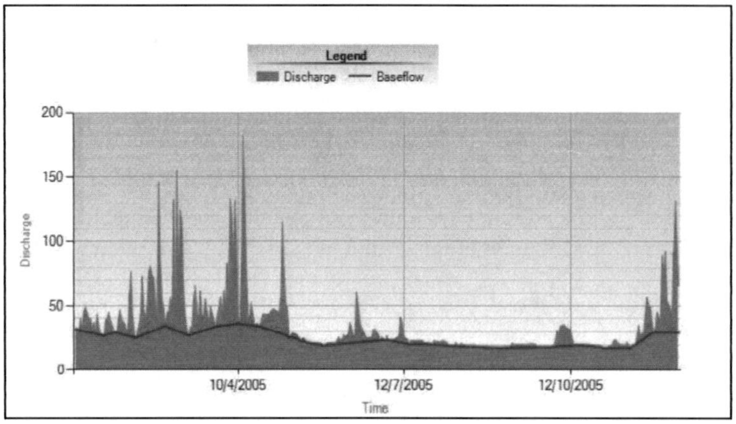

Figure 3. Local minimum method

Furthermore, the Fixed Interval Method uses the lowest discharge in each interval. The number of days in each interval is determined using the formula (2N*Days). N is the number of days after runoff ends. The value of N is obtained empirically from the formula N = A 0.2, where A is the area of the watershed in mile2. This method can be illustrated by a bar chart drawn up until it is in contact with the lowest discharge at each interval. It can be seen in figure 4.

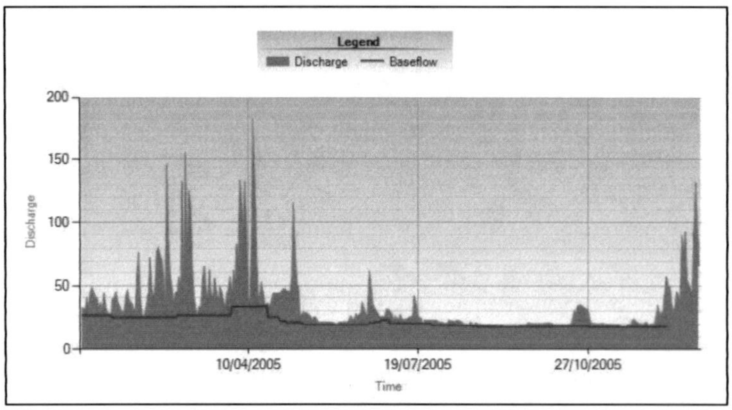

Figure 4. Fixed interval method

2.5. *Baseflow index*

Lvovich first introduced the baseflow index in 1972, and this concept was further developed by the Institute of Hydrology in 1980 to explain the relationship between the geological characteristics of a watershed area and the baseflow. BFI or baseflow index is defined as the ratio between the baseflow volume and the total volume of discharge flow [11]. A watershed with an effluent stream type flow (figure 5), namely a watershed with a high baseflow contribution [12], may have a BFI value approaching 1. In contrast, a watershed with an ephemeral stream or influent stream type (figure 5) will BFI value close to 0 [11].

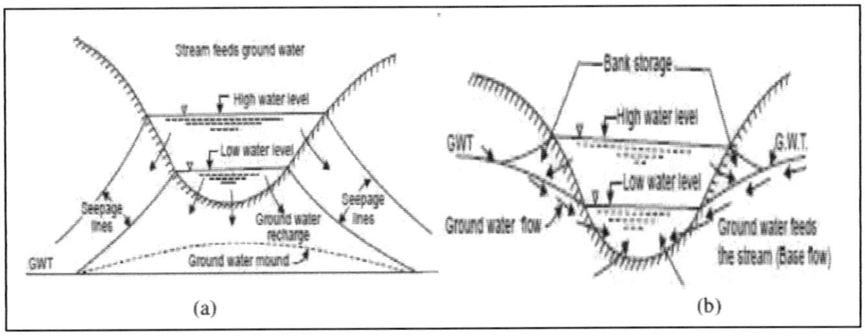

Figure 5. River with the influent flow (a) and effluent flow (b).

2.6. Research methods and stages

This study begins with the observation and selection of continuous daily discharge in each year and rain data to determine the dry/dry season period as the time basis for calibrating the base flow separation model used. At this stage of the research, the calibration process, will choose in the months where the rain is very minimal or close to 0, so it is assumed that the rain that occurs and falls on the ground does not produce runoff, the total discharge (measured discharge) in the river is considered entirely derived from baseflow.

In addition, calibration is a process to determine the parameters in the test method used. The calibration process is carried out during June and October using the "parameter calibration" tool in the BFI 3.0+ software. The calibration process is performed by manually entering the parameter values to produce the appropriate base flow value for each year. Determination of parameter values if the line graph (red) in the dry season coincides with the graph (blue). This can be seen in figure 6 [13].

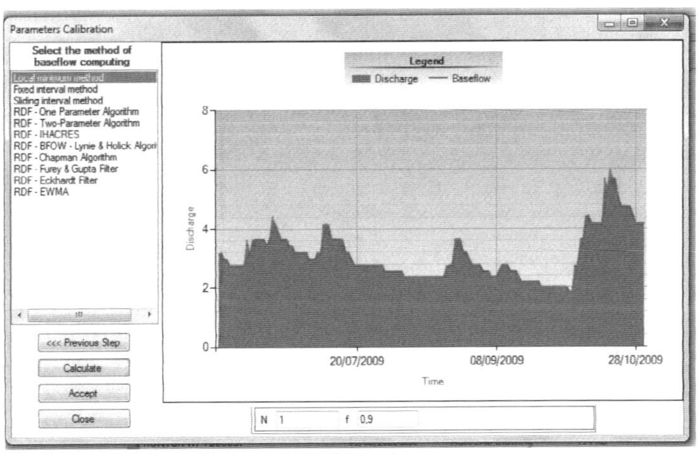

Figure 6. Calibration process

The optimal parameter value for the calibration period is determined from the average parameter value obtained in each year. Baseflow separation Fixed Interval Method using parameters N (time). The next stage is validating the Optimal Parameter values (N value average) obtained from the calibration stage, which is then used for baseflow separation for the Year 2004 – 2013. The final step is evaluating the model's performance by performing statistical tests using the RMSE Method.

Root Mean Square Error (RMSE) is used to determine the error generated during modeling. RMSE is the ratio between the measured discharge and observation baseflow value (assumed from the dry period discharge value). The RMSE equation is expressed by [3]:

$$RMSE = \frac{\sqrt{\Sigma(Qm - Q0)^2}}{n}$$

where:
Qm = total discharge (m3/s)
Q0 = calculated discharge (m3/s)
n = number of samples

3. Result and discussion

3.1. Baseflow calibration period

The calibration process is carried out by manually entering the parameter values to produce the appropriate baseflow value for each year. Determination of parameter values if the line graph (red) in the dry season coincides with the area graph (blue), presented in figure 6. The value of the parameter range (N) is 4 – 6, with the optimal parameter value being 5. Previous studies using the local minimum method resulted in optimum parameter values of N=3 and f= 0.9 [6]. It is different from this research because the minimum local method did not use turning points as a parameter, and N as a time parameter has a different formula.

3.2. Result of base flow separation

Baseflow separation visualization for all-year and dry periods can be seen in figure 7 and figure 8. The parameter value used for the baseflow separation graphical method uses the optimal parameter value (mean) obtained during the calibration period.

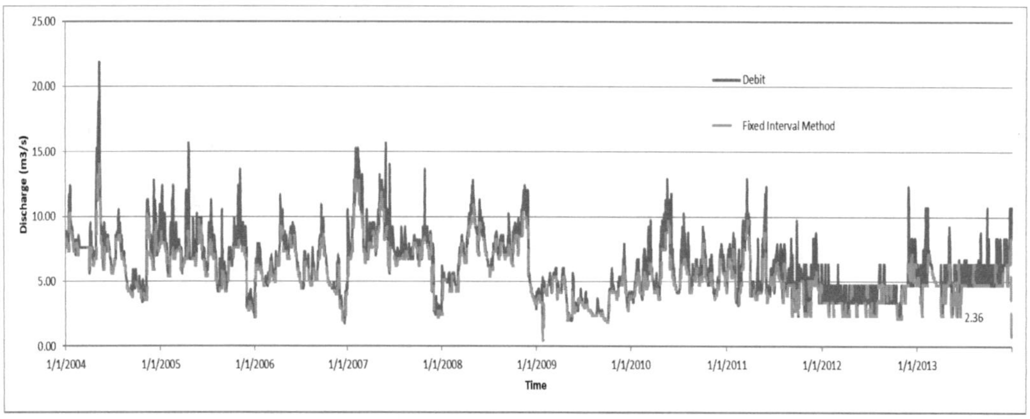

Figure 7. The results of the separation of baseflow in the Miu watershed for the period 2004-2013

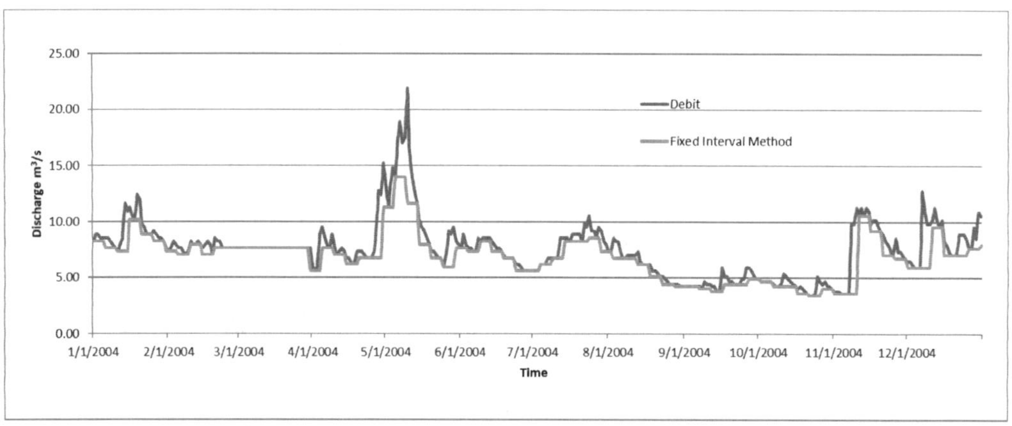

Figure 8. The results of the separation of baseflow in the Miu watershed the period 01 January – 31 December 2004

From figure 8, it can be seen that the discharge value in January – December 2004 has fluctuated. However, in mid-May, there was a significant increase in discharge. A relatively high intensity of rain indicates the increase in discharge at that time. From Juni -October 2004, the discharge and baseflow graphs look close to each other; it can be assumed that Juni -October 2004 is the dry period where the river flow is mostly from baseflow. At that time, there was no rainfall input.

Further, figure 9 shows the baseflow modeling results in the dry period, June – October 2009. From the picture, it is shown that the discharge graph during the dry period relatively looks closely with the baseflow graph, because of almost no rainfall during the dry period. However, in the river, there is still river flow. River flow during the dry season is predicted to contribute to the base flow. The discharge value in the dry season tends to decrease because almost no rain occurs at that time. Meanwhile, during the rainy season, river discharge tends to increase.

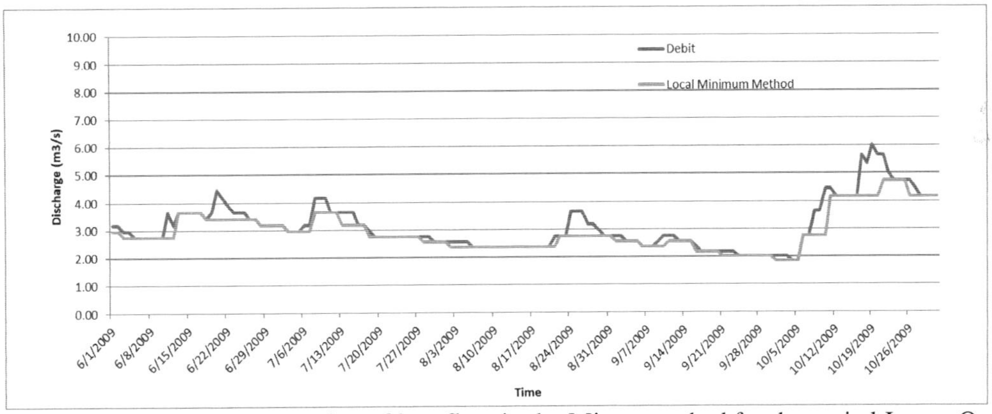

Figure 9. The results of the separation of baseflow in the Miu watershed for the period June – October 2009

Tabel 1. Discharge and baseflow Miu watershed

Discharge (m^3/s)			Baseflow (m^3/s)		
Min	Average	Max	Min	Average	Max
0,48	6,23	21,89	0	5,53	14,00

According to Table 1. The average daily discharge on the Miu River recorded at the SPAS Sungai as a control point is 6.23 m3/s. In comparison, the baseflow value based on the Digital Graphics Method is 5.53 m3/s, the two values are not much different, indicating that the difference could be due to the input of rain with a small intensity. On the other side, there is a significant difference in the maximum values for discharge and baseflow; this is because the input of high-intensity rainfall can cause the maximum discharge recorded at the AWLR station. The maximum values for baseflow are based on the soil's ability and the area's geological structure to respond to rain inputs.

As a comparison, the previous study, which was carried out by Gede Tunas (2007) in the Miu watershed in predicting the magnitude of the dependable flow, the smallest reliable discharge was obtained in mid-November of 5.36 m3/s and the most significant reliable discharge was obtained in mid-May 1 is 7.25 m3/s [14]. It can be concluded that the baseflow value modeled in this study is not much different from the measured daily discharge and the dependable discharge which calculated, hence the assumption that the daily flow rate in the river, especially during the dry season, is a flow contribution from baseflow in the watershed can be fulfilled.

Based on the results of the RMSE value, the method that has better performance is the one that produces an RMSE value close to 0 (zero). Fixed Interval Method results in an RMSE value of 0,164. The RMSE value close to 0 (zero) indicates that the level of data processing error is getting smaller. This method has a good performance in modeling the base flow in the Miu watershed.

Baseflow index is obtained by comparing the base flow and the total discharge. The analysis results show that the average value of BFI in the Miu watershed in 2004 – 2013 is 0.9, while the maximum and minimum values are 1 and 0.15, respectively. The value of the baseflow index gives a view of the availability of discharge in the dry season. According to Tallaksen & Van Lannen, the baseflow index value ranging from 0.9 indicates a permeable watershed. A high baseflow index value describes a watershed with a relatively stable flow type. The watershed can supply water to the river even in the dry season. A low baseflow index value (between 0.15 to 0.2) indicates a relatively impermeable watershed with rapidly dissipating flow rates [8-15]. Based on the research results, it can be concluded that the Miu watershed has high storage characteristics and effluent stream type. This shows that the Miu watershed can maintain river flow during the dry season.

4. Conclusion

Based on the research results, the Fixed Interval Method baseflow range is 0.48 – 12.02 m3/s with an average of 5.43 m3/s. A calibration parameter value of N = 3 and f of 0.9. The BFI value obtained in the Miu Watershed in 2004 – 2013 averaged 0.89, while the maximum and minimum values were 1 and 0.15, respectively. This shows that the Miu watershed can maintain river flow during the dry season. Based on the RMSE criteria, the Fixed Interval Method can be used to separate the base flow from the measured discharge because the RMSE value obtained is close to 0 (zero), which is 0.147 and has an R2 value of 0.751.

References

[1] Frochlich, K, Frochlich, W, Wittenberg, H *Determination of Groundwater Recharge by Baseflow Separation; regional Analysis in Northeast China* IAHS Publ 221, 69-75, 1994

[2] Bokiraiya Latuamury, Studi Tentang Pemisahan Aliran Dasar: Perbandingan Metode Grafis dan Filter Prosiding Seminar Nasional PERTETA, 2015

[3] Indarto, Hydrology – Basic Theory and Examples of Hydrological Model Applications, PT Bumi Aksara, Jakarta, 2010

[4] I Gede Tunas1, Arody Tanga1 and Siti Rahmi Oktavia, Impact of Landslides Induced by the 2018 Palu Earthquake on Flash Flood in Bangga River Basin, Sulawesi, Indonesia, Volume 21, Issue 2, February 2020, pages 190–200

[5] SR Oktavia, OA Mantika, Rugaiyah, Baseflow Separation Study - Local Minimum Method for Miu Watershed Basin, Rekonstruksi Tadulako, Volume 2, Issue 2, September 2021, pages 143-48

[6] M Galib Ishak, I Wayan Sutapa, Armin Basong, and Aditya Dedi, Analysis of Water Availability in Omu Watershed, MATEC Web of Conferences 331, 04003 (2020),

[7] I Indarto, A Ratnaningsih, and S Wahyuningsih, "Calibration of Six Recursive Digital Filters for Baseflow Separation in East Java", ARPN Journal of Engineering and Applied Sciences, vol 12, no10, p 3772, 2017

[8] Szilagyi, J, Parlange, MB, 1997 Baseflow separation based on analytical solutions of the Boussinesq equation Journal of Hydrology 336, 206–17

[9] Saparis, M 2011 Spatial covariation of the relationship between vegetation cover and baseflow recession on the island of Bali Dissertation on Doctoral Program, Faculty of Geography, Universitas Gadjah Mada, Yogjakarta

[10] RA Sloto and MY Crouse, HYSEP: A Computer Program for Streamflow Hydrograph Separation and Analysis, Pennsylvania: US Geological Survey, 1996

[11] Smakhtin, VU, "Low Flow Hydrology: a Review", *Journal of Hydrology*, vol 240, no 3-4, 2001

[12] Raghunath, HM, Hydrology (Principles, Analysis Design) 2nd Edition New Delhi, New Age International (P) Ltd, 2006

[13] Gregor, M (2015) Hydro office User Manual version 2015

[14] I Gede Tunas, Mock Model Parameter Optimization for Calculating Miu River Mainstay Discharge, SMARTek, Vol 5, No 1, 2007

Development of Learning Materials Based on Processed Plastic Waste Products as an Effort to Instill Environmental Care in Elementary School Students

Herlina[1*], **H.P. Melani**[1]

[1]Department of Education Science, Universitas Tadulako, Central Sulawesi, Indonesia

herlina@untad.ac.id

Abstract. This study aims to develop teaching materials based on processed waste products. Using a qualitative approach with Research and Development (R&D) methods, this development research follows the Borg & Gall model. In the development procedure, validation of research products is carried out. The validation process involves two experts, namely material experts and media experts, for one-on-one tests, while fourth grade students as test subjects are a field validation process. Data collection techniques using observation, interviews, questionnaires and learning outcomes tests. Data analysis in the form of descriptive quantitative. The results showed that the assessment of material validators, media and language validators was at 79.3%, with qualitative criteria that could be used with slight revisions. The Trial one-to-one learner reached a percentage score of 85% which was included in the feasible category. small trials groups reached 90% and trials real or large groups reached 95% included in the feasible category. Thus, it can be concluded that the plastic waste recycling learning module for fourth grade elementary school students is feasible to use. The development of learning materials based on processed waste materials can empirically improve attitudes and behavior environmental care in elementary school students.

Keywords: processed plastic waste products, learning materials, modules, environment

1. Introduction

One of the noble mandates contained in the 1945 Constitution is, "Educating the Life of the Nation. "Every human being has the potential/talent of intelligence, it is the responsibility of educators to cultivate and develop systematically. The government's step to realize the 1945 Constitution is to make a law. No. 20 of 2003 concerning SISDIKNAS article 1 point 8 which reads: "Educational levels are stages of education that are determined based on the level of development of students, goals to be achieved, and abilities developed" [1]. Arts, Culture and Crafts (SBdP) subjects are important subjects in schools but are not of great interest to most students. This is also reinforced by the fact that the learning outcomes of students in SBdP subjects are mostly only average, and only a few students are able to get scores that exceed the KKM [2].

Cultural art is a work created from thoughts or expressions of heart that can produce a beautiful work that evokes the feelings of others so that the work created will not be separated from life because it has various aspects that exist within the individual. Learning arts, culture and crafts is an interaction of the educational process in which there is a reciprocal relationship between educators and students. This reciprocal relationship has a certain educational purpose which in the process uses art as an educational medium. This allows SbdP to accommodate the needs of students for creative activities according to their respective abilities [3].

The use of media in learning is very important and beneficial for both teachers and students. But keep in mind that every teacher must pay attention to the characteristics and abilities of each media so that the selection of media is in accordance with the subject matter Children are basically born with creative potential that must be guided and directed in order to develop optimally. With their creative potential, children will always need activities that are conditional with creative ideas [4].

Regarding creativity, [4] states that "creativity is a mental ability and various types of typical human skills that can give birth to a unique, different, original, completely new, beautiful, efficient, targeted, and effective expression". individual life and social life. The formation of creativity in a person is influenced by various factors close to the environment. According to Permatasari, the factors that affect creativity are internal factors that hinder creative behavior such as the influence of habit or habituation, the estimation of other people's expectations, lack of effort or mental laziness, setting their own limits which in fact do not hinder performance. creative thinking, and rigidity or inflexibility in thinking. While Chaplin [5] states that creativity is the ability to produce new forms in art, or in machines, or in solving problems with new methods. The next definition is expressed by Csikszentmihalyi [6], he describes creativity as a product related to finding something, producing something new, rather than gathering skills or practicing knowledge and studying books. As explained in the conclusion of several definitions of creativity by several figures, it is explained that creativity is an individual mental process that causes problems [5].

According to [7], recycling activities or the use of used goods are making toys or objects using used goods that are no longer used, such as matchboxes, bottles, cardboard, plastic, wood and others. The "recycling" activity is by using used goods that are no longer used (such as boxes, bottles, cardboard, plastic, wood, etc.) to make various kinds of objects, depending on the creativity of the creator.

The definition of used/recycled goods as explained by Priharyanto is as follows: "Recycling is solid waste management which consists of sorting, collecting, managing, distributing, and making used goods that have just been returned". [8] explained that an effort to minimize the negative impact of the presence of waste is to apply the 3R principles, namely Reuse, Reduce, Recycle.

Elementary school students scientific community are responsible as well as environmental users. An attitude of caring for the environment must be built from an early age, efforts to instill an attitude of caring for the environment in elementary school students are expected to be the basis for thinking, acting and acting in the future. Instilling an attitude of caring for the environment for students must begin by integrating it in learning, in order to form a character of love for the environment [9]. Planting an attitude of caring for the environment as far as possible is built through holistic learning programs and involves all components, both components in schools and stakeholders [10].

2. Research Method

This research uses a qualitative approach with Research and Development (R&D) research methods. The development model that becomes the reference is the Borg & Gall model. This model consists of 10 (ten) stages which are stated as "research and information gathering, planning, initial product development, preliminary field testing, major product revisions, main field tests, operational product revisions, operational field tests, final product revisions, and dissemination. and implementation" [11]. The stages of Borg & Gall development can be seen in the following figure:

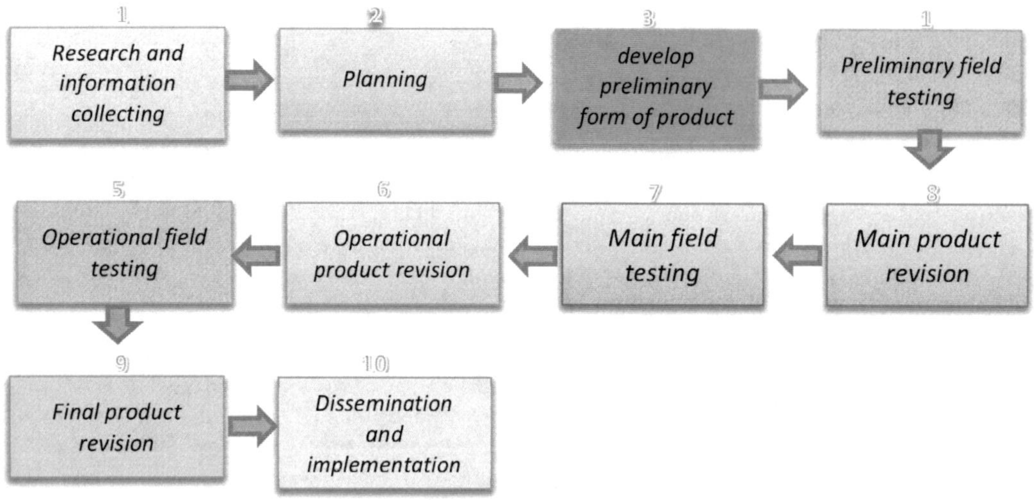

Figure 1. Borg & Gall Development Model

Criteria for evaluating product development in the form of learning modules based on processed waste plastic can be described sequentially. For aspects that are assessed, the instrument and who assesses it can be seen in the following in Table 1.

Table 1. Assessed Aspects, Instruments, Observed Data, and Respondents

Objectives	Aspects Assessed	Instruments	Observed Data	Respondents
Feasibility of material products	product validity	Validation sheet	Student learning module and teacher guide module	Content expert Design expert Language expert
Cultural	effectiveness	Test	learning outcomes	Student
value-	Product	Sheet Observation	learning activities	Observer (teacher)
based	Interesting	Questionnaire	responses	Student
teaching	Absorbed	guide	Student responses and	Teachers and
Using		Interview	teacher responses	students

analysis of qualitative descriptive use to analyze data in the form of notes, suggestions, or comments contained in the validation sheet, observation sheet, and questionnaire. Quantitative descriptive analysis was used to analyze the data in the form of scores from the results of the validation sheet, observation sheet, and questionnaire. There are two types of data in quantitative descriptive data analysis. First, the data on the validity analysis is data that describes the validity of the student learning modules and teacher guide modules that have been developed. The validity of teaching materials obtained from experts [12].
For the feasibility of the media and or the resulting product using the assessment criteria as shown in the following table:

Table 2. Media Feasibility Assessment Criteria

Interval Percentage (%)	Creteria Percentage
81%-100%	Very Eligible
61%-80%	Eligible

41%-60%	Adequate
21%-40%	Less Eligible
<20%	Not Eligible

Students as study subjects at a time the subjects test in the study, especially in the one-to-one learner trial, small group trial and real class trial, refer to the assessment criteria in the following table:

Table 3. Criteria for Attractiveness of Teaching Material Products

No	Percentage of	Qualifications	Description
1	80% - 100%	Attractive	Can be used without revision
2	60% - 79%	Quite Attractive	Can be used with minor revisions
3	50% - 59%	Less Attractive	Cannot be used
4	< 49%	Not Interesting	Forbidden to use

Adapted from Akbar & Sriwiyana [10]

Learning materials are declared interesting to use if they get "quite interesting" and/or "interesting" qualifications.

The effectiveness of teaching material products can be seen from student activities during learning during field trials at least obtaining the "active" criteria. Analysis of learning outcomes obtained from tests given to students in the Competency Test in the product of Using cultural values-based teaching materials. Students are declared complete if they have achieved a score of 70 from a maximum score of 100. The final student learning outcomes are classically declared complete, if the number of students who complete learning is greater than or equal to 75% of the number of students who achieve a score of 70 from ascore maximum of 100.

No	Persentase	Kualifikasi	Keterangan
1	80% - 100%	Very Good	Can be used without revision
2	60% - 79%	Fairly good	Can be used with minor revisions
3	50% - 59%	Not good	Cannot be used
4	< 49%	Not good	Forbidden to use

Table 4. Criteria for Applicability Level of Teaching Material Products

This condition adjusts to the minimum completeness criteria (KKM) for SBdP subjects in fourth grade elementary school students DDI Palu. Data on the applicability of teaching material products was obtained from field trials using questionnaires. Applicable data were obtained from teacher and student questionnaire responses during field trials.

Based on these results, interpretation and decision-making about product quality are then given using the product applicability criteria in Table 5. Language, design experts, and content/material experts. Furthermore, interpretation and decision making about the quality of the development product are given using validity criteria as shown in table 2. Teaching materials are declared valid to be used if they have "quite valid" and/or "very valid" criteria. Validation data is reviewed from the layout and content of the module, collected through questionnaires distributed to students.

3. Result and Discussion

This research was conducted in elementary school. DDI Palu in grade IV. In the early stages of product development, the waste recycling early learning module is a product developed for use in the fourth grade of elementary school. DDI Palu carried out three research processes, namely: a) documentation, b) observation and interviews, and c) questionnaires. This procedure is an initial research in development research which must be preceded by a needs analysis as the first stage in the development of the model used [11]

In the literature study, studies of the literature related to student learning modules, waste recycling In the literature study, a literature study was carried out related to student learning modules, waste recycling learning. From the results of the literature study, these basics are used as guidelines for researchers to develop planned learning module products. The results of the literature study are used to strengthen the theory in developing the product to be produced. The literature search is a scientific step in a research by collecting research data from various written sources [14].

3.1 Development of waste recycling learning module
Production of the Cover of the Waste Recycling Learning Module The cover of the waste recycling learning module is designed using an application Corel Draw X7. The paper used as the cover of the module is 215 mm x 295 mm. On the cover of this module there is a picture of someone painting and there is a title Cover for the Waste Recycling Learning Module The cover of the waste recycling learning module was designed using the Corel Draw X7 application. The paper used as the cover of the module is 215 mm x 295 mm. Mosul components were developed according to Mosul standardization as relevant teaching used for elementary school students [15]. On the cover of this module there is a picture of someone painting and there is a title.

.

Figure 2. cover of learning module

Instructions for Making Waste Recycling Learning Modules Instructions for Making Waste Recycling Learning Modules are used so that the learning modules can be understood by students.

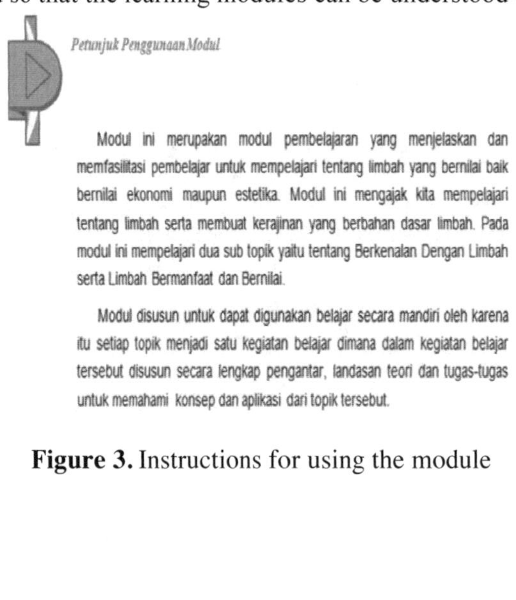

Petunjuk Penggunaan Modul

Modul ini merupakan modul pembelajaran yang menjelaskan dan memfasilitasi pembelajar untuk mempelajari tentang limbah yang bernilai baik bernilai ekonomi maupun estetika. Modul ini mengajak kita mempelajari tentang limbah serta membuat kerajinan yang berbahan dasar limbah. Pada modul ini mempelajari dua sub topik yaitu tentang Berkenalan Dengan Limbah serta Limbah Bermanfaat dan Bernilai.

Modul disusun untuk dapat digunakan belajar secara mandiri oleh karena itu setiap topik menjadi satu kegiatan belajar dimana dalam kegiatan belajar tersebut disusun secara lengkap pengantar, landasan teori dan tugas-tugas untuk memahami konsep dan aplikasi dari topik tersebut.

Figure 3. Instructions for using the module

Production Expected Objectives After Studying the Module The expected goal after studying the module is that what is expected to be achieved

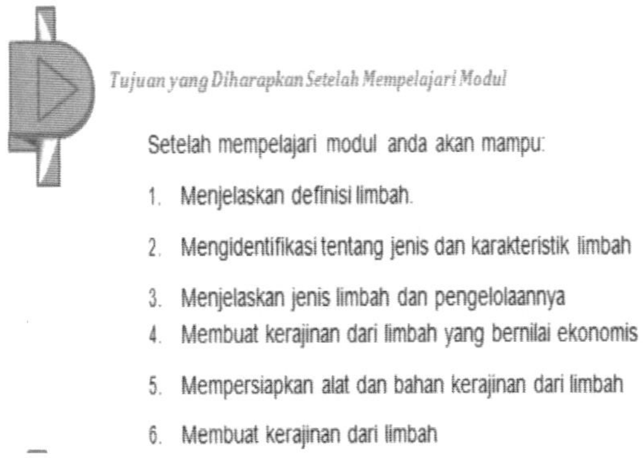
Tujuan yang Diharapkan Setelah Mempelajari Modul

Setelah mempelajari modul anda akan mampu:

1. Menjelaskan definisi limbah.
2. Mengidentifikasi tentang jenis dan karakteristik limbah
3. Menjelaskan jenis limbah dan pengelolaannya
4. Membuat kerajinan dari limbah yang bernilai ekonomis
5. Mempersiapkan alat dan bahan kerajinan dari limbah
6. Membuat kerajinan dari limbah

Figure 4. the expected goals after studying the module

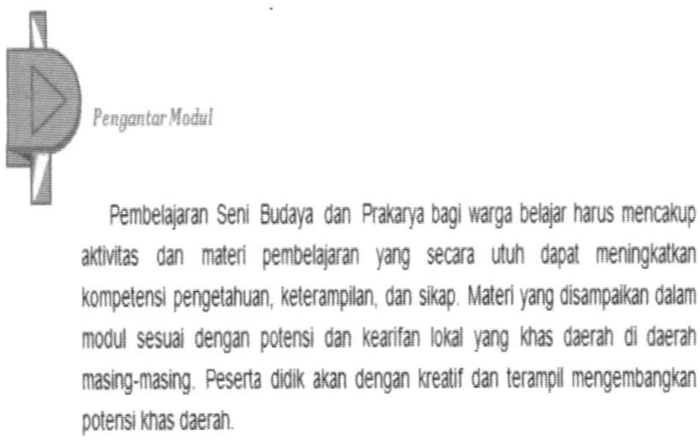
Pengantar Modul

Pembelajaran Seni Budaya dan Prakarya bagi warga belajar harus mencakup aktivitas dan materi pembelajaran yang secara utuh dapat meningkatkan kompetensi pengetahuan, keterampilan, dan sikap. Materi yang disampaikan dalam modul sesuai dengan potensi dan kearifan lokal yang khas daerah di daerah masing-masing. Peserta didik akan dengan kreatif dan terampil mengembangkan potensi khas daerah.

Figure 5. Introduction to the Module

3.2 Description of Learning Module Material

The description of learning materials is always clarified with pictures of processed plastic waste products. The pictures presented in the module are processed waste products made by researchers and are intended as one of the components of the material in the module, and showed of Figur 6.

Figure 6. series of the learning materials

Data on results thief testing research products in the form of learning modules include 3 test components on students and product validation tests by experts. Before being tested on students, the module must be validated by an expert with 3 experts, namely an expert in the field language, an expert in the field of content or learning materials and an expert in the field of learning design. The results of expert or expert validation are as in the following table:

Table 5. Expert validation results

No	Components validated	Expert Validators (%)			Average	Criteria
		Languag e	Content	Design		
1	Compatibility of modules with SK/KD	80	75	79	78	Can be used with a slight revision
2	Compatibility of modules with students	79	79	78	78.7	
3	Interesting modules	80	80	80	80	
4	Readability of the module	81	80	80	80.3	
	Total	80	78.5	79.3	79.3	

In general, the expert validation results state that the product (learning module) can be used with a few revisions, this is in accordance with the criteria used to assess. The module cannot be tested if it does not meet the appropriate criteria according to the results of the expert assessment.

Based on the results of the one-to-one learner trial, the material on how to recycle plastic waste received an assessment of 45 out of a maximum of 50 ratings. The total score has been presented using the percentage eligibility formula and the result is 85%. This data is converted into qualitative data and is included in the "enough" category because it has met the percentage of material eligibility, which is >75%. Even though it is in the sufficient category, the research product still needs to be revised based on the results of the validation carried out by experts. Revisions are made to the module display by adding some pictures to clarify the material, while increasing the attractiveness of the module. The revised module was then piloted in small groups of 8 students. The results of the small group show that 90% of the research products are effective on the ease of material indicators. The results at the small group trial, research products already meet eligibility standards to be tested at the next level. Therefore, the trial was continued with a real class trial. Real class trials or primary care trials in which there are more test subjects than trials small group. The main field trial subjects for making handicrafts in waste processing were 30 students. Based on student assessment, the total score has been presented using the percentage of eligibility formula is 95%. This is converted into qualitative data using the guidelines for the percentage of eligibility for the assessment of the total student instrument and is included in the "very appropriate" category because it has met the percentage of eligibility for the model material, creativity, which is > 75%. Trials results of real class shows that the waste processing-based learning module provides a clear picture that the waste recycling learning module is feasible and very well used for thematic learning for fourth grade elementary school. In this trial, students no longer depend on researchers, students understand the learning module itself, students are active in creating and working using learning modules.

The development of learning materials based on processed plastic waste that was applied at SD DDI Palu, based on the results of the trial, increased student creativity while building an attitude of caring for the environment in students. This proves that attitudes environmental care can be developed through learning, both by integrating concepts and knowledge about the importance of the environment through direct learning media [3] and can also be developed through exemplary methods and giving examples [17]

4. Conclusion

Based on the results of research and discussion it can be concluded that there is; 1) Student assessment of the one-to-one learner trial with a total of 5 students, getting a total score of 85%, or the "appropriate" category because it has met the percentage of material feasibility, which is >75%; 2) The results of the initial product trial revision. Based on the research questionnaire instrument, 2 out of 5 students stated low scores on the learning aspect on the ease of material indicator. The research finally took action against students who experienced these problems by occasionally helping the teacher read the material so that students understood the material themselves; 3) The main field trial is a trial with more test subjects than the initial field trial. The main field trial subjects for making handicrafts in waste processing were 10 students. Based on the student's assessment of the results of the main field trial, the manufacture of handicrafts in processing plastic waste received a total score of 94 out of a total maximum score of 100 assessments. The total score has been analyzed using the percentage eligibility formula and obtained a score of 95%. This is converted into qualitative data by using the guidelines for the percentage of eligibility for the assessment of the total student instrument and is included in the "appropriate" category because it has met the percentage of eligibility for the creativity model material, which is > 75%. Based on the results of the analysis of research data, the development of learning materials based on processed waste materials can empirically improve attitudes and behavior towards environmental care in elementary school students.

Reference

[1] Indonesia, Law No. 20 of 2003 concerning the National Education System. 2003.

[2] YR Lagandesa, "INTEGRATION OF SPORT MOVEMENT IN CULTURAL ARTS AND WORKS LESSONS (SBdP) IN ELEMENTARY SCHOOL," Tadulako J. Sport Sci. Phys. Educ., vol. 8, no. 1, pp. 168–173, 2020.

[3] A. Jihad and A. Haris, Learning Evaluation. Yogyakarta: Multi Pressindo, 2013.

[4] RM Kurnia, Elan, and R. Giyartini, "The Effect of Outdoor Learning Methods on the Formation of Student Creativity in SBdP Learning," Pedadidaktika J. Ilm. educator. School teachers. Basic, vol. 5, no. 4, pp. 240–249, 2018.

[5] Y. Rachmawati and E. Kurniati, Strategies for Developing Creativity in Kindergarten Age Children. Jakarta: Kencana Prenada Media Group, 2010.

[6] I. Trilestari, IMB Setiawan, MW Nurrochsyam, RN Suwarningdyah, and K. Julizar, Creativity Development and Appreciation of Cultural Works. 2017.

[7] Rosdianawati, "Children's Creativity." Center for the Development of Written Teacher Upgrading, Bandung, 2003.

[8] E. Priharyanto,Use Encyclopedia of Popular Knowledge of Paper Recycling and. Jakarta: Jakarta Azka Press, 2008.

[9] MS Yahya, "Integration of Environmental Care Character Education in Learning Activities at SDIT Imam Syafi'i Petanahan Kebumen," Insa. J. Thinker. alternating. Education, vol. 24, no. 2, pp. 232–246, 2019, doi:10.24090/insania.v24i2.3065.

[10] Y. Afriyeni, "Formation of Children's Character to Care for the Environment at Adiwiyata Mandiri School SDN 6 Pekanbaru Yeni Afriyeni Persada Bunda High School Pekanbaru," J. PAUD Lect., vol. 1, no. 2, p. Case. 123-133, 2018.

[11] W. . Borg, MD Gall, and J. . Gall, Applying Educational Research How to Read, DO and Use Research to Solve Problems of Practice, Seventh. New York: Pearson Education, Inc., 2015.

[12] NM Purwanto, Learning Evaluation Principles and Techniques. Bandung: PT Pemuda Rosdakarya, 2010.

[13] Akbar and Sriwijaya, Curriculum Development and Social Science Learning. Yogyakarta: Tiara Annisa, 2011.

[14] BA Habsy, "The Art of Understanding Qualitative Research in Guidance and Counseling: Literature Studies," JURKAM J. Counseling Andi Matappa, vol. 1, no. 2, p. 90, 2017, doi:10.31100/jurkam.v1i2.56.

[15] BP Sitepu, Preparation of Textbooks. Jakarta: Jakarta; Verbum Publishing, 2014.

[16] Y. Lestari, "Instilling the Value of Caring for the Environment in Natural Science Learning," J. Educator. Elementary school, vol. 4, no. 2, pp. 332–337, 2018.

[17] E. Jeramat, H. Mulu, E. Jehadus, and YE Utami, "Inculcating Environmental Care and Responsibility ThroughLearning in Junior High School Students," ScienceJ. Komodo Sci. Educ., vol. 01, no. 02, pp. 24-33, 2019, [Online]. Available: http:ejournal.stkipsantupaulus.ac.id/index.php/jkse.

Sustainable development issues in Togean Islands Tojo Una-Una Indonesia

Muzakkir[1], Suparman[1], M Tallesang[1], I Basir[1], F Bate[1], and M R Fatahilla[1]

[1]Faculty of Economics and Business, Tadulako University, Palu-Indonesia.

pato301008@gmail.com

Abstract. Sustainable development is the main goal of every nation today by reducing the exploitation of the use of natural resources that exceed the capacity and carrying capacity of nature to improve the quality of human life now and in the future. This study aims to identify several issues related to sustainable development goals. The survey was conducted in several sub-districts in the Togean Islands through delivered questionnaires to the community and village and sub-district officials in the archipelago. The results of the study indicate that there are several problems faced by the people of these islands for immediate intervention by the government. This research has a limitation in terms of the number of respondents so that further research can be continued using different approaches. The results obtained from this study can be used as a reference by policy makers in formulating development plans and forms of intervention needed by the people of the Togean Islands, Tojo Una-Una.

1. Introduction

The environmental movement in the 80s led to the emergence of the idea of sustainable development which was defined by the World Commission on Environment and Development in 1987 as development that meets the needs of the present without compromising the ability of future generations to meet their future needs. Sustainable development has been introduced for nearly three decades since being agreed upon in 1992 through the Conference of Rio in Brazil. Nevertheless, the expected outcomes imagined by the concept are neither materialized nor able to provide a significant impact on the sustainability of the global environment. The anthropocentric economic growth paradigm is considered as one of the causes that limit the potential of the concept of sustainable development. Anthropocentrism is an environmental paradigm that considers humans as the center of the universe. In this theory only humans have values and ethics, while nature and its contents are meant for humans to meet their needs. All human obligations to nature and the environment are considered something excessive. This paradigm raises ecological and social problems such as ecosystem degradation, over-exploitation of natural resources and a high level of dependence on fossil energy that causes global climate change. To find solutions to these global environmental problems, a new economic development paradigm is known as a green economy that is expected to improve human welfare without causing environmental and social impacts [2][3].The Togean Islands region is located in a group of islands in Tomini Bay which borders North Sulawesi Province and Gorontalo Province [1]. This area has historical value that still needs to be explored further by researchers. Along with the increasing number of people living in the Togean archipelago, it has consequences for further development. The impact of population growth causes the emergence of social problems, for example, poverty and the level of employment in both the formal and informal sectors [4]. Inadequate supporting facilities and infrastructure have also exacerbated the existing situation. For example, the limited availability of electrical energy can have a significant impact on activities in the fisheries and service sectors.

Sustainable development planning has actually been optimal, but in practice sometimes it cannot be achieved because it is influenced by several factors [1]. Problems in terms of sustainable development are influenced by human resource capacity, budget, and infrastructure conditions. This article is structured systematically, namely an introduction which includes the background of the problem and the research objectives. The literature review in the next section explains the definitions and research results from previous researchers. The approach to writing this article is done by collecting theoretical studies and findings from previous studies obtained from online databases. Furthermore, in the results and discussion section, an approach is carried out, namely collecting the results of the identification of previous findings, describing several things related to the research topic. The final part of this article concludes the findings obtained and recommendations for further research.

2. Literature review

2.1. Poverty

For Indonesia itself, poverty is still a problem that becomes a heavy burden, especially associated with the widening gap between the rich and the poor. As part of the UN member Indonesia is certainly committed to addressing the problem along with the declaration of the SDGs. That means Indonesia is also required to realize the targets set in the UN declaration. The government's efforts to overcome poverty integratively have actually been done since 1995, namely with the issuance of the Inpres Of Disadvantaged Villages. Poverty has established the National Team for The Acceleration of Poverty Reduction (TNP2K). The team is chaired directly by the Vice President. This national effort shows that poverty is still a serious problem. Even the central government has realized the distribution of the first phase of village funds to the village government, about 47 trillion. The village funds have been distributed by the Ministry of Finance (Kemenkeu). After being distributed, the Ministry of Villages, Development of Disadvantaged Regions, and Transmigration (Kemendesa PDTT) is tasked with controlling the priority of the use of desaagar funds in accordance with the Minister's Regulations that have been established. Based on the Regulation of the Minister of Villages, Development of Disadvantaged Regions, and Transmigration Number 21 of 2015 on Prioritization of The Use of Village Funds, village funds in 2016 are used to finance the implementation of village-scale programs and activities in the field of Village Development and Village Community Empowerment. How hard does this poverty problem weigh on the world? The following statistical figures give an idea of the weight of the world shouldering the burden of trouble.

2.2. Sustainable development goals

The concept of the SDGs itself was born at the Conference on Sustainable Development conducted by the United Nations in Rio de Jainero in 2012. The goal to be produced in the meeting is to achieve a universal common goal that is able to maintain a balance of three dimensions of sustainable development: environmental, social and economic [5]. In maintaining the balance of the three dimensions of development, the SDGs have 5 main foundations, namely people, planet, prosperity, peace, and partnership that want to achieve three noble goals by 2030 in the form of ending poverty, achieving equality and tackling climate change. Poverty is still an important and major issue, in addition to two other achievements. To achieve these three noble goals, the following 17 Global Goals are prepared, namely: 1) Without Poverty. There is no poverty of any kind in all parts of the world. 2) Without hunger. No more hunger, achieving food security, improving nutrition, and encouraging sustainable agricultural cultivation. 3) Good Health and Well-Being. Ensuring a healthy life and encouraging the welfare of life for all people at all ages. 4) Quality education. Ensure equality of quality education and improve learning opportunities for everyone, ensure inclusive and equitable education and encourage lifelong learning opportunities for everyone. 5) Gender Equality. Achieving gender equality and empowering mothers and women. 6) Clean Water and Sanitation. Ensure the availability of clean water and sustainable sanitation for everyone. 7) Clean and Affordable Energy. Ensure access to affordable, reliable, sustainable and modern energy sources for everyone. 8) Economic Growth and Decent Jobs. Support sustainable and inclusive economic development, full and productive employment, and decent jobs for everyone. 9) Industry, Innovation and Infrastructure. Build quality infrastructure, drive inclusive and sustainable industry improvement and encourage innovation. 10) Reduce the gap. Reduce inequality both within a country and among the countries of the world. 11) Sustainability of Cities and Communities. Building inclusive, quality, safe, secure cities and settlements. 12) Responsible Consumption and Production. Ensuring the sustainability of consumption and production patterns. 13) Action on climate. Act quickly to combat climate change and its effects. 14) Underwater life. Preserving and maintaining the sustainability of the sea and the life of marine resources for sustainable development development.

3. Research methods

This type of research is conclusive research, which is research that aims to test the relationship between variables. This present study uses descriptive statistical analysis to explain the phenomenon that occurs in the development of sustainability in Kepulauan Togean, Tojo Una Una Regency. Descriptive research is used to draw, explain, and summarize various conditions, situations, phenomena or various research variables according to events as they can be portrayed, interviewed, observed, and that can be expressed through documentary materials. This research is a model of data collection and analysis conducted cross-tabulation. This study analyzed respondents' opinions based on age, education, work and number of family members on the problems of SDGs in Kepualaun Togean. This study used questionnaires distributed to the community in Togean Islands of Tojo Una-Una Regency in the atmosphere of the COVID-19 pandemic. Research enumerators use established protocol standards. The question in this questionnaire is aimed at people who belong to the poor category. The study managed to collect as many as 30 complete questionnaires filled out by respondents from 200 questionnaires distributed. The condition of the Covid-19 pandemic greatly affects the ability of enumerators to reach respondents. The weather conditions and geographical challenges are another reason that the number of respondents who were successfully interviewed was only 15 percent.

4. Results and discussion

4.1. Description of Togean Islands

This type of research is conclusive research, which is research that aims to test the relationship between variables. This present study uses descriptive statistical analysis to explain the phenomenon that occurs in the development of sustainability in Togean Islands, Tojo Una Una Regency. Descriptive research is used to draw, explain, and summarize various conditions, situations, phenomena or various research variables according to events as they can be portrayed, interviewed, observed, and that can be expressed through documentary materials. This research is a model of data collection and analysis conducted cross-tabulation. This study analyzed respondents' opinions based on age, education, work and number of family members on the problems of SDGs in Kepualaun Togean. This study used questionnaires distributed to the community in Togean Islands of Tojo Una-Una Regency in the atmosphere of the COVID-19 pandemic. Research enumerators use established protocol standards. The question in this questionnaire is aimed at people who belong to the poor category. The study managed to collect as many as 30 complete questionnaires filled out by respondents from 200 questionnaires distributed. The condition of the Covid-19 pandemic greatly affects the ability of enumerators to reach respondents. The weather conditions and geographical challenges are another reason that the number of respondents who were successfully interviewed was only 15 percent.

4.2. Demographics

The characteristics of respondents are quite diverse from sides other than gender, for example in terms of the age of respondents who are dominated by respondents with an age range of 30-34 as much as 26.7% and the same in the age range of 40-44 as much as 26.7%. The level of education and employment of respondents was dominated by respondents who were educated in elementary school, which is 50% each.

Table 1. Demographics characteristics.

Characteristics	n	(%)
Gender:		
Women	15	50
Men	15	50
Age:		
30-34	8	26,7
35-39	2	6,7
40-44	8	26,7
45-49	3	10
>50	9	30
Education:		
No School	1	3,3
SD	15	50
SMP	8	26,7
SMA	6	20
Occupation:		
IRT	15	50
Petani	9	30
Nelayan	2	6,7
Penjahit	1	3,3
Buruh Bangunan	2	6,7
Guru	1	3,3
Jenis Bantuan:		
BLT	7	23,3
BNPT	4	13,3
PKH	18	60
UMKM	1	3,3

4.3. Accessibility

The results of data collection show that sea transportation is still a means and infrastructure that can support socio-economic activities of the community in the Island of Togean. Sea transportation access becomes the main transportation of the community to visit other villages or to the sub district capital. The development of accessibility between villages on one island in the Togean Islands showed improvement so that the response to this with a low answer (see the following table).

Table 2. Accessibility.

n=30	Less Urgent		Quite Urgent		Urgent		Very Urgent		Pearson Chi-Square
	n	(%)	n	(%)	n	(%)	n	(%)	
Sea freight access	3	20	10	33,3	14	46,7	3	10	0,881
Access to the district capital	6	20	14	46,7	7	23.3	3	10	0,924
Access to the village	7	23,3	12	40	11	36,7	-	-	0,524

4.4. Electricity and water facilities

The availability of electricity and clean water in Togean Islands is still an urgent issue to build. Limited access to electricity causes people in this region to be unable to move optimally at night. This inadequacy of electricity causes the inability of the community also to store foodstuffs or raw materials for business. Another problem is that clean water is also the most important obstacle in the region. People must pay a lot of money to get clean water. The condition of these two main issues can be seen in the Table below.

Table 3. Electricity and water facilities.

n=30	Less Urgent		Quite Urgent		Urgent		Very Urgent		Pearson Chi-Square
	n	(%)	n	(%)	n	(%)	n	(%)	
24-Hour power grid availability	3	10	5	16,7	18	60	4	(13,3	0,111
Number of villages not electrified	6	20	6	20	16	53,3	1	3,3	0,020
Availability of clean Water	3	10	4	13,3	21	70	2	6,7	0,430
Number of villages not yet clean water	6	20	4	13,3	18	60	2	6,7	0,083
Availability of GENERAL MCK/Latrines	6	20	6	20	16	53,3	2	6,7	0,821
Drainage Development	7	23,3	6	20	17	56,7	-	-	0,648

4.5. *Education*

The education sector is very supportive of improving regional performance because the increase in the number of educated people can increase productivity. Therefore, local governments need to pay attention to the level of education of their people. Based on the data in the table below, shows that school facilities and infrastructure are still very limited, so it needs government intervention to help the sector in order to reduce the dropout rate.

Table 4. Educational facilities.

n=30	Less Urgent		Quite Urgent		Urgent		Very Urgent		Pearson Chi-Square
	n	(%)	n	(%)	n	(%)	n	(%)	
Availability of PAUD in Every Village	-	-	2	6,7	23	76,7	5	16,7	0,326
Availability of Elementary School/MI	-	-	5	16,7	21	70	4	13,3	0,536
School Availability SMP/MTs	9	30	11	36,7	9	30	1	3,3	0,403
Availability of High School/MA/SMK	17	56,7	7	23,3	5	16,7	1	3,3	0,075
Library Availability	9	30	7	23,3	11	36,7	3	10	0,878

4.6. Health

Improving health levels is also the goal of sustainable development goals (SDGs). But the lack of good regional financial management can affect the ability of the government to prepare the necessary health facilities and infrastructure.

Table 5. Health facilities

n=30	Less Urgent		Quite Urgent		Urgent		Very Urgent		Pearson Chi-Square
	n	(%)	n	(%)	n	(%)	n	(%)	
Availability of Health Facilities in Every Village	-	-	3	10	12	40	15	50	0,189
Availability of Health Centres			3	10	12	40	15	50	0,189
Polindes/Posyandu Availability	7	23,3	7	23,3	8	26,7	8	26,7	0,102
Ambulance Availability	10	33,3	9	30	4	13,3	7	23,3	0,125
Availability of Floating Hospital (RS. Ship)	29	96,7	1	3,3	-	-	-	-	0,309

4.7. *Economy*

The increase in the poor population can also be affected by the individual inability of the community to gain access to effort. The addition of market facilities and infrastructure can be a solution to open wide access to individual communities to do business. Conditions that occur in the Togean Islands area, namely market conditions, require development in terms of numbers both at the village and sub-district levels. The condition of facilities and infrastructure in the economic sector in the Togean Islands can be seen in the following Table.

Table 6. Economy.

n=30	Less Urgent		Quite Urgent		Urgent		Very Urgent		Pearson Chi-Square
	n	(%)	n	(%)	n	(%)	n	(%)	
Availability of traditional markets in each village	2	6,7	7	23,3	16	53,3	5	16,7	0,379
Market availability in sub-district	6	20	6	20	16	53,3	2	6,7	0,821
Availability of fish auction Venues	6	20	9	30	13	43,3	2	6,7	0,380
Availability of financial institution s	18	60	5	16,7	6	20	1	3,3	0,601

4.8. *Information and communication*

While the condition of telecommunications networks in the Togean Islands has improved, there are some villages that are still hampered in getting access to information because it is not reached by the network. Ease of getting information can help the government to optimize development. People will also get great benefits if access to information and communication can run smoothly. Easily accessible information has implications for the expansion of society in developing businesses and socializing and gaining access to information in local and global scopes.

Table 7. Information and communication

n=30	Less Urgent		Quite Urgent		Urgent		Very Urgent		Pearson Chi-Square
	n	(%)	n	(%)	n	(%)	n	(%)	
Telephone Network Access Availability	-	-	12	40	12	40	6	20	0,607
Availability of Internet Network Access (Wifi, Hotspot).	3	10	14	46,7	11	36,7	2	6,7	0,237
Availability of TV and Radio Networks	1	3,3	8	26,7	11	36,7	10	33,3	0,574

5. Conclusion

The study obtained data related to the development of sustainable development goals (SDGs). This study only limited the opinions of respondents related to educational health facilities and infrastructure, information and communication, transportation, economy and education. Sea transportation access becomes the main transportation of the community to visit other villages or to the sub-district capital. The number of poverty can also be affected by the individual inability of the community to gain access to effort. The addition of market facilities and infrastructure can be a solution to open wide access to individual communities to do business. Level health is also the goal of sustainable development goals (SDGs). The school infrastructure is still very limited so that government intervention is needed to help the sector in order to reduce the dropout rate. The condition of telecommunications networks in the Togean Islands has improved, but there are some villages that are still hampered in getting access to information because it is not reached by the network. This study has shortcomings in terms of the number of respondents and analytical equipment. However, the results could be used as references for planning the implementation of sustainable development goals (SDGs) in the Togean Islands. Therefore, further research could continue the results of this study with a different research approach with more samples to generalise.

References

[1] Arief, I, and Thahir, H (2020) Community Based Tourism for the Concept of Village Tourism Facility Development in Papan Island IOP Conference Series: Materials Science and Engineering

[2] Boykoff, J, and Mascarenhas, G (2016) Rio 2016: Urban policies and environmental impacts IdeAs

[3] Burchett, K (2016) Anthropocentrism as environmental ethic ProQuest Dissertations and Theses

[4] Laapo, A, Fahrudin, A, Bengen, D G, and Damar, A (2010) Study of Mangrove Areas Characteristic and Suitability for Mangrove Ecotourism Activity in Togean Islands of Togean Islands National Park Jurnal Forum Pascasarjana

[5] Swainson, L, and Mahanty, S (2018) Green economy meets political economy: Lessons from the "Aceh Green" initiative, Indonesia Global Environmental Change

IOP Conf. Series: Earth and Environmental Science 1075 (2022) 012055 doi:10.1088/1755-1315/1075/1/012055

Sustainable marine ecotourism in the togean islands during the covid-19 pandemic

S Hadi[1], H Thahir[1], N Hidayah[2], M I Abdullah[1], F Karim[1], E Rombe[1], and R I Mendi[1]

[1]Faculty of Economy, Tadulako University, Palu, Indonesia
[2]Islamic State University of Datokarama, Palu, Indonesia

suryadihadi@untad.ac.id

Abstract. Tourism is a strategic sector for Indonesia in obtaining foreign exchange. To maintain the sustainability of the tourism sector, sustainable tourism operations both now and in the future are sought to operate in a natural capacity for regeneration and productivity of natural resources in the future. This paper is a literature review by collecting information and conducting a narrative review related to the research topic. Several recommendations can be formulated from this paper regarding the sustainability of ecotourism in the Togean Islands. This study proposes several concepts that can be used as references for further research with a research approach that can produce generalizations.

1. Introduction

Tourism is a strategic sector for Indonesia in earning foreign exchange [1] . Thousands of islands and long coastlines are one of the main assets for Indonesia in developing the tourism sector [2]. In addition, the flora and fauna that are owned by Indonesia are very diverse. Certain animals or plants can only live in Indonesia. The number of tribes that inhabit thousands of islands in Indonesia, is the main capital in enriching the attractiveness of this tourism sector. The Indonesian nation is a nation that is rich with tourism potential to be developed.

To maintain the sustainability of the tourism sector, sustainable tourism operations both now and in the future are sought to operate in a natural capacity for regeneration and productivity of natural resources in the future [1]. Respect for local people, both customs and lifestyle can give a good impression on visitors. Local people feel they get a fair share in the economic benefits of local communities [3].

Tourism management refers to qualitative growth, the intention is to improve welfare, economy and public health, improving the quality of life can be achieved by minimizing the negative impact of non-renewable natural resources. According to Muller [4], five things must be considered in sustainable tourism, namely: 1) healthy economic growth, 2) local community welfare, 3) not changing the natural structure, and protecting natural resources, 4) healthy growing culture of the community, 5) maximize tourist satisfaction by providing good service because tourists in general have a high concern for the environment[1].

The global COVID-19 pandemic in the last two years is very worrying [5]. Indonesia is one of the countries that in 2021 will have a population that is positive for the virus. The spread of this virus is very fast globally so that it greatly affects the mobility of local residents and foreign tourists to visit tourist destinations in the Republic of Indonesia. The number of foreign tourists visiting Indonesia before COVID-19 reached millions, while during the pandemic it was only under five hundred thousand people. However, this pandemic must be responded positively by implementing the health protocols set by the government. The attitude that must be owned by business actors and stakeholders other tourism is to anticipate the major impacts caused by the current pandemic and prepare for post-pandemic mitigation so that tourism in Indonesia can be sustainable [6]. Based on the data that has been collected, it shows that the number of visitors decreased drastically in 2020 in line with the COVID-19 pandemic. There were 408 international visitors visited Togean Islands, while 334 persons were categorized as domestic visitors

Togean Islands National Park is a national park in the Togean Islands located in Tomini Bay, Central Sulawesi (Yulius and Triyono 2011)This park was inaugurated in 2004 and administratively the area is in the Tojo Una-Una Regency. The Togean Islands are a stretch of islands consisting of 6 large islands and about 60 small islands around Tomini Bay. These islands each have biodiversity and are beautiful [8]This beauty is able to attract tourists to come to this area for a tour. This area is also known to have beautiful coral reefs and is not inferior to other coral reefs in Indonesia and abroad [7].

The government's role in tourism development is tasked with making systematic policies and planning [8]For example, the government provides and builds infrastructure to support tourism activities, improves the quality of human resources who work as workers in the tourism sector, and others. The private sector as a business actor has a role in providing tourism supporting facilities. Tourism requires many supporting facilities such as restaurants, accommodation, travel agencies, transportation, and others [9]. While the community as owners and managers can be part of tourist attractions to attract tourists by introducing culture and daily habits that are the uniqueness and characteristics of tourist objects.

Based on the description above, the Togean Islands Region has a large and reliable potential for developing ecotourism in the future. Therefore, the author is interested in describing the above conditions in this paper, so that in the future ecotourism in the Togean Islands can be sustainable.

2. Literature review

2.1. Scope of sustainable tourism

Peeters [11] defines sustainable development as follows: Development of resources (attractions, accessibility, amenitas) tourism that aim to provide optimal benefits to stakeholders and the optimal value for the traveler satisfaction in long-term. Another study, namely Font [12] put forward the definition of sustainability *tourism emphasizing the distinctive characteristics, among others:*

2.1.1. Quality of *sustainable* tourism provides a good experience for visitors while improving the quality of life of the surrounding community and maintaining the quality of the environment.

2.1.2. The sustainability of *sustainable* tourism ensures the continuity of natural resources that underlie the cultural continuity of the surrounding community with a satisfying experience for visitors.

2.1.3. The balance of *sustainable* tourism balances the needs of the tourism industry, environmental advocates, and local communities. Sustainable tourism also emphasizes mutual cooperation between visitors, local communities and tourist attractions.

The main objective of creating a sustainable tourism strategy for a particular region is defined as increasing the number of tourists with the principles of sustainable development [14]. This goal can be achieved by several objectives, such as: All stakeholders participating in the region's tourist development must work together to coordinate their efforts. Regional tourist product inventory in creating tourist goods and marketing efforts, take into account the interests of local populations and the environment. Prospective buyers' marketing evaluations and product perceptions. Creation of the marketing plan's vision, mission, and activity structure during the course of the strategy. Create a regional brand that is consistent with the rest of the country. Develop tools to evaluate progress in implementing strategies. Since sustainability is considered in 3 contexts: economic, environmental and socio-cultural, it is necessary to divide sustainable tourism destinations into 3 aspects [14].

2.2. *Sustainable tourism destinations*

Johnston, [14] stated that the main objective of creating a sustainable tourism strategy for a particular region is defined as increasing the number of tourists with the principles of sustainable development. This goal can be achieved by a number of specific objectives, such as:

1. Coordination of all parties involved in the tourism development of the region.
2. Inventory of regional tourism products.
3. Consideration of the interests of local communities and the environment in shaping tourism products and marketing activities.
4. Marketing assessments and product perceptions by potential buyers.
5. Development of the vision, mission, and activity framework of the marketing plan for the duration of the strategy.
6. Develop the same regional brand.

Develop tools to evaluate progress in implementing strategies.
Since sustainability is considered in 3 contexts: economic, environmental and socio-cultural, it is necessary to divide sustainable tourism destinations into 3 aspects [15]

2.2.1. Economic aspects economic

1. Advantages ensure the viability and competitiveness of regions and businesses to achieve long-term viability;
2. Local prosperity maximizes the economic benefits of tourism for the local community, including the expenditure of tourists in the area;
3. Quality of work increases the quantity and quality of tourism-related work in local communities, including wages, working environment and employment opportunities without discrimination;
4. Social equity ensures fair and equitable distribution of social and economic benefits derived from tourism

2.2.2. Environmental aspects

1. Physical integrity maintains and builds landscape quality, both in urban and rural areas and prevents the formation of ecological and visual pollution;
2. Biodiversity promotes and protects the environment, natural habitats and wildlife, and minimizes the impact of tourism on the environment;
3. Effective waste management minimizes the use of scarce and non-renewable resources in tourism development;
4. Clean environment minimizes water, air, soil pollution, and reduces waste generated by tourists and tour operators

2.2.3. Socio-cultural aspects

1. Community welfare builds community welfare including social infrastructure, access to resources, environmental quality and avoidance of social corruption and resource exploitation;
2. Cultural property maintains and develops the cultural heritage, local culture, customs, and exceptional nature of the host community;
3. Meeting visitor expectations provides a safe and enjoyable travel experience that will meet the needs of tourists and will be available to all?
4. Control of local authorities for planning and decision-making in tourism management by local communities.

2.2.4 Sustainable Development Goals (SDGs)

Sustainable development goals (SDGs) are an agenda that was agreed upon in September 2015 during the 70th General Assembly of the United Nations (UN) in New York, United States of America [15]. The SDGs are designed as a continuation of the Millennium Development Goals (MDGs) [10]. SDGs is a document that becomes a guideline in the framework of development and negotiations of countries in the world [16]. The SDGs are aimed at humans, the planet, and prosperity in order to strengthen world peace with broad freedom and also to overcome poverty [17]. The world has agreed to overcome poverty in any form that exists throughout the world. So the first goal in the 17 SDGs isno poverty [18]. The SDGs agenda has 17 goals as a world agenda for sustainable development for all countries which includes economic growth, social inclusion, and environmental protection. The objectives are as follows:

SDGs 2030 aims to maintain a sustainable increase in the economic welfare of the community, development that maintains the sustainability of community social life, development that maintains the quality of the environment and development that ensures justice and the realization of governance that can maintain an increase in the quality of life from one generation to the next. 2017). The SDGs are expected to be synergized by stakeholders who play a role in the development process in 2016 – 2030. Especially in Indonesia because at the same time, Indonesia also gets a demographic bonus.

3. Methodology

The method used in this paper is qualitative descriptive. The data collected in this paper is qualitative data. Qualitative research usually involves two types of data: primary data and secondary data. Primary data is data obtained directly in the field [15]. The data collection techniques used in the following steps: Author conducted interviews with community leaders and business actors in Tojo Una-Una Regency with academics who have conducted research on tourism in the Togean Archipelago Region; Author made a visit to the Togean Islands-Tojo Una-Una Regency to find out firsthand the state of the area; Author reviews several articles in international and national journals to support the analytical process in writing this scientific paper.

4. Results and discussion

4.1. Togean islands region the togeans

Islands area as an area of the Togean Archipelago National Marine Park (TNL) consists of small islands of up to 60 small islands and 6 large islands. Where the Togean Islands are included in the National Park which is located in the transition zone of thelines Wallace and Weber which is a group of small islands that cross the middle of Tomini Bay. Geographically, the Togean Islands are located in the middle of Tomini Bay with an area of 755.4 km^2. The Togean Islands are bordered to the north by Tomini Bay and Gorontalo Province, to the south by the mainland of Sulawesi Island and the deep sea, to the east by the Maluku Sea and to the west by Tomini Bay and the deep ocean.

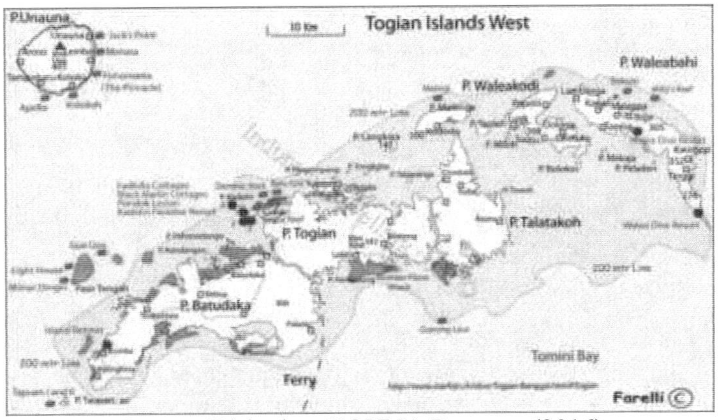

Source: Disbudpar TOUNA Regency (2016)
Figure 1: Map of the Togean Islands

The needs and services of tourist destinations must be supported by four main components with the term "4A" namely attractions, amenities, accessibility, and auxiliary help [16]. Therefore, combining these four factors will increase the number of tourist attractions and build the image of tourists [17]. The Togean Archipelago consists of land and water which is an island ecosystem of terrestrial and marine biodiversity, of which species are endemic and rare that are still natural. The Togean Islands have three different coral environments, namely atoll corals, barrier corals and coastal corals, all of which are habitats for marine flora and fauna.

Marine fauna found in the waters of the Togean Islands includes endemic species (*Paracheilinus togeanensis* and *Ecsenius sp.*), 555 species of molluscs and other rare species such as fish, giant clams (*Tridacna gigas*), turtles (*Tridacna*), green turtles (*squamosa*), turtles green (*Chelonia mydas*), tortoiseshell (*Eretmochelys imbricata*), Laura (*Trochus niloticus*), dugong (Dugong dugong), pilot whale. Natural tourism (nautical tourism) which is an attraction for tourists to visit such as beaches with white sand, jellyfish lakes, views of fighter planes under the sea when *diving*, the uniqueness of Papan Island, and coral reefs. Papan Island is an attraction for tourists because of its uniqueness, namely that there is a bridge over the water made of planks and can connect Papan Island and Malenge Island. Tourists can enjoy the beauty of marine ecotourism in the Togean Islands, such as snorkeling, diving, fishing, sunbathing, enjoying the *sunset* and *sunrise*. Several islands in the Togean Islands, there are cottages with international standards such as Paradise and Black Marlin to be a place to stay for visiting tourists.

4.2. Challenges of sustainable ecotourism development in the togean islands region the

The challenge that needs to be faced is the addition of local superior tourist attractions [18]. The challenges are as follows:

4.2.1. Transportation

Transportation to tourist sites is not optimal because transportation is available only at certain times so that tourists cannot visit tourist sites at any time and need to have information regarding transportation departure schedules. In addition, when tourists go to tourist sites, they must make a stop on the island of Wakai and use other transportation to get to tourist sites. This results in a lack of comfort for tourists to travel, the amount of transportation costs, and it can take time to get to the destination tourist location.

4.2.2. Electricity facilities electricity

Facilities in the Togean Islands have not evenly distributed throughout the Togean Islands and are only available for 18 hours per day. The *resorts* and *cottages* in tourist sites also only rely on generator engines.

4.2.3. Information and communication facilities

Cellular and wifi network services are a challenge in developing ecotourism in the Togean islands because not all cellular networks can be used when visiting, and also wifi facilities have not been provided by *resorts* & *cottages* so tourists are expected to bring their own mobile.

4.2.4. Socio-cultural

The trend in tourism is a journey for change: living like a local, seeking authenticity and transformation [19]. The cultural aspect is one of the destinations for tourists to visit to get to know the uniqueness and authenticity of culture in each region. The readiness of the cultural aspect is very important for every tour, by introducing cultural education in every tourist spot; it is a challenge for tourism managers to attract the attention of local and foreign tourists. The role of local government and society will be an important concern in the process of increasing sustainable tourism.

4.2.5. Local MSMEs

New generation of travelers want experiences that not only offer photo opportunities, but also allow them to "get their hands dirty" and immerse themselves in their destination. Travelers seeking local flavors and wanting to experience a truly authentic destination that makes it unique. As more suppliers and companies tap into the potential of collaborative and shared activities, tourists can access local travel experiences [20]. With the existence of local MSMEs, it will become a tourist attraction to attract tourists to learn and see local products on the tour. Currently, the Togean Islands only have local MSMEs, namely grilled fish, but that does not include local MSMEs because in every place there must be someone who sells grilled fish. The main challenge is to create local tourism products that are characteristic of the Togean Islands. It is the role of the community to form small communities to focus on developing local MSMEs for the use of natural resources and the use of waste to be used as unique works that will become the hallmark of tourist destinations in the Togean islands. Funding and budgets are still the main obstacles, especially related to the limited budget of the government, both central and regional, with the formation of local MSMEs will be the capital to develop tourism in the Togean islands, so that the government and even large companies can look at them to contribute to developing creative tourism in the region.

4.3. Barriers to the development of sustainable ecotourism in the togean islands region

According to Nandi [21], the problems in developing tourist areas in Indonesia are faced by the government as follows: 1) limited facilities & infrastructure that support tourism, 2) limited costs and budget for tourism development, 3) human resources (HR) who do not have high awareness in seeing opportunities and challenges in the tourism sector, 4) local government institutions have not built good coordination with stakeholders in the tourism sector, 5) marketing and tourism promotion programs have not been optimal to increase interest of tourists, both domestic and foreign tourists to visit.
The obstacles in developing sustainable marine ecotourism in the Togean Islands are:

4.3.1. Human resources human

Resources are the main obstacle in developing sustainable ecotourism. The lack of educational facilities in the Togean Islands and the sensitivity of visitors to society and ecotourism are problems. Lack of understanding of the environment and the preservation of community culture is an obstacle for the development of the tourism industry. Bombardment of fish and waste that damage marine ecosystems. Use of underutilized natural resources. Visitors and tourism providers need to work together to create a sense of ownership with an ethic of caution where they visit, visitor perceptions and management of tourist destinations are maintained. The formation of a community in a tourist area opens a wide network. This network is a valuable resource for tourism development in the Togean Islands.

The local community is responsible for finding a culturally, socially, economically and ecologically restorative tourism relationship in the Togean Islands. These problems are the availability of human resources in the Togean Islands, lack of knowledge about sustainable tourism development, and the need for the government's role to provide education to local communities in the Togean Islands. Sustainable tourism development requires the implementation of education and training programs to equip people with knowledge and improve business, skills *vocational* and professional. The training should cover topics on sustainable tourism, hospitality management, as well as other relevant topics.

4.3.2. Infrastructure or Facilities

The development of tourism is supported by the provision of adequate facilities and helping tourists to \enjoy the beauty of nature. An installation that serves as a standard of comfort for tourists on the way from departure to return. The Togean Islands lack services such as transportation, accommodation, restaurants, electricity and clean water. Transportation is the main obstacle and in the Togean Islands, navigation offers departures only twice a day, namely *speedboats* and boats. The mooring facility is only available on one island, which becomes a barrier for new tourists, who must first move to the main island, Wakai, then to the tourist islands. Not all islands have electrical equipment yet, only use electricity independently of generator.

4.3.3. Promotion

Promotion is a one-way flow of information or persuasion to direct a person or organization to an exchange-creating action in marketing [22]. Tourism in the Togean Islands, the problem is the lack of effective promotions carried out so that promotions run effectively, tourism must use marketing strategies for its products, one of which is the promotion mix. According to [23], the promotional mix is a combination of effective strategies including variable advertising, personal selling, and other promotional tools, all of which are designed to achieve the objectives of the sales program. The use of social media is a place to introduce tourist attractions, strategies for using social media are needed to introduce tourist attractions, this is an obstacle because promotions are carried out only through visitors who promote to close relatives or social media from visitors.

4.3.4. Tourism products

One of the obstacles in the development of tourism in the Togean Islands is the development of tourism products. Tourism development focuses on localization and reliability, art, culture as well as the uniqueness and uniqueness of nature, while the Togean Islands tourism product is only in the form of marine tourism and does not have supporting elements to distinguish it from tourism objects in Indonesia. It is also in the interest of tourists to own and maintain. Tourism products are needed because they are sales promotion tools that characterize tourist attractions.

4.4. Marine ecotourism development model

Tourism is a sector that can increase foreign exchange earnings, create jobs, stimulate the growth of the tourism industry, so that it can trigger economic growth [24]. The existence of ecotourism is expected to have a positive impact on the community or local communities around tourist sites. These benefits relate to things that can be enjoyed, such as: economic growth, progress in socio-cultural values, and improving environmental quality [25]. This is related to the *Sustainable Development Goals* (SDGs) which are based on 3 pillars as follows: (1) Economy, (2) Social Culture (3) Environment, so as to stimulate the achievement of the 17 SDGs goals.

There are challenges and obstacles to marine tourism in the Togean Islands and opportunities from tourism development which include: (1) increasing the number of tourists; (2) increasing the length of stay of tourists so that local people have more opportunities to sell local products and services; (3) attracting tourist market share in order to increase purchasing power of local products; (4) increasing the number of tourist purchases through offering a more diverse range of local products; (5) providing local lodging to increase the community's income from the cost of lodging and food and drink. (6) providing guidance services and other services to tourists through community involvement; (7) encourage local attractions (attractions) in the form of local cultural festivals and; (8) the provision of traditional (culinary) food and drinks [26]. So it is necessary to design strategies and policies related to sustainable tourism development. Policies and strategies are important and need to be done for decision making in order to overcome problems related to aspects of marine tourism development [27]. This is done with the aim of stimulating marine tourism to continue to develop from one generation to the next. The strategies and policies that can be carried out in the development of a sustainable tourism industry are as follows:

4.4.1. Improvement of land and sea transportation facilities
The provision of transportation facilities that can be used at any time when tourists will go to tourist sites in accordance with their wishes is important to provide. Procurement of boat access with enough tourist capacity to visit tourist sites but must consider carefully so as not to damage marine ecosystems such as coral. Tourism development uses resources in a sustainable way by avoiding the usage of non-renewable resources. Support for local links during the planning, development, and implementation stages so that a fair share of the gains may be achieved. Tourism operations must focus on features of natural and man-made resources that may be preserved and developed utilizing international criteria and standards.

4.4.2. Adding attractions
Increasing regional capacity in managing marine tourism potential by optimizing current tourism and adding cultural tourism as a supporting factor to attract tourists. The addition of cultural tourism in the Togean Islands can be a distinguishing factor from tourism in other areas because culture is the values that are passed down from generation to generation in the form of solidarity values from generation to generation [28].

4.4.3. Improving human resources
Tourism development must offer quality jobs to local people. Providing quality job opportunities to local communities. Tourism supporting facilities such as hotels, restaurants, and others must be developed and maintained by tourism destination managers and local communities. In addition, education and training for residents as well as easy access for local business actors are needed in realizing local ownership. Furthermore, linkages between business actors and local communities should be pursued to support local ownership. Sustainable tourism development requires the implementation of education and training programs to equip the community with knowledge and improve business, vocational and professional skills. The training should cover topics on sustainable tourism, hospitality management, as well as other relevant topics.

4.4.4. Facilities and infrastructure
Provision of energy and clean water at tourist attractions, so that visitors do not have to think about it when visiting. Community groups and non-governmental organizations, volunteer groups, local governments, tourist associations, corporate associations, and other parties with an impact and interest in tourism activities are among the actors who participate in tourism development. The community and local government play a critical role in ensuring the long-term viability of ecotourism development in the Togean Islands. Recognizing all factors, including the suitable economic, social, cultural, and infrastructure environment, local community awareness, and local government initiatives in local ecotourism development education for the implementation of sustainable tourism in the Togean Islands. Sustainable tourist development is development that is both environmentally and economically sustainable, as well as ethically and socially fair to the community. Sustainable

development is a coordinated and structured endeavour to improve the quality of life by managing the supply, development, use, and upkeep of sustainable resources.

4.4.5. Building a network of domestic and foreign cooperation cooperating

Local communities should monitor or regulate tourism development by participating in the definition of the tourism vision, identifying the resources to be preserved and improved, and setting objectives and strategies for the development and administration of tourist attractions. The community must also be involved in the implementation of previously developed initiatives. The cornerstone to a healthy governance system is active and balanced engagement from the government, the business sector, and the community. As a result, the development of sustainable marine ecotourism is linked not just to environmental challenges, but also to issues of democracy, human rights, and other larger concerns

5. Conclusion

Monitoring and evaluation of sustainable ecotourism development activities to measure their impact should be carried out. The carrying capacity or land capacity that must be considered includes physical, natural, social and cultural carrying capacity. Community goals should be accommodated in tourism activities so that harmonious conditions between visitors/tourists, places and local communities can be realized. Tourism planning should pay great attention to employment opportunities, incomes and improving the health of local people as reflected in development policies. Sustainable tourism development also includes promotion of land use and activities that strengthen local character.

References

[1] Thahir, H. (2018). Increasing the added value of the sustainable tourism business sector based on supply chain management: literature and concept studies. Proceedings of Semkaristek National Seminar on Research and Technology-Based Tourism. 142–7.

[2] Zaman Akhmad, H. dan D. A. (2017). Study of Potential and Development of Sustainable Tourism Strategy on Historical Tourism of Candi Pari. SNTI and SATELLITE Proceedings. 44–50.

[3] Adiati, M. P., & Basalamah, A. (2014). Socio-Cultural Based on Experience and Expectations. Binus Business Review. 80–90.

[4] Müller, H. (1994). The Thorny Path to Sustainable Tourism Development. Journal of Sustainable Tourism.

[5] Antara, Pryanka, A., & Candra, S. A. (2020). The Four Economic Sectors Most Stressed by the Covid-19 Pandemic.

[6] Ministry of Tourism and Creative Economy. (2020). Kemenparekraf/Baparekraf Performance Accountability Report. Kemenparekraf.Go.Id

[7] Yulius, & T. (2011). Island identification based on toponymy rules in the Togean Islands, Central Sulawesi Province.

[8] Arief, I., & Thahir, H. (2020). Community Based Tourism for the Concept of Village Tourism Facility Development in Papan Island. IOP Conference Series: Materials Science and Engineering.

[9] Badarab, F., Trihayuningtyas, E., & Suryadana, M. (2017). Tourism Destination Development Strategy in the Togean Islands, Central Sulawesi Province. THE Journal: Tourism and Hospitality Essentials Journal.

[10] Peeters, P. (2014). Developing a long-term global tourism transport model using a behavioral approach: Implications for sustainable tourism policy making. In Understanding and Governing Sustainable Tourism Mobility: Psychological and Behavioral Approaches.

[11] Wahyuningsih, W. (2012). The Effect of Customer Value on Behavioral Intentions in Tourism Industry. International Research Journal of Business Studies.

[12] Font, X., Tapper, R., Schwartz, K., & Kornilaki, M. (2008). Sustainable supply chain management in tourism. Business Strategy and the Environment.

[13] Günlü, E., Pırnar, I., & Yağcı, K. (2009). Preserving cultural heritage and possible impacts on regional development: Case of Izmir. International Journal of Emerging.
[14] Zaman Akhmad, Henaulu dan Dewi Allina. (2017). Study of Potential and Development of Sustainable Tourism Strategy on Historical Tourism of Candi Pari. SNTI and SATELLITE Proceedings. 44–50
[15] Johnston, T. (2015). Sustainable tourism development. Routledge International Handbook of Sustainable Development. 250–63.
[16] Nasution, S., Nasution, MA, & Damanik, J. (2005). Perceptions of Foreign Tourists on the Quality of Attractions and Attractions (ODTW) of North Sumatra. Journal of Development Studies, 81–96.
[17] Suwena, IK, Widyatmaja, IGN, & Atmaja, M. (2010). Basic knowledge of tourism science. Udayana University Press.
[18] Bagus, I., & Wanda, K. (2018). Visitor Satisfaction (Survey on Trowulan Site Visitors). Journal of Business Administration (JAB), 55(3), 83–91.
[19] Hasiholan, L. (2015). The role of Semarang batik entrepreneurs in promoting local excellence in welcoming MEA 2015. Journal of Contemporary Economics and Business, 1(01).
[20] UNWTO. (2018). Report on tourism and culture synergies.
[21] Nandi, N. (2016). Tourism and Human Resource Development. Journal of Geography of Gea, 33–42.
[22] Rowley, J. (1998). Promotion and marketing communications in the information marketplace. Library Reviews.
[23] Mangold, WG, & Faulds, D. (2009). Social media: The new hybrid element of the promotion mix. Business Horizons. 52(4), 357–65.
[24] Samimi, AJ, Sadeghi, S., & Sadeghi, S. (2011). Tourism and economic growth in developing countries: P-VAR approach. Middle-East Journal of Scientific Research, 10(1), 28–32.
[25] Hermawan, H. (2016). The impact of the development of the Nglanggeran Tourism Village on the local community's economy. Journal of Tourism, 3(2), 105–17.
[26] Tuwo, A. (2011). Coastal and marine ecotourism management: ecological, socio-economic, institutional and regional approaches. International Brilliant.
[27] Nugroho, I., & Dahuri, R. (2004). Regional Development Economic Perspective. Social and Environmental, LP3ES, Jakarta.
[28] Yunus, U., & Tulasi, D. (2012). Batik Semiotics as a medium of communication in Java. Cultura,. 9(2), 143–50.

Implementation of a Sustainable green economy in Indonesia: A Literature Review

R Masdar[1], Husna[1], Jurana[1], R Amborowatie[1], L Meldawaty[1], Tenripada[1], M I Mursali[1], N Naida[1]

[1]Faculty of Economics and Business, Tadulako University, Palu-Indonesia

lucymeldawaty@untad.ac.id

Abstract. This study aims to examine the implementation of a sustainable green economy in Indonesia. The issue of the green economy concept supports the achievement of sustainable development goals, especially in Indonesia. This article was compiled by compiling the results of reports issued by government, nongovernment institutions and universities both from within and outside Indonesia. This article compiles several studies on the development of green investment, green economic growth and strategies for greening the economy. This article cannot be used as a concept that can generalize the application of the green economy because it is limited by the application of the methods in this article. However, the results of the research concepts presented in this article can be a reference for further researchers and policy makers.

1. Introduction

Development that considers natural resources and the environment must be managed well and pay attention to the environment's carrying capacity so that these development assets can benefit society now and in the future. The development idea is built on three pillars: economic, social, and environmental factors, which are broader in scope in terms of implementation because they incorporate pillars such as development governance [1]. Since 1987, The World Commission on Environment and Development has proclaimed the notion of sustainable development at the global level in the "Brundtland Report." Since the 1992 "Earth Summit" in Rio de Janeiro, the concept has grown and been accepted for implementation[2]. The long road since the Earth Summit – more than 20 years – has prompted many parties to study and execute sustainable development, regardless of whether it was in line with the goals when the concept was first proposed. The expected results from sustainable development have not been real, such as in efforts to eliminate poverty, reduce disparities in living standards, consumption and production patterns that are in line with the carrying capacity of ecosystems, sustainable management of renewable resources and agreed methods to achieve the sustainability of renewable resources.

A proper economic development paradigm is required for long-term development. The introduction of a green economy as a new economic development paradigm is projected to promote human welfare while minimizing environmental effect. ecological scarcity or social inequality. Indonesia's green economy concept, in addition to emphasizing the efficient use of resources, also emphasizes the internalization of environmental costs, efforts to alleviate poverty, create decent jobs, and ensure sustainable economic growth. In addition, Indonesia also views the green economy as a vehicle in achieving sustainable development and the MDGs. In principle, the green economy aims to improve people's welfare, provide fair opportunities and minimize environmental damage and economic development in accordance with the carrying capacity of the environment. Green economy can be implemented through low carbon development, natural resource efficiency, and social inclusion. Some examples of programs on natural resource efficiency strategies include sustainable consumption and production, clean production and green industry. In practice, sustainable consumption and production can be used as an integrated approach.

This technique promotes long-term development goals, lowers future economic, environmental, and social costs, and increases competitiveness and reduces poverty. Meanwhile, several programs on socially inclusive strategies include green jobs, green cities and socially responsible investment [3]. Buildings, cities, energy, agricultural fisheries, forests, manufacturing, tourism, transportation, waste, and water are among the 11 areas where development toward a green economy is being prioritized. Economic development in these sectors needs to be balanced with the preparation of several preconditions. Examples of preconditions for the fisheries, transport and water sectors include the removal of subsidies, behavior change, internalization of externalities, land use planning, application of green technology and a shift towards a clean strategy [4]. For the effectiveness of green economy development, it is necessary to apply instruments other than political will. Instruments for green economy development can be in the form of non-economic instruments as well as economic instruments. Non-economic instruments that already exist are usually in the form of regulations and policies to prevent pollution and/or environmental damage. Strategic Environmental Studies (KLHS), spatial planning, environmental quality standards, environmental damage standard criteria, AMDAL, permits, environmental-based budgeting, environmental risk analysis, and environmental audits are all examples of non-economic instruments [5]. In the meanwhile, economic tools are typically in the form of fiscal incentives and disincentives, as well as other sorts. Regional macroeconomic indicators, such as genuine saving, can be used to track the green economy [6]. The challenge for Indonesia in developing a green economy is poverty. With the proportion of poor people still relatively large, the green economy is something that slows growth and increases poverty [7].

The problem in this article is that the implementation of a sustainable green economy is still not getting an idea of the extent of its development. Therefore, the purpose of this study is to identify and find out to what extent the development of sustainable green economy implementation. The systematics of this article consists of an introduction in which the author describes the background of implementing a sustainable green economy in Indonesia. The next section is the method by which the author describes the process of compiling this article. Furthermore, the core part of this article is the results and discussion where the researcher selects articles and reports published by individual researchers and government agencies

2. Literature review

2.1. Sustainable

Sustainable development initiatives have failed to address the issues considered “brown economy” due to the absence of the necessary socioeconomic actors to drive this strategy significantly (Lee et al., 2012). Furthermore, recent economic and environmental disturbances have driven the development or discussion of green economy projects, which are at the core of the debate [3]. The underlying methods such as policy models, assumptions, institutional inertia, and cultural norms that promote increasing unsustainability at all levels be examined [4]. The General Assembly recognizes the concept of a "green economy" as an essential tool for attaining sustainability and poverty alleviation, according to the Rio +20 declaration "The Future We Want” [2]

The purpose of this manifesto is to revitalize national development policies, international cooperation, and aid. This plan is centered on financial growth and is seen as a path to globalization [10]. Insufficient universal principles of what green economy policy measures cover, how to integrate them with national priorities and related goals for economic growth and mission reduction, lack of experience in designing, implementing, and reviewing the costs and benefits of green economy policies, as well as the emergence of various other related terminology, according to Vision [10]. The success of green economy programs and policies in Croatia, Ethiopia, Kenya, Thailand, and Guyana is examined in this article. These countries were chosen to show how green economy initiatives might affect development in developing countries [11]. This article also discusses the many types of national green plans, as well as economic policy alternatives, implementation progress, tools used, and domestic accomplishments. Furthermore, 6 is used to evaluate policy measures that are implemented in different ways in different nations.

Internalization, Incentives, Institutional, Investment, Information, and Inclusion are some of the categories covered by this policy [5], [6]. Furthermore, more educated environmental policy and planning have been identified as a viable road to more socially just and environmentally sound growth, particularly in emerging nations [7]. The Rio+20 agreement recognizes a variety of policy initiatives at the national level, including "regulatory, voluntary, and other."[8]. Furthermore, government activities should consider human welfare and the working environment, resource efficiency and energy conservation, carbon source reduction and emission reduction, technical innovation, and environmental preservation in addition to poverty eradication. A number of developing and developed countries want to take advantage of policy synergies [9], [10]. Leading international organizations such as the United Nations and the World Bank, the OECD, Cosbey, and the International Labor Organization (ILO) have recently published reports that show that sustainable economic policies include complementary concepts and policies that address the three dimensions of development[6].

2.2. *Green economy*

The term "green economy" was originally used in a report published in 1989 that demonstrated how the economy may be a driving force in solving issues such as environmental policy, tropical deforestation, and resource depletion in developing nations [11]. The term "green" with a green political ideology that emphasizes the participation of democracy and grassroots activists in accomplishing environmental goals [12]. Since 2008, the term has continued to focus on how policies respond to an escalating global crisis. most developing and developed societies have embraced this concept in response to today's economic challenges and success stories have been recorded [13]. The United Nations and other non-governmental organizations have also committed to facilitating awareness of the green economy concept. However, there is no agreed definition of a green economy at the international level because this concept has political discourse [12]. This political debate over economic resilience can be viewed as a chance to adopt a radical, robust, and principled vision of sustainable development that provides a normatively appealing and policy-relevant way of structuring the political economy to support sustainable development. Despite the lack of a universally accepted definition of the green economy, recent publications indicate that at least eight different definitions exist. For example, the United Nations Environment Programme (UNEP) defines it as "improving human and social well-being while decreasing environmental consequences and ecological deficits." It's also low-carbon, resource-efficient, and socially responsible [14].

Furthermore, the green economy document is a notion that refers to environmental flexibility as a means of improving the quality of life for all people living in the planet's ecological environment [21]. The International Chamber of Commerce sees it as a pattern of long-term economic growth and environmental stewardship that is fostered jointly while promoting social progress [22]. The concept is defined by the Danish Group 92 as a major systemic transformation process of the current economy that leads to human well-being and equitable access to opportunities for all, while also contributing to environmental and economic integrity so that the planet's carrying capacity is maintained. The Rio+20 conference's goals and themes portrayed the green economy as a lens through which to focus on and exploit chances for concurrently advancing economic and environmental goals [18], [23]. This notion is defined by Ben Glasson [24] as a radical agenda that threatens existing power structures, as well as his fundamental skepticism of future maximizing profit.

The green economy is an extension of the standard economic strategy, which includes, among other things, equitable distribution and environmental quality objectives [15]. Other authors see this concept as a catalyst for creating economic conditions that promote social and environmental justice for the benefit of all humans, nonhuman creatures, nature, and its systems [16]. The green economy concept is defined as investment in economic sectors that promote natural resources, such as renewable energy, low-carbon transportation, energy-efficient buildings, clean technologies, better waste management, improved clean water supply, agriculture, forestry, and sustainable fisheries. natural resources on the planet, and mitigate environmental risks [17]. National policies, as well as the development of international policies and market infrastructure, must all be clearly defined in order to attain this goal. Globally, the policy of applying the green economy concept has grown significantly. Fiscal reform, environmental rules, standards, and certification systems, payments for ecosystem services, integrated natural resource management, and the use of economic tools to support sustainable practices and investments are just a few of these initiatives.

3. Methods

This literature review follows a set of procedures for doing a literature review based on national and international publications or journals on green economic sustainability. Following are some of the steps taken: 1) A literature search was conducted using Google, Google Scholar, and the DIKTI E-Journal. Scopus, ScienceDirect, EBSCOHost, Web of Science, ACM Digital Library, AISel, and Proquest are just a few of the well-known databases that contain scientific papers. To find the content quickly, use keywords in your search. 2) After determining the two existing databases, choose various articles based on several criteria, including peer-reviewed full papers published in national and international journals, empirical studies, research techniques explained, and research connected to a sustainable green economy. 3) The bibliography and nature of research of papers found during the search stage in both databases will be checked. The outcomes of conceptual and empirical study are separated into two groups in the papers obtained. 4) Articles or reports will be examined and classified into numerous areas, including year of publication, methods, outcomes, bibliography, and number of citations. 5) The major topic is broken down into pieces.

4. Results and discussion

4.1. Indonesia's sustainable development

The development of sustainability programs in Indonesia has been applied to the public sector. This study shows that the implementation of economic and non-economic instruments has been announced by BAPPENAS and the sustainable finance initiative has been developed by OJK.

Tabel 1. Indonesia's sustainable development

Author/Institutional Report	Findings
Ministry of Finance Republic of Indonesia, Centre for Climate Change Financing and Multilateral Policy Fiscal Policy Agency	The government determines the direction and other supporting policy frameworks related to sustainable development.
Indonesia Financial Service Authority - OJK (2021)	The sustainable finance initiative developed will integrate seven major components in the ecosystem consist of the following: policy, products, market infrastructure, coordination among related Ministries/Institutions, non-governmental support, human resources, awareness.
BAPPENAS (2013)	The application of non-economic and economic instruments is required

4.2. Policies and initiatives

The form of policies that have been issued by government agencies of the Republic of Indonesia to support the sustainability of the green economy has been in the context of strengthening the role of institutions to encourage the goals of green economy sustainability. The creation of a climate change finance center at the Ministry of Finance is the most prominent example. The following table shows the types of policies pursued by the Indonesian government:

Tabel 2. Policies and initiatives

Author/Institutional Report	Findings
Lilia Endriana, Djoni Hartono, Tony Irawan (2015)[18]	Give special attention to priority sector Participate in CO2 emmisions reduction program Reform public sector Productivity improvement program
OJK (2021)	Green taxonomy development Implementation of environmental, social and governance aspects Real program development
BAPPENAS (2013)	Bio-based economic initiatives Intervensi langsung pemerintah Pendirian National Biodiversity Center Infrastructure establishment, for example: International Center for Interdiciplinary Advance Research Facilities

4.3. Key institutions and organization

The results of the search for documents in this study indicate that government institutions that have implemented green economy sustainability include the Ministry of Forestry, Ministry of Finance, BAPPENAS, and OJK.

Tabel 3. Key institutions and organization

Author/Institutional Report	Findings
Lilia Endriana, Djoni Hartono, Tony Irawan (2015)	Government of Indonesia (GoI)
Direktorat Lingkungan Hidup (2013)	Ministry of Forestry
OJK (2021)	Ministry of Finance
BAPPENAS (2013)	

4.4. Priority sectors

This literature study found that sectors that are the priority for green economy sustainability include forestry, transportation, fisheries, plantations, tourism and others. The government's alignment with sustainable green economy development has been implemented. There are also more priority sectors.

Tabel 4. Priority sectors

Author/Institutional Report	Findings
Makmun (2018)[19]	Forestry, transportation, tourism, Finance

Lilia Endriana, Djoni Hartono, Tony Irawan (2015)	Trade, land transportation, palm oil, forestry, fisheries, textiles, paper, printing, and metal goods, construction, and supporting transportation services are all examples of general services.
Direktorat Lingkungan Hidup (2013)	forestry or finance
OJK (2021)	Agriculture, construction, cities, energy, fishing, forestry, manufacturing, tourism, transportation, waste, and water are all areas that need to be addressed.
BAPPENAS (2013)	Trade, land transportation, palm oil, forestry, fisheries, textiles, paper, printing, and metal goods, construction, and supporting transportation services are all examples of general services.

4.5. Sustainable green economy strategy
The results of the literature search show that the strategy for implementing the green economy has been implemented in several sectors. The participation of indigenous peoples has been optimized in forest conservation.

Tabel 5. Sustainable green economy strategy

Author/Institutional Report	**Findings**
Lilia Endriana, Djoni Hartono, Tony Irawan (2015)[18]	Increasing number and the quality of public facilities, creating better access to public services,
Mohammad Heri Hidayatullah, Azis Nur Bambang, Amirudin (2020)[20]	Improved quality and quantity Empower the community
Direktorat Lingkungan Hidup (2013)[21]	Involving indigenous peoples in forest conservation
OJK (2021) [22]	Make an alignment with the rules of standard setting bodies
BAPPENAS (2013)[23]	Low carbon development Development of environmental services Increasing the use of sustainable biological resources

5. Conclusion
The results of this article search can be seen how far the development of sustainable green economy implementation in Indonesia is. The road map for green economic development began in the 1970s in Indonesia and so far it has been implemented, although not optimally. The Government of the Republic of Indonesia has integrated important things in a sustainable green economy in several policies. Several ministries and institutions have applied efforts to implement this sustainable green economy, namely the ministry of forestry, finance and several other government institutions such as BAPPENAS and OJK. Agriculture, buildings, cities, energy, fishing, forestry, manufacturing, tourism, transportation, waste, and water are the priority sectors that are currently operational. Community involvement is also the main focus of implementing green economy sustainability activities. This study has mapped the implementation of green economy sustainability in Indonesia. However, the weakness of this study is still in the context of literature review so that future researchers are expected to continue the focus of this research with other research methods.

References

[1] B. O. B. Evans and K. Theobald, "POLICY AND PRACTICE LASALA : Evaluating Local Agenda 21 in Europe," *J. Environ. Plan. Manag.*, 2003.
[2] L. Sihombing, "Green Economy dan Konferensi Tingkat Tinggi Rio + 20," *Info Singk. Hub. Int.*, 2012.
[3] K. W. Green, P. J. Zelbst, J. Meacham, and V. S. Bhadauria, "Green supply chain management practices: Impact on performance," *Supply Chain Manag.*, 2012, doi: 10.1108/13598541211227126.
[4] B. Ilić, N. Pavićević, and D. Stojanović, "Sustainability in the example of Indonesia and Serbia by using Green financing management," *Megatrend Rev.*, 2020, doi: 10.5937/megrev2002143i.
[5] N. Urho and V. Niinistö, "Future prospects for enhancing sustainable use of natural resources: The role of international environmental governance after Rio+20," in *International Environmental Law-making and Diplomacy: Insights and Overviews*, 2016.
[6] F. Sindico, "Water Governance in the Aftermath of Rio+20," *International Community Law Review*. 2014, doi: 10.1163/18719732-12341278.
[7] R. A. Coate, "Sustainable development governance," in *International Organization and Global Governance*, 2019.
[8] UNCSD Secretariat, "RIO 2012 Issues Briefs," *Rio +20 United Natl. Conf. Sustain. Dev.*, 2012.
[9] S. Trammell, "Vision 2012," *CFA Inst. Mag.*, 2008, doi: 10.2469/cfm.v19.n4.7.
[10] "UNCSD Rio+20 Focal Point on Managing Climate Change & Adaptation in the Agriculture Sector of Libya," *Int. J. Res. Chem. Metall. Civ. Eng.*, 2016, doi: 10.15242/ijrcmce.iae0316405.
[11] N. D. Bogovic and Z. S. Grdic, "Transitioning to a green economy—possible effects on the croatian economy," *Sustain.*, 2020, doi: 10.3390/su12229342.
[12] E. Lander, "The Green Economy : the Wolf in Sheep's clothing," *Transnatl. Inst.*, 2011.
[13] C. M. Piekarski, L. M. da Luz, L. Zocche, and A. C. de Francisco, "Life cycle assessment as entrepreneurial tool for business management and green innovations," *J. Technol. Manag. Innov.*, 2013, doi: 10.4067/s0718-27242013000100005.
[14] H. Canton, "United Nations Environment Programme—UNEP," in *The Europa Directory of International Organizations 2021*, 2021.
[15] M. I. Setiawan, D. Abdullah, V. N. S. Lestari, and Yuniningsih, "Examining mediating role of environmental performance and green purchasing between green eco design and green information systems with performance, supported renewable energy," *Int. J. Supply Chain Manag.*, 2019.
[16] M. Kennet, "Editorial: Progress in Green Economics: Ontology, concepts and philosophy. Civilisation and the lost factor of reality in social and environmental justice," *International Journal of Green Economics*. 2007, doi: 10.1504/IJGE.2007.013080.
[17] L. Swainson and S. Mahanty, "Green economy meets political economy: Lessons from the 'Aceh Green' initiative, Indonesia," *Glob. Environ. Chang.*, 2018, doi: 10.1016/j.gloenvcha.2018.10.009.
[18] L. Endriana, D. Hartono, and T. Irawan, "Green economy priority sectors in Indonesia: a SAM approach," *Environ. Econ. Policy Stud.*, 2016, doi: 10.1007/s10018-015-0114-5.
[19] Makmun, "Green Economy:Konsep, Implementasi, dan Peranan Kementerian Keuangan," *J. Ekon.*, vol. 19, no. 02, 2011, doi: https://doi.org/10.14203/JEP.19.2.2011.1-15.
[20] A. Hidayatulloh, Mohammad Heri, Bambang, Azis Nur, Amirudin, "The Green Economy Concept as Development Strategy of Cempaka Tourism Village toward Sustainable Tourism Development," *Indones. J. Plan. Dvelopment*, vol. 5, no. 1, 2021, doi: http://dx.doi.org/10.14710/ijpd.5.1.30-37.
[21] D. L. Hidup, "Kumpulan Pemikiran Pengembangan Green Economy di Indonesia (Tahun 2010-2012)," 2013.

IOP Conf. Series: Earth and Environmental Science 1075 (2022) 012056 doi:10.1088/1755-1315/1075/1/012056

[22] T. F. S. Authority, "Roadmap Keuangan Berkelanjutan Tahap 2 (2021-2025), Sustainable Finance Phase 2 (2021-2025), The Future Of Finance," 2021.
[23] Indonesian Ministry of National Development Planning, "DIREKTORAT EVALUASI KINERJA PEMBANGUNAN SEKTORAL KEMENTERIAN PERENCANAAN PEMBANGUNAN/BAPPENAS 2013," 2013.

Entrepreneurial resilience in the hospitality industry during the outbreak covid-19: a case study of the Togean Islands-Tojo Una-Una

A Murad, Farid, Nirwan, E Nainggolan, Nurisky, S Hadi

Faculty of Economics and Business, Tadulako University, Palu-Indonesia.

muhammadalimurad@untad.ac.id

Abstract. Several countries are experiencing economic instability due to the crisis caused by the COVID-19 pandemic, not least Indonesia. This research aims to identify priority aspects that affect the resilience of entrepreneurship in the tourism sector. This research was conducted in the Togean Islands of Tojo Una-Una Regency by interviewing several stakeholders as many as 30 people consisting of business actors. The analysis method used in this study is the analytical hierarchy process (AHP). The results in this study showed that social justice criteria were very prominently chosen by respondents compared to other criteria. This shows the social sensitivity possessed by entrepreneurs during the Covid 19 pandemic. This research requires justification in future research to be used as a generalization in terms of entrepreneurial resilience theory in the face of pandemic risk as it is today.

1. Introduction

Recorded in the second quarter of 2020, economic development slowed and contracted to minus 5.32 percent on an annual basis. The deepest contraction was felt in the household consumption sector (Central Bureau of Statistics, 2020). LIPI conducted a survey by netting 1,548 households spread across 32 provinces taken by a number of respondents who were Working Household status as much as 79.7 percent and Household Businesses as much as 20.3 percent. The survey results show the consequences of the pandemic on the economic survival of households in the face of adversity (Wuryandani, 2020). Business Households and Working Households take advantage of the existence of savings, relics, and or loans.

The resilience and sustainability of the real sector are key to economic success in pandemic times. At the time of this pandemic, MSMEs faced many issues of stalling cash flow and still large expenditures. The role of stakeholders in the development of MSMEs during pandemics is needed. Government assistance is budgeted at 123 trillion but still slow, so the role of OJK is highly expected (CSIS Indonesia, 2020). Other opinions state that the problems faced by capital, marketing, and raw materials, while the government focuses more on the liquidity problems of MSMEs. During this pandemic, the government needs to highlight several things (Wahyu et al., 2020). First, it helps improve the ability of MSMEs to adapt to consumer behavior. Second, the ability to enhance product creativity, innovation, and data management regarding consumers. Third, establish MSMEs in the

IOP Conf. Series: Earth and Environmental Science 1075 (2022) 012057 doi:10.1088/1755-1315/1075/1/012057

procurement of goods for government programs. Fourth, help the ease of access of MSMEs in the digital world and provide access to the internet.

Almost all the nations in the world must face the global pandemic that is COVID-19. Several countries are experiencing negative economic growth due to health protocols that limit people's activities in carrying out their activities. Some of the data collected show some facts. The proportion of the population who carry out work on informal activities is 60.47 percent (77.68 million people), the rest work on other activities (Central Bureau of Statistics, 2020). Employment that experienced an increase occurred in the agricultural sector (2.23 percent), while the sector that experienced the largest decline was the processing industry sector (1.30 percent). The population affected by Covid-19 is 29.12 million people (14.28 percent) consisting of the unemployed due to Covid-19 which is 2.56 million people, not the labor force (BAK) because Covid-19 as many as 0.76 million people, do not work because of Covid-19 which is 1.77 million people, and the working population who underwent a reduction in working hours due to Covid-19 which is 24.03 million people (BPS, 2020).

Uncertainty is always faced by entrepreneurs, so it's important to understand how they can get through tough times. Entrepreneurs often face many obstacles and high stress (Anderson &starnawska, 2008). Invalid information will add to the difficulties that entrepreneurs have to face. Goals and strategies tailored to the dynamics of the conditions faced will guide entrepreneurs in passing the difficulties they face (Fisher et al., 2014). Entrepreneurship resilience is one of the important personal characteristics and can drive entrepreneurial success. Entrepreneurship resilience is the ability of entrepreneurs to be able to survive and quickly solve all difficulties. The ability to survive in the face of crises will be able to help the government to reduce the negative impacts caused by economic crises, natural disasters, and non-natural disasters, such as the COVID-19 pandemic.

Previous empirical research on the influence of entrepreneurial resilience has been less convincing (Ayala &Manzano, 2014). However, research that empirically investigates the influence of entrepreneurial resilience on the success of hotel and restaurant businesses in Indonesia is still under-researched. Research is important given that knowledge of entrepreneurship resilience will be able to contribute to the development and maintaining the continuity of hotel and restaurant businesses that play a role in supporting Indonesia's national income from the tourism sector. Improving the performance of hotel and restaurant businesses has positive implications for sustainable tourism development in Indonesia. The study contributes to sustainable entrepreneurship by linking measures of individual and organizational success to support entrepreneurial resilience. The combination of the two sizes provides a more comprehensive measure of success.

2. Literature review

2.1. *Entrepreneurial resilience*

Entrepreneurial resilience is defined as a multidimensional and multidisciplinary concept that deals with a variety of fields, from physical material properties to psychological behavior, resulting in a variety of theoretical approaches and perspectives. Entrepreneur resilience is a dynamic adaptation process that allows business owners to keep looking ahead to the harsh and destabilizing market situation they face in the market (Ayala &Manzano, 2014). Entrepreneurial resilience can make it easier with entrepreneurs' ability to manage difficult personal and market needs and stability-related difficulties and be future-oriented. There are four indicators in entrepreneurship resilience (Fisher et al., 2016), namely: 1) Actively looking for ways to compensate for the losses experienced 2) Positive will grow in difficult situations 3) Think creatively to change difficult situations 4) Believe you can control yourself in the situation (Ayala & Manzano, 2014)

Entrepreneur Resilience is defined as a multidimensional and multidisciplinary concept that deals with a wide range of fields, from physical material properties to psychological behavior, resulting in a variety of theoretical approaches and analytical perspectives (Ponomarov & Holcomb, 2009). (Ponomarov & Holcomb, 2009). Resilience is a multidimensional construction consisting of a network of favorable attitudes and behaviors (Ayala & Manzano, 2014). Resilience can be thought of as an amalgamation of various personal and behavioral qualities rather than certain characteristics (Ayala & Manzano, 2014).

2.2. *Perception of success*

An extensive network and good relationships with heterogeneous stakeholders can give entrepreneurs the confidence to build learning and experiential resilience so as to provide opportunities for success (Folke et al., 2010; Masten, 2001). Entrepreneurial success is the main benchmark process in an effort to achieve an advantage and competitiveness in the market (Ayala & Manzano, 2014). Success is one indicator that the company has overcome various obstacles in business activities. An entrepreneur's success is a journey, not a goal. The dream you want will make the spirit and motivate yourself. One of the keys is the ability to produce everything that is new and also different from things in the business that already exist today. Although it has risks, entrepreneurs must remain patient to look for opportunities.(Folke et al., 2010; Ayala & Manzano, 2014).

The value of entrepreneurship resilience will give entrepreneurs the ability to face difficult times (Ayala & Manzano, 2014). Entrepreneurship resilience will provide confidence in solving business problems flexibly and agilely. The attitude displayed by employers will have an impact on employees in the form of perseverance in achieving goals (De Jong &Den Hartog, 2007). This entrepreneurial resilience contributes to improving the success of the organization. Entrepreneur success is the acquisition of work accumulated by entrepreneurs is the result of new business creators (Lau et al., 2007; Ayala & Manzano, 2014)

Entrepreneurs are always evolving in uncertain and volatile conditions and strive to find pathways for business growth, especially seeing opportunities to build and manage relationships with external stakeholders (Freeman, 2010; Larson, 1991). Networking ties among companies have an important role, as do individual bonds, especially for managers or entrepreneurs from small companies who work for business growth and are always looking for opportunities (Scuotto et al., 2017; Sefiani et al., 2018).

3. Research methodology

In this study, the authors used this type of descriptive research. Descriptive methods are a method used to describe or analyze a research result but not used to make broader conclusions (Sugiyono, 2018). The study was conducted in Togean Islands, Tojo Una-Una Regency, Central Sulawesi Province. The selection of the location as a research object is due to the Togean Islands being visited by tourists who increase every year, attracted by the beautiful natural beauty, with white sand that lure tourists to come to the attraction. In addition, tourism in the Togean Islands itself has three managers, namely government, private, and also the local community (Sugiyono, 2018).

The population in this study are business people involved in tourism business, ranging from accommodation businesses, restaurants, cafes, food and drink stalls, the government, and the community. The sample from this study is part of the study population. The determination of samples in this study was determined intentionally. The sample in the study amounted to 30 people consisting of tourist entrepreneurs (resorts), cafés, and restaurant or stall entrepreneurs. To see which aspects take precedence in entrepreneurial resilience, a comparison is made between aspects using the Analytical Hierarchy Process (AHP). Basic principles and stages of the AHP are hierarchical, respondent fill, weight calculation or priority vector values, and consistency assessment. The magnitude of the eigenvector determines the calculation of the priority weight of each criterion in each matrix and eigenvalue as its average value. Consistency Index calculation is done to determine the level of consistency to the perception assessment of the respondents. The consistency ratio of respondents in providing perceptions can be declared valid or valid if the consistency ratio is worth 10% or less. If the consistency ratio value is more than 10%, then the consideration is declared invalid or somewhat random so that improvements need to be made.

The Analytical Hierarchy Process (AHP) was developed by Thomas L. Saaty in the 1970s. This method is one of the multi-criterion decision-making models that can help the human frame of mind where logical factors, experience, knowledge, emotions, and feelings are optimized into a systematic process. AHP is a decision-making method developed to give priority to several alternatives when several criteria must be considered and allow decision-makers to organize complex problems into a hierarchical form or an integrated set of levels. Basically, AHP is a method used to solve complex and unstructured problems into groups by organizing the group into a hierarchy, then incorporating numerical values in lieu of human perception in making relative comparisons. With a synthesis, it can be determined which elements have the highest priority. The decision-making in AHP is obtained by comparing alternatives in pairs to be selected using questionnaires. AHP handles a variety of perspectives (criteria) and actions (sub-criteria) with different degrees of importance and translates the overall results into an integrated matrix. The result is to allow managers to know different perspectives from performance assessments and understand the possibility of failure. The steps are as follows:

1. Establish a hierarchy of problems faced
2. Assess criteria and alternatives
3. Determine priorities
4. Logical Consistency. This can be done in the following steps:
   a. Multiply each value in the first column by the priority element, the second column value by the second priority element, and so on.
   b. Sum each line.
   c. The result of the summation of the lines is divided by the priority element. d. Sum the results of the share with the number of elements available. The result is called ƛmax e. ƛmax = the largest eigenvalues of the matrix.
   d. The largest eigenvalue is obtained by summing the result of multiplying the number of columns with the eigenvector. The limit of inconsistency is measured using the consistency ratio (CR), which is the ratio of consistency index (CI) with random consistency value (RI). This value depends on the matrix order n.
   e. Calculate consistency index (CI. The RI value can be seen in Table 1.

Table 1. Random index value

N	2	3	4	5	6	7	8	9	10
RI	0	0,58	0,90	1,12	1,24	1,32	1,41	1,45	1,51

   f. Calculate consistency ratio (CR)
   g. Checking the consistency of the hierarchy, if the value is more than 10%, then the assessment of the judgement data must be corrected. But if the consistency ratio (CI / IR) is less than or equal to 0.1, then the calculation results can be declared correct (Chen et al in Vasakha, 2018)

4. Results and discussion

This study shows that the MYPBB value has the largest weight of criteria value of 0.2422, and the IPRI criteria has the smallest weight of criteria value.

Table 1. Local weight, global weight, and rank

Criteria	**Weight of Criteria**	**Sub Criteria**	**Local Weights of Sub Criteria**	**Global Weights of Criteria**	**Rank**

MY PBB	**0.2422**	Kfin	0.478	0.1158	2
		Plth	0.268	0.0649	5
		Plug	0.105	0.0254	17
		Atif	0.149	0.0361	10
MHPL	**0.156**	MhsPeng	0.307	0.0479	6
		MhPrfL	0.185	0.0289	12
		MhubKL	0.133	0.0207	19
		MhubK	0.2	0.0312	11
		MhunMsyrt	0.175	0.0273	15
Ipri	**0.1419**	TjHdp	0.307	0.0436	8
		KmpKol	0.185	0.0263	14
		Asprsi	0.133	0.0189	20
		Kykin	0.2	0.0284	13
		SpriTual	0.175	0.0248	18
PdMp	**0.1678**	Embroidery	0.478	0.0802	4
		MemOrl	0.268	0.0450	7
		Social	0.105	0.0176	22
		PerLnFisk	0.149	0.0250	16
KTrd	**0.1496**	KepBud	0.666	0.0996	3
		NilBud	0.271	0.0405	9
		KeyBud	0.063	0.0094	23
Ksos	**0.1427**	PerBer	0.875	0.1249	1
		KesSos	0.125	0.0178	21

Based on Table 1, the priority of entrepreneurial resilience starts from the priority and then can be sorted as follows: PerBer, Kfin, KepBud, MemDir, Plth.

5. Conclusion

The results of AHP show that the priority that must be done to increase entrepreneurial resilience during the COVID-19 pandemic is the ability to find meaningful roles. Aspects of social justice become one of the alternatives that must be considered during a pandemic both by entrepreneurs and governments. The next aspect is the financial availability to maintain the company's operations in the face of declining demand for services offered to consumers.

References

[1] Anderson, A R, & Starnawska, M (2008) Research Practices in Entrepreneurship The International Journal of Entrepreneurship and Innovation https://doiorg/105367/000000008786208731
[2] Andriof, J, Waddock, S, Husted, B, & Rahman, S S (2017) Unfolding stakeholder thinking: Theory, responsibility and engagement In Unfolding Stakeholder Thinking: Theory, Responsibility and Engagement https://doiorg/104324/9781351281881
[3] Ayala, J C, & Manzano, G (2014) The resilience of the entrepreneur Influence on the success of the business A longitudinal analysis Journal of Economic Psychology https://doiorg/101016/jjoep201402004
[4] Badan Pusat Statistik (2020) Statistik Indonesia 2020 In Badan Pusat Statistik BPS (2020) Profil Kemiskinan di Indonesia Maret 2020 In Berita Resmi Statistik CSIS Indonesia (2020) LANGKAH PEMBERDAYAAN UMKM DALAM MENGHADAPI COVID-19 CSIS Policy Paper Series PPSECO Department of Economics
[5] De Jong, JP J, & Den Hartog, D N (2007) How leaders influence employees' innovative behaviour European Journal of Innovation Management https://doiorg/101108/14601060710720546
[6] Fisher, R, Maritz, A, & Lobo, A (2014) Evaluating entrepreneurs' perception of success: Development of a measurement scale International Journal of Entrepreneurial Behaviour and Research https://doiorg/101108/IJEBR-10-2013-0157
[7] Fisher, R, Maritz, A, & Lobo, A (2016) Does individual resilience influence entrepreneurial success Academy of Entrepreneurship Journal
[8] Folke, C, Carpenter, S R, Walker, B, Scheffer, M, Chapin, T, & Rockström, J (2010) Resilience thinking: Integrating resilience, adaptability and transformability Ecology and Society https://doiorg/105751/ES-03610-150420
[9] Ghozali, I, & Chariri, A (2014) pendekatan teori akuntansi Research Policy Lau, V P, Shaffer, M A, & Au, K (2007) Entrepreneurial career success from a Chinese perspective: Conceptualization, operationalization, and validation Journal of International Business Studies https://doiorg/101057/palgravejibs8400252
[10] Masten, A S (2001) Ordinary magic: Resilience processes in development American Psychologist https://doiorg/101037/0003-066X563227
[11] Ponomarov, S Y, & Holcomb, M C (2009) Understanding the concept of supply chain resilience The International Journal of Logistics Management https://doiorg/101108/09574090910954873
[12] Scuotto, V, Santoro, G, Bresciani, S, & Del Giudice, M (2017) Shifting intra- and inter organizational innovation processes towards digital business: An empirical analysis of SMEs Creativity and Innovation Management https://doiorg/101111/caim12221 Sefiani, Y, Davies, B J, Bown, R, & Kite, N (2018) Performance of SMEs in Tangier: the interface of networking and wasta EuroMed Journal of Business https://doiorg/101108/EMJB-06-2016-0016
[13] Sugiyono (2018) Metode Penelitian Manajemen (CV Alfabeta (ed); 6th ed) Vasakha, G (2018) ANALISIS PENGENDALIAN SAMPAH MAKANAN PADA WARUNG SARI LAUT DI KOTA PALU

[14] Wahyu, S, Malabay, & Simorangkir, H (2020) Perancangan model sistem informasi e-commerce untuk UMKM di masa pandemik Covid19 Konferensi Nasional Ilmu Komputer (KONIK) Wuryandani, D (2020) Dampak Pandemi COVID-19 Terhadap Pertumbuhan Ekonomi Indonesia 2020 dan Solusinya Info Singkat Bidang Ekonomi Dan Kebijakan Publik Pusat Penelitian Badan Keahlian DPR RI

A conceptual framework for sustainable entrepreneurship in patchouli farming

H Thahir[1], S Hadi[1], F Zahra[1], N Hidayah[1], F Karim[1], R I Mendi[1], M R Fatahilla[1]

[1]Faculty of Economics and Business, Tadulako University, Palu-Indonesia.

suryadihadi@untad.ac.id

Abstract. The world's demand of essential oil has been increased and could contribute to increase the benefits for several countries. Indonesia has potential for patchouli farm and is one of largest exporter countries. Patchouli is one of the sources of essential oil has contributed to fill the high demand. Farmers has the key role in patchouli cultivation must sustain their production without harm the ecosystem. This study aims to identify the theory, experience and perspective of sustainable entrepreneurship in the agribusiness sector, namely patchouli cultivation. This study reviewed previous research on several data bases, including Google Scholar and Researchgate. Several articles related to research topics from reputable journals were collected. The results of this study were carried out succeeded in formulating a research model on the influence of sustainable attitudes, sustainable norms, sustainable behavior, achievement of opportunities, three points of sustainability, on sustainable entrepreneurship. The research model that was formed is recommended for further research. This conceptual paper formulates a model and requires empirical research to generalize the existence of the theory in the future. The research model formulated in this study is still relatively new in the field of sustainable entrepreneurship in patchouli cultivation.

1. Introduction

The current world demand for patchouli oil is quite high and is predicted to continue to increase in the future. Essential oil as a plantation commodity has contributed to increasing foreign exchange for Indonesia. Most of the world's patchouli oil market share and succeeded in placing its position as a major producing country. Patchouli oil, which is also one of the essential oils, is exported to several countries with a volume of 1,200-1,500 tons/year or around 85% of Indonesia's essential oil exports [1]. The end points of export include Singapore, the United States, Spain, France, Switzerland, England, and other countries. Patchouli oil is a binder in the perfume, pharmaceutical, and aromatherapy industries, so far it cannot be replaced by other ingredients [2]. Patchouli oil is often used as a binder or fixative agent and pharmaceutical[3]. Several areas on the Sulawesi island which become the center of patchouli oil production in Indonesia are South Sulawesi, Southeast Sulawesi and Gorontalo. However, another area that is experiencing growth in the number of patchouli plants is Central Sulawesi.

High productivity and profit are the two most important performance indicators in farming analysis. So far, efforts to improve farming performance have mostly been carried out through a cultivation technique approach [4]. Through this approach, farmers are motivated, guided, mobilized, and even directed by the rules to follow all government advice [5]. On the other hand, the role of farmers as the main actors has less than the maximum ability, creativity and creativity [1]. Hisrich [7]

states that entrepreneurship is a dynamic process that is always influenced by environmental factors. Based on this, entrepreneurship is the internal potential of farmers and is used as an alternative approach to improve agricultural performance. According to Rougoor et al. [8] four external environmental factors that play a role in determining farmer entrepreneurship include the physical, social, economic and institutional environment. In addition, other performance indicators are also important to analyze, including output prices, technical efficiency, and competitiveness. Therefore, farming businesses need to improve their performance on an ongoing basis.
Based on the previous findings, there are few researches elaborate the sustainable entrepreneurship in patchouli farmers. Most of the research in patchouli described technical issues. In other words, less of the previous research studied the relationship between sustainable entrepreneurship in patchouli farm. Therefore, it is important to create an understanding among the parties in developing sustainable entrepreneurship in patchouli farming. Failure to address this will lead to future environmental and social problems. Thus, the aim of this paper is to investigate the relationship between the antecedents of farmer entrepreneurship and sustainability. Then, the bottom line of this paper is to provide a conceptual framework for further research.

2. Literature review

2.1. Sustainable entrepreneurship in the agribusiness sector
Sustainability is a consensus to maintain an equitable and harmonious balance of economic, environmental, and social aspects in order to satisfy the demands of the present without hurting the requirements of future generation [9]. Entrepreneurs must innovate in order to lessen the environmental effect of corporate operations. Increasing environmental deterioration is a warning sign for future entrepreneurial ingenuity. Further studies were conducted [10], ensuring future sustainability necessitates innovation, devotion to environmental conservation, and cultural or socioeconomic understanding. current local in environmental dynamics Furthermore, entrepreneurship is growing and progressing toward long-term entrepreneurial growth.

2.2. Sustainability
Existing research demonstrates that a sustainability mindset is required to imagine sustainable entrepreneurial ambitions. Triple Bottom Line (TBL) attitudes and variables were used in this investigation. According to Spagnoli and colleagues [11], Personal values and ideas that drive entrepreneurial ambitions are referred to as sustainable attitudes. Similarly, it comprises a collection of skills to impact entrepreneurial aspirations, which may be viewed as key aspects influencing individual environmental consideration [10] and implementation in sustainability [11]. According to Ajzen [12], Attitude is a factor of behavioral intention and has a beneficial impact on individuals' opinions on entrepreneurial aspirations [13]. Furthermore, it was discovered to be a significant element in pro-environmental behavior intention and play an essential role in driving sustainable entrepreneurship [14]. People who engage in sustainable practices are affected by their positive or negative views, which function as a motivator for businesses to enhance sustainable methods [15]. As a result, while analyzing entrepreneurial intents, it is possible to analyze how confidence in learning influences their attitude toward sustainability in the development of sustainable business models.

3. Method
To discover sustainable entrepreneurship in the agro-industry, this study was developed utilizing a conceptual research technique based on a literature review. This technique consists of numerous processes, including the collecting of related works, the examination of ideas and elements chosen from previous studies, the identification of knowledge gaps, and the discovery of contradictions in the literature for concept generation in new study domains. The reason for this, it was connected to unexplored themes, the literature search was conducted using the keywords sustainable

entrepreneurship and model framework. Several scientific databases, including Science Direct, Scopus, and Google Scholar, are utilized to discover relevant papers from 2000 to 2021 using a restricted range of bespoke queries. An examination of the literature revealed several papers in this field of study that had been published in various journals. These traits, along with a few others, were then included into a framework for sustainable entrepreneurship.

4. Discussion

Understanding sustainable entrepreneurship may be described as the individual identification, development, and exploitation of chances to deliver economic, social, and environmental advantages for future goods and services [21]. Meanwhile, sustainable development may be defined as satisfying future demands that should be grabbed in order to build needs for future generations, connecting sustainable development plans with business models, and evolving into sustainable entrepreneurial synergies. incorporated within a framework of long-term entrepreneurship.

Several scholars have frequently employed the Triple Bottom Line (TBL) idea of sustainable development. A sustainable entrepreneur must foster social cohesiveness in order to address the requirements of individuals and communities across the economy, society, and ecology [27]. Scholars, on the other hand, have incorporated the paradigm while doing important research on sustainable entrepreneurship [26]. Soto-Acosta et al. were not only enlighten or point to sustainable development [25], they also claim that it acts as an effective tool for popular enterprises to achieve so, as well as to play an essential role in exhibiting potential toward agricultural sustainability [26].

Subjective standards that impact one's academic and professional decisions in life are a factor determining a student's business orientation [16]. Based on Bandura's theory of social learning, the existence of norms provides three benefits: learning, inspiration, and motivation to assist individuals establish their goals and attitudes [16]. Previous study has shown that social norms play a role in the process of beginning a business, while students require subjective role models such as parents, friends, specialists, and community leaders who may serve as a source of inspiration for young entrepreneurs [17]. The drivers of the favorable association between sustainability norms and sustainable actions is further confirmed by Sullivan & Meek [18]. Entrepreneurs are encouraged to stimulate sustainable agriculture business ideas through community pressure or inspiration.

Everyone in developing intentions relies on individual control behavior and self-efficacy [19]. Individual behavior is influenced by extrinsic and intrinsic factors along with prosocial motivation [20]. In addition, individuals are also encouraged to take deliberate and practical actions to conserve natural and social resources. For example, individuals tend to behave pro-ecologically, frugally, altruistically, and fairly in terms of positive psychological consequences. Aspects that include positive psychology include satisfaction, self-efficacy, psychological well-being and recovery, happiness, and even pleasure. This allows the preservation of the natural environment and the protection and integrity of society [18]. An entrepreneur can be influenced by social values that stimulate the sustainable behavior of an entrepreneur. Therefore, investigations into the determination to encourage the development of sustainable business models are driven by sustainable entrepreneurial intentions towards the agricultural sector.

Sustainable development in the current unfavorable conditions in society ensuring that future generations can meet their own needs. Creating opportunism in the form of motivation and prior knowledge that emphasizes individual attention to the recognition of sustainable development opportunities that may sustain agriculture, nature, and the environment community and provide development for others [28],[29]. Sustainable entrepreneurship is the discovery, creation and exploitation of opportunities to create future goods and services that are likely to sustain agricultural, natural and communal environments and provide benefits to others. The success of the reality concerning farming production and income is not only determined by the activities of cultivation techniques but is also determined solely by the ability of farmers both from the attitude, knowledge and good life skills of farmers carried out in preparation to farming operations. from planting to marketing the products produced [6].

Sustainable entrepreneurship is a business model that integrates the three domains of sustainable development[20],[22]. The business position was first taken by double bottom line stages and developed into triple bottom lines[10]. Sustainable entrepreneurship creates a profitable business and achieve certain goals of environmental and social objectives[23]. As a result, the creation of an amalgamation of technical learning areas of the value of sustainability for entrepreneurs[23] . For that reason, sustainable entrepreneurs' function as economic agents to initiate, adapt and integrate processes and activities that lead to the development of sustainable and profitable economic opportunities [24]. The proposed research framework can be seen in Figure 1.

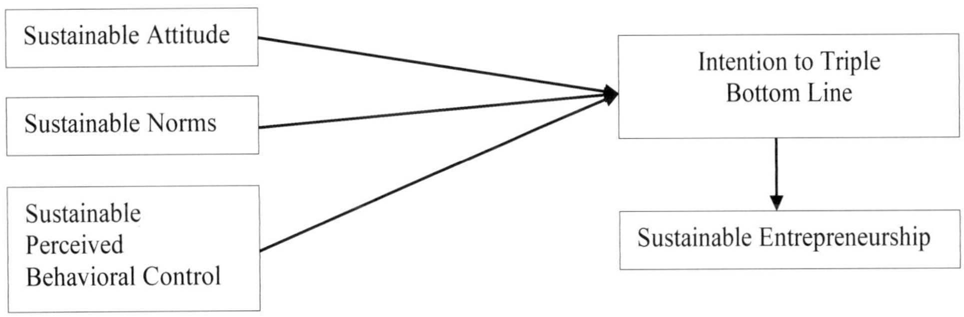

Figure 1. Conceptual framework

5. Conclusion

Sustainable entrepreneurship is not only limited to the creation of products and services, but also includes environmental commitment and equity. This means that the sustainability of a business cannot be separated from support from outside the company. Therefore, the mindset of entrepreneurship is not limited to only how to create new products or services but must also think about environmental and social impacts. The future study of sustainable entrepreneurship will be analyzed on patchouli farmers. The selection of this type of farming is based on a high level of risk and commercial level. Patchouli agricultural commodities are categorized as high risk and require intensive care. The selection of this patchouli commodity is also based on its potential. The area of land that has been developed is very large and has been proven to be able to provide a very large foreign exchange contribution for the nation.

Authors' contributions

HF conceived of the study and coordinated research activities. SH participated in the sequence alignment and drafted the manuscript. FZ participated in design the manuscript. IA participated in collecting and reviewed articles. ER participated in reviewed articles. RI participated in delivered questionnaires. All authors read and approved the final manuscript.

Acknowledgments

This study was conducted under DANA DIPA 2021 Tadulako University.

References

[1] W. Haryudin and S. Suhesti, "KARAKTERISTIK MORFOLOGI, PRODUKSI DAN MUTU 15 AKSESI NILAM," *Bul. Penelit. Tanam. Rempah dan Obat*, 2016, doi:

10.21082/bullittro.v25n1.2014.1-10.

[2] D. P. P. Halimah and M. Dra.Yulfi Zetra, "Minyak Atsiri Tanaman Nilam (Pogostemon cablin B.) Melalui Metode Fermentasi dan Hidrodistilasi Serta Uji Bioaktivitasnya," *Pros. Tugas Akhir Semester Gen.*, 2011.

[3] T. Rihayat, A. Putra, and E. Elwina, "MINYAK NILAM," *J. Sains dan Teknol. Reaksi*, 2016, doi: 10.30811/jstr.v1i1.16.

[4] dan M. H. N. S. Aprilia Budiarti, "Proses Pengambilan Minyak Atsiri Dari Daun Nilam Dengan Pemanfaatan Gelombang Mikro (Microwave)," *J. Tek. Pomits*, 2012.

[5] A. Maryadhi, "Pembuatan Bahan Acuan Minyak Nilam," *Pus. Penelit. Sist. Mutu dan Teknol. Penguji. LIPI*, 2005.

[6] Harunsyah, "Peningkatan Mutu Minyak Nilam Rakyat Melalui Proses Pemurnian," *J. Tekhnologi Politek. Negeri Lhokseumawe*, 2011.

[7] R. D. Hisrich, "The need for marketing in entrepreneurship," *J. Consum. Mark.*, 1992, doi: 10.1108/07363769210035224.

[8] C. W. Rougoor, G. Trip, R. B. m. Huirne, and J. A. Renkema, "How to define and study farmers' management capacity: Theory and use in agricultural economics," *Agric. Econ.*, 1998, doi: 10.1016/S0169-5150(98)00021-8.

[9] T. Strange and A. Bayley, "Sustainable Development - Linking economy, society, environment," *Energy Policy*, 2011.

[10] F. M. Belz and J. K. Binder, "Sustainable Entrepreneurship: A Convergent Process Model," *Bus. Strateg. Environ.*, 2017, doi: 10.1002/bse.1887.

[11] P. Spagnoli, A. Caetano, and S. C. Santos, "Entrepreneurial self-efficacy in Italy: An empirical study from a gender perspective," *TPM - Testing, Psychom. Methodol. Appl. Psychol.*, 2015, doi: 10.4473/TPM22.4.4.

[12] N. F. Krueger, M. D. Reilly, and A. L. Carsrud, "Competing models of entrepreneurial intentions," *J. Bus. Ventur.*, 2000, doi: 10.1016/S0883-9026(98)00033-0.

[13] N. M. Levenburg, P. M. Lane, and T. V. Schwarz, "Interdisciplinary Dimensions in Entrepreneurship," *J. Educ. Bus.*, 2006, doi: 10.3200/JOEB.81.5.275-281.

[14] A. Fayolle and B. Gailly, "The impact of entrepreneurship education on entrepreneurial attitudes and intention: Hysteresis and persistence," *J. Small Bus. Manag.*, 2015, doi: 10.1111/jsbm.12065.

[15] Z. Liu, J. Xu, and B. T. Han, "Small- and medium-sized enterprise post-disaster reconstruction management patterns and application," *Nat. Hazards*, 2013, doi: 10.1007/s11069-013-0657-3.

[16] M. M. Nauta, A. M. Saucier, and L. E. Woodard, "Interpersonal influences on students' academic and career decisions: The impact of sexual orientation," *Career Dev. Q.*, 2001, doi: 10.1002/j.2161-0045.2001.tb00962.x.

[17] S. Karimi, H. J. A. Biemans, T. Lans, M. Chizari, and M. Mulder, "Effects of role models and gender on students' entrepreneurial intentions," *Eur. J. Train. Dev.*, 2014, doi: 10.1108/EJTD-03-2013-0036.

[18] D. M. Sullivan and W. R. Meek, "Gender and entrepreneurship: A review and process model," *Journal of Managerial Psychology*. 2012, doi: 10.1108/02683941211235373.

[19] H. Van Auken, P. Stephens, F. L. Fry, and J. Silva, "Role model influences on entrepreneurial intentions: A comparison between USA and Mexico," *Int. Entrep. Manag. J.*, 2006, doi: 10.1007/s11365-006-0004-1.

[20] F. Tilley and W. Young, "Sustainability entrepreneurs: Could they be the true wealth generators of the future?," *Greener Manag. Int.*, 2006, doi: 10.9774/gleaf.3062.2006.au.00008.

[21] L. E. Schlange, "Stakeholder identification in sustainability entrepreneurship: The role of managerial and organisational cognition," *Greener Manag. Int.*, 2006, doi: 10.9774/GLEAF.3062.2006.au.00004.

[22] T. McEwen, "Ecopreneurship as a Solution to Environmental Problems : Implications for College Level Entrepreneurship Education," *Int. J. Acad. Res. Bus. Soc. Sci.*, 2013.

[23] D. Y. Choi and E. R. Gray, "The venture development processes of 'sustainable' entrepreneurs," *Manag. Res. News*, 2008, doi: 10.1108/01409170810892127.
[24] B. Mupfasoni, A. Kessler, and T. Lans, "Sustainable agricultural entrepreneurship in Burundi: drivers and outcomes," *J. Small Bus. Enterp. Dev.*, 2018, doi: 10.1108/JSBED-03-2017-0130.
[25] P. Soto-Acosta, D. M. Cismaru, E. M. Vătămănescu, and R. S. Ciochină, "Sustainable entrepreneurship in SMEs: A business performance perspective," *Sustain.*, 2016, doi: 10.3390/su8040342.
[26] W.-L. Koe, R. Omar, and I. A. Majid, "Factors Associated with Propensity for Sustainable Entrepreneurship," *Procedia - Soc. Behav. Sci.*, 2014, doi: 10.1016/j.sbspro.2014.04.009.
[27] J. K. Binder and F. M. Belz, "Sustainable entrepreneurship: What it is," in *Handbook of Entrepreneurship and Sustainable Development Research*, 2015.
[28] J. Elkington, "Partnerships from cannibals with forks: The triple bottom line of 21st-century business," *Environ. Qual. Manag.*, 1998, doi: 10.1002/tqem.3310080106.
[29] S. Schaltegger, M. Beckmann, and K. Hockerts, "Collaborative entrepreneurship for sustainability. Creating solutions in light of the UN sustainable development goals," *Int. J. Entrep. Ventur.*, 2018, doi: 10.1504/IJEV.2018.092709.

AUTHOR INDEX